Advanced Structural Wind Engineering

Yukio Tamura • Ahsan Kareem
Editors

Advanced Structural Wind Engineering

Springer

Editors
Yukio Tamura
Wind Engineering Research Center
Tokyo Polytechnic University
Atsugi, Kanagawa, Japan

Ahsan Kareem
NatHaz Modeling Laboratory
University of Notre Dame
Notre Dame, IN, USA

ISBN 978-4-431-54336-7 ISBN 978-4-431-54337-4 (eBook)
DOI 10.1007/978-4-431-54337-4
Springer Tokyo Heidelberg New York Dordrecht London

Library of Congress Control Number: 2013943687

Printed on acid-free paper

Springer is part of Springer Science+Business Media (www.springer.com)

Preface

The effect of wind is a very important issue in the building and structural engineering field. It is said that almost 70–80 % of economic losses due to natural disasters in the world are caused by extreme winds and related water hazards. The risk of future disasters continues to escalate with population shifts towards urban centers and the impending threat of their increased intensity and frequency as hypothesized by potential climate change. Urbanization has also led to the deterioration of regional and global environmental quality with a far-reaching impact on public health. Not only extremely strong wind events or other natural hazards of short duration, but also serious environmental problems should be treated as long-lasting hazards causing severe disasters in human society. This calls for a sustainable society that emphasizes reduced energy consumption and improved environmental quality.

The over-arching vision of the Tokyo Polytechnic University–Wind Engineering Group (TPU-WEG) is to build a sustainable urban environment that is resilient to wind-related events and is in harmony with regional local climates. Its focus is on developing an integrated education and research program that covers a wide spectrum of problems to address wind-related challenges of the next frontiers in urban regions of Asia and beyond.

In order to create efficient and comfortable environments, and to satisfy people's basic demand for iconic symbols in cities, buildings are becoming taller and slenderer, and bridges and roof structures are becoming longer and lighter, thus making them more vulnerable to wind. The importance of wind engineering is significantly increasing, but few universities have curriculums in this field.

Since July 2006, the TPU-WEG has been organizing the International Advanced School (IAS) on Wind Engineering with the intention of covering this lack of relevant advanced professional training.

Thirty-seven world-eminent professors and researchers were invited as lecturers to the past nine IAS sessions: Chris Baker, Shuyang Cao, Qingyan Chen, Chiiming Cheng, Tadeusz Chmielewski, Richard de Dear, David Etheridge, Richard Flay, Yaojun Ge, Jamie Hernandez, John Holmes, Ahsan Kareem, Michael Kasperski, Shinsuke Kato, Youngduk Kim, Greg Kopp, Prem Krishna, Takashi

Kurabuchi, Kenny Kwok, Sangjoon Lee, Chris Letchford, Xiaofeng Li, Rheo Lim, Masaru Matsumoto, Robert Meroney, Akashi Mochida, William Nazaroff, Siva Parameswaran, Michael Reyes, Matthew Santamouris, Partha Sarkar, Michael Schatzmann, Giovanni Solari, Ted Stathopoulos, Charles Weschler, Youlin Xu, and Lingmi Zhang. They collaborated with three lecturers from TPU-WEG—Masaaki Ohba, Ryuichiro Yoshie, and me—to make the IAS sessions fruitful and successful.

The IAS lectures dealt with wind climates, wind-induced disasters, structural aerodynamics, wind loads, and various wind environmental problems. The lectures provided necessary basic knowledge as well as the latest, state-of-the-art information. The IAS also provided a platform for exchanging and sharing information through discussions between leading researchers in the world and young people new to the field, so that serious wind-related problems regarding wind hazard risk due to strong winds, urban air pollution, and increase of environmental load will be solved by young leaders in the future. On behalf of TPU-WEG and the TPU Global Center of Excellence (COE) program members, I would like to express our sincere gratitude to all lecturers and participants for their great contributions and achievements.

Fourteen lecture notes in the structural wind engineering field prepared for the past IAS sessions were rearranged for publication in this textbook, *Advanced Structural Wind Engineering*. A textbook titled *Advanced Environmental Wind Engineering* is also planned for publication, to be based on the IAS lecture notes in the environmental wind engineering field.

I am confident that these two textbooks will be useful to students, engineers, and researchers who work in relevant scientific research or design topics, and will enable them to contribute to the development and construction of sustainable urban environments and hence to achieve wind hazard resilient cities.

Finally, I would like to express my gratitude to all the chapter authors for their cooperation in the creation of these books that provides a valuable contribution to the common goal of the structural and environmental wind engineering.

Kanagawa, Japan Yukio Tamura

Contents

Contributors

Enrica Bernardini NatHaz Modeling Laboratory, University of Notre Dame, Notre Dame, IN, USA

Shuyang Cao Tongji University, Shanghai, China

Richard G.J. Flay University of Auckland, Auckland, New Zealand

Yaojun Ge State Key Laboratory for Disaster Reduction in Civil Engineering, Tongji University, Shanghai, China

John Holmes JDH Consulting, Mentone, VIC, Australia

Ahsan Kareem NatHaz Modeling Laboratory, University of Notre Dame, Notre Dame, IN, USA

Michael Kasperski Department of Civil and Environmental Engineering Sciences, Ruhr-Universität Bochum, Bochum, Germany

Gregory A. Kopp Boundary Layer Wind Tunnel Laboratory, University of Western Ontario, London, ON, Canada

Kenny C.S. Kwok Institute for Infrastructure Engineering, University of Western Sydney, Penrith, NSW, Australia

Chris Letchford Department of Civil and Environmental Engineering, Rensselaer Polytechnic Institute, Troy, NY, USA

Ning Lin Department of Civil and Environmental Engineering, Princeton University, Princeton, NJ, USA

Megan C. McCullough NatHaz Modeling Laboratory, University of Notre Dame, Notre Dame, IN, USA

Seymour M.J. Spence NatHaz Modeling Laboratory, University of Notre Dame, Notre Dame, IN, USA

Theodore Stathopoulos Department of Building, Civil and Environmental Engineering, Concordia University, Montreal, QC, Canada

Yukio Tamura Wind Engineering Research Center, Tokyo Polytechnic University, Atsugi, Kanagawa, Japan

Hiroshi Tanaka Department of Civil and Environmental Engineering, University of Ottawa, Ottawa, ON, Canada

Chapter 1
Strong Winds and Their Characteristics

Shuyang Cao

Abstract The time-dependent aerodynamic forces experienced by a structure immersed in the air flow relate to the wind properties directly. Understanding of the strong wind characteristics is very important for safe and serviceable design. This chapter focuses on strong winds and their characteristics. Wind climates that may bring strong wind, for instance monsoons, tropical cyclones, thunderstorms and so on, are briefly described together with the underlying hydrodynamics. Variation of wind speed with height above ground and turbulent structure inside the atmospheric boundary layer, including turbulence intensity and scale, gust factor, peak factor, decay factor of the coherence function, probability distribution function, power spectrum, and their variations with wind speed, are discussed in detail. In addition, Typhoon Maemi is referred as an example of strong wind event to exhibit the wind characteristics of strong winds.

Keywords Hydrodynamics • Mean velocity • Strong winds • Turbulence characteristics • Typhoon maemi

1.1 Introduction

Wind is basically caused by the temperature gradient of the atmosphere due to variable solar heating of the earth's surface. It is initiated, in a more immediate sense, by density difference or pressure gradient between points of equal elevation. Mechanical effect (velocity shear effect) and temperature effect (buoyancy effect) are the sources of turbulence production of atmospheric flow. Variation of solar heating with time makes the atmospheric boundary layer at different metrological conditions, i.e., stable, unstable or neural conditions, which can be defined by

S. Cao (✉)
Tongji University, Siping Road 1239, Shanghai 200092, China
e-mail: shuyang@tongji.edu.cn

Y. Tamura and A. Kareem (eds.), *Advanced Structural Wind Engineering*,
DOI 10.1007/978-4-431-54337-4_1, © Springer Japan 2013

gradient Richardson number or Richardson numbers in variable formations. Generally, a negative Richardson number corresponds to an unstable condition in which the mechanical and convective effects co-exist, while a positive Richardson number means a stable condition with mechanical effect only. A temperature inversion, i.e. an increase in temperature with height near the ground, can lead to pollution such as smog being trapped close to the ground. Generation of strong winds is closely related with the global atmospheric circulations, and also related with smaller scale temperature differences. However, in the case of strong wind, the buoyancy effect is usually weaker than the mechanical effect and can be neglected. Therefore, in the structural wind engineering field, usually a neutral condition is assumed.

1.2 Strong Winds and Hydrodynamics

1.2.1 Strong Wind Climates

There are many wind climates that might result in strong wind, e.g. monsoons, frontal depressions, tropical cyclones (hurricanes, typhoons, and cyclones), thunderstorms (downburst and microburst), tornadoes, devils, gravity winds (katabatic winds), lee waves and so on.

Monsoons are seasonal winds, and may be considered as large-scale sea breezes, due to seasonal heating and the resulting development of a thermal low over a continental landmass. The monsoon effects are developed most strongly in Asia. The influenced area of monsoons may be larger, and its sustained time may be longer than in typhoon. They are characterized by warm, rainy summer monsoons and cold, dry winter monsoons.

Frontal depressions are boundaries separating two masses of air of different densities, and are the principal cause of strong winds. Cold fronts may feature narrow bands of thunderstorms and severe meteorological phenomena, and may on occasion be preceded by squall lines which contain heavy precipitation, hail, frequent lightning, strong straight line winds, and possibly tornadoes and waterspouts.

Tropical cyclone is intense cyclonic storm which occurs over the tropical oceans, mainly in summer and early autumn. It is known as typhoon in the Far East, cyclone in the region of Australia and the Indian Ocean, and hurricane in the America, with different definitions. For example, a typhoon is a tropical cyclone whose maximum mean wind speed near the center is greater than 17 m/s. The diameters of the tropical cyclones are between 100 and 2,000 km, usually of the order of several hundred kilometers. The depth of the atmosphere involved is about 10–12 km. From the ratio of the diameter to the thickness, it is understandable that the entire rotating body of a tropical cyclone is like a "compact disc". A roughly circular "eye", or "hot tower", is formed in the center of the storm. The air inside the eye is relative dry and light, and rises slowly near the perimeter of the eye. Outside the eye wall, there is a vortex in which warm, moist air is convected at high altitude and forms tall convective clouds. Spiral rain-band clouds are often noticed in the outside of the vortex.

Thunderstorms are small disturbances in horizontal extent, compared with extra-tropical depressions and tropical cyclones, but they are capable of generating severe winds. Thunderstorms result from the rapid upward movement of warm, moist air. As the air moves upward, its cools, condenses, and forms cumulonimbus clouds. Thunderstorms are responsible for the development and formation of many severe weather phenomena: downburst winds, large hailstones, flash flooding, tornadoes and waterspouts. In downburst winds, a strong downdraft reaches the ground, and produces a strong wind for a short period of time—perhaps 5–10 min.

The tornadoes are observed as funnel-shaped vortex in the thunderclouds. It is the most destructive of wind storms. A tornado is typically of the order of 300 m in diameter and moves with respect to the ground with speeds of the order of 30–100 km/h in a distance approximately 15-km long before dissipating, producing a long narrow path of destruction.

Gravity winds are winds that flow downhill under the pull of gravity. They occur mostly in mountainous or glacial regions. When the slopes are particularly steep, the cold air can gather tremendous kinetic energy giving very strong winds. Lee waves are atmospheric standing waves. The most common form is mountain waves.

1.2.2 Strong Wind Hydrodynamics

The surface of the earth exerts upon the moving air a horizontal friction force to retard the flow, forming an atmospheric boundary layer with a vertical velocity profile through the effects of turbulent mixing. It is the wind regime within the atmospheric boundary layer that is of direct interest to the designers of civil engineering structures. The depth of the atmospheric boundary normally ranges in the case of neutrally stratified flows from a few hundred meters to several kilometers, depending on the wind intensity, roughness of terrain and angle of latitude (Simiu and Scanlan 1986). Within a synoptic boundary layer, the wind speed increases with elevation; its magnitude at the top of the boundary layer is often referred as the gradient speed. Outside the boundary layer, the wind flows approximately with the gradient speed along the isobars. The forces acting on the air inside the atmospheric boundary layer consist friction force, pressure gradient force, Coriolis force and centrifugal force.

The friction force acts to prevent movement the air. It is opposed to the direction of movement, and perpendicular to Coriolis force.

The pressure gradient force is caused by pressure difference, which acts from high pressure region to low pressure region. It is usually responsible for accelerating a parcel of air, resulting in wind. Actually it is not a "force" but the acceleration of air due to pressure difference (a force per unit mass).

Coriolis force is a "force" that acts upon a moving body in an independently rotating system. Coriolis effect is caused by the Earth's rotation. In the northern hemisphere, the Coriolis force is to the right of an observer facing the flow direction. This results in anti-clockwise rotation of wind around a low pressure center.

Centrifugal force is an outward "force" away from the center of rotation. It expresses the effects of inertia that arise in connection with rotation.

The above forces co-exist and govern the movement of air flow, however with different contributions depending on the wind climate and elevation.

1.2.2.1 Geostrophic Wind and Gradient Wind

For air flows well above the surface of the earth where friction free wind (FFW) can be assumed, the pressure gradient force, the Coriolis force and the centrifugal force are the dominate forces. Geostrophic wind is a theoretical wind that is directed parallel to perfectly straight isobars, in order words, geostrophic wind would result from an exact balance between the Coriolis force and the pressure gradient force. Thus, the geostrophic wind speed U_G is defined by the next equation;

$$-\frac{dp}{dn} + \rho_a U_G f_c = 0 \tag{1.1}$$

where ρ_a is the air density, f_c is the Coriolis parameter expressed by $f_c = 2\Omega \sin\phi = 1.454 \times 10^{-4}\sin\phi$ (rad/s), Ω is the earth's rotational speed, and ϕ is the latitude. In the northern hemisphere, the Coriolis force acts to the right of the direction of motion. Therefore, high pressure is always to the right of an observer moving with the flow. It is opposite in the southern hemisphere.

Once the flow develops curvature (as for example near the center of a tropical cyclone), centrifugal forces are generated, which direct from the center of curvature and equals to $\rho_a U_g^{\,2}/r$, where r is the radius of the curvature of the isobars. Around a low pressure center (a cyclone), the centrifugal force acts in the same direction as the Coriolis force (opposite to the pressure gradient force). This results in anti-clockwise rotation of wind around a low pressure center in the northern hemisphere. In this situation, the gradient wind speed U_g is defined by the following equation:

$$-\frac{dp}{dr} + \rho_a U_g f_c + \rho_a \frac{U_g^{\,2}}{r} = 0 \tag{1.2}$$

The gradient wind speed U_g is smaller than geostrophic wind speed U_G when the air curves anti-clockwise around a low pressure centre, otherwise it is larger than geostrophic wind speed when the air curves clockwise around a high pressure center.

Inside a tornado or typhoon eye where the radius of the curvature of an isobar is small, wind speed is high and pressure gradient force is large, Coriolis force is small compared with the centrifugal force and can be neglected. Thus the wind is balanced between the pressure gradient force and centrifugal force, and rotates around a low pressure center. This wind is called cyclostrophic wind, whose velocity is defined by the following equation

$$\frac{1}{\rho}\frac{\partial p}{\partial n}=\frac{v^2}{r} \tag{1.3}$$

1.2.2.2 Atmospheric Boundary Layer

Near the ground, the frictional force due to the ground surface becomes significant and FFW cannot be assumed. The friction force is perpendicular to Coriolis force, and these two forces shall be balanced with pressure gradient force. This means that the wind direction cannot be parallel to the isobars, and shall turn to the left in the northern hemisphere. The effects of the friction force become more significant at a lower altitude, and the inclination angle of the wind direction from the isobar becomes larger near the ground. Together with the reduction of wind speed near the ground, wind direction tends to move anti-clockwise and the tip of the velocity vector traces a spiral that is called Ekman spiral.

Inside the atmospheric boundary layer, the flow is balanced among pressure gradient force, Coriolis force and friction force because the centrifugal force is usually small, thus,

$$2\Omega V\sin\varphi=-\frac{\partial P}{\partial x}+\frac{1}{\rho}\frac{\partial\tau_x}{\partial z} \tag{1.4}$$

$$2\Omega U\sin\varphi=-\frac{\partial P}{\partial y}+\frac{1}{\rho}\frac{\partial\tau_y}{\partial z} \tag{1.5}$$

where $\tau_x=-\rho\overline{u'w'}$ and $\tau_y=-\rho\overline{v'w'}$ are the Reynolds stress. Considering the geostrophic wind speed (U_G, V_G) can be expressed as

$$U_G=-\frac{1}{\rho f_c}\frac{\partial P}{\partial y} \tag{1.6}$$

$$V_G=\frac{1}{\rho f_c}\frac{\partial P}{\partial x} \tag{1.7}$$

following equation can be obtained:

$$-f_c(V-V_G)=\frac{1}{\rho}\frac{\partial\tau_x}{\partial z} \tag{1.8}$$

$$f_c(U-U_G)=\frac{1}{\rho}\frac{\partial\tau_y}{\partial z} \tag{1.9}$$

With the assumption of Reynolds stress, $\tau=-\rho\overline{uw}=\rho\varepsilon\frac{\partial U}{\partial z}$ (ε : Kinetic Eddy Viscosity), which links the Reynolds stress with the mean velocity profile, the mean velocity may be approximately calculated.

1.3 Statistical Description of Wind Speed

Wind speed that varies randomly in time can be assumed statistically as the sum of a mean component and a fluctuating component. In other words, the wind velocity $U(t)$ is composed of a mean component U, in the x axis direction, and fluctuating components (u, v, w) in three orthogonal directions (x, y, z). The mean wind speed U is defined as the average of the fluctuating velocity $U(t)$ over the averaging time T. The averaging time of the mean wind speed is 10 min in Japan and ISO4354, and 1 h in the UK. The National Hurricane Center in the US uses a 1 min averaging time for reporting relatively long-lasting wind, named "sustained wind". The fluctuating wind speed structures can be described by kinds of statistics, like turbulence intensity and scale, gust factor/peak factor, power spectrum, spatial-/temporal- correlations and so on. These features of atmospheric turbulence are of interest in various wind engineering applications.

In this chapter, Typhoon Maemi, which passed Miyakojima Island of Okinawa Prefecture, Japan, on September 11, 2003, is referred as an example of strong wind event to exhibit the wind characteristics of strong winds. The details of this typhoon and the measurements are available at Cao et al. (2009).

One of the key assumptions for analyzing the wind is that the mean wind is in fact steady. If this is not the case, then more sophisticated processing will be necessary to remove the turbulent part of the signal, as otherwise the trend in the mean wind may bias the sample mean away from the true mean. For example, a 10-min mean measured during the passage of a sharp eye wall wind maximum of a fast moving by storm may reduce the amplitude of the true mean wind maximum (WMO 2010). In practice, averaging periods are chosen as a compromise between minimizing sampling errors, reducing the errors due to non-stationarity. Figure 1.1 is an example to extract the time dependent mean velocity from a non-stationary velocity recode by Empirical Mode Decomposition (EMD), with which any complicated data set can be decomposed into a finite but often small number of intrinsic mode functions (IMF). This decomposition is adaptive and applicable to nonlinear and non-stationary processes because it is based on the local characteristic time scale of the data (Xu and Chen 2004).

Generally, the mean wind estimate is always of critical importance and should be based on the longest practical interval that can be regarded as stationary. In practice, the 10-min average generally satisfies this requirement. The World Meteorological Organization (WMO) standard for estimating the mean wind is the 10-min average.

1.3.1 Mean Wind Speed

1.3.1.1 Mean Wind Speed in General Atmospheric Boundary Layer

As already mentioned in the previous paragraph, friction forces forms ABL. Friction forces or the velocity shear generate vortices, resulting in large wind fluctuation near the ground. ABL, roughly speaking, contains the interfacial layer,

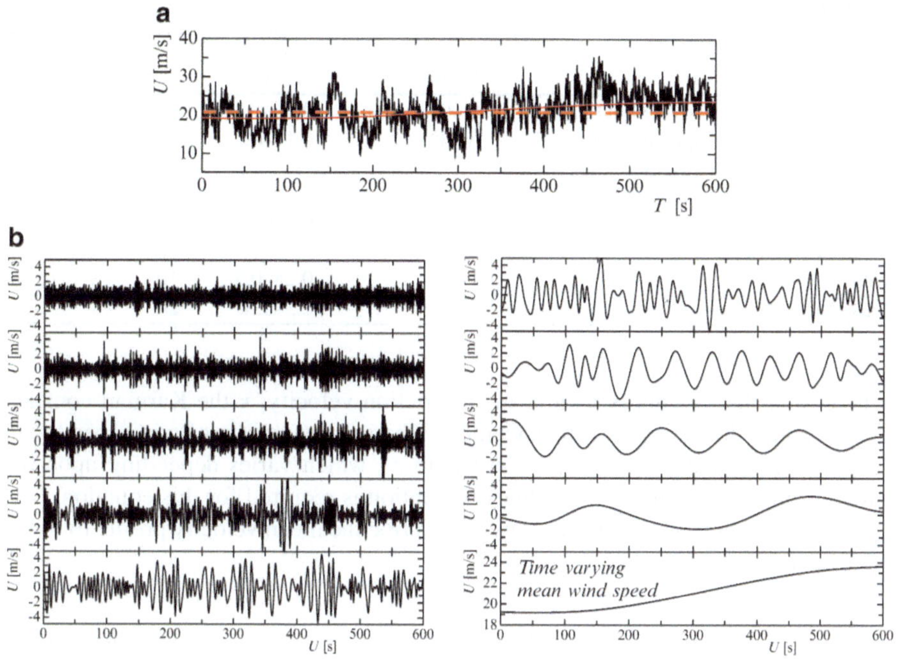

Fig. 1.1 Example of extracting the time dependent mean velocity from a non-stationary velocity recode by Empirical Mode Decomposition (EMD). (**a**) 10-min non-stationary velocity record; (**b**) EMD components

surface layer and outer layer. The height of ABL is defined as the gradient height Z_g, and the friction free gradient wind speed U_g can be assumed at above Z_g. The Reynolds stresses play important roles in this process of momentum transfer near the ground.

Strong wind condition is usually concerned in a thermally neutral condition resultant from sufficient mixing of air in structural wind engineering field. In the surface layer, Reynolds stress is roughly constant and taken at $\tau_0 = \rho_a u_*{}^2$, where u_* is the friction velocity. A good mathematical model to describe the wind speed profile in strong wind condition is the logarithmic law. In practical wind load codes and standards, the log law, the power law and Deaves & Harris model are used to describe the wind speed profile.

According to the log law, the mean wind speed profile is defined as:

$$U = \frac{u_*}{\kappa} \ln\left(\frac{z-d}{z_0}\right) \tag{1.10}$$

Table 1.1 Parameters for mean wind speed profile

Terrain	Roughness length z_0 (m)	Power-law index α	Zero-plane displacement d (m)
Sea, mudflats, snow covered flat land, etc.	0.000–0.003	0.1–0.13	0
Open flat countryside, fields with crops, fences and few trees, etc. (meteorologically standard condition)	0.003–0.2	0.14–0.2	0.1
Dense woodland, domestic housing, suburban area	0.2–1	0.2–0.25	5
City	1–2	0.25–0.3	10
Large city center	2–4	0.3–0.5	20

where z_0 is the roughness length, u_* the friction velocity, κ the Karman constant ($\approx$0.4) and d the zero-plane displacement. The increasing rate of wind speed with height is governed by the roughness length z_0, which varies depending upon the ground roughness. Uniform roughness condition is required for this equation.

The power law is an empirical law to describe the mean velocity profile to fit the field data,

$$U = U_r \left(\frac{z - d}{z_r} \right)^{\alpha} \tag{1.11}$$

where U_r is the wind speed at reference height z_r, α is the power law index. This expression is more tractable for structural designers in wind load estimation or response analysis.

Deaves and Harris (1981) developed a more complicated semi-empirical model to approximate the profile throughout ABL as follows:

$$U = \frac{u_*}{\kappa} \left[\ln\left(\frac{z - d}{z_0} \right) + 5.75 \left(\frac{z - d}{Z_g} \right) - 1.875 \left(\frac{z - d}{Z_g} \right)^2 - \frac{4}{3} \left(\frac{z}{Z_g} \right)^3 + \frac{1}{4} \left(\frac{z}{Z_g} \right)^4 \right] \tag{1.12}$$

where the gradient height Z_g is defined:

$$Z_g = \frac{u_*}{6 f_c} \tag{1.13}$$

Australian/New Zealand Standard (AS/NZS 1170.2) (2002) specified a profile based on Deaves and Harris model.

The values of roughness length z_0, power-law index α, and zero-plane displacement d vary with the development of the boundary layer. Table 1.1 shows the field data of roughness length and power-law index measured at different terrain categories.

Table 1.2 Parameters to describe the mean velocity profile in APEC-WW economies

			Minimum		Basic		Maximum	
Economy	Law	N	α_{min}	δ_{min}	α_b	δ_b	α_{max}	δ_{max}
Australian and New Zealand	D&H	4						
Canada	Power	3	0.14	270	0.14	270	0.36	400
China	Power	4	0.12	300	0.16	350	0.30	450
Hong Kong	Power	1	0.11	500	0.11	500	0.11	500
India	Power	4	0.10	250	0.14	270	0.34	500
Indonesia	Power	4	0.09	213	0.11	274	0.20	457
Japan	Power	5	0.10	250	0.15	350	0.35	650
Korea	Power	4	0.10	250	0.15	300	0.33	500
Malaysia	Power	4	0.12	250	0.15	300	0.30	500
Philippines	Power	4	0.09	213	0.11	274	0.20	457
Singapore	Log	1						
Taiwan (Chinese Taipei)	Power	3	0.15	300	0.15	300	0.32	500
Thailand	Power	3	0.14	270	0.14	270	0.36	600
United States	Power	3	0.11	210	0.15	270	0.25	360
Vietnam	Power	3	0.07	250	0.09	300	0.14	400

Table 1.2 summarizes the parameters related to the mean velocity profiles specified by the current codes in Asia-Pacific economies (Ge et al. 2010). Majority codes adopt power law to describe the mean velocity profile. However, deviations of gradient height and power index among the APEC economies are obvious.

1.3.1.2 Mean Wind Speed in Tropical Cyclones

Vertical profile of mean wind speed during tropical cyclones is of interest in wind resistant design in cyclone prone regions. Consistent results seemed have been reached that the winds in tropical cycles away from their eyes can be represented by ordinary atmospheric boundary layers, although some reports indicated higher turbulence intensities for tropical cyclone winds. However, there is still insufficient understanding of mean wind profiles near the eye-wall region where the wind speed reaches maximum, because few successful field measurements are available due to difficulty of measurement. Franklin et al. (2000) compared the wind-speed profile near the eye wall with those at the outer vortex regions within 200 miles (320 km) of the cyclone center. Figure 1.2 shows that the wind-speed profiles of both eye wall winds and outer vortex winds show same tendency, i.e., the mean wind speed increased with height, reached its maximum value at a certain height, and then slightly decreased with height up to several thousand kilometers. It is also shown in Fig. 1.2 that the maximum wind speed in the outer vortex region appears at a somewhat higher elevation and is not as pronounced as that in the eye wall. Giang et al. (2007) reviewed the wind speed profiles in tropical cyclones, and showed in Fig. 1.3 that wind speed profiles near the ground can be approximated by logarithmic-law or power-law at both the eye wall and outer vortex regions.

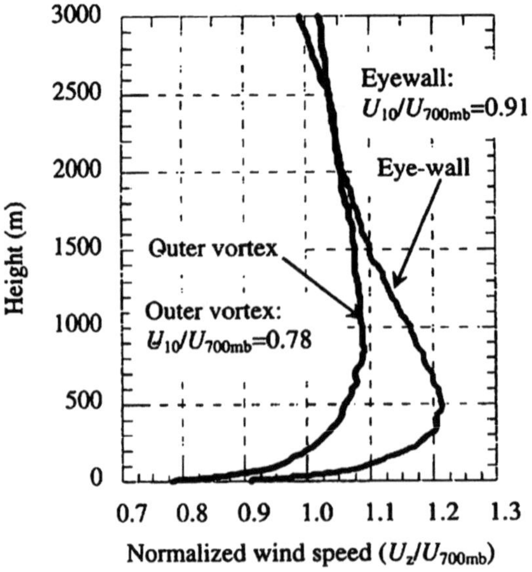

Fig. 1.2 Mean wind-speed profiles of storms at eye wall and outer vortex regions (Franklin et al. 2000)

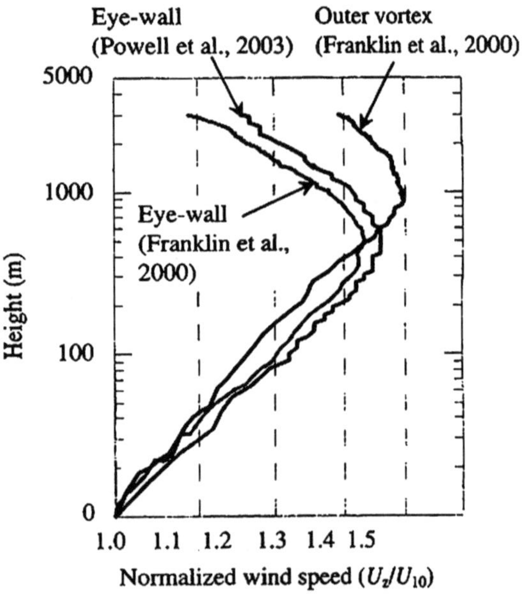

Fig. 1.3 Log-law expression of mean wind-speed profiles

It was also concluded that there is no significant difference between the wind-speed profiles in the eye-wall and outer vortex regions.

There is discussion on the gradient height in tropical cyclones where the wind speed reaches maximum. Amano et al. (1999) measured Typhoon T9612 in Okinawa by Doppler sodars and showed that the gradient height may possibly decrease to lower than 100 m. This result was referred by Holmes (2001) who characterized the wind speed profile in hurricanes and proposed the mean wind speed profiles compressed with a steep logarithmic-type profile up to a height (60–200 m), above which the flow has nearly constant mean wind speed. The result of low gradient velocity height in tropical cyclone was doubted by Giang et al. (2007) who considered that the wind speeds reported by Amano were too low for typhoon wind, and even a thermally neutral condition could not necessarily be assumed. Alternatively, Giang et al. (2007) suggested that the average feature of the gradient heights above the sea is basically around 500–700 m for the eye-wall region and 900 m or higher for the outer vortex region.

A simple Rankine vortex model that consists of the rigid rotating part near the center and an induced free vortex part at surrounding area is widely utilized to describe the horizontal structure of mean velocity of a tropical cyclone. The wind speed increases in proportion to the distance r form the center in the rigid rotating part, and decreases in proportion to the reciprocal of the distance $1/r$ in the surrounding area. Due to the superimposition effect of the rotating velocity and the moving velocity of the tropical cyclone, the wind speed in the right semicircle tends to be higher than the left semicircle in the Northern hemisphere.

1.3.1.3 Mean Wind Speed Profile in Downbursts

It may not be appropriate to define a mean wind speed, but using filtering techniques or conditional averaging it is possible to identify a time varying mean wind speed representing the environmental flow and a fluctuating component due to thermal and mechanical turbulence. The basic structure of a downburst is that of a low-level jet where a maximum velocity is reached at between 50 and 100 m, above which the mean wind speed reduces slowly to the environmental flow (Letchford and Kwok 2004). There are some field data measured by radars and tower anemometer in Australia and the US to give insights into the structure of downbursts. Physical simulations by Letchford et al. (2002) and Mason et al. (2003) demonstrated that the transient flow is best simulated by a roll vortex generated by a pulsed flow rather than the steady wall jet impacting on the ground. The maximum wind speeds occur very near the point of impact, typically within one diameter of the downdraft which has been shown to range from 1 to 2 km in diameter. Oseguera and Bowles (1988) proposed a model of the mean wind speed profiles in a steady wall jet, which the horizontal wind speed component is expressed by:

$$U = \left(\frac{\lambda R^2}{2r}\right)\left\{1 - \exp\left[-\left(\frac{r}{R}\right)^2\right]\right\}\left[\exp\left(-\frac{z}{z^*}\right) - \exp\left(-\frac{z}{\varepsilon}\right)\right] \tag{1.14}$$

where r is the radial coordinate from the center of the wall jet, R the characteristic radius of the jet'shaft', z is the height above the ground, z^* is a characteristic height

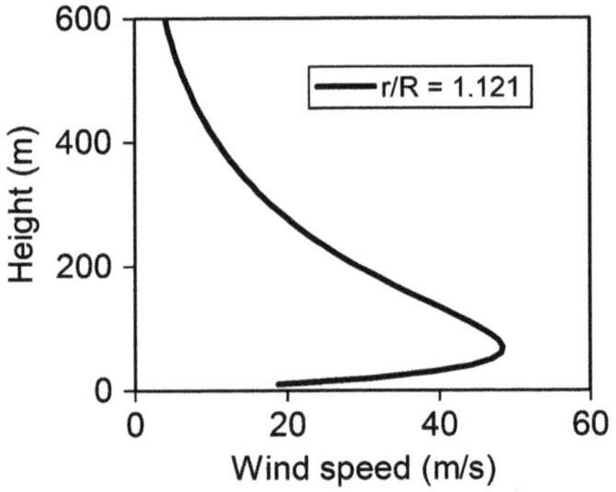

Fig. 1.4 Mean wind speed profile in a downburst (Holmes 2001)

Table 1.3 Fujita scale for tornadoes (Letchford and Kwok 2004)

Fujita scale	Damage description	Approximate wind speed (m/s)
F0	Light	18–32
F1	Moderate	33–50
F2	Considerable	51–70
F3	Severe	71–91
F4	Devastating	92–115
F5	Incredible	116–141

out of the boundary layer, ε is a characteristic height in the boundary layer, λ is a scaling factor, with dimensions of $[s]^{-1}$. Figure 1.4 shows an example of the wind speed profile at the radius of maximum winds (Holmes 2001).

1.3.1.4 Mean Wind Speed in Tornadoes

Tornadoes produce the strongest winds, and the peak gust can be more than 100 m/s. There are two types of tornados: a cold tornado accompanied by a cold front and a warm tornado often accompanied by a tropical cyclone or a developed depression. When a typhoon is approaching, the warm air flows into the front of a typhoon, and the atmosphere becomes unstable. A developed cumulonimbus can be the matrix of the tornadoes. Incidentally, tornados and downbursts are both born from cumulonimbus, it is often difficult to identify them from damage.

Tornado damage is categorized by the Fujita scale from F0 to F5 with approximately 90 % of all operated tornadoes being F2 or less (Letchford and Kwok 2004). The approximate wind speeds corresponding to the Fujita scales are given in Table 1.3. Tornadoes can be modeled as a vertical Rankine vortex surrounded by a free vortex similar to the one of a tropical cyclone.

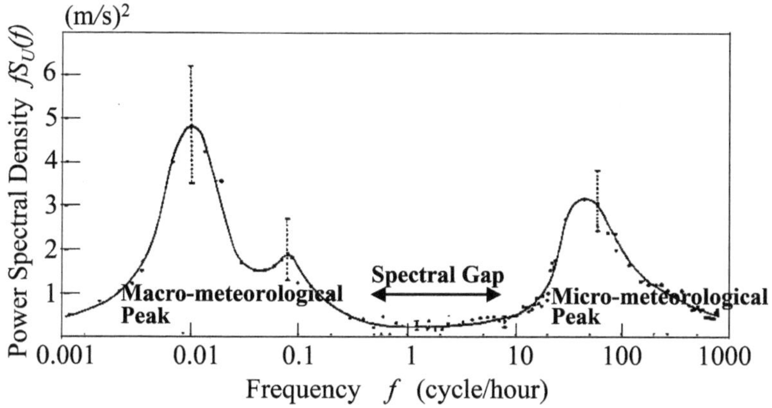

Fig. 1.5 Van der Hoven spectrum of wind speed in a wide frequency range

1.3.2 Wind Speed Fluctuations

Figure 1.5 shows the famous power spectrum by van der Hoven (1957), in which there are two clear peaks. One is so called the "macro-meteorological peak", which shows energy concentration for a period of around 4 days. The other is called the "micro-meteorological peak", which has an energy peak for around 1 min. The former is related to the movement of a large-scale weather system due to global circulations. The latter is related to the turbulence caused by topographic effects, terrain roughness, obstacles around the site and so on. The frequency range between the two peaks contains almost no energy. This range is called the "spectral gap", which separates the two different frequency ranges of the wind fluctuation. The wind fluctuation in the micro-meteorological peak should be taken into account in evaluating dynamic wind forces in building design. However, the wind fluctuation in the low frequency range including the macro-meteorological peak can be taken as the variation of the mean wind speed in building design.

1.3.2.1 Turbulence Intensity

The simplest descriptor of atmospheric turbulence is turbulence intensity. It is defined as the ratio of the standard deviation of velocity fluctuation to the mean velocity:

$$I_u = \frac{\sigma_u}{U},\ I_v = \frac{\sigma_v}{U},\ I_w = \frac{\sigma_w}{U} \tag{1.15}$$

where $\sigma_u{}^2 = \frac{1}{T}\int_0^T u^2 dt,\ \sigma_v{}^2 = \frac{1}{T}\int_0^T v^2 dt,\ \sigma_w{}^2 = \frac{1}{T}\int_0^T w^2 dt$

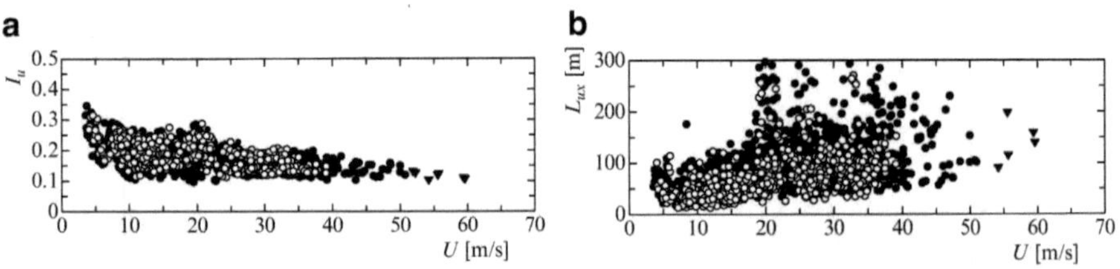

Fig. 1.6 Variations of turbulence intensity and turbulence scale with mean velocity. (**a**) Turbulence intensity; (**b**) Turbulence scale

Standard deviation of velocity fluctuation, σ, can also be obtained by:

$$\sigma^2 = \int_{-\infty}^{\infty} S(f)df \tag{1.16}$$

where $S(f)$ is the power spectral density of the wind speed component. Turbulence intensity is also an important parameter, in both estimating buffeting responses of buildings and interpreting fluid–structure interaction. Major wind load codes specified the vertical profile of turbulence intensity, which varies with height as well as terrain category.

Figure 1.6a shows the variation of longitudinal turbulence intensities, I_u, with mean wind speed obtained during typhoon Maemi. It can be seen that I_u decrease with mean wind speed and remain almost constant when the wind speed becomes high, say 30 m/s. The ratios of the standard deviations among the turbulence components obtained during typhoon Maemi by the sonic anemometers are $\sigma_u/\sigma_v/\sigma_w = 8/1.5/1.0$, which is comparable to the general relations obtained in the neutral surface boundary layer, $\sigma_u/\sigma_v/\sigma_w = 1.9/1.5/1.0$ by Teunissen (1970), or $\sigma_u/\sigma_v/\sigma_w = 2.0/1.5/1.0$ by Counihan (1975).

1.3.2.2 Turbulence Scale

The turbulence scale of the u-component in the longitudinal direction x is an average scale of wind speed fluctuation u, and is defined as:

$$L_{ux} = \int_0^{\infty} R_{uu}(x)dx \tag{1.17}$$

where $R_{uu}(x)$ is the spatial correlation coefficient of u, defined as:

$$R_{uu}(x) = \overline{u(r,t)u(r+x,t)} \tag{1.18}$$

Here, $(\overline{})$ means the operation of temporal averaging.

Turbulence scale can be calculated from the following equation when Taylor's hypothesis of frozen turbulence is applicable.

$$L_{ux} = U \int_0^{\infty} R_{uu}(\tau) d\tau \tag{1.19}$$

where $R_{uu}(\tau)$ is the auto-correlation coefficient of $u(t)$, defined as:

$$R_{uu}(\tau) = \overline{u(r,t)u(r,t+\tau)} \tag{1.20}$$

The turbulence scale varies greatly in the atmospheric boundary layer and the method of calculating it also significantly influences the results. Turbulence scale is often calculated from the following equation, which is deduced from the Wiener-Khintchine relation of the auto-correlation function $R_{uu}(\tau)$ and power spectral density $S_u(f)$.

$$L_{ux} = US_u(0)/2\sigma_u^{\,2} \tag{1.21}$$

As space has three orthogonal directions, even for the u-component, three turbulence scales L_{ux}, L_{uy}, and L_{uz} can be defined. In the isotropic turbulence field, the turbulence scale in the longitudinal direction is twice that in the lateral and vertical directions, that is to say, $L_{ux} = 2L_{uy} = 2L_{uz}$. At points near the ground with a velocity shear, $L_{ux} \approx 3L_{uz}$ is noticed. Similar to turbulence intensity, major wind load codes specified the vertical profile of turbulence scale also, which varies with height as well as terrain category. Figure 1.6b shows that the turbulence scale L_{ux} increases with mean wind speed, although it obviously deviate.

1.3.2.3 Gust Factor and Peak Factor

The gust factor G_u is the ratio of the maximum peak gust with a specified period, say, 10 min, to the mean wind speed.

$$G_u = \frac{U_{\max}}{U} = \frac{U + u_{\max}}{U} \tag{1.22}$$

Peak factor g_u is the ratio of the maximum wind speed fluctuation to the standard deviation of the wind speed fluctuation, that is:

$$g_u = \frac{u_{\max}}{\sigma_u} \tag{1.23}$$

The following relation between gust factor and peak factor can be established:

$$G_u = 1 + \frac{u_{\max}}{U} = 1 + g_u \frac{\sigma_u}{U} \tag{1.24}$$

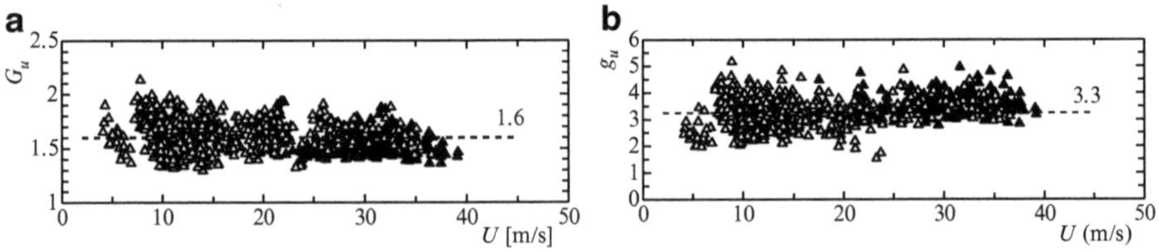

Fig. 1.7 Dependence of gust factor and peak factor on mean wind speed. (**a**) Gust factor; (**b**) Peak factor

Figure 1.7 illustrates the dependences of gust factor and peak factor with mean wind speed measured at typhoon Maemi. The average value of gust factor and peak factor for all data is 1.6 and 3.3 individually, which is almost unchanged with wind speed. However, they deviate greatly.

Inside the existing framework of WMO practice for tropical cyclone conditions, the averaging period for the maximum sustained wind speed is either 1, 3 or 10 min, depending on the tropical cyclone regional association panels and committees. Furthermore, the averaging period to estimate the gust wind speed was not harmonized among nations and international organizations. Obviously a shorter period of averaging will produce values that are greater or smaller than the mean value obtained with a longer averaging period. The gust factor, which represents the ratio of the highest mean wind speed of a shorter averaging period to the mean wind speed of a longer period of observation, is one kind of averaging conversion factor between the two kinds of mean velocities with different averaging periods. This conversion factor is more important at the tropical cyclone condition for the design of buildings and other structures because a structure as a whole and individual elements of that structure have resonance frequencies that response to winds of varying periods. The effects of averaging time and sample duration on estimation of gust wind speed had received intensive attention over several decades and there had been many researches on the wind averaging conversion factors including the gust factor of tropical cyclones. Table 1.4 summarized the wind speed conversions factors under tropical cyclone conditions focusing on the open ocean and coastal situations, which includes the recommendation of WMO and the data of Typhoon Maemi. The tropical cyclone conditions at four exposure categories, i.e., in-land (roughly open terrain), off-land (offshore winds at a coastline), off-sea (onshore winds at a coastline) and at-sea (>20 km offshore) were compared.

1.3.2.4 Power Spectrum Representing Micro-Meteorological Wind Speed Fluctuation

Wind speed randomly fluctuates both in time and in space. This can be represented by the phenomena in which various sizes of eddies are conveyed by the mean flow. A large-size eddy results in a long-period (low-frequency) fluctuation, and a

Table 1.4 Wind speed conversions factors under tropical cyclone conditions (Data at brackets are the recommendations of WMO)

Exposure at + 15 m + 10 m (WMO)		Reference period	Gust factor $G_{\tau,To}$				
			Gust duration (s)				
Class	Description	To (s)	3	60	120	180	600
Off-land	Offshore winds at acoastline	3600	1.49 (1.60)	1.23 (1.22)	1.18 (1.15)	1.16 (1.12)	1.10 (1.06)
		600	1.36 (1.52)	1.13 (1.16)	1.08 (1.09)	1.06 (1.06)	1.00 (1.00)
		180	1.29 (1.44)	1.07 (1.10)	1.02 (1.04)	1.00 (1.00)	–
		120	1.26 (1.42)	1.04 (1.08)	1.00 (1.00)	–	–
		60	1.23 (1.36)	1.00 (1.00)	–	–	–
Off-sea	Onshore winds at a coastline	3600	1.54 (1.45)	1.29 (1.17)	1.23 (1.11)	1.19 (1.09)	1.11 (1.05)
		600	1.43 (1.38)	1.17 (1.11)	1.11 (1.05)	1.08 (1.03)	1.00 (1.00)
		180	1.35 (1.31)	1.09 (1.05)	1.03 (1.00)	1.0 (1.00)	–
		120	1.32 (1.28)	1.06 (1.03)	1.0 (1.00)	–	–
		60	1.28 (1.23)	1.00 (1.00)	–	–	–

small-size eddy a short-period (high-frequency) fluctuation. Distribution of turbulence energy with frequency is described by a power spectral density function (usually abbreviated to "power spectrum").

When only the micro-meteorological peak is picked up, the power spectrum schematically consists of three parts: production range, inertial range, and dissipation range. In the production part at the low-frequency end, the turbulence is generated as large eddies due to instabilities of the mean wind. In the inertial range next to the production range, the large eddies break up and transfer their momentum to smaller eddies, then yet smaller ones, and so on. In the dissipation range at the high-frequency end, the eddies become so small that viscosity becomes predominant and their energy is dissipated as heat. There are many mathematical forms to define the power spectrum representing the micro-meteorological wind speed fluctuation. The Fichtl and McVehill model gives a general mathematical expression, where the power spectrum is defined as:

$$\frac{f\,S_u(f)}{\sigma_u^2} = \frac{4f^*}{(1+\alpha\,f^{*\beta})^{\frac{5}{3\beta}}} \tag{1.25}$$

where

$$f^* = \frac{fL_{ux}}{U},\quad \alpha = 1.5\frac{4^\beta}{b^\beta},\quad b = \frac{1.5^{\frac{1}{\beta}}\beta\,\Gamma\left(\frac{5}{3\beta}\right)}{\Gamma\left(\frac{1}{\beta}\right)\Gamma\left(\frac{2}{3\beta}\right)}$$

and $\Gamma(\cdot)$ is Gamma function. It can be found that

$$S_u(f) \propto f^{-\frac{5}{3}} \tag{1.26}$$

when $f \to \infty$. This is Kolmogorov's hypothesis of local isotropy in the inertial sub-range where the power spectrum follows a power of −5/3 law.

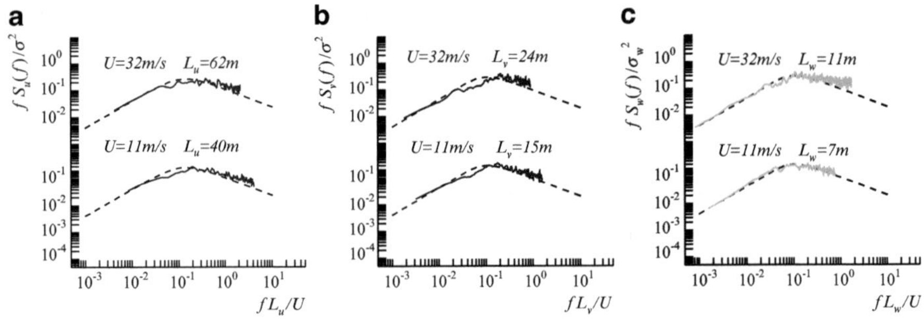

Fig. 1.8 Power spectra of three wind speed components measured by sonic anemometers (*dashed lines*, Karman type). (**a**) Longitudinal; (**b**) Lateral; (**c**) Vertical

It becomes the familiar power spectra such as the Kaimal type when $\beta = 1$, the Panofsky type when $\beta = 5/3$, and the Karman type when $\beta = 2$. The Karman-type spectrum is expressed as follows:

$$\frac{fS_u(f)}{\sigma_u^2} = \frac{4\left(\frac{fL_{ux}}{U}\right)}{\left[1 + 70.8\left(\frac{fL_{ux}}{U}\right)^2\right]^{\frac{5}{6}}} \tag{1.27}$$

Many structural codes adopt the Karman-type spectrum as the expression of wind spectrum.

Figure 1.10 shows the power spectra of the longitudinal, lateral and vertical velocity components of moderate and strong wind measured by sonic anemometers at typhoon Maemi. The power spectra can be well described by Karman-type spectra. There are discussions on the power spectra of wind speed fluctuations during a typhoon. Tamura et al. (1993) shows a case study of power spectra of the wind speed fluctuations measured before and after the eye of a strong typhoon passed over the site, in which both spectra have two peaks. The peak in the lower frequency region corresponds to the wind speed variation similar to that of Fig. 1.8, and the peak in the higher frequency region is the micro-meteorological peak. Meanwhile, many field measurements showed that power spectrum has no significant difference in this case of a typhoon and a frontal depression.

1.3.2.5 Cross-Correlation Functions Between Wind Speed Fluctuation Components

Cross-correlation functions between fluctuation components of wind speeds:

$$\sigma_{uv}^2 = \overline{uv}, \ \sigma_{vw}^2 = \overline{vw}, \quad \text{and} \quad \sigma_{uw}^2 = \overline{uw} \tag{1.28}$$

are also of interest. Their non-dimensional forms are cross-correlation coefficients:

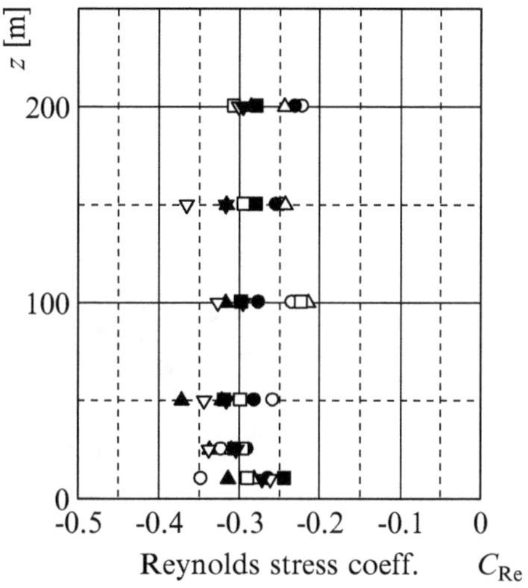

Fig. 1.9 Distribution of Reynolds stress coefficient measured at a 200 m meteorological tower

$$\rho_{uv} = \frac{\sigma_{uv}^2}{\sigma_u \sigma_v}, \ \rho_{vw} = \frac{\sigma_{vw}^2}{\sigma_v \sigma_w}, \quad \text{and} \quad \rho_{uw} = \frac{\sigma_{uw}^2}{\sigma_u \sigma_w} \tag{1.29}$$

and range between -1 and 1.

The cross-correlation functions multiplied by air density ρ_a with a negative sign are Reynolds stresses, which represent stresses due to turbulent flow components that transport momentum acting on the fluid lump in addition to the stress due to the mean flow component. The Reynolds stresses acting on the surface normal to the x-axis are:

$$\tau_{xx} = -\rho_a \sigma_{uu}^2, \ \tau_{xy} = -\rho_a \sigma_{uv}^2, \quad \text{and} \quad \tau_{xz} = -\rho_a \sigma_{uw}^2 \tag{1.30}$$

If there is a wind speed profile where the u-component wind speed increases with height, the 2nd quadrant "ejection" and 4th quadrant "sweep" motions becomes dominant due to the organized structure of the ABL, resulting in a negative Reynolds stress. Figure 1.9 illustrates the Reynolds stress coefficient obtained at a 200 m tower at the Japan National Meteorological Institute at neutral condition. The Reynolds stress coefficient falls in the regime from -0.4 to -0.2 (Cao et al. 2001).

1.3.2.6 Coherence of Wind Speed Fluctuations

Frequency-dependent correlation of wind speed fluctuations u_i and u_j at two separated points i and j can be described by a function known as "coherence", which is the normalized magnitude of the cross spectrum $S_{ui\,uj}(f)$ defined as follows:

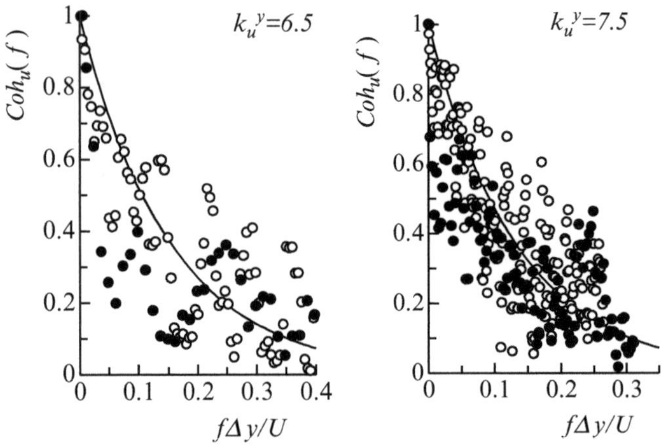

Fig. 1.10 Coherence function at different wind speed obtained by typhoon Maemi (*white circle*: $\Delta y = 6$ m, *black circle*: $\Delta y = 14$ m)

$$Coh(f) = \frac{|S_{ui\,uj}(f)|^2}{S_{ui}(f)S_{uj}(f)} \tag{1.31}$$

The square root of the coherence (root-coherence) can be a cross-correlation coefficient of wind speeds u_i and u_j for a frequency component f. The root-coherence is often approximated by an exponential function of separation distance and frequency:

$$\sqrt{Coh(f)} \approx \exp\left(-\frac{\sqrt{k_y^2 y^2 + k_z^2 z^2}}{\sqrt{U_1 U_2}} f\right) \tag{1.32}$$

where k_y and k_z are empirical constants used to fit the measured data, which depend on many factors including height, surface roughness, stability level and so on.

Figure 1.10 shows examples of the root-coherence of fluctuating wind speeds (u-component) at two different velocities. Significant change of the decay factor with mean wind speed was not observed at typhoon Maemi.

1.3.2.7 Probability Distribution

The probability distribution of fluctuating wind speed is usually assumed to follow a Gaussian distribution. It is interesting to know whether or nor the probability density function of fluctuating wind speed during a strong typhoon also follows a Gaussian distribution. Figure 1.11 shows examples of probability density functions of the wind speed at several mean wind speeds, $U = 8, 21, 32$ m/s, measured by the

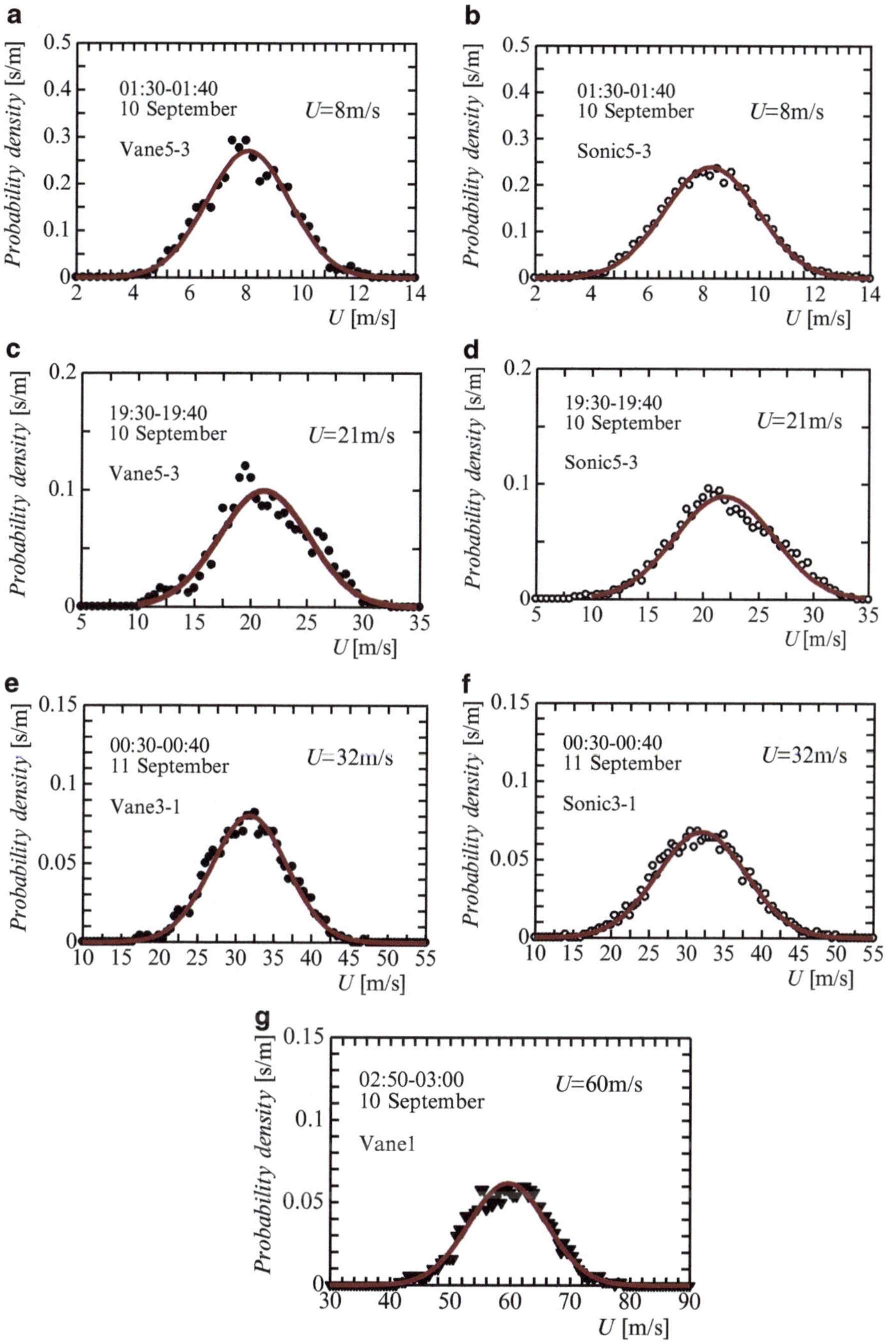

Fig. 1.11 Probability density function of wind speed (*black circle*: Vane; *white circle*: Sonic). (**a**) At 8 m/s (Vane); (**b**) At 8 m/s (Sonic); (**c**) At 21 m/s (Vane); (**d**) At 21 m/s (Sonic); (**e**) At 32 m/s (Vane); (**f**) At 32 m/s (Sonic); (**g**) At 60 m/s (Vane)

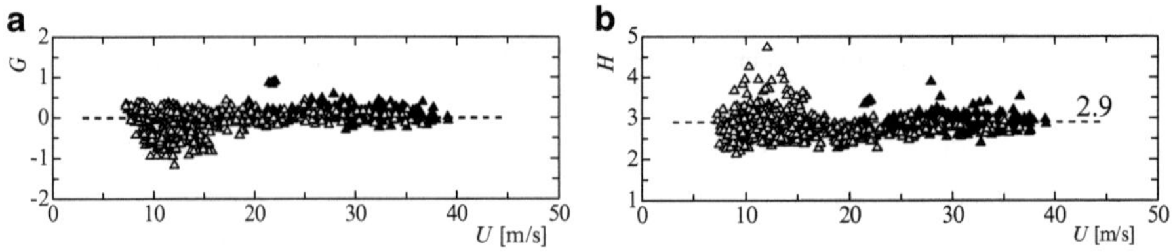

Fig. 1.12 Variations of skewness and kurtosis with mean speed measured by sonic anemometers. (**a**) Skewness; (**b**) Kurtosis

sonic and vane anemometers. It is found that they agree approximately with a standard Gaussian distribution at these wind speeds. The probability density function of fluctuating wind speed at a very high wind speed $U = 60$ m/s is shown in Fig. 1.11e. This also fits a Gaussian distribution.

Figure 1.12 shows the variations of skewness G and kurtosis H with mean speed measured by sonic anemometers. Skewness and kurtosis are independent of wind speed and wind direction. Averaged values of skewness and kurtosis are $G = 0$ and $H = 2.9$, which are close to the values $G = 0$ and $H = 3.0$ for a standard Gaussian distribution.

1.4 Topographic Effects

Topography has a considerable effect on mean and gust wind speeds, mostly notably by accelerating flow over the crests of hills. Topography can also cause shelter effects, such as in valleys and in the lees of hills or ridges. Escarpments, cliffs, ridges, hills and valleys are common topographical features, among which escarpments, ridges and hills are usually considered in wind engineering applications. Considerable research has been undertaken for boundary layer flows over moderate topography.

In codes and standards, the speed-up ratio or topographic multiplier is defined as:

$$\frac{\text{Wind speed at height } z \text{ above the feature}}{\text{Wind speed at height } z \text{ above the flat ground}}$$

Figure 1.13 shows the mean wind speed-up ratio for the flow over a ridge obtained by a wind tunnel test using a split-fiber probe. The speed-up ratio becomes largest at the crest near the ground, and reduces at a higher altitude and with distance from the crest. This is a typical feature of the flow over a ridge. For shallow escarpments and ridges having an upwind slope angle less than 17° with flow normal to their upstanding face, the flow first decelerates from the approaching wind speed to the minimum value near the upwind foot of the slope. Then, it accelerates to a maximum near the crest, and then decelerates again downwind of

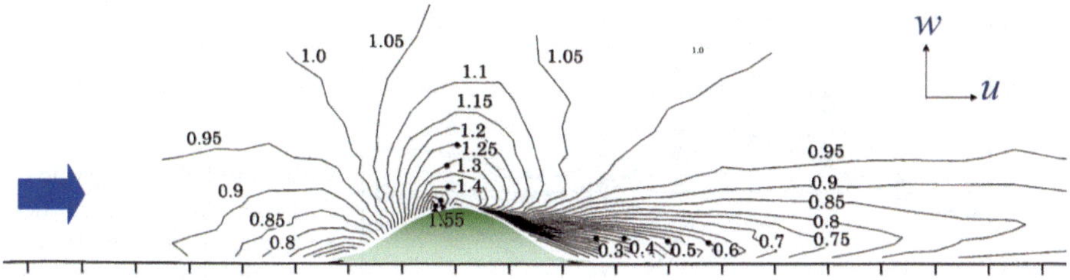

Fig. 1.13 Topographic effects at ridge (Speed-up ratio by wind tunnel test, Tsuchiya et al. 1999)

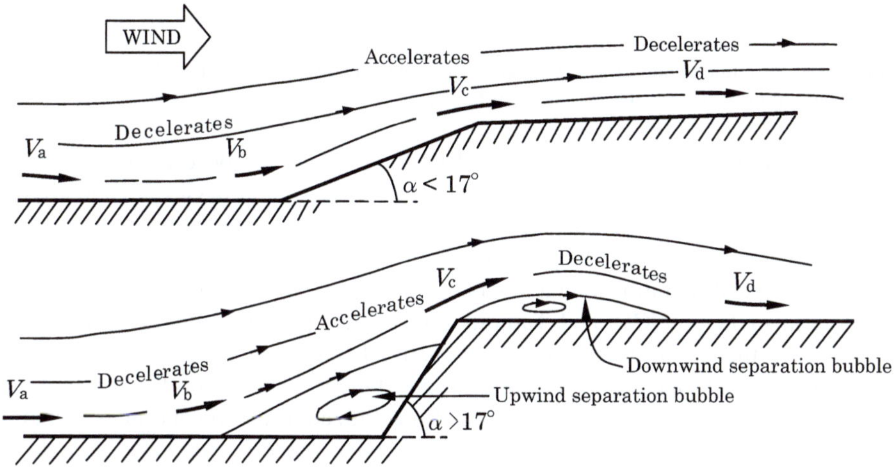

Fig. 1.14 Flow over escarpments (Cook 1985)

the crest (Cook 1985). For steep escarpments and ridges having an upwind slope angle larger than 17°, the flow separates upwind of the slope face and downwind of the crest, forming separation bubbles as shown in Fig. 1.14. The effects of terrain roughness and surface roughness of the feature itself are also recognized.

ASCE 7-02 (2002), Australian/New Zealand Standard (AS/NZS 1170.2) (2002), and AIJ-RLB (2004) have codified topographic effects for hills, ridges and escarpments.

ASCE 7-02 (2002) defines a topographic factor K_{zt} for wind load calculation, and the speed-up ratio for gust wind speed can be expressed as follows:

$$\sqrt{K_{zt}} = 1 + K_1 K_2 K_3 \tag{1.33}$$

K_1 is the factor giving the maximum speed-up effect depending on the shape of the topographic feature, K_2 is the factor for reduction in speed-up effect depending on horizontal distance from the crest, K_3 is the factor for reduction in speed-up effect depending on height. The topographic factor K_{zt} can take the effect of terrain roughness into account.

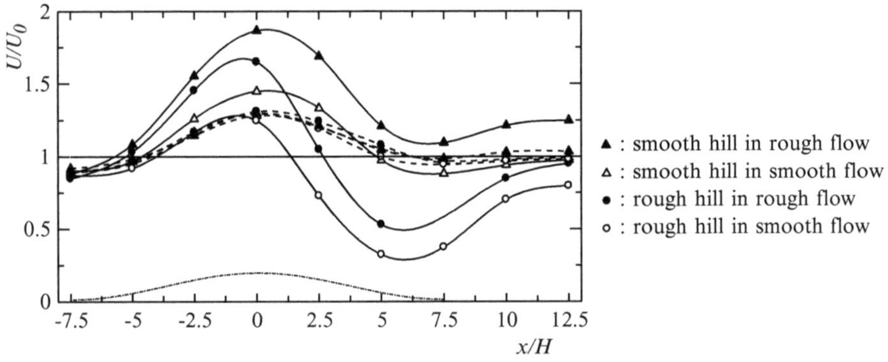

Fig. 1.15 Variation of normalized wind velocity in the streamwise direction (*solid lines*: the inner layer height; *broken lines*: the middle layer height)

Topography factor specified in AIJ-RLB (2004) is a speed-up ratio, which reflects the change in mean wind speed that occurs as wind passes at right angles over escarpments or ridge-shaped topography.

The magnitude of speed up ratio is influenced by many factors. Figure 1.15 shows the speed-up ratio along a low two-dimensional hill at inner layer height and middle layer height (Cao and Tamura 2007). The speed-up ratio is greater than unity when there is a sudden roughness change from rough to smooth condition.

The effects of topography also appear in turbulence, and distort the shape of the turbulence spectra. However, they are not recognized as significant unless separation does not occur. The standard deviation of wind speed fluctuation is thought to be almost constant unless separation occurs. AIJ-RLB (2004) has specified the distortion effects not only of the mean wind speed but also of the turbulence intensity due to topographic features.

References

AIJ-RLB-2004 (2004) Recommendations for loads on buildings. Architectural Institute of Japan, Japan (in Japanese)

Amano T, Fukushima H, Ohkuma T, Kawaguchi A, Goto S (1999) The observation of typhoon winds in Okinawa by Doppler sodar. J Wind Eng Ind Aerodyn 83:11–20

ASCE 7-02 (2002) Minimum design loads for buildings and other structures. American Society of Civil Engineers, Reston

Australian/New Zealand Standard (AS/NZS 1170.2) (2002) Structural design actions, Part 2: wind actions. Standards Australia & Standards, New Zealand

Cao S, Nishi A, Hirano K, Ozono S, Miyagi H, Kikugawa H, Matsuda Y, Wakasugi Y (2001) An actively controlled wind tunnel and its application to the reproduction of the Atmospheric Boundary Layer. Bound-Layer Meteorol 101(1):61–76

Cao S, Tamura T (2007) Effects of roughness blocks on atmospheric boundary layer flow over a two-dimensional low hill with/without sudden roughness change. J Wind Eng Ind Aerodyn 95:679–695

Cao S, Tamura Y, Kikuchi N, Saito M, Nakayama I, Matsuzaki Y (2009) Wind characteristics of a strong typhoon. J Wind Eng Ind Aerodyn 97:11–21. doi:10.1016/j.jweia.2008.10.002
Cook NJ (1985) The designer's guide to wind loading of building structures Part 1. Butterworths, London
Counihan J (1975) Adiabatic atmospheric boundary layers—a review and analysis of data from the period 1880–1972. Atmos Environ 9:871–905
Deaves DM, Harris RI (1981) A mathematical model of the structure of strong winds. C.R.I.A. Report 76. Construction Industry Research and Information Association, London
Franklin JL, Black ML, Valde K (2000) Eye wall wind profiles in hurricanes determined by GPS drop wind sondes, extracted from Le et al. (2007)
Yaojun Ge, Xinyang Jin, and Shuyang Cao (2010) Comparison of APEC wind loading codification and revision of Chinese National Code. In: Proceedings of 6th workshop on regional harmonization of wind loading and wind environmental specifications in Asia-Pacific economies, Korea
Holmes JD (2001) Wind loading of structures. Spon Press
van der Hoven I (1957) Power spectrum of horizontal wind speed in the frequency range from 0.0007 to 900 cycles per hour. J Meteorol 14:160–164
Giang TG, Tamura Y, Cao S, Matsui M (2007) Wind-speed profiles in tropical cyclones. J Wind Eng 4(1):39–48
Letchford CW, Mans C, Chay MT (2002) Thunderstorms, their importance in wind engineering—a case for the next generation wind tunnel. J Wind Eng Ind Aerodyn 90:1415–1433
Letchford CW, Kwok K (2004) Simulation and flow measurement of atmospheric boundary layer flows. In: State-of-the-art wind tunnel modeling and data analysis techniques for infrastructure and civil engineering application, Croucher Advanced Study Institute, Hong Kong
Mason M, Letchford CW, James DL (2003) Pulsed jet simulation of a thunderstorm downburst. In: Proceedings of 11ICWE, Lubbock, 2249–2256
Oseguera RM, Bowles RL (1988) A simple analytic 3-dimensional downburst model based on boundary layer stagnation flow, NASA Technical Memorandum 100632. National Aeronautics and Space Administration, Washington
Powell MD, Vickery PJ, Reinhold TA (2003) Reduced drag coefficient for high wind speeds in tropical cyclones. Nature 422:279–283
Simiu E, Scanlan R (1986) Wind effects on structures, 2nd edn. Wiley, New York
Teunissen HW (1970) Characteristics of the mean and turbulence in the planetary boundary layer. UTIAS Review No. 32
Tsuchiya M, Kondo K, Sanada S (1999) Effects of micro-topography on design wind velocity—characteristics of wind velocity amplification around various topographies. Summaries of Technical Papers of Annual Meeting, Architectural Institute of Japan, 119–120
Tamura Y, Shimada K, Hibi K (1993) Wind response of a tower (Typhoon observation at the Nagasaki Huis Ten Bosch Domtoren). J Wind Eng Ind Aerodyn 50:309–318
WMO (2010) Guidelines for converting between various wind averaging periods in tropical cyclone conditions. World Meteorological Organization, WMO/TD-No.1555, Geneva
Xu YL, Chen J (2004) Characterizing non-stationary wind speed using empirical mode decomposition. J Struct Eng ASCE 130(6):912–920

Chapter 2
Estimation of the Design Wind Speed

Michael Kasperski

Abstract The modern approach to the specification of the design wind speed favours the explicit format which directly presents the design value of the wind speed instead of hiding the value behind the product of the characteristic wind speed and a partial factor. This approach leads to better reliability consistency in case of highly variable geographic conditions and it avoids underestimations for systems with unfavourable non-linear characteristics. Some leading codes therefore have adopted this concept. This chapter presents tentative target values of the exceedance probability of the design wind speed for the ultimate limit state with reference to the design working life considering six importance classes for buildings and structures. Additionally, tentative target values are given for the serviceability limit state with reference to a single year. For the practical application, strategies are discussed in regard to the quality check of data, the sampling of the appropriate ensembles and the estimation of the design value considering an appropriate confidence interval. Since statistical uncertainties remain large even for several decades of observations, a new approach is presented which leads to consistent values of the design wind speed. Additionally, the complex topic of directionality is addressed on the example of storms induced by strong frontal depressions.

Keywords Best estimate • Design wind speed • Design working life • Directionality • Structural class

M. Kasperski (✉)
Department of Civil and Environmental Engineering Sciences, Ruhr-Universität Bochum, 44780 Bochum, Germany
e-mail: michael.kasperski@rub.de

Y. Tamura and A. Kareem (eds.), *Advanced Structural Wind Engineering*,
DOI 10.1007/978-4-431-54337-4_2, © Springer Japan 2013

2.1 Introduction

2.1.1 Terminology

Design wind speed in a narrower sense means the design value of the wind speed which is used to evaluate the wind resistance of a structure in the ultimate limit state. Together with the design value of the aerodynamic coefficient and a deterministic value of the air density the design wind speed leads to the design wind load as follows:

$$w_{des} = \frac{1}{2} \cdot \rho \cdot {v_{des}}^2 \cdot c_{des} \tag{2.1}$$

w_{des}: design wind load; v_{des}: design value of the wind speed; c_{des}: design value of the aerodynamic coefficient; ρ: air density

The explicit format in Eq. (2.1), i.e. the direct specification of the design value, origins from earthquake engineering and has first been adopted in wind engineering in the Australian Standard in 1989 (AS 1170). It has been adopted in the recently revised ISO-document in 2009 (ISO 4354 2009), it is used in the actual ASCE (ASCE SEI 7 10) and of course is still implemented in the latest Australian/New Zealand Standard (AS/NZS 1170). As alternative approach to the traditional format, which hides the design value behind the product of a characteristic value and a corresponding partial factor, the explicit format has two advantages: it leads to better reliability consistency in case of highly variable geographic conditions and it avoids underestimations for systems with unfavourable non-linear characteristics as they occur e.g. for the resonant part of the along wind response of structures which increases rather with v^3 than with v^2. Furthermore it meets the basic demand to transparency and allows the designing engineer to identify that the wind resistant design is based on very rare events which are unlikely to occur in the anticipated lifetime of buildings and structures which is used as reference period for specifying appropriate target values of the exceedance probability of the design event.

Design wind speed in a broad sense also includes those values of the wind speed which are used in the analysis of the different serviceability limit states. Basically, standards distinguish between reversible and irreversible limit states. As reference period, usually a single year is used.

At least for the specification of the ULS-design wind speed, larger extrapolations are required. Then, two types of uncertainties have to be considered: Statistical uncertainties as they are to occur due to the confined ensemble of observations, and uncertainties due to inadequate knowledge of the process e.g. in regard to long term trends as they might be induced by the climatic change.

Basically, uncertainties may lead to over- or under-estimations of the true design value. While over-estimations may lead to unnecessary expensive structures, under-estimations may lead to structures which have a lower reliability than intended. The latter case can be avoided by introducing an appropriate one-sided confidence interval, i.e. the probability of erroneously estimating a too low value of the design wind speed is limited to an acceptable low value. This error probability is the complementary to the confidence. In structural engineering, typically a confidence of 75 % is applied when estimating the design value of the resistance based on tests (EN 1990). The same target can be adopted when estimating design values of actions.

2.1.2 *Basic Reliability Demands*

The basic reliability demands can be specified with reference to a single year or with reference to the projected or design working life which is the period for which the structure is intended to perform its required functions. The latter approach allows a more economic design of the build environment. Typical values for the design working life are 1–5 years for temporary structures, 10–40 years for industrial structures and buildings, 60–80 years for residential buildings and 100–150 years for bridges. Most of today's codes do not allow considering the design working life as an explicit design variable; instead, they often consider implicitly a general value of 50 years and relate the design event to a single year.

Basically, target values with reference to the design working life can be translated to corresponding annual exceedance probabilities as follows:

$$p_1 = 1-(1-p_L)^{1/L} \tag{2.2}$$

p_L: exceedance probability in the design working life; p_1: annual exceedance probability; L: design working life in years

It is reasonable to classify structures in regard to the consequences of a failure in terms of risk to life, economy and social fabric. On the basis of modern Standards like ISO 22111-2007: *Bases of design of structures—General requirements* (ISO 22111 2007), the following classes can be distinguished:

0	Structures presenting no hazard to life, e.g. free field solar parks
1	Structures presenting a low degree of hazard to life and other properties, e.g. agricultural buildings without regular human occupancy or commercial green houses without public access
2	Structures presenting an ordinary degree of hazard to life, e.g. office buildings, commercial buildings, residential buildings, factories
3	Buildings which as whole contain people in crowds, e.g. high-rise buildings, stadia and concert halls
4	Buildings or structures having a special post disaster function, e.g. hospitals, schools, bridges and transmission lines
5	Buildings which are designed to offer shelter for people during extreme storm events, e.g. hurricane or tornado shelters

The target value of the exceedance probability of the design wind load for structural safety is specified in the following for the design work life (Table 2.1). For the serviceability limit state respective target values are presented with reference to the period of 1 year.

The influence of the design working life on the yearly exceedance probability of the design wind speed for the ultimate limit state is presented in Table 2.2 for values

Table 2.1 Tentative target values of the exceedance probability of the design wind load for the ultimate limit state (reference period: design working life) and serviceability limit state (reference period: 1 year)

Structural class	0	1	2	3	4	5
ULS in lifetime	0.4	0.2	0.1	0.05	0.025	0.005
SLS in 1 year	0.2	0.1	0.05	0.02	0.01	0.002

Table 2.2 Influence of the design working life on the yearly exceedance probability of the design wind load for the ultimate limit state

	Structural class					
L (years)	0	1	2	3	4	5
1	1/2.5	1/5	1/10	1/20	1/40	1/200
2	1/4.5	1/10	1/20	1/40	1/80	1/400
5	1/10	1/25	1/50	1/100	1/200	1/1,000
10	1/20	1/45	1/95	1/200	1/395	1/1,995
20	1/40	1/90	1/190	1/390	1/790	1/3,990
50	1/100	1/225	1/475	1/975	1/1,975	1/9,975
80	1/160	1/360	1/760	1/1,560	1/3,160	1/15,960
100	1/200	1/450	1/950	1/1,950	1/3,950	1/19,950

of the design working life from 1 to 100 years. For a design working life of 50 years, the target values of the yearly exceedance probability range from 1/100 for class 0 to about 1/10,000 for class 5.

For the evaluation of the wind resistance of existing buildings it may be necessary to extend the list of structural classes to buildings having specific cultural and historical values. These buildings have to be conserved for future generations, i.e. the intended future working life is much larger than 100 years. A first tentative value for the yearly target exceedance probability for this structural class is 1/20,000.

For the specification of the two design values—the design wind speed and the design value of the aerodynamic coefficient—only one demand is available which is meeting the specified target exceedance probability of the design wind load. Consequently, for a specific problem, there is an infinite number of solutions. A more general approach, which is valid for all kind of structures, specifies a common design value for the wind speed and the individually required design value of the aerodynamic coefficient. Then, the specified target values for the exceedance probability of the design wind load can be used as target values for the design wind speed as well.

2.2 Data Basis

2.2.1 Basic Information Provided by National Weather Services

Meteorological observations form the data basis for the statistical analysis of the extreme wind climate. Although already in 1873 the International Meteorological Organization was founded, still today in different countries different information on wind is sampled. As a matter of fact, the sampled information not always meets the needs of wind engineers. In Germany, per day 24 hourly mean wind speeds are given together with their corresponding wind direction. Additionally, per day the largest gust wind speed and the time of its occurrence is listed. However, the wind direction of the gust is not available. In India, wind speeds are sampled as 10-min mean values. Per hour, only one value is given which is the last 10-min mean before the full hour. The wind direction is given only for each third hour. However, the largest gust wind speed per day comes with its direction.

The German Weather Service (Deutscher Wetterdienst DWD) provides data for the wind speed with a step width of 0.1 m/s. While today's modern ultrasonic sensors are able to provide high quality data for this fine resolution, cup anemometers counting e.g. three pulses per full rotation may not be able to resolve the above step width. Each pulse corresponds to a specific length, which is different for different instruments. A typical length is 0.12 m, i.e. one pulse per second means 0.12 m/s, two pulses per second 0.24 m/s and three pulses per second 0.36 m/s. If these results are re-sampled with a step width of 0.1 m/s, the result for 0.36 is sampled in the 0.4 m/s class. The same sampling error occurs for wind speed levels of 0.9, 1.5, 2.1 m/s and so on. This distortion can be seen in the plot of the frequency of wind speeds in 0.1 m/s classes. Older data are influenced by the above sampling error, while more recent data are sampled correctly. In the overall count, the neighbouring classes 0.4, 1.0, 1.6 m/s and so on are over-represented. The sampling error disappears if the class width is increased to 0.6 m/s, since 0.6 is an integer multiple of the basic length of 0.12 m. A similar effect is obtained when the wind speed is sampled in different dimensions, e.g. first in knots and later in km/h. Since knots and km/h have no integer common multiple, the sampling error can not be removed but only seems to vanish when the class width is increased.

2.2.2 *Quality Control and Corrections*

The quality of the statistical analysis of the meteorological observations depends on the quality of the input data. The two basic demands to the data set are completeness and consistency, i.e. over the observation period there should be no longer gaps of missing data, and the conditions around the measuring station should remain constant. Supposing that storms in consecutive meteorological years are independent allows using data which consist of non-consecutive meteorological years, i.e. gaps of full meteorological years are acceptable. Thereby, a meteorological year is defined as a twelve month cycle which provides a meaningful comparison of the data. Basically, the meteorological year may differ from the calendar year. In case of wind data, the corresponding meteorological year can be called aeolian year (Kasperski 2007). For Germany, an aeolian year has to include a complete winter storm season and a complete summer thunderstorm season. An appropriate start of the German aeolian year then is given with the 1st of May.

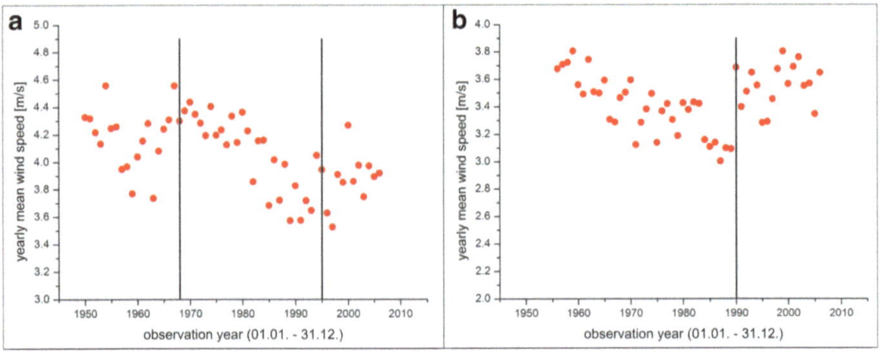

Fig. 2.1 Typical examples of disturbed data sets. (**a**) Hamburg-Fuhlsbüttel; (**b**) Saarbrücken/Ensheim

Considerable violations of the second demand usually can be detected in the time history of the yearly means as shown in Fig. 2.1 for two German stations. For Hamburg, the yearly means show a clear decreasing trend, indicating the growth of the city over the years. Therefore, the mast has been relocated in 1968 and in 1995. At Saarbrücken, there was a relocation of the mast in 1990. Before the relocation, the data show a decreasing trend.

In Germany, undisturbed stations are scarce. A correction of the disturbed data set can be obtained by fitting a trend to the observed trace of the yearly means and multiply all observed wind speeds with the inverse of the trend. This method requires additionally an anchoring value which translates the effects of the surrounding roughness to the standard reference roughness of flat open country. Since the roughness effects usually are different for different wind directions, the correction has to be applied to each sector individually. For many German stations, the German Weather Service (Traup and Kruse 1996) has produced a snap shot of the surrounding roughness considering roughness changes in a distance of up to 10 km. These values, which represent the roughness in the mid 1990s, can be used as anchor points. Figure 2.2 shows as an example the surrounding roughness for Hamburg.

2.2.3 *Parent Distribution*

The lower range of wind speeds often is called the parent distribution, suggesting that strong storms are born from small winds (“*they sow the wind and reap the whirlwind*”). However, a closer look to the different driving mechanisms behind e.g. a sea-breeze and a fully developed winter storm prohibits extrapolating from the parent distribution to the seemingly corresponding extremes.

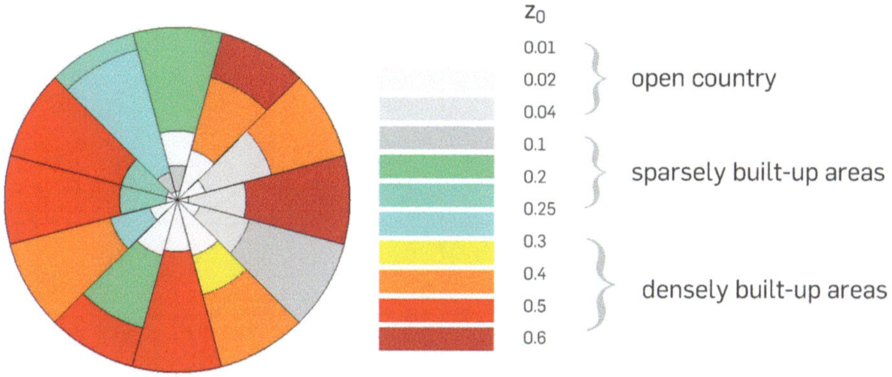

Fig. 2.2 Roughness around the meteorological station Hamburg Fuhlsbüttel, radius 3 km

Basically, the observed hourly wind speeds have to be understood simply as a measure of the velocity of moving air. A widely accepted model for the relative frequency of hourly mean wind speeds is given with the Weibull distribution. Its probability density and corresponding cumulative probability distribution are given as follows:

$$f(x) = \frac{k}{x_0} \cdot \left(\frac{x}{x_0}\right)^{k-1} \cdot \exp\left(-\left(\frac{x}{x_0}\right)^{k}\right) \tag{2.3}$$

$$F(x) = 1 - \exp\left(-\left(\frac{x}{x_0}\right)^{k}\right) \tag{2.4}$$

with k: shape parameter; x_0: scale parameter

If a local wind climate is characterized by considerable periods of calms, the above approach has to be applied to the non-zero hourly means. It is important to note, that the Weibull distribution is an efficient expression to describe the observed relative frequencies of wind speeds; however, it is not the "true" distribution of "parent" winds. This is illustrated based on the observed wind speeds at the meteorological station Düsseldorf airport for the years from 1952 to 1999 in Fig. 2.3. A first solution is obtained from a least-square fit of the counted relative frequencies in a linear range for f(x). This solution is an appropriate model for wind speeds from 0 to 10 m/s. The model is less appropriate for higher wind speeds up to 15 m/s. A second fitting method uses the number of hours above a threshold value and performs the least-square fit in the log-scale for (1-F(x)). The resulting solution is an appropriate model for wind speeds in the range from 7.5 to 15 m/s; however it fails in the lower wind speed range.

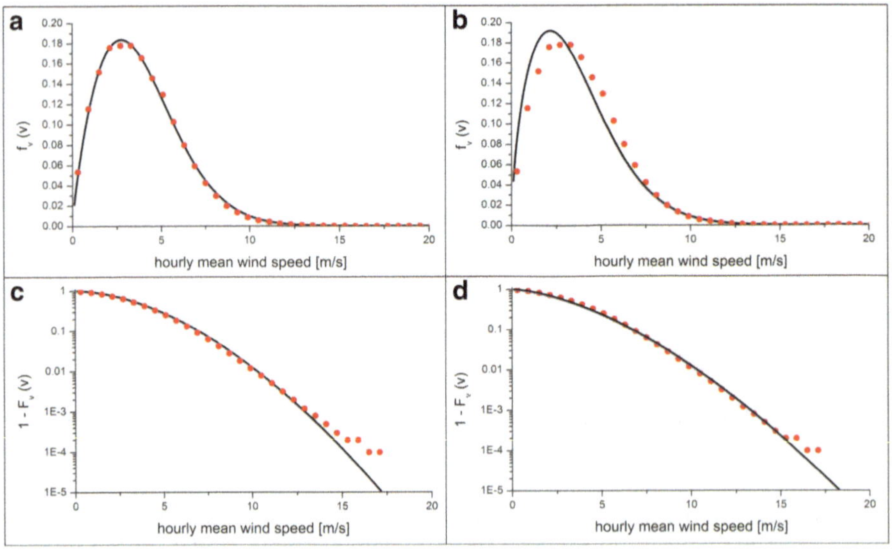

Fig. 2.3 Fitting the Weibull distribution to the "parent" wind climate at Düsseldorf airport (Kasperski 2009a). (**a** and **c**) $x_0 = 4.36$ m/s· k = 1.78; (**b** and **d**) $x_0 = 3.97$ m/s· k = 1.60

The mean value m and the standard deviation σ of the Weibull-distributed process x are connected to the describing parameters x_0 and k as follows:

$$m = x_0 \cdot \Gamma\left(1 + \frac{1}{k}\right)$$
$$\sigma = x_0 \cdot \left(\Gamma\left(1 + \frac{2}{k}\right) - \Gamma^2\left(1 + \frac{1}{k}\right)\right)^{1/2} \tag{2.5}$$

The data in Fig. 2.3 are obtained as an average over 48 years. Supposing that the "parent" distribution is sufficiently described by the Weibull approach, the inter-annual variability of e.g. the annual mean wind speed can be obtained based on simulations. Figure 2.4 shows the respective results in terms of the non-exceedance probability of the random annual mean wind speed assuming $x_0 = 4.36$ m/s and k = 1.78. The coefficient of variation of the annual mean wind speed is 0.006, using an average over 365.25·24 = 8,766 independent hours per year. The observed coefficient of variation is larger by a factor of 12.8. Therefore, the trace of the observed non-exceedance probabilities of the random annual wind speeds deviates drastically from the theoretically expected values. Obviously, the "parent" distribution is not completely described by a simple Weibull model. Furthermore, the strong frontal depressions which induce the storm conditions at Düsseldorf

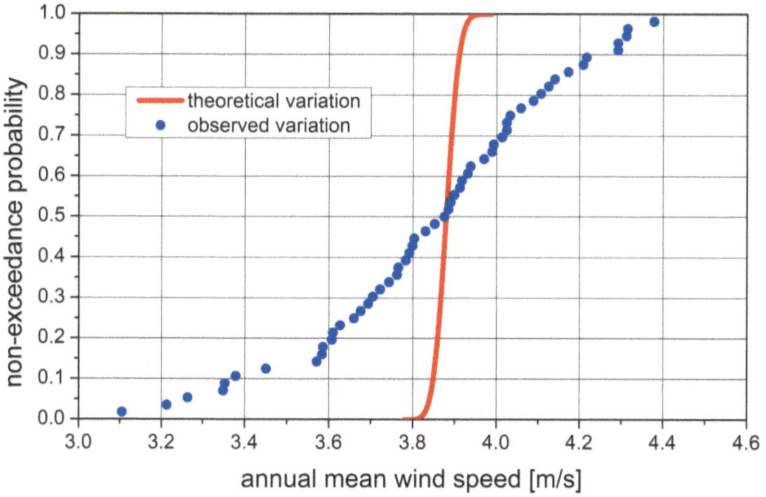

Fig. 2.4 Probability distribution of the annual mean wind speed—theoretical range assuming that the hourly wind speeds are Weibull-distributed and observed range (Düsseldorf airport 1951–1999)

show a clear seasonal variation. More than 90 % of the storms occur in the months from October to March, and there is no reason why the "parent" wind speeds from the summer can be assumed to influence the winter storms.

2.2.4 Sampling Ensemble of Extremes

The intuitive strategy for sampling the ensemble for the extreme value statistics uses yearly extremes typically based on the western calendar year. The basic advantage of this approach lies in its simplicity. All statistical information is already provided with reference to a single year. The only problem which might occur is obtained for a storm blowing at the turn of the year. Without additional control, this event is counted twice. The problem can be avoided by introducing an appropriate meteorological year which starts already outside the storm season. Figure 2.5 shows two typical examples of the seasonal distribution of storms induced by strong frontal depressions.

Although yearly extremes form an easy access to extreme value statistics, there are some possible shortcomings to consider. Many extreme wind climates around the world have as specific feature that storms tend to occur in families or clusters. Then, in 1 year—or better in one storm season—several storms are observed, while in other years the observed largest wind speed remains relatively small and does not correspond to strong storm conditions. Obviously, yearly extremes are not able to sample the complete set of strong storm events. In other words: the set of yearly

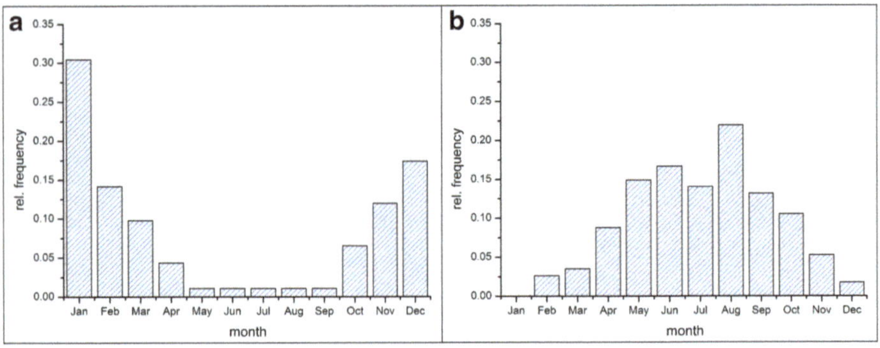

Fig. 2.5 Seasonal variation of the relative frequency of strong frontal depressions (data for Neumayer III from (König-Langlo 2006)) (**a**) Düsseldorf, Germany, 1952–2007; (**b**) Neumayer III, Antarctica, 1982–2004

extreme wind speeds may contain events which have nothing to do with strong storm conditions, while other events with strong storm conditions are ignored. The statistical analysis based on this inappropriate set may lead to biased results.

Furthermore, yearly extremes are not sufficient to evaluate the exceedance probability of the design wind load. Since the extreme aerodynamic coefficient is a random variable, the exceedance of the design wind load amplitude might occur for the second or third largest storm of one season (Cook 1990).

A more consistent approach is obtained by sampling all independent events above a specific threshold value. The threshold value has to be large enough to ensure that only strong events are sampled and it has to be small enough to give sufficient number of events for the statistical analysis. Sampling of independent events is easy in case of short storm system like e.g. thunderstorms, supposing they do not occur shortly before midnight and are not registered with two gust wind speeds, one before and one after midnight. For storms induced by strong frontal depressions, the threshold value usually is exceeded for several hours. Only the strongest hour can be collected in the ensemble, since the additional storm hours are not independent events. Additionally, this storm type has as characteristic feature that it might take a deep breath, i.e. there might occur a phase in the storm where wind speeds decrease and later increase again (Fig. 2.6). Although wind speeds during this phase may drop below the threshold value, the freshening-up wind speeds still belong to the same event. Strictly speaking, independent events can only be identified taken a closer look to large scale weather maps. As a fair approach, however, it has become common practice to assume a new event if the gap of small wind speeds exceeds say 12 or 24 h.

Tropical cyclones show a similar behaviour, however, for a slightly different time scale. The example in Fig. 2.6 shows measured mean values in terms of 15-min means for hurricane Wilma (Florida Coastal Monitoring 2007). Clearly, more than one 15-min period has to be considered for the accumulation of the exceedance probability of the design wind load.

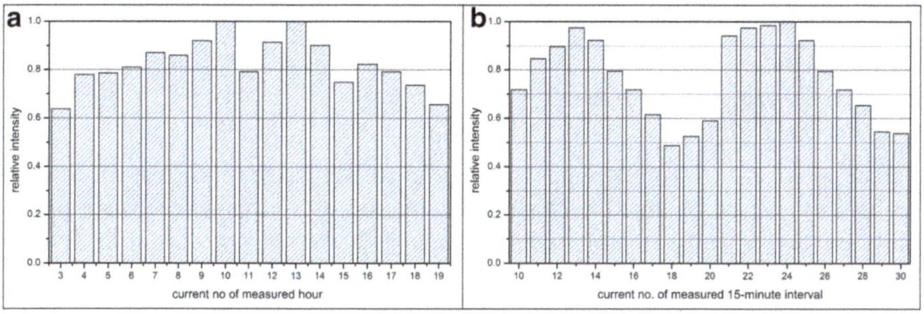

Fig. 2.6 Decrease of wind speeds and freshening up for two storm examples (Kasperski 2009a). (**a**) Frontal depression, Düsseldorf airport, 1954; (**b**) hurricane Wilma, Florida, USA, 2005

Table 2.3 Storm types and appropriate time scale

Storm type	Time scale
Storms induced by strong frontal depressions (h)	1
Overshooting gust fronts in frontal depressions (s/min)	30/1
Tropical cyclones (min/min)	10/15
Thunderstorms (s/min)	30/1
Gravity winds (h)	1
Other local storms	unknown

Extreme value theory is valid only for independent events having the same physical origin. Unfortunately, this does not mean that an extreme value analysis will not work on a set of data which contains—literally speaking—apples and pears, or on sets of data which artificially subdivide the ensemble of one physical process. However, results obtained from such distorted statistics may be considerably biased, and conclusions drawn from these results may be wrong. It is important to note that subdividing the ensemble of one storm into independent ensembles for different wind sectors is inappropriate.

So the first step in the analysis of any extreme wind climate is to set up independent ensembles for each independent storm phenomenon which occurs (Gomes and Vickery 1978). It is important to note that for each storm type a different time scale applies. Probably the largest time scale is obtained for gravity winds which in the Antarctic may last for several days. The smallest time scale belongs to thunderstorms which lead to high wind speeds at a specific location for only a few minutes. Table 2.3 summarizes the design-decisive storm types and recommended basic time scales to describe the time history of the individual events. Tornadoes are not included in this list since usually structures will not be designed to withstand the enormous wind speeds in such storms.

2.3 Basic Statistics

2.3.1 Combination Rules

For the approach based on extremes of independent events, the yearly non-exceedance probability can be obtained from the combination of the non-exceedance probability per event and the probability of having N storms per season as follows (Kasperski 2007):

$$p(v \leq v_{ref} |\ year) = \sum_{N=0}^{\infty} p(N) \cdot p(v \leq v_{ref})^N \tag{2.6}$$

$p(N)$: probability of N storms per year; $N = 0, 1, 2, 3, 4, \ldots$
$p(v \leq v_{ref})$: non-exceedance probability per independent event

The exceedance probability of the design wind load for more than one storm type can be obtained as the complementary to the event that the design wind load is not exceeded in any of the contributing storm types:

$$p_{year}(w > w_{des}) = 1 - \prod_{j} p_{year,j}\ (w \leq w_{des}) \tag{2.7}$$

j: storm type, e.g. frontal depressions, gust fronts, thunderstorms

Finally, the exceedance probability of the wind speed in a specific sector Φ is obtained as:

$$p(v > v_{ref}|\Phi) = p(v > v_{ref}) \cdot f_{rel}(\Phi) \tag{2.8}$$

$p\ (v > v_{ref})$ exceedance probability based on the whole ensemble of extremes of the considered storm type

$f_{rel}(\Phi)$ relative frequency of extremes from the considered storm type in the chosen sector

2.3.2 Generalized Extreme Value Distribution

The generalized extreme value (GEV) distribution usually can be applied to the trace of the yearly non-exceedance probabilities obtained from Eq. (2.6). The GEV distribution combines the three families of extreme value distributions which are:

- Type I: Gumbel distribution
- Type II: Fréchet distribution
- Type III: Reverse Weibull distribution

There are two options for the notation of the GEV. The first option makes use of the first two moments of the distribution and the cumulative probability distribution is obtained as:

$$F(x) = \exp\left[-\left(f_1 - f_2 \cdot \frac{x-m}{\sigma}\right)^{1/\tau}\right] \tag{2.9}$$

m: mean value; σ: standard deviation; τ: shape parameter

The coefficients f_1 and f_2 depend on the actual shape parameter τ and are given as follows:

$$f_1 = \Gamma(1+\tau) \tag{2.10}$$

$$f_2 = \sqrt{\Gamma(1+2\cdot\tau) - {f_1}^2}$$

Γ: Gamma function

The second notation is based on the location parameter μ and the scale parameter s:

$$F(x) = \exp\left[-\left(1 - \tau \cdot \frac{x-\mu}{s}\right)^{1/\tau}\right] \tag{2.11}$$

For positive shape parameters, type III distributions are obtained. This family of distributions has a finite upper tail, which can not be exceeded with probability of 1. The maximum value is given as:

$$x_{max} = m + \sigma \cdot \frac{f_1}{f_2} \tag{2.12}$$

Negative shape parameters lead to the family of the type II distributions. These distributions have a finite lower tail. For $\tau = 0$, the extreme distribution type I is obtained, i.e. in the limit Eq. (2.9) becomes:

$$F(x) = \exp\left[-\exp\left(-\left[\gamma + \frac{\pi}{\sqrt{6}}\frac{x-m}{\sigma}\right]\right)\right] \tag{2.13}$$

Γ: Euler constant = 0.577216

The alternative notation is as follows:

$$F(x) = \exp\left[-\exp\left(-\left(\frac{x-\mu}{s}\right)\right)\right] \tag{2.14}$$

The type I distribution is unlimited to both sides.

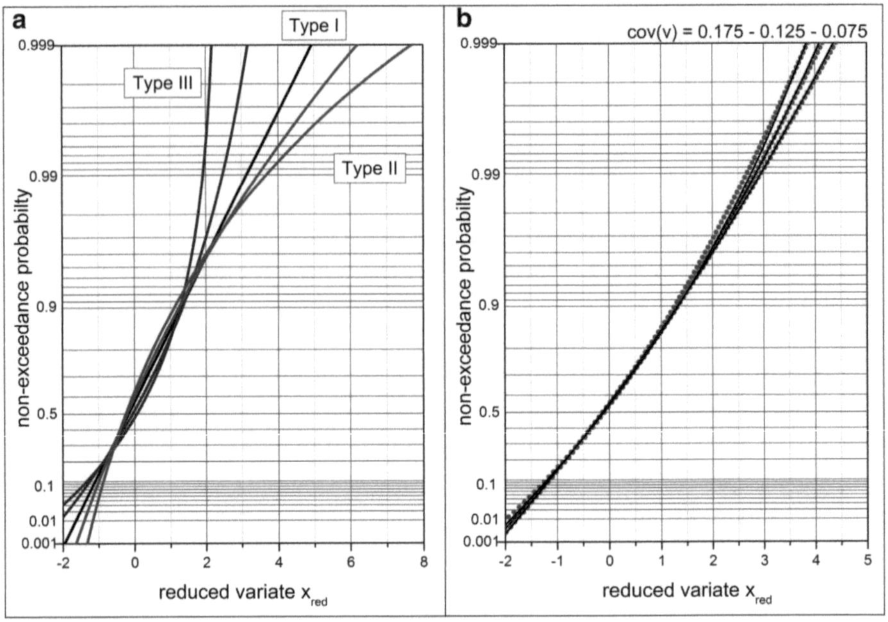

Fig. 2.7 Traces of the extreme value limit distributions in a Gumbel probability paper (Kasperski 2009a). (**a**) Traces for v following type I, II or III; (**b**) traces for v^2 following type I

The distributions of type II and III each form a family of curves with specific characters. Compared to the type I distribution, they show as special feature a certain curvature when plotted in Gumbel probability paper (Fig. 2.7). Generally, a probability paper is a graph paper with one axis specially ruled to transform the distribution function of a specified function to a straight line when it is plotted against the variate as the abscissa. While the curves for the type II distribution bend in a concave shape in respect to the axis of the reduced variate, the curves corresponding to type III show a distinct convex character. These two types are separated by the type I distribution which appears in the plot as a straight line.

In an alternative approach, which also has been adopted in the Eurocode, it is assumed that x^k follows a type I distribution. Especially for $k = 2$ applied to wind speeds there seems to be a physical justification for this concept, since the velocity pressure is obtained from squaring the wind speed. In Fig. 2.8, the traces of the non-exceedance probabilities of v^2 following a type I distribution are shown for three different variation coefficients of v, namely 0.075, 0.125 and 0.175. All three traces, which are marked by dots, show a clear convex curvature; in other words: if v^2 follows the type I distribution, this may indicate that v follows the type III distribution. A respective fit is shown in Fig. 2.7 as solid lines. The differences between these two approaches are quiet small, and fitting a type III distribution for v basically may cover the trace as good as fitting a type I distribution for v^2.

2.3.3 Generalized Pareto Distribution

The Generalized Pareto distribution (GPD) can be used as theoretical expression for the non-exceedance probability of the wind speeds per independent event (Holmes and Moriarty 1999). The cumulative probability distribution is given as follow:

$$F(x) = 1 - \left[1 - \frac{x - x_s}{s} \cdot k\right]^{1/k} \qquad (2.15)$$

x_s: threshold value; s: scale parameter; k: shape parameter

In case of a positive shape factor, the distribution has a finite upper tail which can not be exceeded. The corresponding largest value is given as:

$$x_{max} = x_s + s/k \qquad (2.16)$$

For $k = 0$, the exponential distribution is obtained:

$$F(x) = 1 - \exp\left(-\frac{(x - x_s)}{s}\right) \qquad (2.17)$$

The close relation between GPD and GEV can be seen in the relation of the two shape parameters which is:

$$k = \frac{1}{\tau} \qquad (2.18)$$

2.3.4 Identification of the Describing Parameters of the Probability Distributions

There are basically four different approaches to identify based on observations the unknown parameters of the probability distribution of the underlying process. The first approach is called method of moments. Depending on the number of parameters of the distribution, the first two or three central moments are required. Using for the GEV the notation of Eq. (2.9) requires the calculation of the ensemble mean and standard deviation:

$$m = \frac{1}{N}\sum_{i=1}^{N} x_i, \quad \sigma^{\frac{1}{2}} = \left(\frac{1}{N-1}\sum_{i=1}^{N}(x_i - m)^2\right) \qquad (2.19)$$

The curvature parameter τ is connected to the skewness γ_1 as follows:

$$\gamma_1 = \frac{\Gamma(1 + 3 \cdot \tau) - 3 \cdot \Gamma(1 + \tau) \cdot \Gamma(1 + 2 \cdot \tau) + 2 \cdot \Gamma^3(1 + \tau)}{\left(\Gamma(1 + 2 \cdot \tau) - \Gamma^2(1 + \tau)\right)^{3/2}} \qquad (2.20)$$

The skewness is obtained from the third central moment μ_3:

$$\gamma_1 = \frac{\mu_3}{{\sigma_x}^3}, \text{ with } \mu_3 = \frac{N}{(N-1)\cdot(N-2)} \sum_{i=1}^{N} (x_i - m_x)^3 \qquad (2.21)$$

The parameters of the GPD are obtained from the ensemble mean and standard deviation as follows:

$$s = 0.5 \cdot m \cdot \left(1 + \left(\frac{m}{\sigma}\right)^2\right), \; k = 0.5 \cdot \left(1 - \left(\frac{m}{\sigma}\right)^2\right) \qquad (2.22)$$

It has to be noted that m in Eq. (2.22) relates to the mean value of the exceedances of the threshold value. The method of moments leads to biased estimations, i.e. on average the method will not lead to the "true" parameters.

The second basic approach is the least-square fit in the corresponding probability paper. For this approach, the fundamental equation of the cumulative probability distribution has to be transformed to obtain a linear expression for the observation variable. First step is the estimation of the non-exceedance probability based on the sorted and ranked observations. The lowest value gets the rank 1; the largest value the rank N, where N is the ensemble size. There are numerous recommendations on how to estimate the non-exceedance probability or the plotting position, respectively, of each observation. The general expression for the plotting position is obtained as follows in terms of the relative frequency of non-exceedance:

$$f_{rel}\,(x \leq x_i) = \frac{i\text{-}\alpha}{N + 1 - 2\cdot\alpha} \qquad (2.23)$$

Typical values for α are in the range from 0 to 0.5. In Gumbel's original concept (Gumbel 1958), α is set to zero. Later, Gringorten (Gringorten 1963) recommends that for plots in the Gumbel probability paper $\alpha = 0.44$ should be used. Basically, the estimated values for the design wind speed will decrease if α increases.

It has to be noted that the estimated parameters and therefore the estimated design wind speed may differ from the "true" values if the ensemble size is not large. For observation periods of only a few decades, the random scatter in the estimated design wind speed may be larger than the differences obtained from different α-values. This is shown in Fig. 2.8 on the example of a wind climate with one storm per year where the extreme wind speeds follow the type I extreme value distribution with m = 16.0 m/s and σ = 2.0 m/s. As design target, the 99.9 %-fractile is used. Based on 100,000 simulations for 30 years, the parameters are estimated using the least square approach in Gumbel probability paper.

For $\alpha = 0.44$, as median the "true" design wind speed is obtained, i.e. the "true" design wind speed equals the 50 %-fractile value. For all other α-values, the method is biased, i.e. the median is too large for smaller α-values and too small for larger α-values.

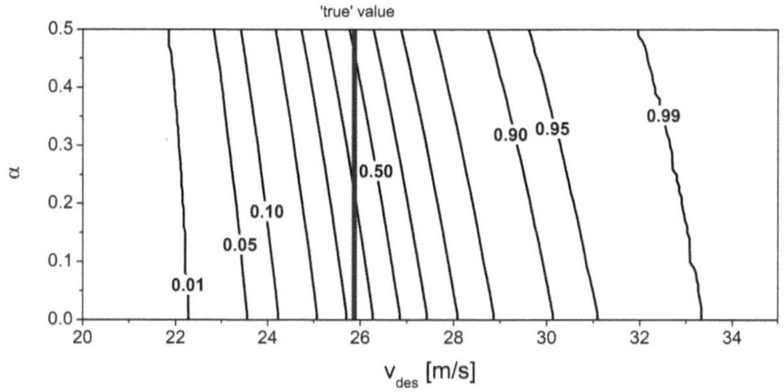

Fig. 2.8 Non-exceedance probabilities of the estimated design wind speed for different α-values; wind climate with one storm per year, extremes follow the type I distribution with m = 16.0 m/s, σ = 2.0 m/s, p_{target} = 0.999, $v_{des,\ true}$ = 25.87 m/s

Basically, for all non-zero values of the curvature parameter τ, there exists a corresponding probability paper for the GEV. This approach will lead for each chosen τ-value to a solution in terms of the two characteristic parameters. The different solutions for each τ-value can be compared based on the deviations in a reference paper. It is reasonable to choose as reference the Gumbel probability paper. Finally, the best fit is obtained as the solution leading to the smallest sum of the squared deviations in the reference paper.

The least-square approach for the GPD can be obtained with a similar approach, i.e. in an appropriate reference paper the sums of the squared deviations are obtained for different pairs of characteristic parameters. Starting point for the iterative strategy is the solution from the method of moments, in the vicinity of this solution, other pair of scale and shape factor are evaluated in regard to the error sum. The pair with the smallest error sum becomes the least- square solution.

The third method to identify the unknown parameters is the maximum likelihood method. For the GEV, the likelihood function is obtained as the product of the corresponding values of the density function f_x for each observation x_i:

$$L\ (m,\sigma,\tau) = \prod_{i=1}^{N} f_x(x_i, m, \sigma, \tau) \stackrel{!}{=} \text{Max.} \tag{2.24}$$

Due to the monotony of the logarithm, the position of the maximum of L is the same as for the logarithmic likelihood function l, i.e. instead of using Eq. (2.24), the following simpler expression is obtained:

$$l\ (m,\sigma,\tau) = \ln\left(\prod_{i=1}^{N} f_x(x_i, m, \sigma, \tau)\right) = \sum_{i=1}^{N} \ln(f_x(x_i, m, \sigma, \tau)) \stackrel{!}{=} \text{Max.} \tag{2.25}$$

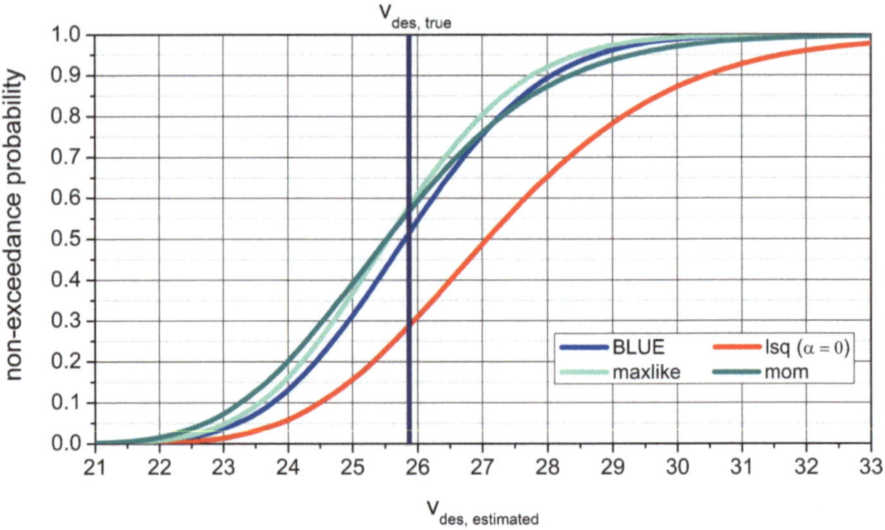

Fig. 2.9 Random scatter in the estimated design wind speed applying different estimation methods

The logarithmic likelihood function becomes maximum if the partial derivates to the unknown parameters m, σ and τ vanish. For the Gumbel distribution using the notation in Eq. (2.9), the basic system of equations becomes:

$$\begin{aligned} &\frac{n}{s} - \frac{1}{s} \cdot \sum_{i=1}^{N} \exp\left(-\frac{x_i - \mu}{s}\right) = 0 \\ &\frac{n}{s} + \frac{1}{s} \cdot \sum_{i=1}^{N} \left(\frac{x_i - \mu}{s}\right) + \frac{1}{s} \cdot \sum_{i=1}^{N} \left(\frac{x_i - \mu}{s}\right) \cdot \exp\left(-\frac{x_i - \mu}{s}\right) = 0 \end{aligned} \tag{2.26}$$

The solution of this non-linear system of equations has to be obtained by iterations. A similar approach can be obtained for the GPD.

The fourth approach is known as the Best Linear Unbiased Estimator (BLUE). It estimates the characteristic parameters as a weighted sum of the ranked observations, specifying individual sets of weighting factors for each characteristic parameter. For the type I extreme value distribution, respective coefficients are published in (Balakrishnan and Chan 1992) for an ensemble size up to 30, where the coefficients relate to the notation in Eq. (2.14):

$$\mu = \sum_{i=1}^{N} x_i \cdot A_i, \quad s = \sum_{i=1}^{N} x_i \cdot B_i \tag{2.27}$$

In Fig. 2.9, the four methods (least-square with $\alpha = 0$, method of moments, maximum likelihood and BLUE) are compared based on the non-exceedance

Table 2.4 Mean value and coefficient of variation of the estimated design wind speed, error probabilities for the estimated design wind speed being smaller than the "true" design wind speed and adjusting factor to meet the error probability of 25 %

Method	Blue	Maxlike	Mom	lsq ($\alpha = 0$)
Mean value (m/s)	25.87	25.62	25.70	27.33
c.o.v.	0.0652	0.0653	0.0789	0.0887
Error probability	0.515	0.579	0.568	0.288
Adjusting factor	1.048	1.057	1.066	1.009

probability of the estimated design wind speed for random ensembles of 30 observation assuming that the extreme wind climate follows the Gumbel distribution with mean = 16 m/s and standard deviation = 2 m/s. Additionally, in Table 2.4, the mean value and the standard deviation of the estimated design wind speed are summarized. Only the BLUE-estimations are unbiased, for the original Gumbel concept, on average, considerable overestimations are obtained, for the method of moments and the method of maximum likelihood, slight underestimations are obtained. The smallest coefficient of variation (c.o.v.) in the estimated design wind speeds is obtained for BLUE, closely followed by the method of maximum likelihood. The largest c.o.v. is obtained for the least-square approach using $\alpha = 0$. From this point of view, BLUE is—as indicated by its name—the best of the four discussed estimation methods. However, the estimations based on the BLUE-method lead to an error probability of 51.5 %, i.e. with probability of 51.5 % the estimated design wind speed will be smaller than the "true" design wind speed. In regard to this criterion, the best solution is obtained with the original Gumbel approach, i.e. the least square fit for $\alpha = 0$, which leads to "only" 28.8 % error probability.

The question arises which target should be used for the error probability. Basically, the smaller the value of the error probability the larger will be the probability of an overestimation, i.e. the design becomes more and more uneconomic for decreasing target values of the error probability. As compromise for the two conflicting demands, the Eurocode dealing with *Bases of Design in Structural Engineering* recommends as confidence 75 % in its section *Design by testing*, i.e. the target error probability is 25 % for estimating from tests the design value for the resistance. To meet this demand, an adjusting factor can be applied which simply shifts the four traces to the higher wind speeds. The adjusting factor is obtained as the ratio of the "true" design wind speed and the 25 %-fractile value of the estimated design wind speed. If the appropriate adjusting factors are applied, all four curves will go through the anchor point $v_{des,\ true}$ and $p = 0.25$. It is worth mentioning that for the original Gumbel approach the required adjusting factor is close to unity (Table 2.4), i.e. the original Gumbel approach with $\alpha = 0$ can be applied without any further considerations.

Figure 2.10 shows the four adjusted curves which all meet the target error probability of 25 %. Obviously, the method of moments and the least-square fit lead to larger overestimations compared to the maximum likelihood method and the

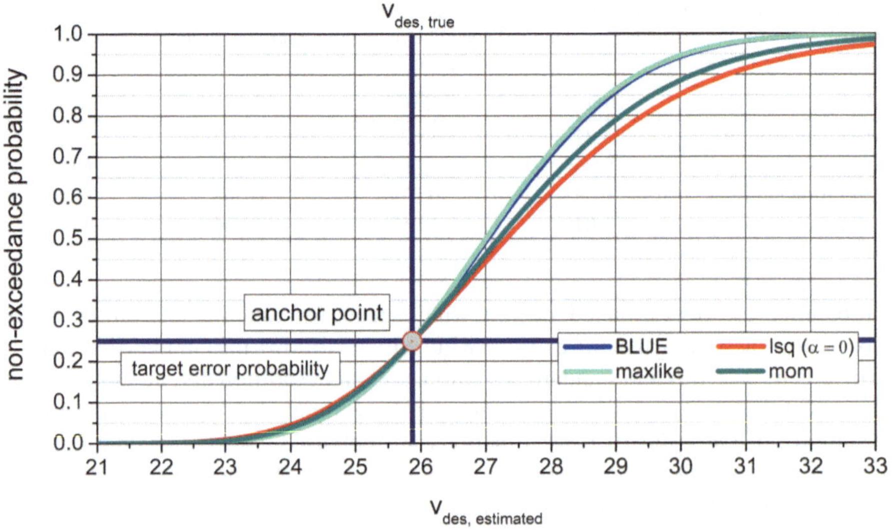

Fig. 2.10 Random scatter in the estimated design wind speed with adjusted estimation methods

BLUE-method. The overestimations obtained with the BLUE-method are slightly larger than those obtained with the maximum likelihood method, i.e. the best method again is not obtained with the BLUE-method.

It is important to note that the appropriate adjusting factor only can be obtained if the "true" probability distribution of the underlying process is known. Usually, the observed ensemble will not reveal this secret, i.e. strictly speaking it remains unknown, if the "true" distribution is type I, II or III. As reasonable compromise, ISO 4354: 2009 recommends to use the type I distribution, thus trying to balance the demand to safety and economy.

2.3.5 Statistical Stability of the Identified Parameters

The statistical stability of the estimated parameters can be studied by simulations. In the following, a wind climate is assumed which follows the GPD with $v_s = 14.1$ m/s, $s = 2.0$ m/s and $k = -0.1$. Per season, the average number of storms is given with $\lambda = 2$, assuming that the random number of storms per season follows the Poisson distribution. In Fig. 2.11, the probability densities of estimated λ, s and k values are shown based on 10^6 independent runs for each individual combination, simulating 20, 30 and 50 years observation periods. As a matter of fact, none of the parameters can be identified with sufficient statistical stability. The deviations of the estimated values from the "true" values are large enough to affect the estimation of the design wind speed considerably.

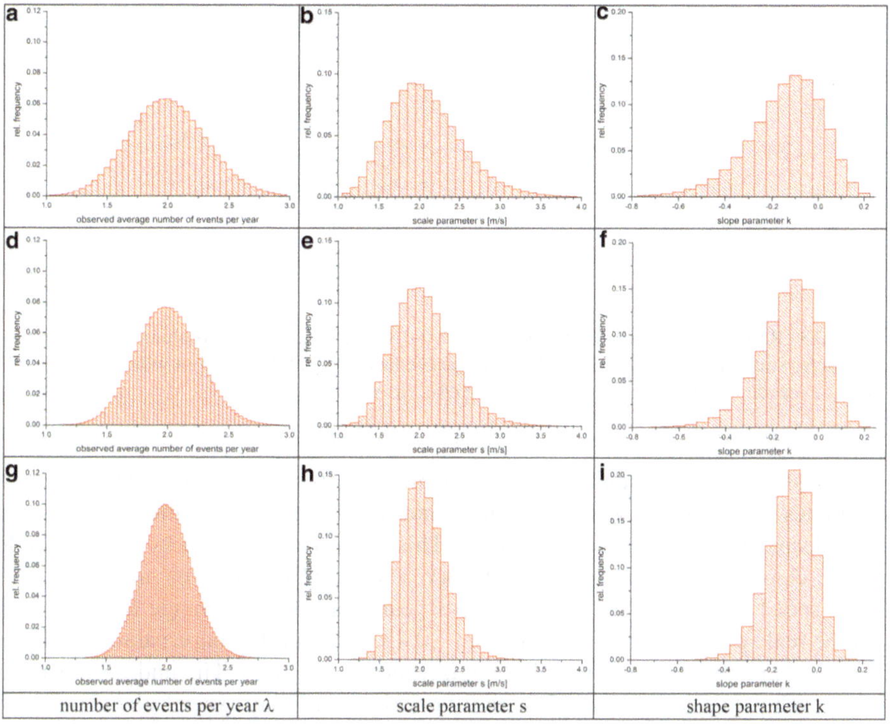

Fig. 2.11 Statistical stability of estimated parameters of the extreme wind climate (Kasperski 2009b). (**a–c**) T_{obs} = 20 years; (**d–f**) T_{obs} = 30 years; (**g–i**) T_{obs} = 50 years

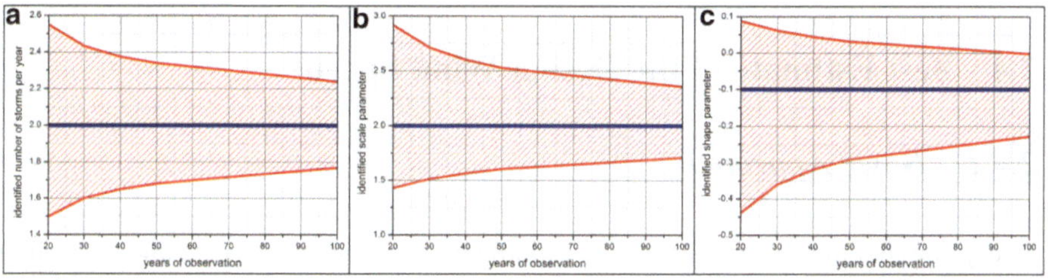

Fig. 2.12 Influence of the observation period on the 90 %-confidence interval of the statistical parameters of the extreme wind climate (Kasperski 2009b). (**a**) Number of events per year λ; (**b**) scale parameter s; (**c**) shape parameter k

In Fig. 2.12, the 90 %-confidence intervals are shown for the three parameters λ, s and k. Although the confidence intervals become smaller with increasing observation period, even for 100 years the possible deviations from the "true" values remain large. For observation periods of only a few decades, the analysis is hardly able to identify the basic type of distribution (type III or type II).

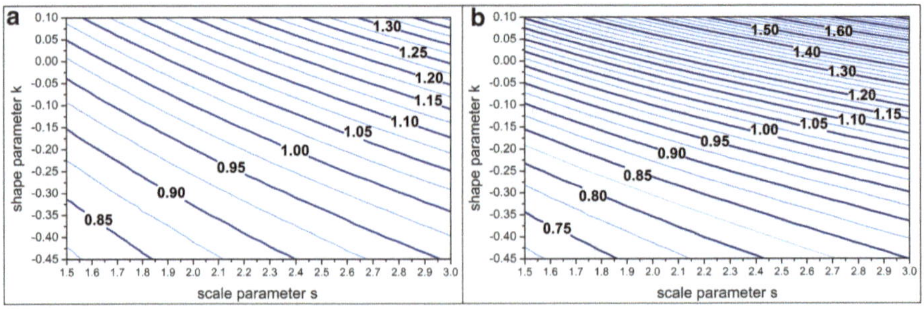

Fig. 2.13 Ratio of the estimated to the "true" wind speed for different combinations of s and k (characteristic wind speed v_k with 0.02 yearly exceedance probability, design wind v_d with 0.001 yearly exceedance probability) (Kasperski 2009b). (**a**) v_k; (**b**) v_d

The statistical uncertainties in the estimated values of λ, s and k lead to uncertain estimates for extrapolated values of the wind speed. The uncertainties will increase with decreasing exceedance probability. This is shown in Fig. 2.13 on the example of the characteristic wind speed with $p(v > v_k = 1/50$ per year) and the design wind speed with $p(v > v_d = 1/1{,}000$ per year) considering a range from 1.5 to 3 for the scale parameter and a range from −0.45 to 0.1 for the shape parameter. Even for the characteristic wind speed, the uncertainties are large. The underestimations may reach values larger than 15 %; the overestimations may exceed a factor of 1.35. For the design wind speed, the deviations become even larger. Underestimations up to 25 % may occur; overestimations may reach a factor of 1.8.

2.4 Best Estimate of the Design Wind Speed

2.4.1 Basic Approach

Accepting that the "true" parameters cannot be identified leads to the understanding that virtually every wind climate may hide behind the observed parameters, however, with different probabilities. The probability p, that a specific triple (λ, s, k) randomly leads to the observed triple (λ_{obs}, s_{obs}, k_{obs}), can be obtained from simulations, i.e. for a wide range of triples (λ, s, k) the number of cases are counted which randomly lead to the observed triple. Then, the number of random matches for a specific triple is divided by the total number of random matches. This relative frequency becomes a good estimate of the probability if the number of simulations is sufficiently large (Kasperski 2009b).

In the next step, for each triple (λ, s, k), a design wind speed can be estimated. These design wind speeds are sorted in ascending order together with the corresponding probabilities. These two results then allow constructing a probability distribution of the uncertain design wind speed:

$$p(v_{des} \leq v_{des,\,i}) = \sum_{j=1}^{i} p_j\left(v_{des}\left(\lambda_j,\ s_j,\ k_j\right)\right) \tag{2.28}$$

with p_j: individual probability of random match for the specific triple (λ_j, s_j, k_j)

The final design wind speed in terms of the best estimate can be obtained based on a chosen confidence level. Applying the recommended 75 % confidence from the Eurocode, identifies the best estimate as the 75 % fractile value of the probable design wind speeds.

Since usually the number of storms and the intensity of storms can be treated as independent, the simulations can be performed in two independent steps, i.e. one simulation for getting the probabilities of random matches for the number of storms and a second simulation for getting the random matches for s and k. The two results then are combined in the following product:

$$p_j\left(v_{des}\left(\lambda_j,\ s_j,\ k_j\right)\right) = p\left(\lambda_j\right) \cdot p\left(s_j,\ k_j\right) \tag{2.29}$$

The above approach can be extended to consider long term trends e.g. in the number of independent events λ and the scale factor s. Since the available observation periods usually are relatively short, the "true" shape of trends can hardly be identified. Therefore, a fair approach introduces linear trends for the observation period and for the extrapolation (Kasperski 2010).

2.4.2 *Example of Application*

The proposed method is applied to the extreme wind climate at Düsseldorf, Germany, which is governed by strong frontal depressions. As representative value, the largest hourly mean wind speed of independent events is used. In Fig. 2.14, the identified probability densities and distributions are shown. These results can be used directly for the estimation of the design wind speed. As target value, an exceedance probability of 5 % in the projected working life is chosen. The resulting design values and the corresponding yearly exceedance probabilities are summarized in Table 2.5 for different values of the working life L for class 3 buildings.

In Table 2.6, the influence of the building class is shown for the lifetime L = 50 years. Changing the design working life from 10 to 50 years leads to 14 % increase of the wind loads. The wind loads for a class 4 building and a class 2 building differ by 10 % assuming 50 years design working life.

Figure 2.15 presents the probability densities of the triples (λ, s, k) randomly leading to the observed values, assuming that the number of storms per year and the

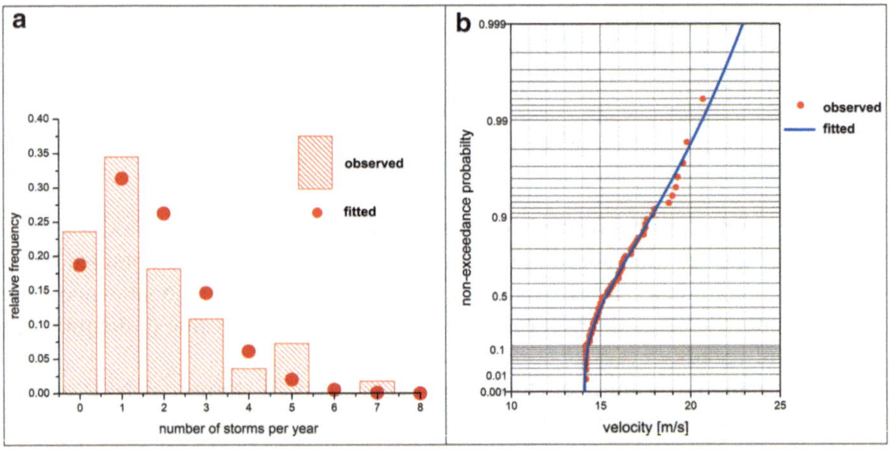

Fig. 2.14 Observed parameters of the extreme wind climate at Düsseldorf airport (1952–2007) (Kasperski 2009b). (**a**) $\lambda_{obs} = 92/55$; (**b**) $v_{s,\ obs} = 14.1$ m/s, $s_{obs} = 1.803$ m/s, $k_{obs} = -0.1063$

Table 2.5 Design wind speeds based on the identified parameters, class 3

L (years)	10	20	50	80
p_{yearly} $(v > v_d)$	1/195	1/390	1/975	1/1,560
v_d (m/s)	21.90	22.55	23.34	23.71

Table 2.6 Design wind speeds based on the identified parameters, L = 50 years

Class	0	1	2	3	4	5
p_{yearly} $(v > v_d)$	1/100	1/224	1/375	1/975	1/1,975	1/9,975
v_d (m/s)	21.20	22.03	22.72	23.34	23.89	25.03

intensity of the storms are statistically independent. The full simulation involves 10^6 independent runs each corresponding to the sub-period of 55 years. The resulting possible range for λ is from 1.18 to 2.27. For the scale parameter, the simulation identifies a range from about 1.0 to 3.5 m/s, while the shape factor probably lies between −0.5 and +0.25.

The cumulative probability distributions for the design wind speeds are shown in Fig. 2.16 for different values of the design working life and for different classes. The best estimates of the design wind speed are obtained as the 75 %-fractile. The respective values are summarized in Tables 2.7 and 2.8. Additionally, the resulting wind loads are compared to values from the "classical" approach.

There are large differences between the direct estimations based on the observed wind climate parameters and the new approach. This is mainly due to the large uncertainties in the shape factor. For a class 3 building, the best estimate of the design wind load is about 18–26 % larger than the wind load estimated with the observed parameters. The influence of the building class leads to a similar range of differences.

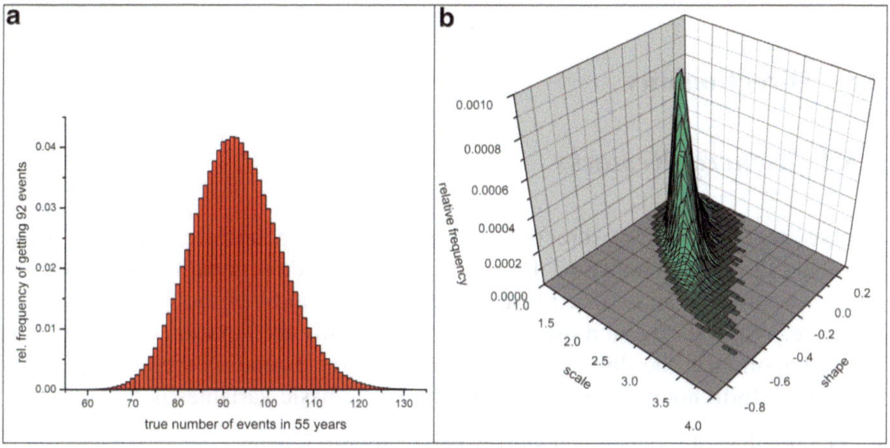

Fig. 2.15 Randomness in the identified parameters (Kasperski 2009b). (**a**) Probability density for observing 92 events in 55 years for different λ-values; (**b**) joint-probability density for getting the observed parameters s_{obs} and k_{obs} from an ensemble with 92 events

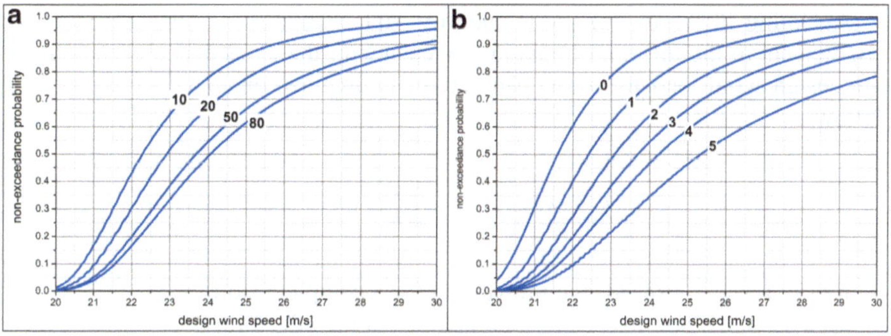

Fig. 2.16 Cumulative probability distributions for the design wind speed. (**a**) Variation of lifetime L, class 3. (**b**) variation of building class, L = 50 years

Table 2.7 Best estimates of the design wind speed and comparison to the wind loads obtained with the "classical" approach, class 3

L (years)	10	20	50	80
v_d (m/s)	23.74	24.71	25.98	26.62
Wind load increase (%)	+17.5	+20.1	+23.9	+26.1

Table 2.8 Best estimates of the design wind speed and comparison to the wind loads obtained with the "classical" approach, L = 50 years

Class	0	1	2	3	4	5
v_d (m/s)	22.74	23.93	24.98	25.99	26.95	29.08
Wind load increase (%)	+15.1	+18.0	+20.9	+23.0	+27.3	35.0

2.5 Directionality

2.5.1 General Remarks

A consistent analysis of directional effects requires a holistic approach which considers the directionality of the wind climate, the directionality of the aerodynamic coefficients and—if applicable—the directionality of the resistance of the structure. While for the non-directional design usually the duration of a storm is of minor importance, the directional approach requires extending the analysis to the additional storm hours. Then, the following list of basic variables is obtained:

- The number of independent storms per 12-month season
- The maximum hourly mean wind speed per independent storm
- The corresponding mean wind direction
- The duration of an independent storm in terms of the number of hours above the storm threshold value
- The relative intensity of the additional storm hours
- The probability of a direction change in the different storm hours

Assuming that—in the vast majority—the extremes of the aerodynamic coefficients in the different hours are independent, the exceedance probability of the design wind load for a complete storm is obtained analogue to Eq. (2.7) as follows:

$$p(w > w_{des}) = 1 - \prod_{k=1}^{m} p_k(w \leq w_{des}) \tag{2.30}$$

with $p(w > w_{des})$: exceedance probability for a complete storm
$p_k(w \leq w_{des})$: non-exceedance probability in the kth storm hour
m: duration of the storm

2.5.2 Maximum Wind Speed Per Storm

The wind climate of Western Europe is mainly governed by strong frontal depressions which form over the Northern Atlantic and travel westwards. Therefore, the extreme wind speeds induced by this storm type show a distinct directionality, i.e. hourly wind speeds at storm level are only to be expected for a confined sector of wind directions. From the clearly origin of this storm type it can be excluded that e.g. a winter storm is blowing from eastern directions.

An appropriate probabilistic model featuring an upper and lower limiting value is given with the Beta distribution. The probability density of the Beta distribution is given as follows:

$$f(x) = \frac{(x-a)^{r-1} \cdot (b-x)^{t-1}}{(b-a)^{r+t-1} \cdot B(r,t)} \quad \text{for } a \leq x \leq b \tag{2.31}$$

a: lower bound; b: upper bound; r, t: exponents
The constant B is defined as:

$$B(r,t) = \frac{\Gamma(r) \cdot \Gamma(t)}{\Gamma(r+t)} \tag{2.32}$$

For the general case, a closed form solution for the cumulative probability distribution F(x) does not exist, i.e. F(x) has to be obtained from a numerical integration of f(x). If the two bounds a and b are known, the exponents r and t can be obtained from the mean value and the standard deviation of x as follows:

$$r = \frac{h_1 \cdot {h_2}^2 - (1+h_1)^2}{(1+h_1)^3} \quad \text{and } t = r \cdot h_1$$
$$\text{with } h_1 = \frac{m-a}{b-m} \quad \text{and } h_2 = \frac{b-a}{\sigma} \tag{2.33}$$

Otherwise, the four parameters have to be identified based on a trial-and-error scheme applying e.g. a least-square approach.

The observed duration of storms and therefore the observed probability of direction change are biased parameters, since weaker storms are artificially cut at both sides. Therefore, it is reasonable to confine the analysis of the directionality to only the strongest of the observed independent storms. For the following example of the strong wind climate at Düsseldorf, only the 30 strongest storms have been used.

Figure 2.17 shows the mean wind direction for the largest wind speed during an independent storm in terms of the cumulative Beta distribution and the corresponding density in a polar-plot. At Düsseldorf, storms are confined to the sector from 180° to 330°. The mean wind direction of storms is 240°.

Based on Eq. (2.8), the directional design value of the wind speeds can be obtained in terms of the wind speed having a specific exceedance probability in a specific range of wind directions. In Table 2.9, the directional design wind speeds with an annual exceedance probability of 1/1,000 are summarized for six sectors applying the wind climate model as shown in Fig. 2.17.

Since for the sector from 330° to 180°, storms induced by strong frontal depressions are not be expected with a probability of unity, the design wind speed for these sectors is 14.1 m/s. It is important to note that this design value corresponds to storm types which are characterized by large hourly means, i.e. this design wind speed does not cover thunderstorms.

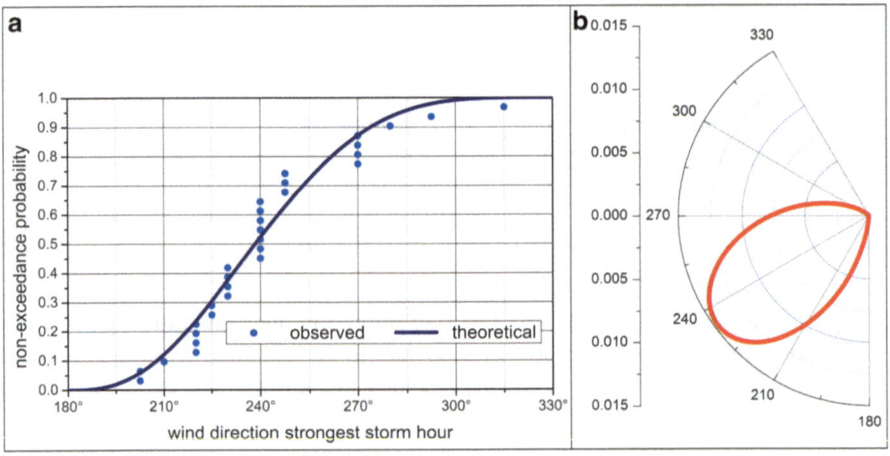

Fig. 2.17 Directionality of the largest wind speed during an independent storm event (Düsseldorf 1952–2007: m = 240°, σ = 25°, r = 3.04, t = 4.59). (**a**) Cumulative probability distribution; (**b**) probability density

Table 2.9 Directional design wind speeds for Düsseldorf airport

Sector Φ	p (Φ)	p_{year} ($v_{des, Φ}$)	$v_{des, Φ}$ (m/s)
180–195°	0.01995	0.94987	19.35
195–225°	0.28522	0.99649	22.25
225–255°	0.41871	0.99761	22.61
255–285°	0.23295	0.99571	22.06
285–315°	0.04275	0.97661	20.28
315–330°	0.00042	0	14.1
All	1.0	0.999	23.35

2.5.3 Duration of Storms and Relative Intensity of the Additional Storm Hours

In Fig. 2.18, the observed durations for the 30 strongest storms are plotted versus the largest hourly mean wind speed during each independent storm. The scatter plot indicates an increasing trend, i.e. the duration of the storm increases with increasing intensity of the storm. The corresponding correlation is about 0.5. For an ensemble size of 30 events, the observed absolute value of the correlation is not random with a probability of 0.993. Extrapolating this relation to the level of the design wind speed having an annual exceedance probability of 1/1,000 leads to an expected duration of 14 storm hours.

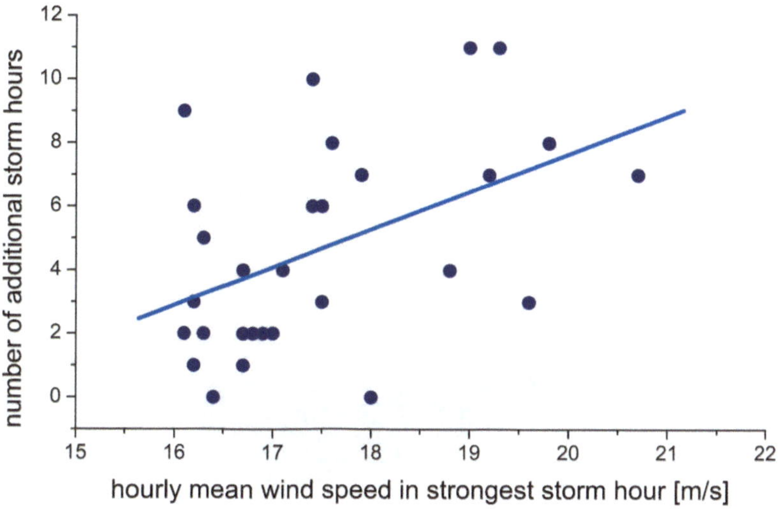

Fig. 2.18 Duration of the 30 strongest storms plotted versus the hourly mean wind speed in the strongest hour (Düsseldorf 1952–2007)

The additional storm hours in each independent storm can be sorted in descending order, their wind speed levels can be normalized to the largest hourly mean wind speed in the respective storm. This leads to the relative intensities of the additional storm hours. For a consistent probabilistic model, the relative intensity of the 3rd strongest storm hour may not be larger than the relative intensity of the 2nd strongest storm hour. This can be achieved by describing the relative intensities in a cascaded model, i.e.:

$$v_k = v_{max} \cdot \prod_{j=1}^{k} a_j \tag{2.34}$$

v_{max}: strongest wind speed in an independent storm; a_j: relative intensity in j-strongest storm hour

Basically, the relative intensities are in the range from 0 to 1. The mean value of the cascaded relative intensity increases with k:

$$a_{rel}(k) = 0.9548 + 0.0027 \cdot k \tag{2.35}$$

Figure 2.19 shows the cascaded relative intensities based on 542 storm hours from the 30 strongest storms at Düsseldorf. Thereby, the first additional storm hour is the 2nd strongest storm hour in an independent storm. The larger k becomes, the smaller will be the effective range of the cascaded relative intensities. Basically, the information in Fig. 2.19 allows an extrapolation clearly beyond the expected duration of 14 h for the design storm event.

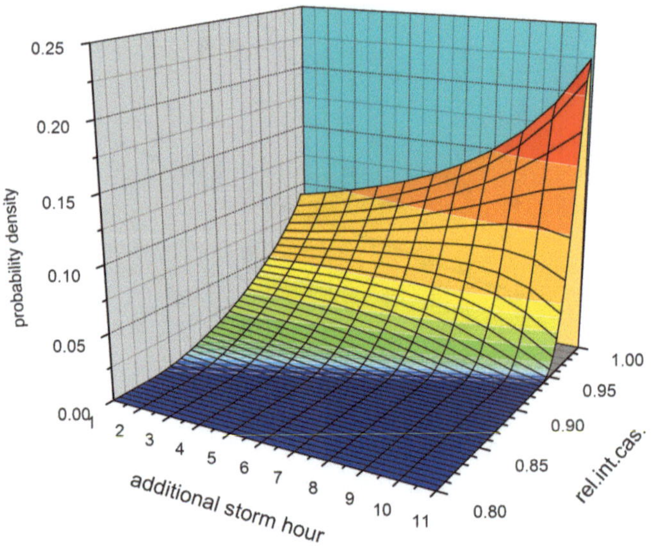

Fig. 2.19 Probability density of the cascaded relative intensity of the additional storm hours

2.5.4 Direction Change in the Additional Storm Hours

From the 30 strongest storms, altogether 542 h are available to describe the probability of the direction change in the k-strongest hours of a storm. Basically, the probability density of the wind direction in the additional storm hours can be described by the Beta distribution. For the example Düsseldorf, all distributions are bounded in [180°, 330°].

For the second strongest hour, the probability densities are shown in Fig. 2.20 for the three major contributing storm sectors 210°, 240° and 270°, which contain more than 90 % of the strong storms. If the strongest storm hour occurs for 240°, the second strongest storm hour is likely to occur in the same sector. The probability of a direction change smaller than ±5° is about 60 %. If in the strongest hour the wind blows from 210°, the second strongest storm hour occurs in the same direction with a probability of about only 20 %. Finally, if the wind direction for the strongest storm hour is 270°, the second strongest storm hour will have the same direction with less than 10 %. Basically, the storm tends to turn clockwise, if the wind direction in the strongest storm hour is smaller than 240°. For storms blowing in their strongest hour from wind directions larger than 240°, the probability density of the direction change is skewed in the counter-clockwise direction.

The mean value of the direction change is more or less independent of the rank k. However, the standard deviation increases with increasing k for the central sectors, i.e. generally the probability distribution becomes broader for higher k-values.

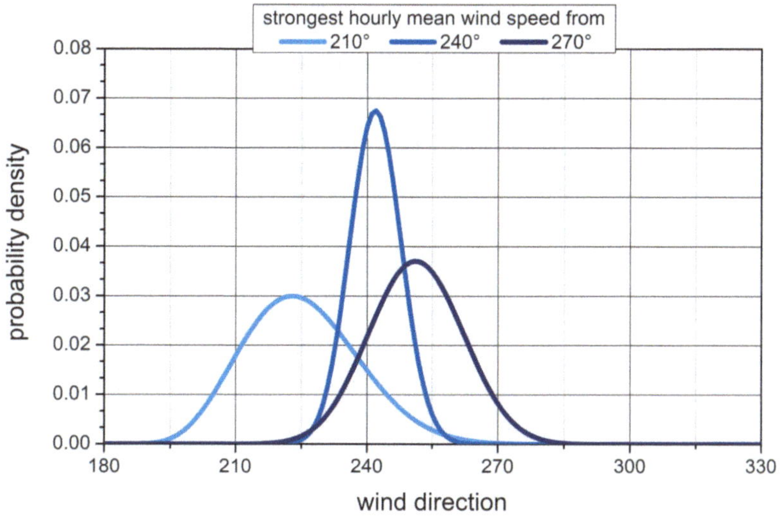

Fig. 2.20 Probability density of direction change in the second strongest storm hour (Düsseldorf 1952–2007)

References

AS 1170.2-1989 (1989) Australian Standard – minimum design loads on structures, Part 2: wind loads. Standards Australia, Sydney

AS/NZS 1170.2:2011 (2011) Australian/New Zealand Standards: structural design actions - wind actions. Standards Australia, Sydney

ASCE/SEI 7-10 (2010) Minimum design loads for buildings and other structures. American Society of Civil Engineers, New York

Balakrishnan N, Chan PS (1992) Order statistics from extreme value distribution II: best linear unbiased estimates and some other uses. Commun Stat Simulation Comput 21 (4):1219–1246

Cook NJ (1990) The designer's guide to wind loading of building structures, Part 2 Static structures. Butterworths, London. ISBN 0-408-00871-7

EN 1990 (2002) Basis of structural design. European Committee for Standardization, Brussel

Florida Coastal Monitoring Program (2007) http://fcmp.ce.ufl.edu/ accessed May 2007

Gomes L, Vickery BJ (1978) Extreme wind speeds in mixed climates. J Ind Aerodyn 2:331–344

Gringorten II (1963) A plotting rule for extreme probability paper. J Geophys Res 68 (1963):813–814

Gumbel EJ (1958) Statistics of extremes. Columbia University Press, New York

Holmes JD, Moriarty WW (1999) Application of the generalised Pareto distribution to wind engineering. J Wind Eng Ind Aerodyn 83:1–10

ISO 22111-2007 (2007) Bases for design of structures – general requirements. International Organisation for Standardisation, Geneva

ISO 4354-2009 (2009) Wind actions on structures. International Organisation for Standardization, Geneva

Kasperski M (2007) Design wind loads for a low-rise building taking into account directional effects. J Wind Eng Ind Aerodyn 95:1125–1144

Kasperski M (2009a) Specification of the design wind load - a critical review of code concepts. J Wind Eng Ind Aerodyn 97:335–357

Kasperski M (2009) Estimation of the design wind speed based on uncertain parameters of the wind climate. In: Proceedings of APCWE-7, Taipei, 8–12 November 2009

Kasperski M (2010) A new approach for the specification of the design wind speed considering long-term trends in the wind climate. Global Environ Res 13(2):161–168

König-Langlo G (2006) http://www.pangaea.de/PangaVista?count=100&query=Continuous+meteorological+observations+Neumayer+-Gernandt accessed December 2006

Traup S, Kruse B (1996) Winddaten für Windenergienutzer Selbstverlag des Deutschen Wetterdienstes DWD, Offenbach am Main

Chapter 3
Bluff Body Aerodynamics

Richard G.J. Flay

Abstract The aerodynamic characteristics of bluff bodies differ substantially from streamlined bodies, and an understanding of bluff body aerodynamics is essential to make progress in understanding wind engineering. Streamlined bodies like aircraft wings have a rounded nose, a thin profile, and a sharp trailing edge. Their wakes are small and for small angles of attack, the lift force developed is considerably greater than the drag force. On the other hand, bluff bodies have a large separated wake, with unsteady flow, and the drag force is comparable with the lift force. It is necessary to understand the size and nature of these forces to ensure that engineered designs are fit for purpose under wind action.

Keywords Along-wind • Bluff body • Cross-wind • Drag • Lift • Separated flow • Turbulent flow • Unsteady wake • Wind engineering

3.1 Introduction

This section introduces the concepts of drag and lift (Cengel et al. 2008). Drag is the aerodynamic force produced by a flowing fluid over a body and is in the direction of the flow. Lift is the aerodynamic force developed by a flowing fluid in a direction normal to the flow direction, e.g. vertically upwards for an aircraft flying at constant speed and direction in a horizontal plane, but is horizontal in the case of a tower, and is usually referred to as the cross-wind force.

R.G.J. Flay (✉)
University of Auckland, Private Bag 92019, Auckland 1142, New Zealand
e-mail: r.flay@auckland.ac.nz

Y. Tamura and A. Kareem (eds.), *Advanced Structural Wind Engineering*,
DOI 10.1007/978-4-431-54337-4_3, © Springer Japan 2013

3.2 Drag and Lift

3.2.1 Drag

Drag is the force a flowing fluid exerts on a body in the flow direction. A stationary fluid exerts a pressure force normal to the surface of an immersed body. Relative motion between the surface and the fluid causes a tangential shear force due to the viscosity in real fluids and the no-slip condition at the surface. The drag force is due to the combined effects of the components of pressure and shear forces acting on each elemental area of the body in the flow direction, as illustrated in Fig. 3.1.

If one examines the elemental area *dA* illustrated in Fig. 3.1, it can be seen that the drag and lift forces resulting from the normal and shear forces on the element are given by Eqs. (3.1) and (3.2) respectively.

$$dF_D = -PdAcos\theta + \tau_w dAsin\theta \tag{3.1}$$

$$dF_L = -PdAsin\theta + \tau_w dAcos\theta \tag{3.2}$$

In general, both skin friction and pressure contribute to the drag and lift. A flat plate with negligible thickness normal to the flow has only pressure drag. Wings are streamlined and thin, and their drag also is primarily from skin friction, except when the angle of attack is high and the flow is separated.

Bluff bodies are not streamlined, and so the drag is primarily pressure drag, due to the higher pressure at the front of the body than at the rear (often referred

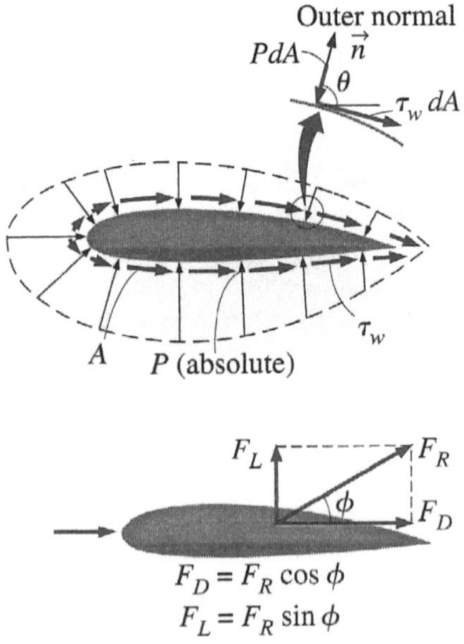

Fig. 3.1 Normal and shear forces acting on an element of a wing-like shape (after Cengel et al. 2008)

to as the base). In fact one tries to reduce drag by increasing the pressure on the downstream parts of cars, aircraft etc. This is because pressure drag results from the non-recovery of pressure on the downstream side of bodies. The pressure drag becomes most significant when the flow separates and creates a very low pressure region on the base or leeward side of a body.

Friction drag is due to the "stickiness" or viscosity of fluids, and also depends on the viscosity coefficient of the fluid, the boundary layer profile gradient $\frac{dU}{dy}$ at the surface, where U is fluid speed parallel to the surface, and y is the direction normal to the surface. Hence skin friction drag is a strong function of viscosity, and depends on whether the boundary layer is laminar or turbulent.

Figure 3.2 illustrates different drag cases. The top diagram shows a flat plate parallel with the flow, and the drag on the plate is primarily from skin friction. The centre image shows the flat plate rotated so that it is normal to the flow. In this case the drag is primarily pressure drag, as the pressure on the leeward face is much less than on the windward face. The lower diagram shows streamlines around a circular cylinder, where the drag results from skin friction and pressure.

Drag may be reduced by streamlining, and this may be of importance in some wind engineering applications. It is always considered in the aeronautical industry, since fuel consumption is proportional to drag. In wind engineering it is more problematic, as the wind can come from any direction, and so any streamlining device may have to be self-aligning. This approach is sometimes used on yachts where it is very important to minimise drag in serious racing such as the America's Cup.

The drag coefficient of a circular cylinder at $\mathrm{Re} = 4 \times 10^4$ is about 1.2, and for $\mathrm{Re} > {\sim}5 \times 10^5$ it is about 0.3, whereas the drag coefficient on a streamlined aerofoil of the same projected frontal area is considerably less. NACA carried out an experimental investigation of streamlining (Abbott and von Doenhoff 1959). Their results are shown in Fig. 3.3. Note that the abscissa is D/L, and so the left hand end of the figure corresponds to a flat plate, and the right hand end corresponds to a 40 % thick aerofoil. Note that since the reference area in the drag coefficient is the frontal area DB, a good way to interpret Fig. 3.3 is to imagine an aerofoil that has constant thickness D, and variable chord L. The contributions to the drag from pressure and skin friction are illustrated. Hence the friction drag reduces as the shape changes from long and thin to short and fat, whereas the pressure drag increases. C_D has a minimum value of about 0.06 at $D/L = 0.27$, which is substantially less than a circular cylinder of the same diameter.

3.2.2 Flow Separation

Fluid flow will separate from a surface if it curves too much, thus creating a region of adverse pressure gradient—where the pressure increases in the downstream direction. This makes it difficult to design "diffusers" which expand in area in the downstream direction without having the flow separate. The flow will separate from

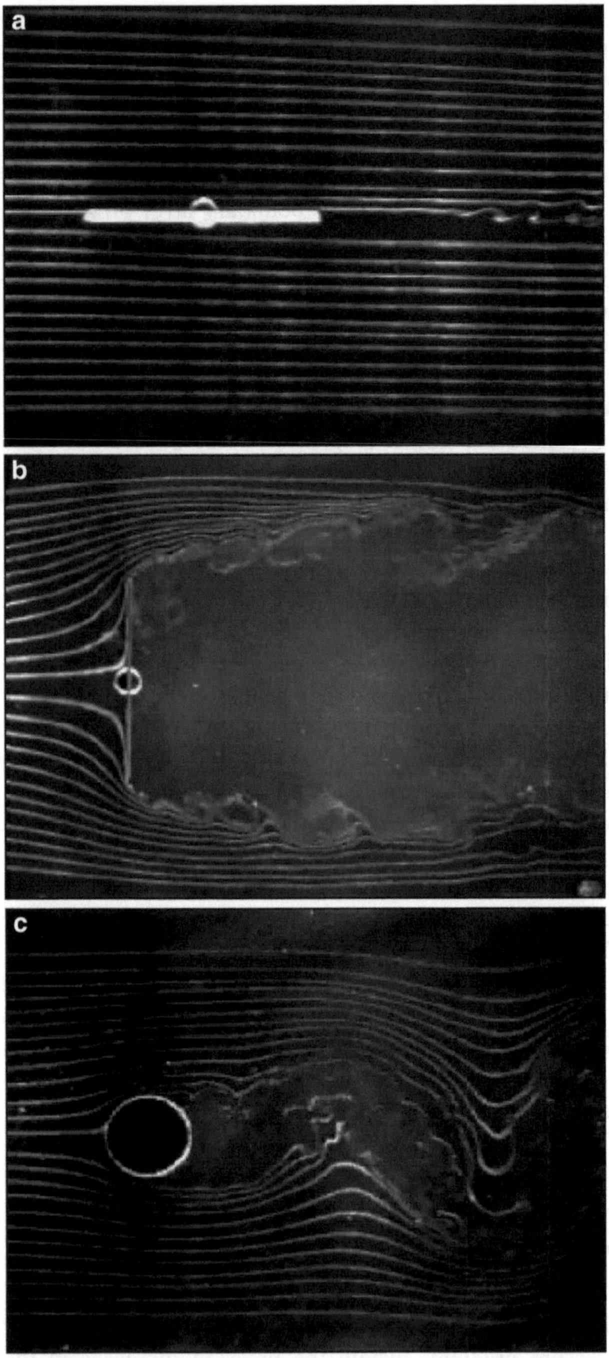

Fig. 3.2 Flow visualisation of several drag cases. (**a**) *Top*: flat plate parallel with the flow; (**b**) *Centre*: flat plate normal to the flow; (**c**) *Bottom*: circular cylinder (after Cengel et al. 2008)

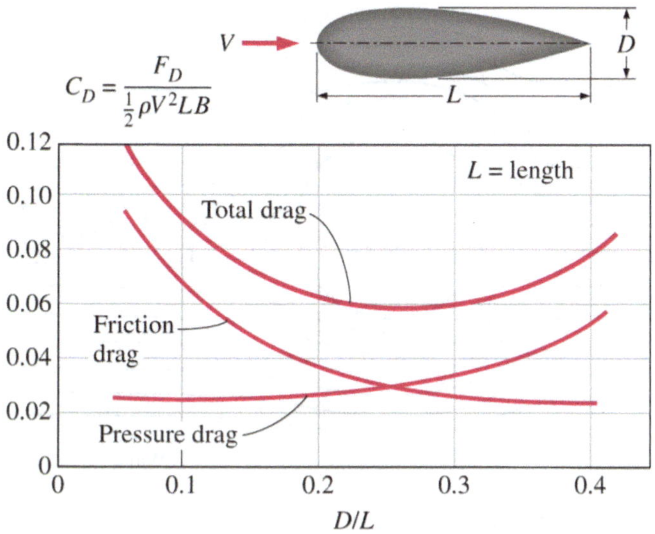

Fig. 3.3 The variation of friction, pressure, and total drag coefficients of a streamlined strut with thickness-to-chord ratio for Re $= 4 \times 10^4$. C_D is made dimensionless using the frontal area *DB*, where *B* is the width of the body normal to the page (after Cengel et al. 2008)

the top of a wing, where the pressure is increasing in the downstream direction, not underneath it, where the pressure is fairly uniform.

Separation depends on: Re, surface roughness, and turbulence. The separated region behind a bluff body is called the wake. Separation will occur for a wing at high angles of attack (above about 15°) and this is often referred to as stall. Figure 3.4 shows the separated region behind a circular cylinder, and behind a wing at a high angle of attack.

It is interesting to observe how the shape of cars has changed over the decades in order to reduce the drag coefficient and consequently the fuel consumption. In the 1920s, cars were very box-like, and had drag coefficients of about 0.8 based on frontal area. In the 1950s the drag coefficient dropped to 0.5. A typical drag coefficient in the 2000s is about 0.3. These improvements in drag have come about by having more rounded shapes at the front to reduce separation, streamlining aspects like the door edges, window joints, and rear-view mirrors etc.

3.3 Pressures on Bluff Bodies

3.3.1 Smooth Flow Over a Streamlined Body

For streamlined bodies like wings, the boundary layer thickness, δ, is very small, and outside δ, the flow is irrotational (see Fig. 3.5). Prandtl (1905) first came up with the idea of subdividing flow into viscous and non-viscous regions. The external

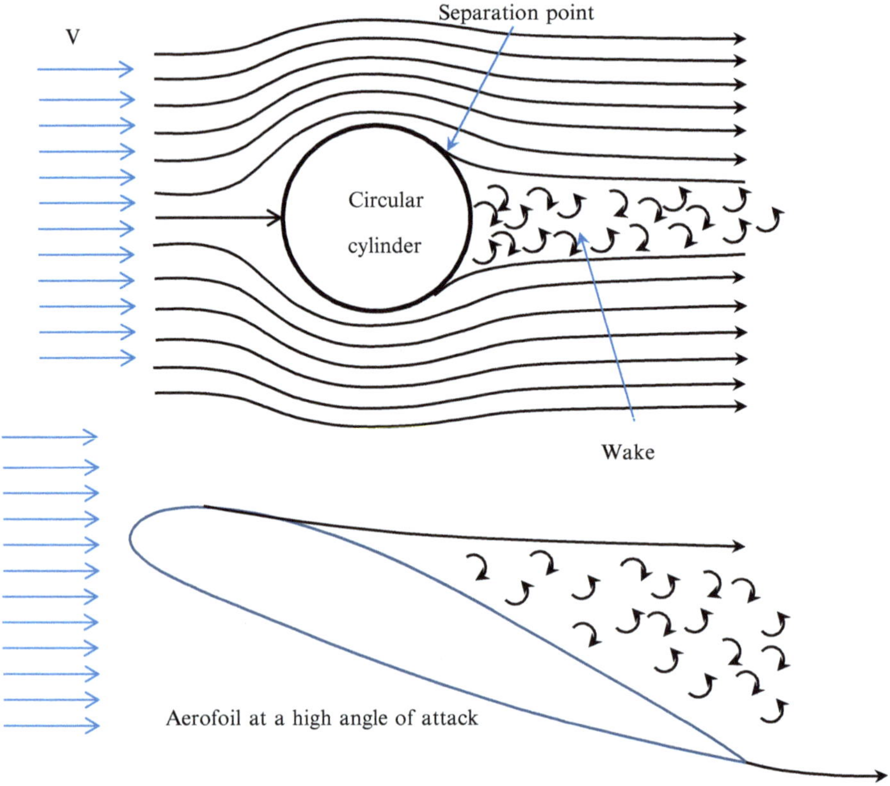

Fig. 3.4 Separation behind a circular cylinder and above an aerofoil at a high angle of attack

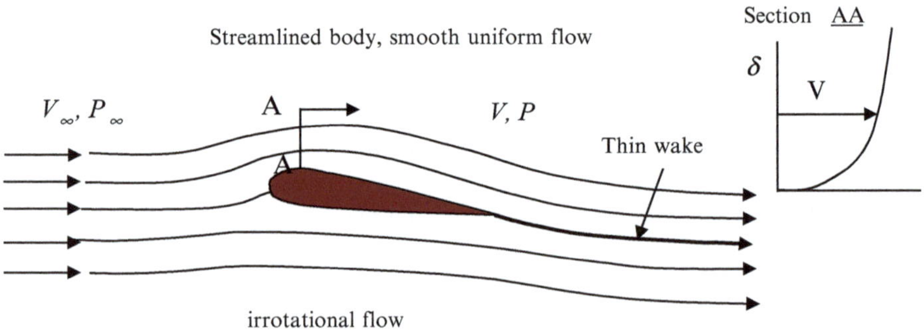

Fig. 3.5 Flow over a streamlined body

flow can therefore be analysed using the Bernoulli equation, and only within δ does viscosity need to be accounted for, and this makes streamlined body analysis much simpler than in wind engineering with bluff bodies where turbulent viscosity is important everywhere. The drag of streamlined bodies is mainly due to skin friction as there is good pressure recovery on the leeward side of well-designed aerofoils.

The Bernoulli equation is:

$$P + \frac{1}{2}\rho V^2 = P_\infty + \frac{1}{2}\rho V_\infty^2 = P_o = const. \tag{3.3}$$

A pressure coefficient can be obtained by making the pressure difference the subject, and then making it dimensionless by dividing by the reference dynamic pressure, as shown in Eqs. (3.4) and (3.5).

$$P - P_\infty = \frac{1}{2}\rho V_\infty^2 - \frac{1}{2}\rho V^2 \tag{3.4}$$

$$Cp = \frac{P - P_\infty}{\frac{1}{2}\rho V_\infty^2} = 1 - \left(\frac{V}{V_\infty}\right)^2 \tag{3.5}$$

P_∞ is the freestream static pressure (a datum), $\frac{1}{2}\rho V_\infty{}^2$ is the freestream dynamic pressure (a scaling parameter), where ρ is the fluid density and V_∞ is the freestream velocity.

Hence if we can solve the potential flow to get $\left(\frac{V}{V_\infty}\right)^2$, then Cp can be calculated, and the load on the aerofoil can be found.

3.3.2 Smooth Uniform Flow Over a Bluff Body

Figure 3.6 shows a rectangular prism immersed in a uniform stream of fluid. In this situation, it is evident that although the onset flow is smooth and without turbulence, the body produces a large wake, which is unsteady and fluctuates from side to side. Although the Bernoulli equation can be used in the flow far from the wake to connect the local pressures and velocities to the freestream values, this is not the case in the wake, where the sum of the static and dynamic pressures are less than the freestream values. This loss in mechanical energy of the fluid has been converted to turbulence. Also evident in Fig. 3.6 is a separation bubble on the windward corner of the lower sidewall. Here the mean flow direction is reversed on the body surface, and "backflow" would be observed on the surface of a building at this location. Further downstream the flow reattaches. Hence is it clear that the flow in Fig. 3.6 is much more complicated and unsteady than that shown in Fig. 3.5 over a streamlined shape.

However, real flows are more complicated than the situation shown in Fig. 3.6. Wind flows are turbulent and the mean speed increases with height above the ground. This is discussed in the next section.

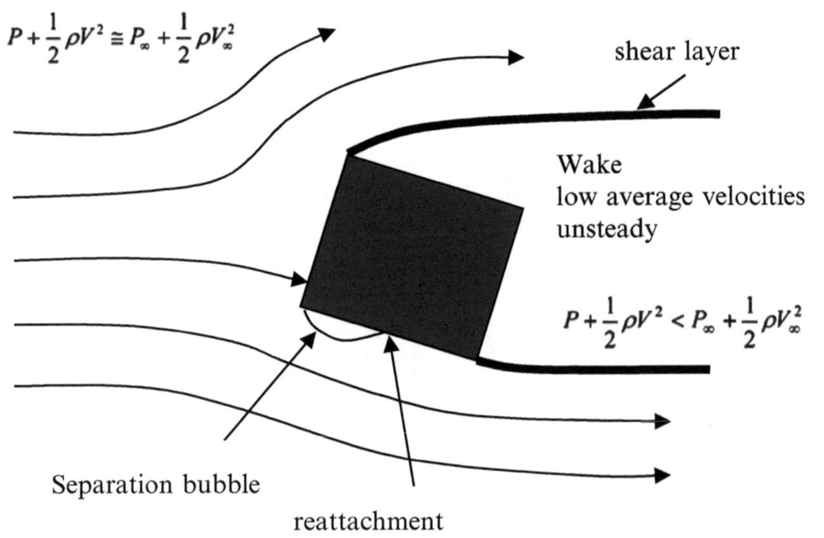

Fig. 3.6 Bluff body in smooth uniform flow

3.3.3 Turbulent Approach Flow, Bluff Three-Dimensional Body

Figure 3.7 shows a schematic diagram of a boundary layer flow onto a three-dimensional building-like shape. One possible realisation of the instantaneous velocity profile is sketched, along with the mean profile which could be obtained after averaging for a period that was long compared with the turbulent fluctuations; perhaps say 20 min in full-scale, or 1 min in the wind tunnel. Such a complicated flow, which varies in time and space, will cause fluctuating pressures at all points on the surface. The pressure time history at a point such as that marked "A" in the diagram would have the appearance illustrated in Fig. 3.8.

Figure 3.8 shows that the wind has a range of eddy sizes and frequencies. There is a peak pressure, a minimum pressure, and a mean pressure that can be defined for the period.

It is evident that with such a signal several pressure coefficients can be defined, as there are several pressure values that may be of interest, and several dynamic pressures that may be used as reference values. Equation (3.6) defines the mean pressure using the mean dynamic pressure as the reference. Equation (3.7) defines the peak pressure using the mean dynamic pressure. Equation (3.8) defines the peak pressure using the peak dynamic pressure. Equation (3.9) defines the minimum pressure using the mean dynamic pressure. Equation (3.10) is probably not very realistic, but it defines the

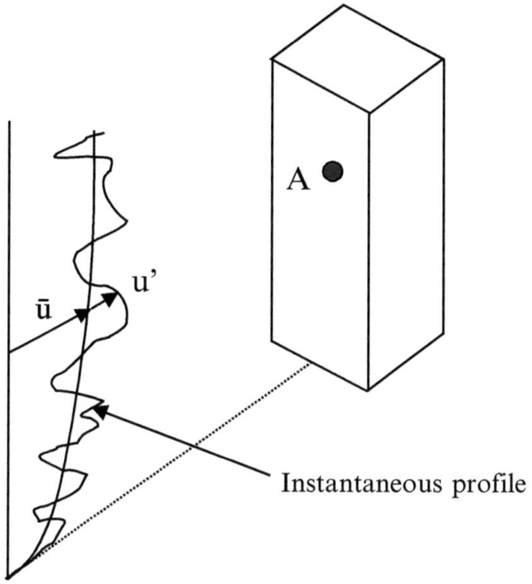

Fig. 3.7 Turbulent approach flow onto a 3-D bluff body

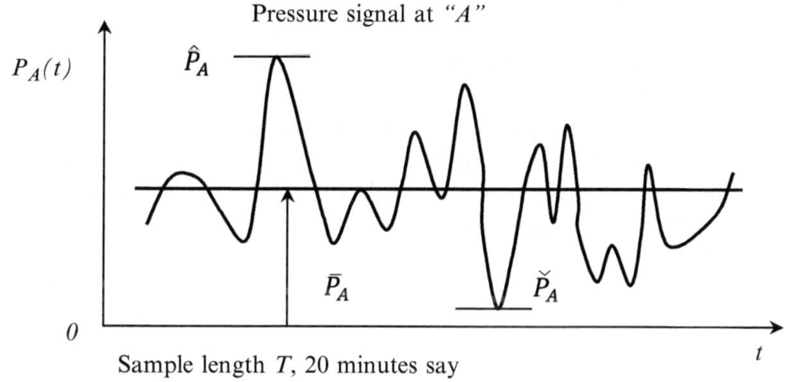

Fig. 3.8 Schematic diagram showing pressure fluctuations at point "A" on the front of the bluff body shown in Fig. 3.7

minimum pressure using the minimum dynamic pressure, which actually could be close to zero, and would result in very large values of the pressure coefficient.

$$\bar{P}_A - P_\infty = \bar{C}p\frac{1}{2}\rho\bar{V}^2 \tag{3.6}$$

$$\hat{P}_A - P_\infty = \hat{C}p\frac{1}{2}\rho\bar{V}^2 \tag{3.7}$$

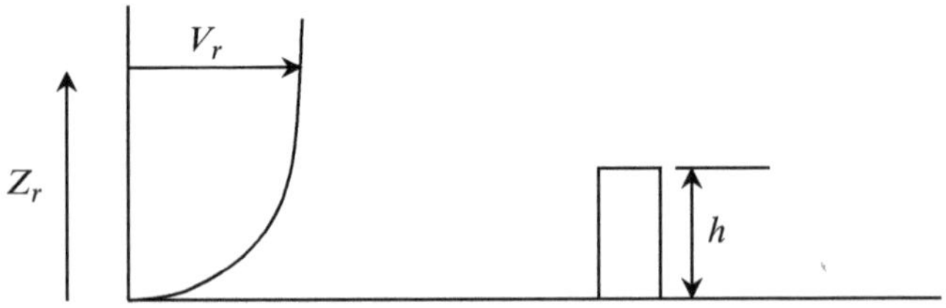

Fig. 3.9 Schematic diagram of a wind tunnel setup, showing a reference wind speed at the top of the boundary layer

$$\hat{P}_A - P_\infty = \hat{\hat{C}}p\frac{1}{2}\rho\hat{V}^2 \tag{3.8}$$

$$\breve{P}_A - P_\infty = \breve{C}p\frac{1}{2}\rho\bar{V}^2 \tag{3.9}$$

$$\breve{P}_A - P_\infty = \breve{\breve{C}}p\frac{1}{2}\rho\breve{V}^2 \tag{3.10}$$

In the wind-tunnel one also has to be careful about defining reference quantities. Figure 3.9 shows a schematic diagram of a wind tunnel set up, where V_r is the reference wind speed at the height Z_r near or at the top of the boundary layer.

In most wind tunnel setups, the reference dynamic pressure is based on a measurement at the top of the boundary layer well away from the influence of the model, and then pressure coefficients corresponding to the mean, maximum and minimum pressures can be defined according to Eqs. (3.11), (3.12), and (3.13). At The University of Auckland, this is upstream of the turntable and usually about 500 mm above the floor, which for a scale of 1:400 corresponds to a full-scale height of 200 m, which is approximately the height of the restaurant in the Skytower in Auckland City.

$$\bar{P} - P_\infty = \bar{C}p\frac{1}{2}\rho\bar{V}_r^2 \tag{3.11}$$

$$\hat{P} - P_\infty = \hat{C}p\frac{1}{2}\rho\bar{V}_r^2 \tag{3.12}$$

$$\breve{P}_A - P_\infty = \breve{C}p\frac{1}{2}\rho\bar{V}_r^2 \tag{3.13}$$

It is usually convenient to relate the pressure coefficients to a speed at the top of the building, so the subject building can be removed and a Pitot-static probe or other wind speed sensor placed at its top, and a (linear) relationship found between the reference speed, V_r, and V_h at the top of the building. Then by substituting for V_h in

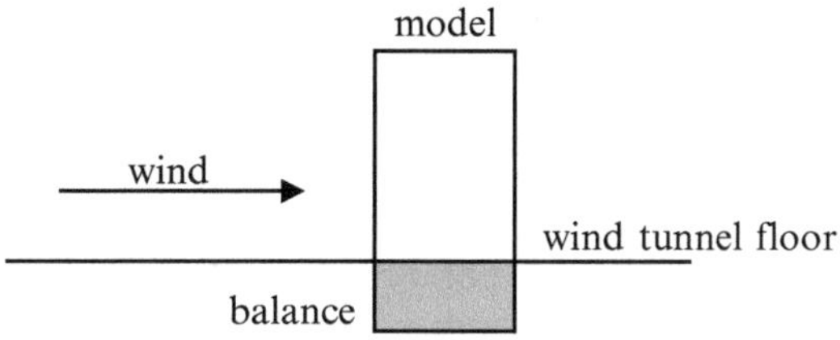

Fig. 3.10 Schematic diagram of a model on a force balance in a wind tunnel

the above "wind tunnel" equations, even more pressure coefficients can be defined, e.g. Eq. (3.14) for the mean pressure.

$$\bar{P} - P_\infty = \bar{C}p_h \frac{1}{2}\rho \bar{V}_h^2 \tag{3.14}$$

Pressure coefficients based on the speed at the top of the building are useful as test engineers know what kinds of numbers to expect. For example on the front of a model building the mean pressure coefficient may be about 0.8, and the peak pressure coefficient about 1.2. On the sides and rear the values will be somewhat negative.

Force coefficients are also required. This is because wind tunnel balances measure forces, as shown in Fig. 3.10, and force coefficients are usually relatively insensitive to the actual test speed. These are conveniently orientated (usually) in directions parallel and normal to the mean wind direction, the drag and lift directions respectively. Drag and lift coefficients are defined in terms of an undisturbed freestream dynamic pressure in Eqs. (3.15) and (3.16) respectively.

$$C_D = \frac{F_D}{\frac{1}{2}\rho V_\infty^2 A} \tag{3.15}$$

$$C_L = \frac{F_L}{\frac{1}{2}\rho V_\infty^2 A} \tag{3.16}$$

Now that the general features of turbulent flows have been introduced, some of the important bluff body aerodynamics concepts are presented.

3.4 The Effect of Reynolds Number on Bluff Body Forces and Pressure Distributions

The Reynolds number is the ratio of inertia forces to viscous forces, and is defined as follows.

$$\mathrm{Re} = \frac{\text{inertia forces}}{\text{viscous forces}} \propto \frac{\text{mass} \times \text{acceleration}}{\text{shear stress} \times \text{area}} = \frac{\rho L^3 \frac{V_\infty}{T}}{\mu \frac{V_\infty}{L} L^2} = \frac{\rho L^2}{\mu T} = \frac{\rho L^2}{\mu \frac{L}{V_\infty}}$$

$$= \frac{\rho V_\infty L}{\mu} = \frac{V_\infty L}{\nu}.$$

L = characteristic length, V_∞ = characteristic velocity, μ = dynamic viscosity, ν = kinematic viscosity = μ/ρ. The characteristic time T is equal to characteristic length/characteristic velocity, and ν is approximately 15×10^{-6} m^2/s for air at 21° C.

A typical Reynolds number for a building might be: 40 m/s × 40 m/ $15 \times 10^{-6} = 10^8$. For a chimney of 10-m diameter it will be about 3×10^7.

The Reynolds number is important in determining the extent to which boundary layers are laminar or turbulent. This in turn determines their ability to withstand an adverse pressure gradient, or their sensitivity to separate from the surface of a body. In the wind tunnel, the model scale is typically about 1/400 of full-scale, and the test speed is typically 1/3 of the full-scale design speed. This means that the wind tunnel Reynolds number is only about 1/1,200th of full scale, i.e. for a building it will be of the order of 10^5, and for a chimney it will be of the order of 2×10^4.

It is evident that the Reynolds number is very much smaller in the wind tunnel than in full-scale, and one may be concerned that this would give laminar boundary layers over surfaces much more than in full-scale. It turns out that this is not much of a problem, as the natural wind is very turbulent, so as long as there is sufficient turbulence in the wind tunnel, there is not expected to be large regions of laminar boundary layers on the moderately rough surfaces of typical building models. One might also be concerned that the flow could separate differently in the wind tunnel compared with full-scale. It turns out that this too is not a problem for sharp-edged shapes like buildings, as their sharp corners promote separation, and the flow separates in the wind tunnel like it does in full-scale. However, for curved shapes, this is not usually the case, and testing model chimneys in the wind tunnel is fraught with difficulty.

Figure 3.11 shows the drag coefficients of circular cylinders and spheres for a large range of Reynolds numbers. It is evident in Fig. 3.11 that for a smooth cylinder, the drag coefficient is about 1.2 for the Reynolds number range 10^3 to 2×10^5 and then it shows a rather sudden drop to 0.3 at a Reynolds number of 5×10^5 after which it gradually increases again. Thus a full-scale chimney with a Reynolds number in excess of 10^7 will be above the Cp dip (or critical Reynolds number) region, and in the so-called supercritical region, whereas a model in a wind tunnel, with a Reynolds number of 10^4 will be well below the critical Reynolds number, and in the sub-critical region.

This rather unusual behaviour of the drag coefficient can be explained by the behaviour of the boundary layer on the surface of the cylinder, which results in the flow regimes shown schematically in Fig. 3.12. At low Reynolds numbers the boundary layer is laminar and is very vulnerable to the adverse pressure gradient on the rear of the cylinder, and separation occurs at $\theta = 82°$ producing a large low pressure wake and a drag coefficient of about 1.2. At higher Reynolds numbers the boundary layer becomes turbulent before this location and is better able to withstand the adverse pressure gradient, and so it does not separate until $\theta = 120°$, resulting in a much smaller wake, a higher pressure on the rear (base), and a reduction in the drag coefficient to 0.3.

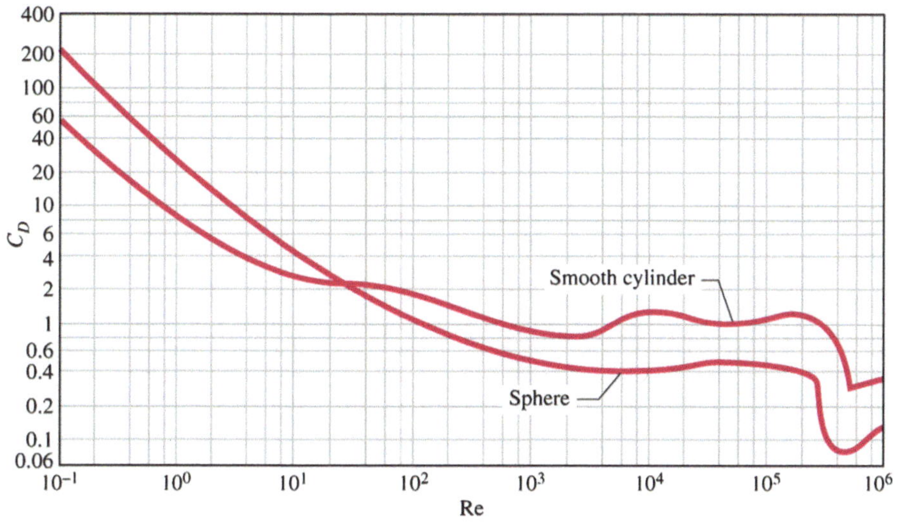

Fig. 3.11 Drag coefficient for smooth spheres and circular cylinders as a function of Reynolds number (after Cengel et al. 2008)

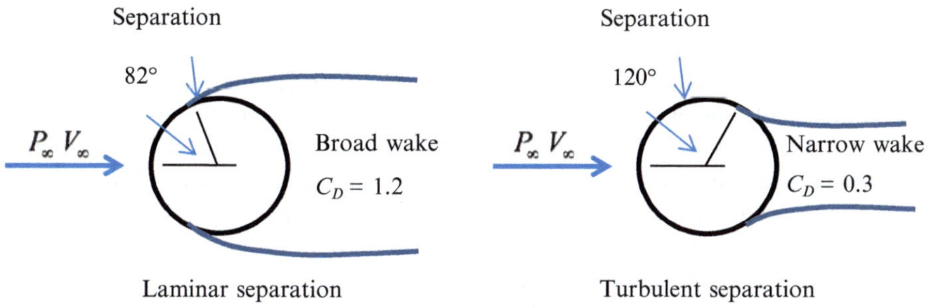

Fig. 3.12 Flow past a circular cylinder showing laminar and turbulent separation

When the flow is turbulent, or the surface of a sphere or circular cylinder is roughened, then these effects shown in Figs. 3.11 and 3.12 still occur, but the Reynolds number that they occur at changes. The reason for this is because turbulence and surface roughness cause early transition of the boundary layer on the surface from laminar to turbulent. This effect is illustrated rather dramatically in Fig. 3.13, which shows the drag coefficients for spheres over a smaller range or Reynolds numbers but for a range of different surface roughnesses. As the surface roughness increases, the critical Reynolds number where the drop in drag occurs is reduced. Hence one can emulate higher Reynolds number behaviour by roughening the surface. Turbulence in the onset flow produces the same effect (Son et al. 2010). Because of this effect, so-called Turbulence Spheres were used historically to determine the turbulence levels in wind tunnels (Platt 1937) by measuring the drag

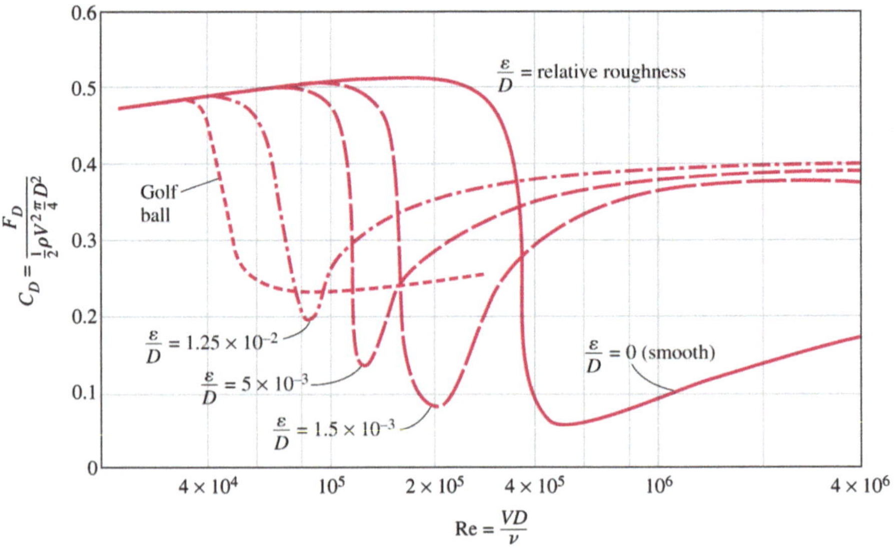

Fig. 3.13 The effect of surface roughness on the drag coefficient of a sphere (after Cengel et al. 2008)

and determining the drag coefficient. A wind tunnel with very low turbulence would have a critical Reynolds number of about 3×10^5, whereas the drop in drag occurs at progressively lower values of Reynolds numbers as the turbulence in the wind tunnel flow is increased. It is therefore extremely important to have low turbulence in aeronautical wind tunnels which are used for simulating aircraft travelling in a low turbulence environment. Such low turbulence levels are not important in boundary layer wind tunnels for wind engineering studies where turbulence is deliberately introduced into the flow stream to simulate the natural wind.

The drag coefficients shown in Fig. 3.11 result from integrations of the pressure distributions on the cylinder surface in the streamwise direction. Figure 3.14 shows pressure distributions measured on circular cylinders for a range of Reynolds numbers, and it is clearly evident that they vary quite remarkably depending on Reynolds number.

Referring to Fig. 3.14, line (a) for $\mathrm{Re} = 1.1 \times 10^5$ is at a sub-critical Reynolds number, line (b) for $\mathrm{Re} = 6.7 \times 10^5$ is at the bottom of the drag curve, and line (c) for $\mathrm{Re} = 8.4 \times 10^6$ is well above the critical Reynolds number. Line (a) has the most negative pressure coefficient on the leeward side of the cylinder, whereas line (b) corresponding to the lowest drag has the most positive pressure coefficient. It is clearly evident that none of the measured pressure distributions agree particularly well with the theoretical potential flow pressure distribution shown as line (d).

The result of this is, as noted previously, that it is not usually wise to test rounded shapes in wind tunnels at Reynolds numbers that are much lower than full-scale, unless it is done very carefully, as the pressure distributions on the body, and the

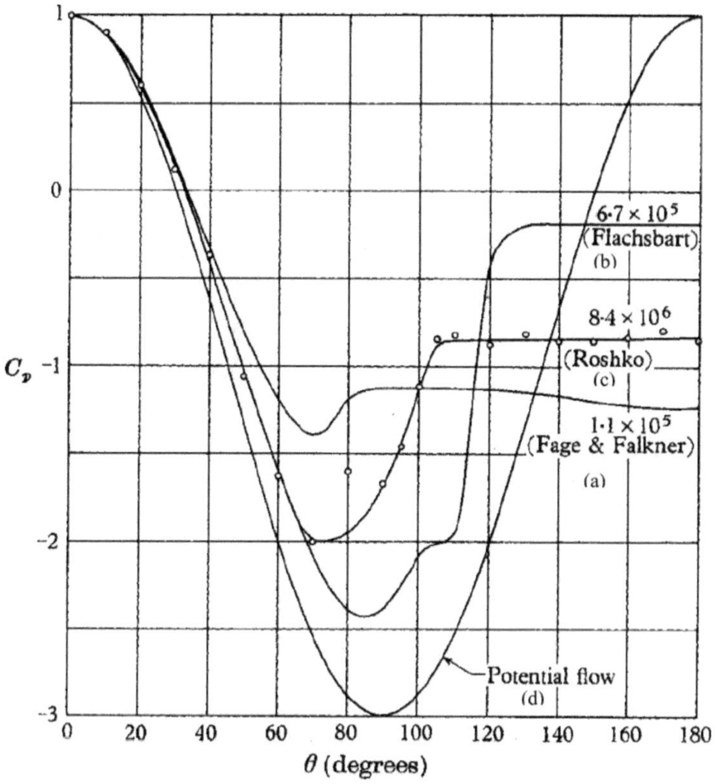

Fig. 3.14 Influence of Reynolds Number on the pressure distribution around a circular cylinder (after Roshko 1961)

resulting forces may be far from those that exist in reality. If such tests need to be undertaken, then one should attempt to test at the highest Reynolds number possible.

3.5 The Effects of Turbulence on Bluff Body Forces

It is common knowledge that in order for wind tunnels to give the correct answers in building aerodynamic studies it is necessary for the flow structure to be properly simulated. Why is this? The reason is that turbulence affects the flow in two ways. One reason is not particularly important, and one is very important.

It is well known that turbulence causes early transition of boundary layers from laminar to turbulent. This generally increases drag, but also can have the beneficial effect of causing flow separation to be reduced. In the atmosphere, the flow is at very large Reynolds numbers and one would expect all boundary layers on

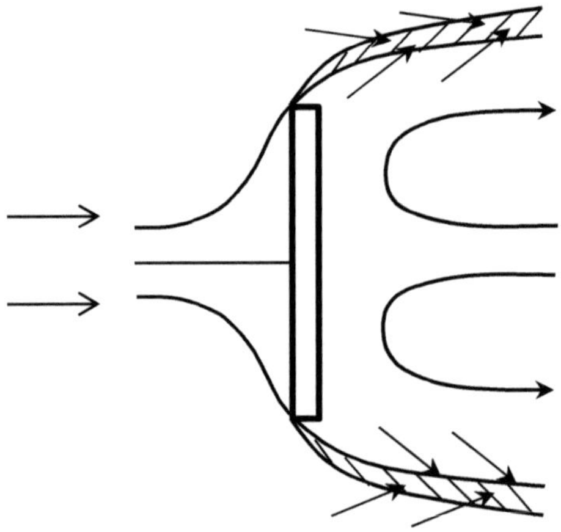

Fig. 3.15 Entrainment of freestream and wake fluid into the shear layers

buildings etc. to be turbulent. Thus in the wind tunnel one should have turbulent flow to get turbulent boundary layers, as mentioned in the previous section. However, whether or not the boundary layers change to turbulent at just the right place will not make too much of an impact on the overall loads, because they are mainly generated through pressure drag, i.e. differences in pressures between the front and back faces. There are special problems with chimneys, as already mentioned due to the variation of the separation point with Reynolds number.

The other effect of turbulence on the flow is much more important. The amount of turbulence in the flow affects the way that wakes develop behind objects, and this has a very strong impact on the resulting pressures on the sides and rear of these objects. Thus turbulence has to be modelled correctly in the wind tunnel to get the wakes right, and therefore to get the correct base pressures.

Figure 3.15 (Gartshore 1973) shows diagrammatically the wake developing behind a plate orientated normal to the flow direction. The freestream and wake fluid are being entrained into the shear layers. Gartshore argued that turbulence in the fluid enhances the entrainment into these shear layers, and this can affect the loads by affecting the radius of curvature of the shear layers.

Figure 3.16 shows how the wakes differ when objects are put in flow with and without turbulence. (Gartshore 1973) explains that small scale freestream turbulence can influence the drag, fluctuating lift, base pressure coefficient and galloping behaviour of bluff bodies by increasing the rate of entrainment of the shear layers shed from them. This is described in more detail in Sects. 3.5.1 and 3.5.2.

3.5.1 The Effects of Turbulence on Drag

In the flow around the objects depicted in Fig. 3.16, the freestream turbulence, after severe distortion by the mean flow field produces the increased mixing and entrainment by its presence in the shed shear layers. This increased rate of entrainment of

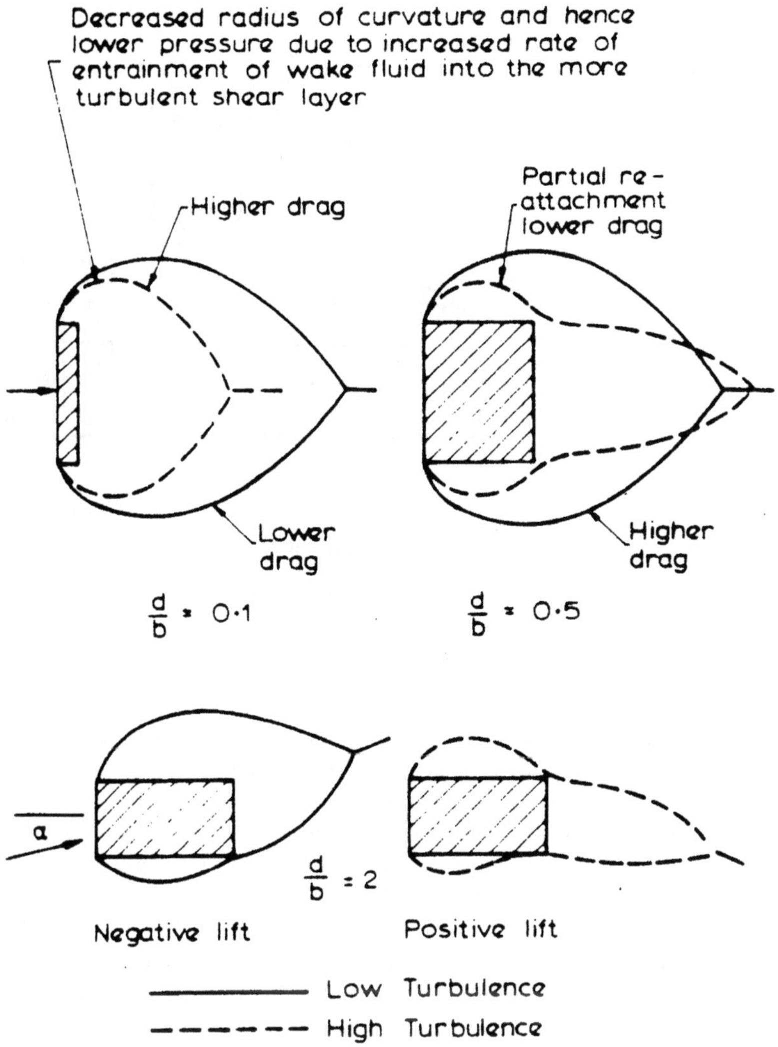

Fig. 3.16 Sketches of expected effect of turbulence on dividing streamlines as it affects drag and lift of rectangular sections (after Gartshore 1973)

wake fluid into the shear layer affects its shape and causes it to curve more sharply—the radius of curvature is reduced. This affects the base pressures quite dramatically, as increased curvature requires an increased radial pressure gradient, and since the freestream pressure cannot change, the pressure inside the shear layer therefore reduces. For the thin plate shown in Fig. 3.16 in high turbulence (top left), the increased curvature gives lower pressures in the wake and therefore the drag increases. For thicker plates (top right), this increased curvature may cause the flow

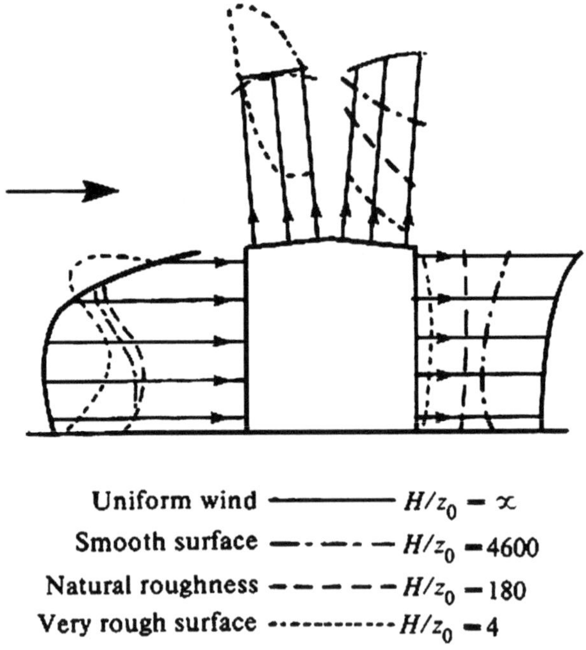

Fig. 3.17 Effect of surrounding ground roughness on the pressure distribution over a house, where H is height of house and z_0 is the roughness length (after Scruton 1981)

to reattach onto the "after-body" as shown. This makes the wake narrower and the shear layer straighter with lower curvature, and has the effect of making the wake pressure closer to ambient (i.e. higher than it otherwise would have been), and so the drag is reduced. It is very important that the right flow curvature of the shear layers is produced in wind tunnel simulations so that they will reattach (or not) in an analogous way as in full scale, and so that therefore the resulting base pressures and corresponding forces are simulated correctly, as depicted in Fig. 3.17 which shows the results of measurements on a house-like model in various levels of turbulence.

In Fig. 3.17 the roughness length is dependent on the surrounding surface roughness and is obtained by fitting the measured velocity profile to a logarithmic law. It can be seen that this length ratio, H/z_0 (called the Jensen number) has a strong influence on the resulting pressure distribution, and thus it is important that it is the same in a wind tunnel simulation as in the real flow in order for the simulated pressures to be representative of the real situation.

3.5.2 The Effects of Turbulence on Lift

The lower sketches in Fig. 3.16 are important because they show how turbulence can affect the reattachment when the flow is at an angle of attack to the object. The position of reattachment can have a large effect on the lift, and so turbulence

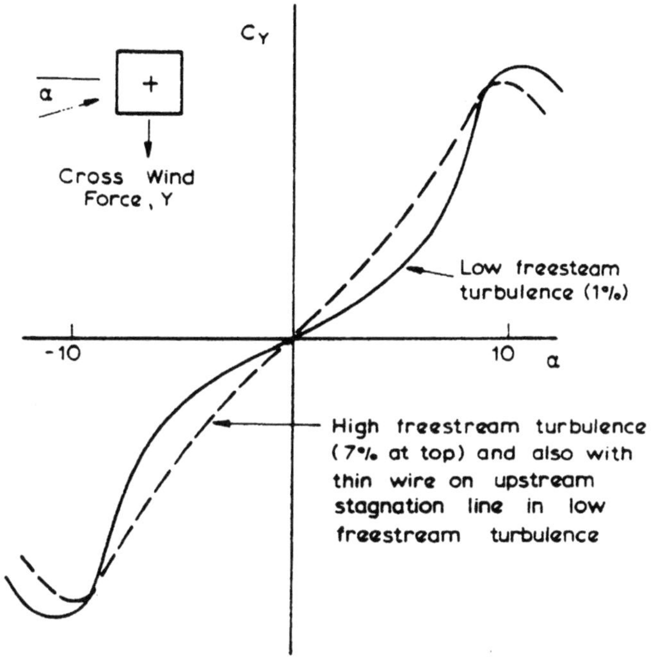

Fig. 3.18 Effect of freestream turbulence and small scale turbulence (due to a thin wire) on the transverse force characteristics of a slender square sections tower (after Melbourne 1975)

therefore affects the ability of a bluff body to generate a cross-wind force. It can even change the sign of the cross-wind force. Such sign changes are very important as they may lead to unstable cross-wind motion like "galloping", where the cross-wind force is in phase with the velocity, and thus is able to feed energy into the motion of the structure.

Figure 3.18 shows the cross-wind force developed by a square cross-section in turbulent and non-turbulent flows. It can be seen that at a given angle of attack, C_y is larger in the turbulent flow compared to the non-turbulent flow. In this case, the cross-wind force was large enough for the model to gallop in turbulent flow, whereas it was stable in low turbulent flow. Note in Fig. 3.18 that motion downwards produces a positive angle α and thus positive $C_{y,}$ and thus the motion has the conditions to allow galloping to occur if there is not enough damping in the system.

For bluff shapes with longer after-body lengths (like the cross-section of some bridges) the bodies can be stable in turbulent flow, but unstable in laminar flow. This is why bridges are often tested in smooth flow as well as turbulent flow, because it can be a worse case and may occur. This is because rather low turbulence flows can occur in stable atmospheric conditions above cold water in rivers, so model testing bridges in flows with low turbulence is very sensible.

3.6 The Effect of Corner Radius

Corner radius on bluff body shapes is very important, as it can help or hinder the way in which the flow goes past a corner. Sharp corners encourage separation, and the flow will separate there no matter what the Reynolds number. This is the reason that boundary layer wind tunnels can be used for the majority of wind engineering investigations at Reynolds numbers that are low by three orders of magnitude. However, when the corner is rounded, the flow can become sensitive to the Reynolds number, as discussed in Sect. 3.4 and shown in Fig. 3.19 from (Scruton 1981) adapted from (Delany and Sorensen 1953).

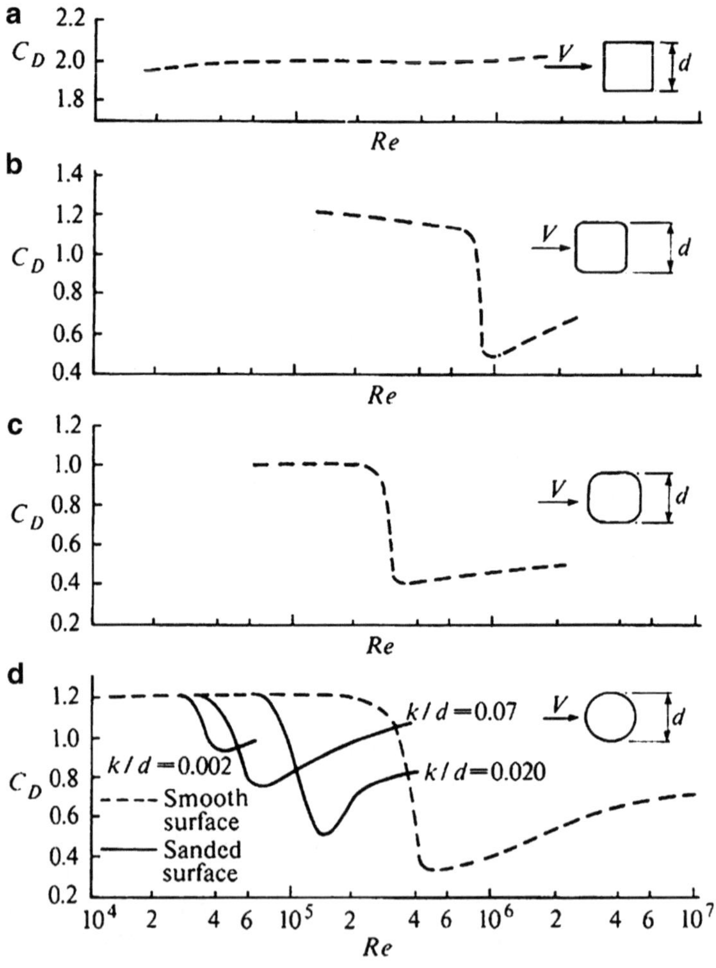

Fig. 3.19 Influence of Reynolds number, corner radius ratio r/d, and surface roughness k/d, where k is grain size of sand, on the values of C_D for prisms of square section and circular cylinders (aspect ratio $= \infty$): (**a**) $r/d = 0.021$, (**b**) $r/d = 0.167$, (**c**) $r/d = 0.333$, and (**d**) $r/d = 0.5$ (circular section) (after Scruton 1981, adapted from Delany and Sorensen 1953)

The top diagram with sharp corners shows that the drag is insensitive to Reynolds number. As the corner radius is increased, it can be seen that the drag reduces when the Reynolds number is increased above a certain value, and that this value gets lower with increased radius. This figure provides evidence for the observed rounding of the front of modern cars, as it reduces the drag. The bottom diagram shows the effect of surface roughness on the circular cylinder drag curve and is similar to Fig. 3.13 which shows the effect of surface roughness on the drag of a sphere. In wind engineering, increasing the surface roughness of a curved surface can be used to "simulate" a higher Reynolds number for wind tunnel testing, but this approach should be used with care.

3.7 Along-Wind and Cross-Wind Excitation, and Vortex Shedding

It is very convenient for wind engineering purposes to decompose the wind excitation and motion of structures into along-wind and cross-wind. This is because of the different excitation mechanisms that apply. The along-wind excitation results primarily from pressure drag, and is affected greatly by the oncoming turbulence. This is covered in Chap. 6. The cross-wind excitation comes from more sources. It depends on the oncoming turbulence, as well as the building shape, the wake, and the building motion. It is commonly observed that the wake in the river flow behind a bridge support consists of eddies that seem to turn one way and then the other. This is known as a von Karman vortex street. This kind of flow pattern is shown in the visualisation of Fig. 3.20. It illustrates flow at low Reynolds numbers, which is why the separation occurs at about 82° from the stagnation point. Vortices rotating in opposite directions are clearly evident. Alternating vortices are also shed from two-dimensional shapes at the much higher Reynolds numbers associated with buildings and towers etc. in the natural wind.

These vortices also exist behind tall two-dimensional shapes in wind flow, although they cannot be seen, and give rise to "lift" forces normal to the mean wind direction, i.e. the cross-wind direction. They arise because of the "circulation" Γ, that they produce around the shape, and the lift force, L can be shown to be related to circulation by the following very important equation $L = -\rho U \Gamma$. This is the same phenomenon that produces the down-force on a tennis ball when a good player puts on a lot of top spin. These forces become more important in wind engineering in smooth flow with tall slender shapes like chimneys when the wake may be correlated for several diameters along its height. They can also be an issue with tall buildings, and designers go to some effort to try to break up these vortices, so that the cross-wind excitation forces, and therefore the subsequent motions, are reduced.

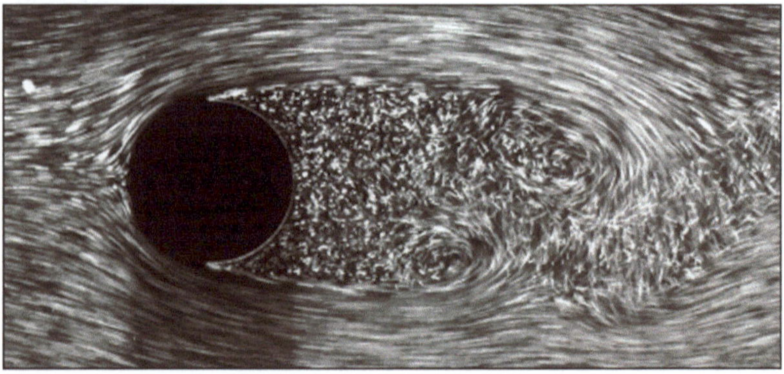

Fig. 3.20 Laminar boundary layer separation with a turbulent wake; flow over a circular cylinder at Re = 2000 (after Cengel et al. 2008)

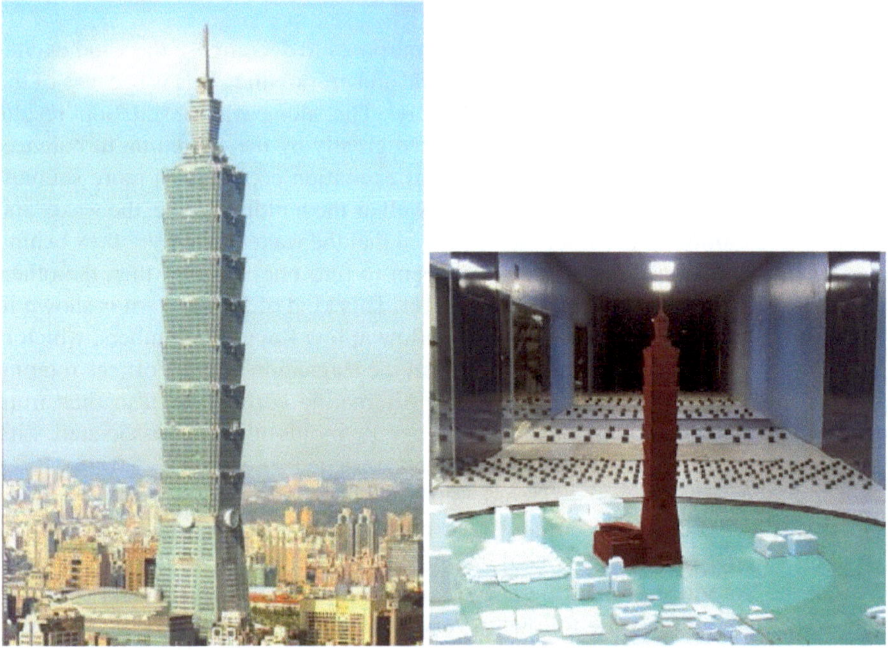

Fig. 3.21 *Left*: Photograph of Taipei 101 tower, and *right*: photograph of Taipei 101 tower wind tunnel model (after Irwin 2008)

There are some good examples of practical methods to break up the wake from the Consultancy RWDI, published in (Irwin 2008, 2009) and shown in Figs. 3.21, 3.22, and 3.23. Figure 3.21 shows a clever external shape that not only helps to break up the wake, but is sympathetic to the traditional architectural style in the region. Figure 3.23 shows clearly how the top of the building has been made very

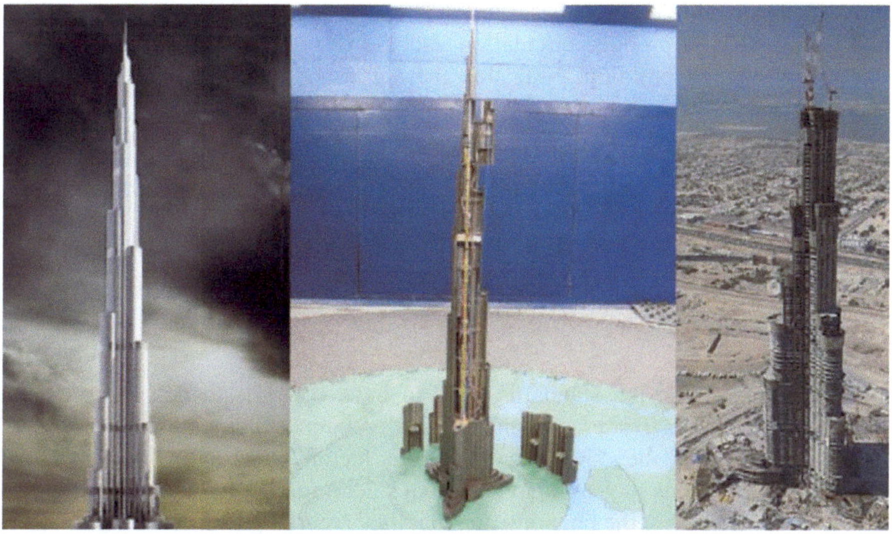

Fig. 3.22 Burj Kalifa—future view, aeroelastic model and construction photograph (after Irwin 2009)

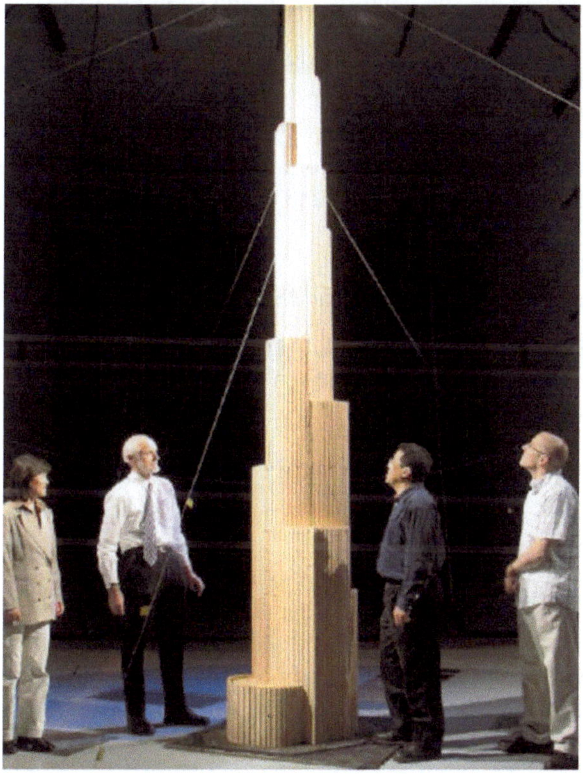

Fig. 3.23 1:50 scale model of top portion of Burj Kalifa for high Reynolds number testing (after Irwin 2009)

three-dimensional in shape in order to break up the wake so that it is more random and is less able to produce circulation and the associated cross-wind forces on this extremely tall building. Note that the spirals on chimney stacks serve the same purpose.

3.8 How the Australian and New Zealand Wind Loading Standard: AS/NZS 1170.2 Deals with "Bluff Body Aerodynamics"

Wind loading codes and standards need to have methods to determine design pressures and loads from component sizes, reference velocities and pressure coefficients. Ideally, they should have various factors which adjust the pressures due to various reasons such as the presence of shielding, and these factors should be based on the physics of the phenomenon that they are trying to account for. The NZ and Australian standards specify a regional peak gust wind speed V_R for a particular region and return period (R years) for a height of 10 m, and terrain roughness category 2, i.e. rural terrain. This regional speed is modified by various M coefficients to obtain the site wind speed, as shown in Eq. (3.17).

$$V_{sit,\beta} = V_R M_d \left(M_{z,cat} M_s M_t \right) \tag{3.17}$$

where

V_R = regional gust wind speed (m/s) for annual probability of exceedance of $1/R$
M_d = wind directional multipliers for the eight cardinal directions (β)
M_s = shielding multiplier
M_t = topographic multiplier
$V_{sit,\beta}$ is the wind speed calculated for the site, where direction β is N, NE, E, SE etc.

Thus Eq. (3.17) deals with the building height and features of the site and climate which have an influence on the local design wind speed. The pressure is determined using Eq. (3.18)

$$p = (0.5\rho_{air}) \left[V_{des,\theta} \right]^2 C_{fig} C_{dyn} \tag{3.18}$$

where

p = design wind pressure acting normal to a surface, Pa
ρ_{air} = density of air, which shall be taken as 1.2 kg/m^3
$V_{des,\theta}$ = building orthogonal design wind speeds, usually $\theta = 0, 90, 180, 270°$, and is determined from $V_{sit,\beta}$ and depends on the orientation of the building with respect to the cardinal wind directions

C_{fig} = aerodynamic shape factor; C_{dyn} = dynamic response factor (= 1.0, except when the building is wind sensitive)
$V_{des,\theta}$ = peak gust wind speed, and is in relation to the direction of the faces of the building.

The aerodynamic shape factor, C_{fig} is determined from an external pressure coefficient and various K factors.

$$C_{fig} = C_{p,e} K_a K_c K_l K_p \tag{3.19}$$

where

$C_{p,e}$ = external pressure coefficient
K_a = area reduction factor
K_c = combination factor
K_l = local pressure factor
K_p = porous cladding reduction factor.

The pressure factors account for some of the physics of wind flow over bluff bodies. The external pressure coefficient is a function of location on a structure or building, and takes into account knowledge of the mean value of pressure coefficients in various areas. For example, for a typical tall building the external pressure coefficients on the front, side and rear can have values of 0.8, −0.65 and −0.5 respectively, thus being positive on the front, but negative elsewhere. The area reduction factor K_a acts to lower the loads on structural support members dealing with large areas, and helps to account for the lack of correlation of wind gusts over large areas. It is equal to 1.0 for areas less than 10 m^2, reducing to 0.8 for areas in excess of 100 m^2. The combination factor reduces the loading when the load effect results from pressures acting on several surfaces, and allows for the fact that the peak pressures and suctions do not occur on all faces simultaneously. The local pressure factor K_l takes account of the intensity of small eddies on certain regions of building to produce higher pressures and suctions. For example it can have a value 3.0 for windward regions of wall and roof edges. This is where local separations can cause high suctions. The porous cladding reduction factor lowers the pressure coefficient for porous cladding where the flow can leak through the wall thus reducing the pressures and suctions. Hence it has been shown that understanding bluff body aerodynamics has enabled the writers of codes and standards to develop economical calculation processes that generally try to embody the physics of the wind loading process in the calculation procedures.

3.9 Conclusions

Thus it is evident from the foregoing that the fluctuating pressures acting bluff bodies immersed in highly turbulent sheared onset flows with large pulsating wakes that swing around from side to side, are very complicated, and so the study of bluff body aerodynamics is challenging and interesting. Much of the understanding that has been presented in this chapter has been gained by dedicated researchers undertaking extensive measurements of pressures and loads in the field on real structures and buildings, and also on detailed models in boundary layer wind tunnels. Computational fluid dynamic analysis has also contributed significantly to understanding bluff body aerodynamics, and is growing in importance. Thus a large body of knowledge exists which can enable specialists to make good estimates of loads and pressures on a vast range of shapes based on experience and published results. However, such is the complication of the subject that it is common for large structures and buildings to be wind tunnel tested to obtain the best possible estimates of pressures and loads for design purposes, and through such tests, and continuing research, the understanding of the nature of bluff body aerodynamics will continue to improve decade by decade.

References

Abbott HI, Von Doenhoff EA (1959) Theory of wing sections. Dover Publications, New York, US

Cengel Y, Cimbala J, Turner R (2008) Fundamentals of thermal-fluid sciences (SI units), 3rd edn. McGraw-Hill, Europe, Middle East and Africa. ISBN 9780071266314

Delany NK, Sorensen NE (1953) Low-speed drag of cylinders of various shapes, N.A.C.A. Tech. Note. 3038. Wash

Gartshore IS (1973) The effects of free stream turbulence on the drag of rectangular two-dimensional Prisms, Engineering science research report, BLWT-4-1973 Vol 73-74. Boundary Layer Wind Tunnel Laboratory, University of Western Ontario, London

Irwin PA (2008) J Wind Eng Ind Aerodyn 96:701–712

Irwin PA (2009) J Wind Eng Ind Aerodyn 97(7–8):328–334

Melbourne WH (1975) Cross-wind response of structures to wind action. In: Eaton KJ (ed) Proceedings of the fourth international conference on wind effects on structures. Cambridge University Press, London, pp 343–359

Platt RC (1937) Turbulence factors of N.A.C.A. wind tunnel as determined by sphere tests, NACA report No. 558

Prandtl L (1905) In: Krazer A (ed) Verhandlungen des dritten internationalen Mathematiker-Kongresses in Heidelberg 1904, Teubner, Leipzig, Germany, p. 484. English trans. in Early Developments of Modern Aerodynamics, Ackroyd JAK, Axcell BP, Ruban A1 (eds) Butterworth-Heinemann, Oxford, (2001), p 77

Roshko A (1961) Experiments on the flow past a circular cylinder at very high Reynolds number. J Fluid Mech 10(3):345–356

Scruton C (1981) An introduction to wind effects on structures. Oxford University Press, Oxford

Son K, Choi J, Jeon W, Choi H (2010) Effect of free-stream turbulence on the flow over a sphere. Phys Fluids 22(4):1–7

Chapter 4
Long-Span Bridge Aerodynamics

Yaojun Ge and Hiroshi Tanaka

Abstract In this Chapter, the formulation of aerodynamic forces of bridge decks is introduced with static force components, quasi-steady and unsteady aerodynamic force, and transient forces, and followed by various types of wind induced responses of bridges, including aerodynamic instability of flutter and galloping, buffeting vibration and vortex induced oscillation.

Keywords Aerodynamic force • Buffeting • Flutter • Galloping • Vortex excitation

4.1 Introduction

Wind has always had a strong influence upon man's activity in both favorable and unfavorable ways. Hot summer days become unbearable if there is no gentle breeze under green shade; but the same extent of breeze can be a life-threatening addition to a freezing weather. Development of oversea trade was made possible from early days of human history because of winds. At the same time they were fully aware that these ships would be wretched by a strong storm if they were unlucky. It is, hence, not surprising to find that wind has been an important factor since ancient days for the design of houses and towns and also in terms of ventilation and cooling particularly in warm and humid areas.

Y. Ge (✉)
State Key Laboratory for Disaster Reduction in Civil Engineering, Tongji University, Shanghai 200092, China
e-mail: yaojunge@tongji.edu.cn

H. Tanaka
Department of Civil and Environmental Engineering, University of Ottawa, Ottawa, ON, Canada

Y. Tamura and A. Kareem (eds.), *Advanced Structural Wind Engineering*,
DOI 10.1007/978-4-431-54337-4_4, © Springer Japan 2013

Table 4.1 John Smeaton's table of wind force (1759)

Velocity of the wind		Perpendicular force on one foot aero in pounds avoirdupois	Common appellations of the force of winds
Miles/hour	Feet/second		
1	1.47	0.005	Hardly perceptible
2	2.93	0.020	Just perceptible
3	4.40	0.044	
4	5.87	0.079	Gentle pleasant wind
5	7.33	0.123	
10	14.67	0.492	Pleasant brisk gale
15	22.00	1.107	
20	29.34	1.968	Very brisk
25	36.67	3.075	
30	44.01	4.429	High wind
35	51.34	6.027	
40	58.68	7.873	Very high
45	66.01	9.963	
50	73.35	12.300	A storm or a tempest
60	88.02	17.715	A great storm
80	117.36	31.490	A hurricane
100	146.70	49.200	A hurricane that tears up trees, carries buildings before it, etc.

4.1.1 First Turning Point to Mean Wind Loading

The earliest presentation of practical knowledge on wind for engineering design purposes was probably not until that of John Smeaton (1724–1792), the first "civil engineer", presented to the Royal Society of London in 1759 as the Table of Wind Force. The table is something like shown in Table 4.1 and the listed force can be summarized by a simple equation.

$$P = 1.2v^2 \tag{4.1}$$

Where P is the wind induced pressure (Pa or N/m^2) and v is the wind speed normal to the surface (m/s). This is equivalent to assume the drag coefficient of $C_D = 2.0$. Note that C_D for a 2D flat plate placed normal to the flow is approximately 1.9.

Smeaton's pioneering work, however, was not given enough respects as it deserves. It may not have contributed very much for better understanding and recognition of wind loading by engineers (Biggs 1961).

4.1.2 Second Turning Point to Gust Wind Loading

After the turn of the nineteenth century, wrought iron as a new and more reliable structural material started being used extensively. It resulted in design and construction of much greater structures than before. A remarkable progress of applied mechanics was also in support of this leap. The need for wind loading information became acute in this period. The impact of the collapse of the Tay Bridge should be understood in this context. There was no other incident that gave such a grave impact to design engineers for the recognition of wind loading on structures than this failure.

The Firth of Tay Bridge was the world longest railway bridge of the day with the total length of 3 km. It consisted of 84 spans of wrought iron trusses, 13 of which were 70–75 m truss spans crossing 27 m above water. It was designed by a renowned Scottish railway engineer Thomas Bouch (1822–1880) who was knighted by Queen Victoria because of this work. Unfortunately, the bridge was blown down by a strong gale in a stormy night of December 29, 1879, only 2 years after it was opened, with a passenger train on it claiming 75 lives. The wind speed in that evening was said to be 30–35 m/s, which was less than the design wind speed of 36 m/s. The main reason of the collapse was gust wind speed or gust wind loading.

The success of Forth Railway Bridge and Eiffel Tower, on the other hand, the first two greatest steel structures, completed in 1887 and 1889 respectively, were remarkable milestone established by excellent engineers. Forth Bridge is a 2×488 m suspension bridge, the longest span bridge in that time, and Eiffel Tower is a 300 m tall structure, the tallest man made structure in the world. These great successes can also be attributed to gust wind loading, which was firstly included in the design of the structures. They were followed by the days of huge suspension bridges and skyscrapers particularly in USA through the first half of the twentieth century.

4.1.3 Third Turning Point to Dynamic Wind Loading

The Tacoma Narrows Bridge was the world's third longest span bridge of the day and designed by Leon Moisseiff, who had been progressing the use of the deflection theory for the past quarter of a century and one of the highly regarded bridge designers in USA those days. Even 4 months prior to the opening of the bridge, there were vibrations observed quite frequently which called the attention of Prof. F.B. Farquharson of University of Washington as well as the others. As he observed, the bridge was firstly found frequently doing vertical bending vibrations in 1st and 2nd symmetric modes very light wind of 3–6 m/s. The maximum double amplitude was said to be up to 0.6 m and 0.8 m in these modes respectively. On the day of the final failure, November 7, 1940, the bridge oscillated for some time in the 7th symmetric and 8th asymmetric modes with the wind of 18–22 m/s. When the wind speed was about 19 m/s, suddenly the 1st asymmetric torsional motion started and became violent. Lateral motion also increased. It is believed that at this time a weak centre-tie connecting the main cables and the deck were snapped.

The most important matter to be noted of this incident was the fact that the bridge was destroyed not by static wind loading but by dynamic excitation. The bridge was supposed to withstand the static wind loading up to the design wind

Table 4.2 Bridge collapse due to wind action

No.	Year	Name of bridge	Span (ft)	Wind action	Designer	Country
1	1818	Dryburgh Abbey	260	Static	John and William Smith	Scotland
2	1821	Union	499	Static	Samuel Brown	England
3	1834	Nassau	245	Static	Lossen and Wolf	Germany
4	1836	Brighton Chair Pier	255	Static	Samuel Brown	England
5	1838	Montrose	432	Static	Samuel Brown	Scotland
6	1839	Menai Strait	580	Static	Thomas Telford	Welsh
7	1852	Roche-Beruard	641	Static	Le Blanc	France
8	1854	Wheeling	1,010	Static	Charles Ellet	USA
9	1864	Lewiston-Queenston	1,041	Static	Edward Serrell	USA
10	1879	Firth of Tay	246	Static	Thomas Bouch	Scotland
11	1889	Nigara-Clifton	1,260	Static	Samuel Keefer	USA
12	1940	Tacoma Narrows	2,800	Dynamic	Leon Mosseiff	USA

speed, which was more than twice the speed the bridge was actually destroyed. It was because of the induced vibration which was not considered at the design stage. Looking back the history in Table 4.2, there had been some incidents before in which bridges vibrated due to wind. However, apparently the dynamics of bridges had never become this much serious topic for bridge engineers.

The collapse of Tacoma Narrows Bridge in 1940 opened a new horizon to the structural design against wind action. Of course this was not the end of the static problems but growing need of dynamics was inevitable with the growing span and height of the structures. Development of new lighter and stronger materials, reduction of damping due to progress of welding techniques, more "optimized" design supported by the development of electronic computers and calculation techniques, everything seems to be making structures more vulnerable to dynamic excitations. The question of course was what kind of vibrations need to be considered and how they could be predicted. Later von Karman collaborated with Farquharson to do wind tunnel investigation of the bridge and tried to attribute the cause of vibration in both modes to vortex excitation. Bleich, another famous bridge engineer, tried to explain it by the analogy to the aerofoil flutter theory with some additional aerodynamic force terms. Modification of flutter theory was also attempted by Klöppel of Germany and Selberg of Norway. Hirai of Japan found the bridge failure as the lateral buckling due to lack of torsional rigidity of the stiffening frame (Bleich 1949a; Klöppel 1963; Selberg 1963; Hirai 1942).

Following this incident, the contribution of the aeronautical engineers was significant. The developing theories in flutter instability provided the theoretically interesting frame work to explain the bridge failure. Disagreement of observation with flutter theory even more spurred rather intensive research works on the matter at various institutions. Amongst them, the most important serious of study were carried out by a team of Farquharson, Vincent and von Karman towards the rebuilding of the New Tacoma Bridge. A similar series of study also took place at the National Physical Laboratory in Teddington, England, under the

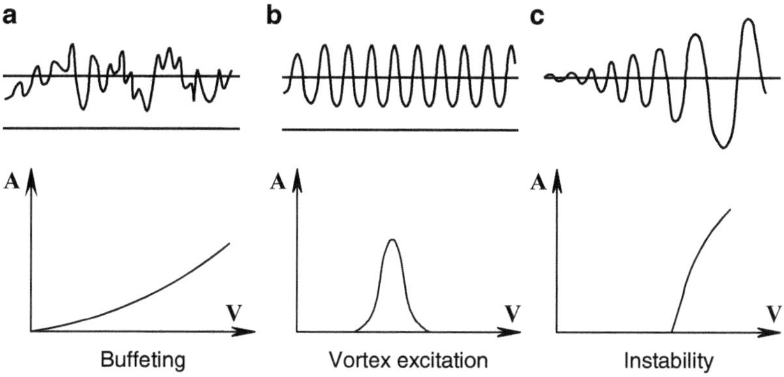

Fig. 4.1 Various types of wind induced response. (**a**) Buffeting; (**b**) Vortex excitation; (**c**) Instability

leadership of Kit Scruton. The results were applied to the design of both Forth and Severn suspension bridges (Davenport 1977; Aynsley et al. 1977; Moisseiff and Lienhardt 1933).

Apart from flutter or galloping instability shown in Fig. 4.1c, there are also two kinds of wind induced vibrations, stochastic buffeting and vortex excitation, shown in Fig. 4.1a, b, which should be considered in bluff-body aerodynamics and its application in bridges.

4.2 Formulation of Aerodynamic Forces

4.2.1 Static Force Components

When a 3D structure is exposed to an air flow, three force components and three moment components can be generally considered as indicated in Fig. 4.2 and these components are termed as

Corresponding to these, the dimensionless force and moment coefficients are sometimes defined as follows:

$$C_F = F/(QBW) \quad F = D, S \,\&\, L \quad \text{and} \quad Q = \rho U^2/2 \tag{4.2}$$

$$C_M = M/\left(QB^2W\right) \quad M = M_R, M_P \,\&\, M_Y \quad \text{and} \quad Q = \rho U^2/2 \tag{4.3}$$

In many wind engineering problems, it is not necessary to consider all six of these components, or even if it is necessary, 2D alternatives could be considered as convenient mathematical models to simplify the idea and make practical analyses possible.

Force components:
F_D: drag force
F_S: side force
F_L: lift force

Moment components:
M_R: rolling moment
M_P: pitching moment
M_Y: yawing moment

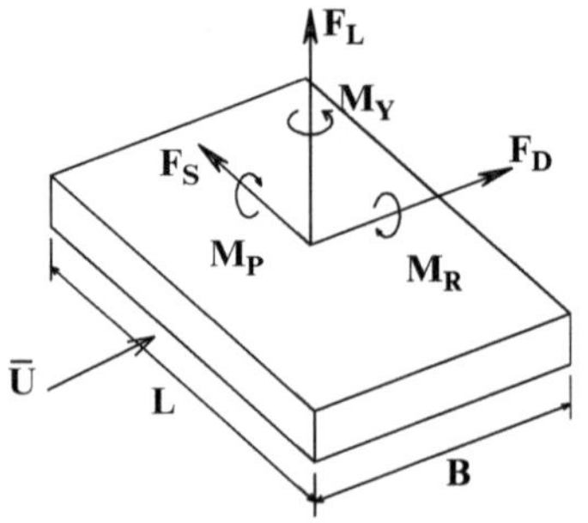

Fig. 4.2 Aerodynamic force components

The strip theory assumption, originally introduced for aerofoil and now often used for bridges, is a typical example of it. Since bridges are extended only in one direction and the primary concern is its behavior when wind comes perpendicular to its longitudinal axis, instead of looking at the whole structure, often its 2D slice (or strip) of unit thickness cut off by two planes in mean wind direction is considered. The idea is the same as the plane strain analysis in the theory of elasticity.

For this case, there are only three components need to be considered; lift force F_L, drag force F_D and pitching moment M_P, since each span wise station is considered as though it were a portion of an infinite span bridge with uniform span wise properties. Corresponding to these, the displacements taken into consideration are (h_x, h_z, α_y). In the following sections, only these three directions are counted for the formulation of aerodynamic force components and the displacements are written as (p, h, α) for simplicity.

4.2.2 Quasi-Steady Aerodynamics

Another approximation often introduced is the quasi-steady approximation which is to say that the aerodynamic forces at any time depend only on the instantaneous position of the body at that particular moment. In other words, the temporal memory effect or the history of the motion in the aerodynamic model can be ignored.

The strip theory approximation discussed above is unambiguous and its meaning is generally accepted. However, this is not true for the quasi-steady approximation. It is an acceptable assumption for relatively high wind speed case but clearly unacceptable, for example, in case of vortex excitations.

When the quasi-steady approximation is applied, three aerodynamic force components are given simply by

$$L = QB \cdot C_L \quad D = QB \cdot C_D \quad \text{and} \quad M = QB^2 \cdot C_M \tag{4.4}$$

where Q is the dynamic pressure considering the relative wind speed and is given by

$$Q = (\rho/2) \cdot (U + u - dp/dt)^2 \tag{4.5}$$

and the force and moment coefficients, as their first approximation is given by

$$C_j = C_j(\alpha) + \left(dC_j/d\alpha\right) \cdot \alpha_{rel} \tag{4.6}$$

where U and u are the mean wind speed and the turbulence speed, respectively, and α_{rel} is the relative angle of attack, and is given by

$$\alpha_{rel} = \alpha - \frac{\dot{h} + nB\dot{\alpha} - w}{U + u - \dot{p}} \approx \alpha - \frac{\dot{h} + nB\dot{\alpha} - w}{U} \tag{4.7}$$

where n is a factor representing the position of the point of application of the aerodynamic lift force. For a thin aerofoil at low reduced frequency, $n \approx 0.75$. Hence the force components can be expressed as follows:

$$\begin{aligned} L &= \bar{L} + L_f + L_b \\ D &= \bar{D} + D_f + D_b \\ M &= \bar{M} + M_f + M_b \end{aligned} \tag{4.8}$$

where the first terms are static components given by

$$\begin{aligned} \bar{L} &= \left(\rho U^2/2\right) \cdot B \cdot C_L(\alpha) \\ \bar{D} &= \left(\rho U^2/2\right) \cdot B \cdot C_D(\alpha) \\ \bar{M} &= \left(\rho U^2/2\right) \cdot B^2 \cdot C_M(\alpha) \end{aligned} \tag{4.9}$$

the second terms are motion-dependent components which are

$$\begin{aligned} L_f &= \left(\rho U^2/2\right)B\left[C'_L\alpha - \left\{2C_L\dot{p} + C'_L\left(\dot{h} - nB\dot{\alpha}\right)\right\}/U\right] \\ D_f &= \left(\rho U^2/2\right)B\left[C'_D\alpha - \left\{2C_D\dot{p} + C'_D\left(\dot{h} - nB\dot{\alpha}\right)\right\}/U\right] \\ M_f &= \left(\rho U^2/2\right)B^2\left[C'_M\alpha - \left\{2C_M\dot{p} + C'_M\left(\dot{h} - nB\dot{\alpha}\right)\right\}/U\right] \end{aligned} \tag{4.10}$$

and the last terms, buffeting force terms are represented by

$$\begin{aligned} L_b &= (\rho UB/2)\{2C_L \cdot u(t) + C'_L \cdot w(t)\} \\ D_b &= (\rho UB/2)\{2C_D \cdot u(t) + C'_D \cdot w(t)\} \\ M_b &= (\rho UB/2)\{2C_M \cdot u(t) + C'_M \cdot w(t)\} \end{aligned} \tag{4.11}$$

The first terms relate only to the static deformation and can be set aside for the present analysis. C' means the slope of C at the vicinity of the static angle. If the cross section is nearly symmetric, C_L, C_M and C'_D are close to zero and hence

$$
\begin{aligned}
L(t) &\approx \frac{\rho U^2}{2} \cdot B \cdot \frac{dC_L}{d\alpha} \cdot \left[\alpha - \frac{\dot{h} - nB\dot{\alpha}}{U}\right] + \frac{\rho UB}{2} \cdot \frac{dC_L}{d\alpha} \cdot w(t) \\
D(t) &\approx -\rho U^2 BC_D \cdot \dot{p} + \rho UB \cdot C_D \cdot u(t) \\
M(t) &\approx \frac{\rho U^2}{2} \cdot B^2 \cdot \frac{dC_M}{d\alpha} \cdot \left[\alpha - \frac{\dot{h} - nB\dot{\alpha}}{U}\right] + \frac{\rho UB^2}{2} \cdot \frac{dC_M}{d\alpha} \cdot w(t)
\end{aligned} \tag{4.12}
$$

4.2.3 Unsteady Aerodynamic Force Coefficients

The motion-dependent force components can be expressed more generally, including the virtual mass effects, as follows:

$$
\begin{aligned}
L_f &= L_h h + L_{\dot{h}}\dot{h} + L_{\ddot{h}}\ddot{h} + L_\alpha \alpha + L_{\dot{\alpha}}\dot{\alpha} + L_{\ddot{\alpha}}\ddot{\alpha} + L_p p + L_{\dot{p}}\dot{p} + L_{\ddot{p}}\ddot{p} \\
D_f &= D_h h + D_{\dot{h}}\dot{h} + D_{\ddot{h}}\ddot{h} + D_\alpha \alpha + D_{\dot{\alpha}}\dot{\alpha} + D_{\ddot{\alpha}}\ddot{\alpha} + D_p p + D_{\dot{p}}\dot{p} + D_{\ddot{p}}\ddot{p} \\
M_f &= M_h h + M_{\dot{h}}\dot{h} + M_{\ddot{h}}\ddot{h} + M_\alpha \alpha + M_{\dot{\alpha}}\dot{\alpha} + M_{\ddot{\alpha}}\ddot{\alpha} + M_p p + M_{\dot{p}}\dot{p} + M_{\ddot{p}}\ddot{p}
\end{aligned} \tag{4.13}
$$

The coefficients $L_{\dot{h}}, L_{\ddot{h}}, \ldots, M_p$ are generally called aerodynamic derivatives and are given as functions of reduced velocity or reduced frequency

$$V_r = U/(\omega B) \tag{4.14a}$$

$$K = 1/V_r = \omega B/U \tag{4.14b}$$

The analytical expressions for the derivatives have been obtained only for the case of a simple harmonic motion of a flat plate in coupled heave-pitch motion with infinitesimal amplitudes and zero mean angle of attack. Another simple case analytically given is the lift force on a flat plate induced by a sudden stepwise position change, $h = 1$ and $\alpha = 1$, when it is exposed to a uniform air flow with zero angle of attack. For other cases, the aerodynamic derivatives need to be determined experimentally. Usually a simple harmonic oscillatory motion is assumed. For this case, the first, fourth and seventh terms can be put together with the third, sixth and ninth terms in Eq. (4.13), respectively, as

$$L_x x + L_{\ddot{x}}\ddot{x} = L_x x + L_{\ddot{x}}(-\omega_x^2 x) = (L_x - \omega_x^2 L_{\ddot{x}})x = L'_x x \tag{4.15}$$

Consequently, the motion-dependent force also known as self-excited aerodynamic force can be expressed through six state variables and eighteen aerodynamic parameters. The following expression with aerodynamic derivatives by Scanlan has been widely accepted:

$$
\begin{aligned}
&KH_1^*(K)\cdot\frac{\dot{h}}{U}+KH_2^*(K)\cdot\frac{B\dot{\alpha}}{U}+K^2H_3^*(K)\cdot\alpha+K^2H_4^*(K)\cdot\frac{h}{B}+KH_5^*(K)\cdot\frac{\dot{p}}{U}\\
&\quad+K^2H_6^*(K)\cdot\frac{p}{B}\\
&KP_1^*(K)\cdot\frac{\dot{p}}{U}+KP_2^*(K)\cdot\frac{B\dot{\alpha}}{U}+K^2P_3^*(K)\cdot\alpha+K^2P_4^*(K)\cdot\frac{p}{B}\\
&\quad+KP_5^*(K)\cdot\frac{\dot{h}}{U}+K^2P_6^*(K)\cdot\frac{h}{B}\\
&KA_1^*(K)\cdot\frac{\dot{h}}{U}+KA_2^*(K)\cdot\frac{B\dot{\alpha}}{U}+K^2A_3^*(K)\cdot\alpha\\
&\quad+K^2A_4^*(K)\cdot\frac{h}{B}+KA_5^*(K)\cdot\frac{\dot{p}}{U}+K^2A_6^*(K)\cdot\frac{p}{B}
\end{aligned}
\tag{4.16}
$$

where $K = \omega B/U$. For the case of flat plate aerodynamic derivatives are given as follows:

$$
\begin{aligned}
&H_1^* = -\frac{\pi}{k}\cdot F(k)\quad H_2^* = -\frac{\pi}{4k}\cdot\left[1+F(k)+\frac{2G(k)}{k}\right]\\
&H_3^* = -\frac{\pi}{4k^2}\cdot[2F(k)-kG(k)]\quad H_4^* = -\frac{\pi}{2}\cdot\left[1+\frac{2G(k)}{k}\right]\\
&A_1^* = \frac{\pi F(k)}{4k}\quad A_2^* = \frac{\pi}{16k}\cdot\left[-1+F(k)+\frac{2G(k)}{k}\right]\\
&A_3^* = \frac{\pi}{8k^2}\cdot\left[\frac{k^2}{8}+F(k)-\frac{kG(k)}{2}\right]\quad A_4^* = -\frac{\pi G(k)}{4k}\\
&P_j^* = 0\quad (j = 1,2,3,4)
\end{aligned}
\tag{4.17}
$$

in which *F(k)* and *G(k)* are the real and imaginary parts of the Theodorsen's function defined by using the Hankel functions as follows:

$$
C(k) = F(k) + i\cdot G(k) = \frac{H_1^{(2)}(k)}{H_1^{(2)}(k) + i\cdot H_0^{(2)}(k)}
\tag{4.18}
$$

and $k = K/2$ is used as the reduced frequency here.

4.2.4 Transient Forces

The definition of these aerodynamic derivatives, however, assumes that the body in motion is performing a simple harmonic coupled oscillation, just like the case of an aerofoil flutter analysis, with infinitesimally small amplitudes. Hence they are not directly applicable to the description of any transient motion, which can be obtained by a convolution of any forcing function with indicial admittances.

When a thin two-dimensional aerofoil of the chord length B is placed in a uniform flow U and the angle of attack α is suddenly given as stepwise increase from zero, only a half of the final lift force would become effective immediately and another half of the lift force will be gained asymptotically, or

$$L(\tau) = \pi\rho U^2 \mathrm{B}\alpha \cdot \Phi(\tau) \tag{4.19}$$

where $\tau = 2Ut/B > 0$ is dimensionless time. $\Phi(\tau)$ is called the Wagner's function and is approximated by

$$\Phi(\tau) = 1 - 0.165 \cdot \exp(-0.0455\tau) - 0.335 \cdot \exp(-0.300\tau) \tag{4.20}$$

The Wagner's function is related to the Theodorsen's function by the inverse Laplace transform as

$$\Phi(\tau) = L^{-1}[C(-is)/s] \tag{4.21}$$

This relationship implies that the indicial lift and steady state sinusoidal lift terms form a Laplace transform pair. Since the Laplace transform of an exponential function is given by a rational form, or

$$L[e^{at}] = 1/(s - a) \tag{4.22}$$

It has been established that the unsteady aerodynamic force terms in general can be approximated by a series of partial fraction forms of the Laplace variable (Fung 1969; Karpel 1981; Roger 1977). The most common form of the approximating functions currently used for each unsteady generalized force coefficient is the Roger's formulations given as follows:

$$[A_p] = [Q_1] + [Q_2]p + [Q_3]p^2 + \sum_{j=1}^{N} \frac{p}{p + \gamma_j}[R_j] \tag{4.23}$$

Another fundamental formulation is when the lift force is induced by a suddenly applied constant upward gust $w_o(\tau)$, which is given by

$$L(\tau) = \pi\rho U^2 B \cdot (w_o/U) \cdot \Psi(\tau) \tag{4.24}$$

where $\tau = 2Ut/B$ is again the dimensionless time and $\Psi(\tau)$ is the indicial admittance and is called the Küssner's function for this case. Küssner function is somewhat similar to Wagner function except it starts from zero at time zero and approaches asymptotically its steady-state value of unity when $\tau \to 0$. If the gust is of an arbitrary vertical velocity distribution $w(\tau)$, the lift becomes a convolution, or Duhamel's integral, involving the derivatives of Küssner function, which is equivalent to the impulse response function, as

$$L(\tau) = \pi\rho U^2 B \cdot \int_0^\infty w(\tau\text{-}s)\Psi'(s)ds \tag{4.25}$$

In particular, if the gust variation takes a sinusoidal form, $w(s) = W_o\exp(iks)$, where $k = \omega B/(2U)$, the lift force becomes

$$L(s) = \pi\rho U^2 B \cdot (w(s)/U) \cdot \Theta(k) \tag{4.26}$$

where $\Theta(k)$ is the Sears' function which is related to the Küssner's function by

$$\Theta(k) = \int_0^\infty \Psi'(s)\exp(-iks)ds = ik\int_0^\infty \Psi(s)\exp(-iks)ds \tag{4.27}$$

The concept of convolution integral explained above becomes an import tool for the time domain analysis.

4.3 Aerodynamic Instability

4.3.1 Concept of Aerodynamic Instability

Consider, as an example, a pitching motion in Fig. 4.3. The equation of motion can be as follows: (Fung 1969; Blevins 1990)

$$J\ddot{\phi} + C\dot{\phi} + K\phi = \left(\rho U^2/2\right) \cdot B^2 \cdot C_M(\phi, \dot{\phi}, \ldots) \tag{4.28}$$

As a simple expression, if $C_M = a\phi + b\dot{\phi}$

$$\ddot{\phi} + 2\varsigma_s\dot{\phi} + \omega_T^2\phi = \frac{\rho U^2 B^2}{2J} \cdot \left(a\phi + b\dot{\phi}\right) \tag{4.29}$$

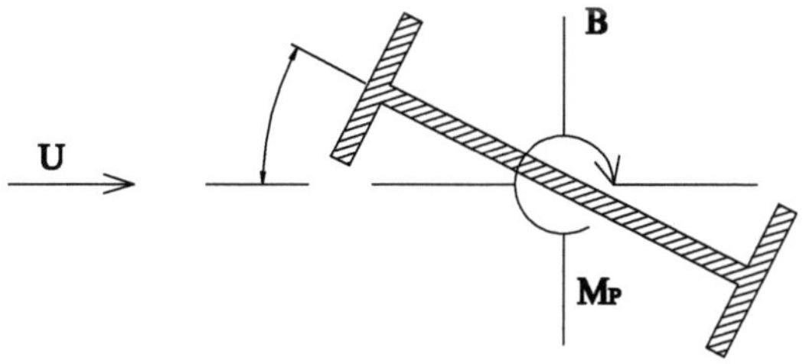

Fig. 4.3 Aerodynamic instability model

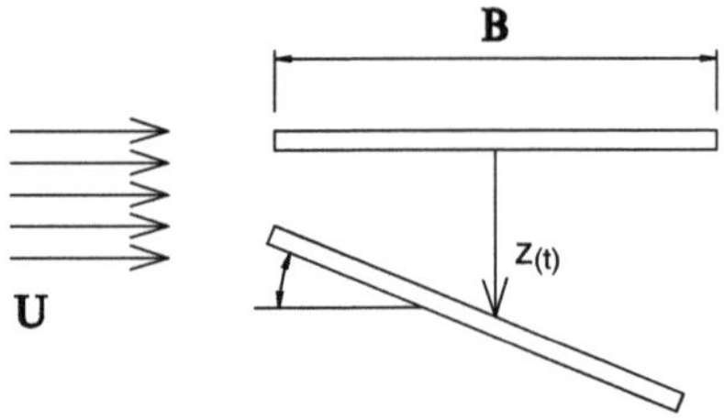

Fig. 4.4 Displacement of wing flutter

where $\omega_T = (K/J)^{\frac{1}{2}}$, $\varsigma_s = C/(2J\omega_T)$

$$\ddot{\phi} + 2(\varsigma_s + \varsigma_a)\omega_T\dot{\phi} + \omega_T^2(1 - S_a)\phi = 0 \tag{4.30}$$

where $\varsigma_a = -\frac{\rho U^2 B^2}{4\omega_T J} \cdot b(U_r)$ $S_a = -\frac{\rho U^2 B^2}{4J\omega_T^2} \cdot a(U_r)$

a and b are found to be functions of the reduced velocity $U_r = U/(B\omega_T)$.

$\varsigma_s + \varsigma_a < 0 \rightarrow$ negative damping $\rightarrow$ flutter

$S_a > 1 \rightarrow$ negative stiffness $\rightarrow$ divergence

4.3.2 2D Strip Theory of Aerofoil Flutter

Though a dynamic failure of aircraft wings caused by aeroelastic phenomena was observed from the early days of flight, the real development of the non-stationary aerofoil theory did not occur till the 1920s. The numerical calculation of the aerodynamic force on an oscillating flat plate was first given by Birnbaum (1922) as an application of Prandtl's Theory of bound vortices. Through the following decade, the analysis of unsteady aerodynamic farces on an oscillating two-dimensional plate attracted significant interests of aerodynamicists such as Wagner, Glauert, Küssner, Duncan and Collar and the most complete solution to this problem was presented by Theodorsen in 1935. Similar solutions were also developed by Küssner and Schwarz, Cicala, Schmieden, Ellenberger but the solution by Theodorsen has been most extensively used (Bisplinghoff et al. 1955; Dowell et al. 1989).

The problem of wing flutter as shown in Fig. 4.4 is formulated as follows:

$$\begin{aligned} \ddot{z} + \omega_z^2(1 + i \cdot 2\varsigma_z)z &= L/m \\ \ddot{\phi} + \omega_\phi^2(1 + i \cdot 2\varsigma_\phi)\phi &= M_p/J \end{aligned} \tag{4.31}$$

The unsteady aerodynamic forces are

$$\begin{aligned} L &= (\rho U^2/2) \cdot B \cdot C_L(\dot{z}, \ddot{z}, \phi, \dot{\phi}, \ddot{\phi}) \\ M_p &= (\rho U^2/2) \cdot B^2 \cdot C_M(\dot{z}, \ddot{z}, \phi, \dot{\phi}, \ddot{\phi}) \end{aligned} \tag{4.32}$$

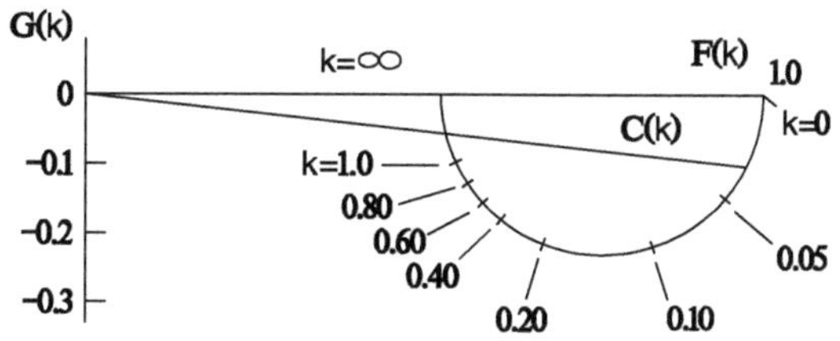

Fig. 4.5 Vector diagram of C(k)

where

$$C_L = 2\pi \cdot \left[\left(\phi + \frac{\dot{z}}{U}\right) \cdot C(k) + \frac{B\dot{\phi}}{4U}\{1 + C(k)\} + \frac{B\ddot{z}}{4U^2}\right]$$

$$C_M = \frac{\pi}{2} \cdot \left[\left(\phi + \frac{\dot{z}}{U}\right) \cdot C(k) - \frac{B\dot{\phi}}{4U}\{1 - C(k)\} - \frac{B^2\ddot{\phi}}{32U^2}\right] \tag{4.33}$$

in which

$$C(k) = \frac{H_1^{(2)}(k)}{H_1^{(2)}(k) + i \cdot H_0^{(2)}(k)} = F(k) + i \cdot G(k) \tag{4.34}$$

is called the Theodorsen function as shown in Fig. 4.5 and $k = \pi fB/U$.

By assuming the steady-state solution of the equations as

$$z(t) = Ze^{i\omega t} \& \phi(t) = Fe^{i\omega t} \tag{4.35}$$

the flutter conditions (U_F, ω_F) can be determined.

Theodorsen's analysis was based on the potential flow theory which it is assumed that the flow does not separate from the solid surface of the body. Also it assumes a coupled simple harmonic vibration with infinitesimally small amplitudes.

4.3.3 Application of Flutter Theory

The collapse of the original Tacoma Narrows Bridge (1940) was, from the time immediately following the incident, frequently compared to galloping of iced cables or flutter of aircraft wings. Bleich tried to analyses the accident as a flutter by applying Theodorsen's aerodynamic formulation to the bridge and found the critical speed thus calculated was considerably higher than that of the Tacoma Narrows Bridge. It was obvious that the aerofoil flutter coefficients were not applicable directly to more aerodynamically bluff bridge sections. Bleich tried to cover this defect by modifying the Theodorsen's expression and considered additional terms corresponding to the effect of vortex formation from the leading edge of the deck but was not very successful. Pugsley commented at this point that experimentally determined aerodynamic coefficients rather than Theodorsen's may be of more help in future. He was right (Bleich 1949b; Scanlan 1978).

Ironically, the use of streamlines shallow box girders as a suspension bridge stiffening girder has become quite popular inspired by successful application of them for both Severn and Lillebelt crossings in the late 1960s. For these cases, the flow separation is much less and Bleich's original calculation with Theodorsen's forces can actually provide a reasonably good approximation. Because of a tedious process required for its calculation in pre-computer days, various simplifications of flutter calculation were considered. Well-known formulae by Selberg, Rocard etc. were proposed in those days (Selberg 1957).

$$\frac{U_F}{Bf_\phi} = 3.7 \cdot \left[\frac{mr}{\rho B^3} \cdot \left\{1 - \left(\frac{\omega_z}{\omega_\phi}\right)^2\right\}\right]^{\frac{1}{2}} \tag{4.36}$$

where m = mass per unit length of the plate, r = radius of gyration and $f_\phi = \omega_\phi/2\pi$.

4.3.4 Experimental Determination of Unsteady Lift Force

Experimental determination of the unsteady lift force on wings has been carried out since the 1930s particularly when the flow separation is involved. There are basically two ways to do it: one is to make a direct measurement of aerodynamic force components by strain gauges, for example, when the body is forced to do a specific motion and another is to calculate the force indirectly from the flow-induced motion of the model.

In absence of any analytical means to determine unsteady aerodynamic forces on various aerodynamically bluff non-streamlined cross-sections for most of the civil engineering applications, these methods have been widely applied. Most of these measurements have been done in smooth air flow and the force coefficients are assumed to be given by linear combinations of displacements and their first derivatives. Some attempts have been made to investigate the effects of turbulence (Ukeguchi et al. 1966; Scanlan and Sabzevari 1967; Forsching 1968; Tanaka and lto 1969; Scanlan and Tomko 1970; Morimitsu et al. 1973; Scanlan and Lin 1978; Tanaka and Davenport 1979; Hnston 1987; Poulsen 1990; Sarkar et al. 1991, 1993; Yamada et al. 1991; Jakobsen and Hjorth-Hansen 1993; Jensen and Nielsen 1995).

As a mathematical formulation of unsteady aerodynamic forces, the proposed format by Scanlan is most widely employed in civil engineering applications:

$$\begin{aligned} C_L &= KH_1^*(K) \cdot \frac{\dot{z}}{U} + KH_2^*(K) \cdot \frac{B\dot{\phi}}{U} + K^2H_3^*(K) \cdot \phi + K^2H_4^*(K) \cdot \frac{z}{B} \\ C_M &= KA_1^*(K) \cdot \frac{\dot{z}}{U} + KA_2^*(K) \cdot \frac{B\dot{\phi}}{U} + K^2A_3^*(K) \cdot \phi + K^2A_4^*(K) \cdot \frac{z}{B} \end{aligned} \tag{4.37}$$

where reduced frequency is defined as $K = \omega B/U = 2\ k$. For the case of flat plate aerodynamics, the above defined derivatives become as follows:

$$\begin{aligned}
H_1^* &= -\frac{\pi}{k}\cdot F(k) & H_2^* &= -\frac{\pi}{4k}\cdot\left[1+F(k)+\frac{2G(k)}{k}\right] \\
H_3^* &= -\frac{\pi}{4k}\cdot\left[\frac{2F(k)}{k}-G(k)\right] & H_4^* &= \frac{\pi}{2}\cdot\left[1+\frac{2G(k)}{k}\right] \\
A_1^* &= \frac{\pi F(k)}{4k} & A_2^* &= \frac{\pi}{16k}\cdot\left[-1+F(k)+\frac{2G(k)}{k}\right] \\
A_3^* &= \frac{\pi}{8k}\cdot\left[\frac{k}{8}+\frac{F(k)}{k}-\frac{G(k)}{2}\right] & A_4^* &= -\frac{\pi G(k)}{4k}
\end{aligned} \tag{4.38}$$

There is no comprehensive summary of measured aerodynamic derivatives readily available though there have been many measurements earned out over the last four decades. Most of the measurements were done in smooth air flow and linear aerodynamic forces are assumed. The effects of turbulence and nonlinearity of forces have been studied for very limited cases so far.

The calculation process to determine flutter conditions using measured aerodynamic coefficients is the same as before. Since the aerodynamic force terms are given as linear combinations of the displacements and their first derivatives by assuming a set of simple harmonic solutions with the same frequency at a certain wind speed, the complex homogeneous equation will be given as a condition to have a non-trivial solution set to the equations of motion. Hence

$$C_R(\omega_F, U_F) + i\cdot C_I(\omega_F, U_F) = 0 \tag{4.39}$$

Thus the flutter conditions, w_F and U_F, are given by solving this equation.

4.3.5 3D Flutter Analysis

A two-dimensional analysis described above is applicable to a relatively single case such as a straight decked suspension bridge for which two clearly identified vibration modes in vertical bending and torsion are the same in form across the span. However, in any other cases, it becomes necessary to consider the contribution of three-dimensional mode shapes of the entire structure.

Suppose that three participating modes of vibration of the bridge deck are defined by

$$\begin{aligned}
h(y,t) &= \phi(y)\cdot h(t)\cdot B \\
p(y,t) &= \Psi(y)\cdot p(t)\cdot B \\
\alpha(y,t) &= \theta(y)\cdot\alpha(t)
\end{aligned} \tag{4.40}$$

in vertical bending, lateral sway and pitching motion, respectively, y is the span wise coordinate taken along the bridge, $h(t)$, $p(t)$ and $\alpha(t)$ are dimensionless generalized coordinates. The generalized equations of motion can be given by

$$
\begin{aligned}
m_h B \cdot \left(\ddot{h} + 2\varsigma_h \omega_h \dot{h} + \omega_h^2 h\right) &= \int_L L(h,p,\alpha) f(y) dy \\
m_p B \cdot \left(\ddot{p} + 2\varsigma_p \omega_p \dot{p} + \omega_p^2 p\right) &= \int_L D(h,p,\alpha) \Psi(y) dy \\
J_\alpha \left(\ddot{\alpha} + 2\varsigma_\alpha \omega_\alpha \dot{\alpha} + \omega_h^2 \alpha\right) &= \int_L M(h,p,\alpha) \Theta(y) dy
\end{aligned} \tag{4.41}
$$

where the generalized inertia terms are given by

$$
m_h = \int_L m(y)\phi^2(y)dy \quad m_p = \int_L m(y)\Psi^2(y)dy \quad J_\alpha = \int_L J(y)\theta^2(y)dy \tag{4.42}
$$

and all integrations shall be done over the bridge span.

In this formulation, new terms are required for the aerodynamic forces since three directions are now considered. They can be as follows

$$
\begin{aligned}
L &= QB \cdot \left[KH_1^* \frac{\dot{h}}{U} + KH_2^* \frac{B\dot{\alpha}}{U} + K^2 H_3^* \alpha + K^2 H_4^* \frac{h}{B} + KH_5^* \frac{\dot{p}}{U} + K^2 H_6^* \frac{p}{B}\right] \\
D_f &= QB \cdot \left[KP_1^* \frac{\dot{p}}{U} + KP_2^* \frac{B\dot{\alpha}}{U} + K^2 P_3^* \alpha + K^2 P_4^* \frac{p}{B} + KP_5^* \frac{\dot{h}}{U} + K^2 P_6^* \frac{h}{B}\right] \\
M_f &= QB^2 \cdot \left[KA_1^* \frac{\dot{h}}{U} + KA_2^* \frac{B\dot{\alpha}}{U} + K^2 A_3^* \alpha + K^2 A_4^* \frac{h}{B} + KA_5^* \frac{\dot{p}}{U} + K^2 A_6^* \frac{p}{B}\right]
\end{aligned} \tag{4.43}
$$

where $Q = \rho U^2/2$. The generalized forces are then

$$
\begin{aligned}
\int_L L(h,p,\alpha) \cdot \phi(y) dy &= \left(\rho B^3 \omega^2/2\right) \left[H_h^* q_{11} \cdot h + H_p^* q_{12} \cdot p + H_\alpha^* q_{13} \cdot \alpha\right] \\
\int_L D(h,p,\alpha) \cdot \Psi(y) dy &= \left(\rho B^3 \omega^2/2\right) \left[P_h^* q_{21} \cdot h + P_p^* q_{22} \cdot p + P_\alpha^* q_{23} \cdot \alpha\right] \\
\int_L M(h,p,\alpha) \cdot \theta(y) dy &= \left(\rho B^4 \omega^2/2\right) \left[A_h^* q_{31} \cdot h + A_p^* q_{32} \cdot p + A_\alpha^* q_{33} \cdot \alpha\right]
\end{aligned} \tag{4.44}
$$

in which

$$q_{11} = \int\limits_L \phi^2(y)dy \;\; q_{22} = \int\limits_L \Psi^2(y)dy \;\; q_{33} = \int\limits_L \theta^2(y)dy$$
$$q_{23} = q_{32} = \int\limits_L \Psi(y)\cdot\theta(y)dy \;\; q_{13} = q_{31} = \int\limits_L \theta(y)\cdot\phi(y)dy$$
$$q_{12} = q_{21} = \int\limits_L \phi(y)\cdot\Psi(y)dy \tag{4.45}$$

and

$$H_h^* = H_4^* + i\cdot H_1^* \;\; H_p^* = H_6^* + i\cdot H_5^* \;\; H_\alpha^* = H_3^* + i\cdot H_2^*$$
$$P_h^* = P_6^* + i\cdot P_5^* \;\; P_p^* = P_4^* + i\cdot P_1^* \;\; P_\alpha^* = P_3^* + i\cdot P_2^*$$
$$A_h^* = A_4^* + i\cdot A_1^* \;\; A_p^* = A_6^* + i\cdot A_5^* \;\; A_\alpha^* = A_3^* + i\cdot A_2^* \tag{4.46}$$

Setting up the flutter condition as

$$h(t) = h_0 e^{i\omega t} \;\; p(t) = p_0 e^{i\omega t} \;\; \alpha(t) = \alpha_0 e^{i\omega t} \tag{4.47}$$

with the equation of motion leads to the following:

$$2\mu_h\left[(\omega_h/\omega)^2 - 1 + i\cdot 2\varsigma_h\omega_h/\omega\right]\cdot h_0 = H_h^* q_{11}\cdot h_0 + H_p^* q_{12}\cdot p_0 + H_\alpha^* q_{13}\cdot\alpha_0$$
$$2\mu_p\left[(\omega_p/\omega)^2 - 1 + i\cdot 2\varsigma_p\omega_p/\omega\right]\cdot p_0 = P_h^* q_{21}\cdot h_0 + P_p^* q_{22}\cdot p_0 + P_\alpha^* q_{23}\cdot\alpha_0$$
$$2\nu_\alpha\left[(\omega_\alpha/\omega)^2 - 1 + i\cdot 2\varsigma_\alpha\omega_\alpha/\omega\right]\cdot \alpha_0 = A_h^* q_{31}\cdot h_0 + A_p^* q_{32}\cdot p_0 + A_\alpha^* q_{33}\cdot\alpha_0 \tag{4.48}$$

where the following mass parameters are introduced:

$$\mu_h = m_h/\left(\rho B^2\right) \;\; \mu_p = m_p/\left(\rho B^2\right) \;\; \nu_\alpha = J_\alpha/\left(\rho B^4\right) \tag{4.49}$$

The critical flutter speed U_F and the flutter frequency ω_F are decided to give a non-trivial solution set to the equations. Once the eigen frequencies and corresponding mode shapes are calculated by applying a dynamic analysis, the flutter equation can be established for any selected combination of three mode shapes, one for each direction of vertical, lateral and torsion, as follows:

$$\begin{vmatrix} 2\mu_h\left[\lambda_1^2X^2 - 1 + i\cdot 2\varsigma_h\lambda_1 X\right] - H_h^* q_{11} & -H_p^* q_{12} & -H_\alpha^* q_{13} \\ -P_h^* q_{21} & 2\mu_p\left[\lambda_2^2X^2 - 1 + i\cdot 2\varsigma_p\lambda_2 X\right] - P_p^* q_{22} & -P_\alpha^* q_{23} \\ -A_h^* q_{31} & -A_p^* q_{32} & 2\nu_\alpha\left[X^2 - 1 + i\cdot 2\varsigma_\alpha X\right] - A_\alpha^* q_{33} \end{vmatrix} = 0 \tag{4.50}$$

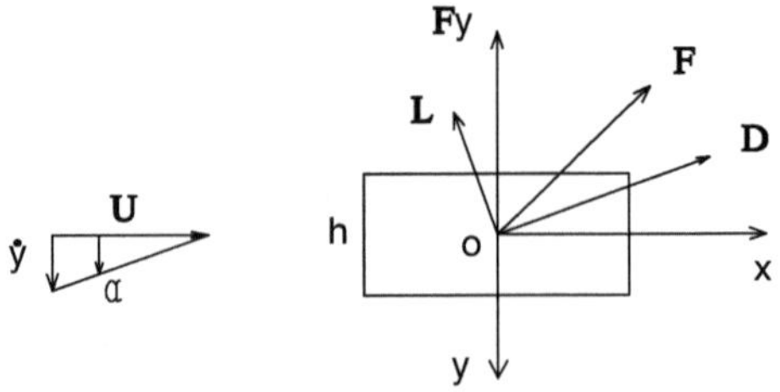

Fig. 4.6 Galloping force components

where $\lambda_1 = \omega_h/\omega_\alpha$, $\lambda_2 = \omega_p/\omega_\alpha$, $X = \omega_\alpha/\omega$ and ζ_j ($j = h, p, \alpha$) are the structural damping assumed as a fraction of critical in each of three vibration modes.

The 3D flutter analyses are mainly based on the idea of multi-mode superposition in which a dynamic coupling between the natural modes takes place through the self-excited aerodynamic forces. Some natural modes which are assumed to participate in flutter instability are chosen to be included in flutter analyses. Theoretically, the mode combination is only an approximate expression of the flutter mode, but in reality this approximate expression will have very good precision if enough natural modes can be included in analysis. In order to develop a more comprehensive and accurate procedure to analyze flutter mode analysis, the author presented a full-mode participation method in 1999 which is basically different from the multi-mode participation method and can contain all natural modes in flutter analysis. With using the inverse form of flutter equation, the full-mode method adopts the most effective eigen-solution technique, the inverse vector iteration with QR transformation, to solve flutter eigenvalues and critical wind speeds of the system composed of the structure and the wind flow (Ge 1999; Ge and Tanaka 2000).

4.3.6 Galloping

Another type of wind induced, self-excited structural vibration is often very violent, strongly nonlinear. Drastic effects of turbulence have been pointed out. In many ways, galloping exhibits its unique aspect (Blevins 1990; Parkinson and Brooks 1961; Novak 1969).

Consider a solid tower, for example, of height H. A cross-section of it at the elevation z is shown in Fig. 4.6. The lateral force (F_y) in the direction of the body motion is

$$F_y(\alpha) = \left(\rho U^2/2\right)hH \cdot C_{Fy}(\alpha) \tag{4.51}$$

Since the lateral force is taken positive in the direction of $-y$, $F_y = -(L\cos\alpha + D\sin\alpha)$ and the lateral force coefficient is

$$C_{Fy}(\alpha) = -(C_L + C_D \cdot \tan\alpha) \cdot \sec\alpha \tag{4.52}$$

and

$$\frac{dC_{Fy}}{d\alpha} = -\left\{\frac{dC_L}{d\alpha} + C_D \cdot \left(1 + 2\tan^2\alpha\right)\right\} \cdot \sec\alpha - \left(C_L + \frac{dC_D}{d\alpha}\right) \cdot \tan\alpha \cdot \sec\alpha \tag{4.53}$$

The lateral force excites the vibration when $dC_{Fy}/d\alpha$ at $\alpha = 0$ is negative, or

$$[dC_L/d\alpha + C_D]_{\alpha=0} < 0 \tag{4.54}$$

which is called the Den Hartog's criterion.

Substituting $\alpha = \tan^{-1}(\dot{y}/U)$

$$C_{Fy}(\alpha) = \sum_{i=odd}^{m} A_i \cdot \left(\frac{1}{U} \cdot \frac{dy}{dt}\right)^i + \sum_{j=even}^{k} A_j \cdot \left(\frac{1}{U} \cdot \frac{dy}{dt}\right)^j \cdot \frac{\dot{y}}{|\dot{y}|} \tag{4.55}$$

Equation of motion is

$$m\ddot{y} + c\dot{y} + ky = F_y(t) \tag{4.56}$$

or

$$\ddot{y} + 2\varsigma\omega\dot{y} + \omega^2 y = \left(\rho U^2/2m\right)hH \cdot C_{Fy}(\alpha) \tag{4.57}$$

Introducing new dimensionless parameters as

$$\begin{aligned} &Y = y/h \quad u_r = U/(\omega h) \ \text{ and } \ \mu = \rho h^2 H/(4m) \\ &\ddot{Y} + \omega^2 \dot{Y} + F(\dot{Y}) = 0 \end{aligned} \tag{4.58}$$

where

$$F(\dot{Y}) = 2\varsigma\omega\dot{Y} - 2\mu \cdot \left[\sum_i A_i \omega^{2-i} \cdot \frac{\dot{Y}_i}{u_r^{i-2}} + \sum_j A_j \omega^{2-j} \cdot \frac{\dot{Y}_j}{u_r^{j-2}} \cdot \frac{\dot{Y}}{|\dot{Y}|}\right] \tag{4.59}$$

From the energy principle

$$W = \int_0^H \int_0^T F(\dot{Y}) dx dY(t) = 0 \tag{4.60}$$

Hence

$$\frac{1}{V} = \sum_{r=1}^{s} A_r \cdot B_r \cdot C_r \cdot \left(\frac{a}{V}\right)^{r-1} \tag{4.61}$$

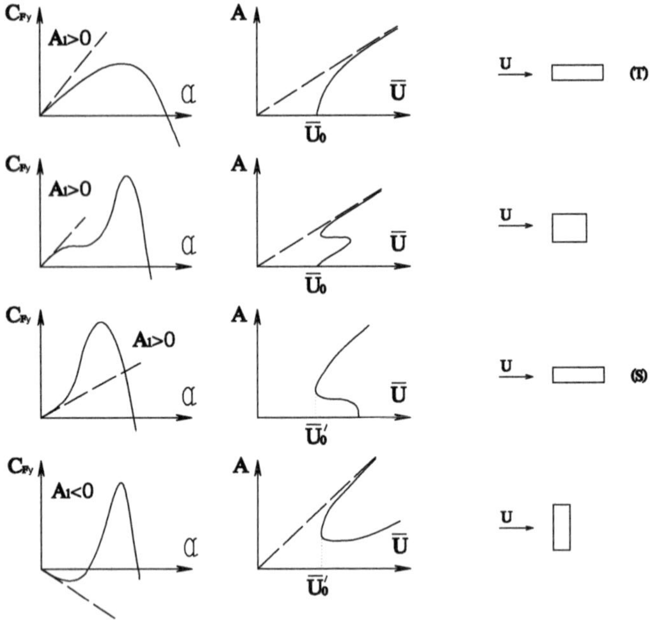

Fig. 4.7 Typical values of A_1

in which

$$V = \mu u_r/\varsigma \quad a = \mu Y_{\max}/\varsigma$$
$$B_r 2 \cdot i!!/(i+1)!! \ \ (i = odd) \ \text{ or } \ (4/\pi) \cdot j!!/(j+1)!! \ \ (j = even)$$
$$C_r = \int_o^H v^{2-r}(z)|y(z)|^{r+1}dz \Bigg/ \left\{ \int_0^H y^2(z)dz \right\} \tag{4.62}$$

$v(z)$ = dimensionless profile of $U(z)$, s = the maximum power index at the expansion of C_{Fy}.

The critical speed for instability is decided by

$$V_{cr} = \varsigma/(\mu C_1 A_1) \ \text{ or } U_{cr} = \varsigma \cdot \omega_i \cdot h/(\mu A_1) \tag{4.63}$$

The typical values of A_1 are shown in Fig. 4.7.

4.3.7 Torsional Flutter

Analysis of torsional instability considering a SDOF system should be parallel to SDOF galloping. However, its quasi-steady analysis is not so straight-forward because of the difficulty in taking the effective angle of attack.

Analyses were attempted by Modi and later by Nakamura and Mizota and well summarized by Blevins (Blevins 1990; Modi and Slater 1972; Nakamura and Mizota 1975).

In order to avoid any possible ambiguity, it makes more sense to use unsteady aerodynamic forces. Only relevant terms are A_2^* and A_3^* of the pitching moment. Hence, the equation of motion is

$$\ddot{\alpha} + 2\varsigma_\alpha \omega_\alpha \dot{\alpha} + \omega_\alpha^2 \alpha = \left(\rho U^2/2J\right)B^2 \cdot \left[KA_2^*(B\dot{\alpha}/U) + K^2 A_3^* \alpha\right] \tag{4.64}$$

where, as before, the aerodynamic coefficients A_2^* and A_3^* assumed to be functions of reduced frequency $K = \omega B/U$. If no coupling with any other modes are assumed, $\omega = \omega_\alpha$ and hence the critical condition for the onset of instability is define by

$$A_2^* = (4J_{\varsigma\alpha})/\left(\rho B^4\right) \tag{4.65}$$

4.4 Buffeting Analysis

4.4.1 Random Vibration Analysis

A structural system can be often approximated by a SDOF system or by a combination of SDOF systems by applying the modal analysis. Once this model is established and the external force is given by an explicit form $p(y, t)$, the response can be calculated at least in principle. However, for the case of buffeting induced by wind turbulence, the situation is more complicated since the wind excitation force cannot take generally any analytical forms. It is given in terms of its statistical properties at the best.

There are basically two different approaches that can be taken for this case. One way is to perform the response prediction in frequency domain by applying Fourier transform technique. As long as the structure is well defined mathematically, the response can be calculated in terms of its statistical properties such as the mean, rms (root-mean-square) and peak response for a given observation period. This is a standard method of random vibration analysis and is quite efficient as long as the system is linear and the excitation is considered to be a steady state process.

Another way is to carry out the calculation in time domain by expressing the excitation forces as time series which simulate the given statistical characteristics. This is a standard method in Earthquake Engineering since the ground motion and resulted excitation are basically transient processes and steady state solutions by the application of frequency domain analysis does not make too much sense. Even in Wind Engineering field, there have been an increasing number of attempts using this method in recent years particularly for the consideration of structural and aerodynamic nonlinearity.

In this chapter, the buffeting analyses are mostly in frequency domain. First, some additional mathematical tools are introduced. Consider a SDOF system characterized by eigen frequency and inherent damping as follows:

$$f_0 = \frac{\omega_0}{2\pi} = \frac{(k/m)^{1/2}}{2\pi}$$
$$\varsigma_s = \frac{c}{2m\omega_0} = \frac{c/2}{(km)^{1/2}} \tag{4.66}$$

Its free vibration is expressed by

$$x(t) = \exp(-\varsigma_s\omega_0 t)[A\sin(\omega_D t) + B\cos(\omega_D t)]$$
$$\omega_D = \omega\left(1 - \varsigma_s^2\right)^{1/2} \tag{4.67}$$

and the forced vibration under any known excitation force $F(t)$ is

$$x(t) = \int_0^t F(s) \cdot h(t - s)ds \tag{4.68}$$

where $h(t)$ is the Impulse Response Function (IRS) defined by

$$h(t) = (m\omega_D)^{-1} \cdot \exp(-\varsigma_s\omega_0 t) \cdot \sin\omega_D t \tag{4.69}$$

$h(t)$ corresponds to the response of the system when $F(t) = \delta(t)$. When $F(t) = 1(t)$, the response is expressed by

$$x(t) \approx \{1 - \exp(-\varsigma_s\omega_0 t) \cdot \cos\omega_D t\}/k = A(t) \tag{4.70}$$

where $A(t)$ is called Indicial Admittance (IA) and is related to IRS by

$$h(t) = dA/dt + A(0)\delta(t) \tag{4.71}$$

Duhamel's integral (convolution) can be expressed by using IA as

$$x(t) = \int_0^t F(t - s) \cdot A'(s)ds \tag{4.72}$$

When the excitation is single harmonic or F(t) = exp(iωt)

$$x(t) = \left(e^{i\omega t}/k\right)\left[1 - (\omega/\omega_0)^2 + i \cdot 2\varsigma_s\omega/\omega_0\right]^{-1} = F(t) \cdot H(\omega)/k \tag{4.73}$$

$H(\omega)/k$ is called the Frequency Response Function (FRF). FRF and IRF consist a Fourier Transform pair, or

$$H(\omega)/k = F\{h(t)\} \& h(t) = F^{-1}\{H(\omega)/k\} \tag{4.74}$$

where $\omega = 2\pi f$.

Response to a random loading can now be formulated as follows. First of all, the input–output relationship is established by Eq. (4.68) as

$$x(t) = \int_{0}^{t} F(t-\theta)\cdot h(\theta)d\theta \tag{4.75}$$

Obviously the mean or static response is simply given by

$$\bar{y} = \bar{x}/k \tag{4.76}$$

The auto-covariance function of the response signal x(t) is

$$\begin{aligned} R_x(\tau) &= \int_{-\infty}^{\infty} x(t)\cdot x(t+\tau)dt = \int_{-\infty}^{\infty}\left[\int_{-\infty}^{\infty} F(t-\theta)\cdot h(\theta)d\theta\right]\left[\int_{-\infty}^{\infty} F(t+\tau-\theta)\cdot h(\theta)d\theta\right]dt \\ &= \int_{-\infty}^{\infty}\int_{-\infty}^{\infty} R_F(\tau+\theta_1-\theta_2)\cdot h(\theta_1)h(\theta_2)d\theta_1 d\theta_2 \end{aligned} \tag{4.77}$$

By taking the Fourier transform of Eq. (4.77)

$$\begin{aligned} S_x(\omega) &= \int_{-\infty}^{\infty} R_x(\tau)\cdot \exp(-i\omega\tau)d\tau \\ &= \int_{-\infty}^{\infty} \exp(-i\omega\tau)\cdot \int_{-\infty}^{\infty}\int_{-\infty}^{\infty} R_F(\tau+\theta_1-\theta_2)\cdot h(\theta_1)h(\theta_2)d\theta_1 d\theta_2 d\tau \\ &= \int_{-\infty}^{\infty} h(\theta_1)\cdot \exp(-i\omega\theta_1)d\theta_1 \cdot \int_{-\infty}^{\infty} h(\theta_2)\cdot \exp(-i\omega\theta_2)d\theta_2 \\ &\cdot \int_{-\infty}^{\infty} R_F(\tau+\theta_1-\theta_2)\cdot \exp\{-i\omega(\tau+\theta_1-\theta_2)\}d\tau \\ &= \frac{H(\omega)}{k}\frac{H^*(\omega)}{k}\int_{-\infty}^{\infty} R_F(\theta)\cdot \exp(-i\omega\theta)d\theta = \frac{|H(\omega)|^2}{k^2}\cdot S_F(\omega) \end{aligned} \tag{4.78}$$

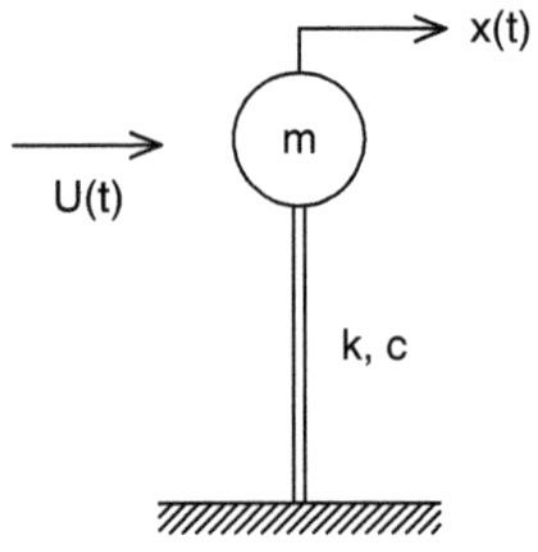

Fig. 4.8 SDOF point-like structure

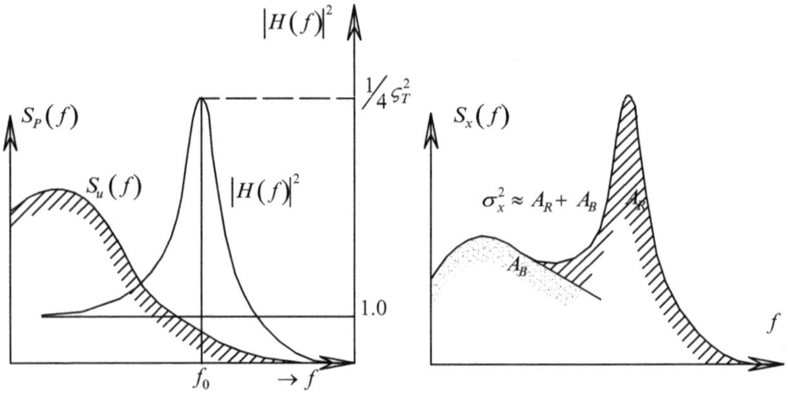

Fig. 4.9 Spectra of drag force and along-wind response

Thus the spectra of input and output are related. By integrating Eq. (4.78) the mean-square response can be calculated as follows:

$$\sigma_x^2 = k^{-2} \cdot \int_{-\infty}^{\infty} |H(\omega)|^2 \cdot S_x(\omega) d\omega \tag{4.79}$$

4.4.2 Application to SDOF Systems

Consider the along-wind response of a SDOF point-like structure given in Fig. 4.8, where the drag force (Fig. 4.9) is defined by

$$P(t) = \left(\rho U^2/2\right) A C_D \tag{4.80}$$

By putting

$$\begin{aligned}U(t) &= U + u(t)\\ P(t) &= C_D(\rho A/2)U^2\{1 + 2\cdot u(t)/U\} = \bar{P} + p(t)\\ p(t) &= 2\bar{P}\cdot[u(t)/U]\cdot X_a\left(f\sqrt{A/U}\right)\\ S_p(f) &= (2\bar{P}/U)^2|X_a(f)|^2\cdot S_u(f)\end{aligned} \tag{4.81}$$

where X_a and $S_u(f)$ are called the aerodynamic admittance function and the velocity spectrum, respectively.

Response $X(t) = X + x(t)$ is evaluated as follows

$$\begin{aligned}S_x(f) &= [|H(f)|/k]^2\cdot S_p(f)\\ |H(f)|^2 &= \left[\left(1-\Omega^2\right)^2 + (2\varsigma_T\Omega)^2\right]^{-1}\\ \Omega &= \omega/\omega_0\end{aligned} \tag{4.82}$$

and

$$\left(\frac{\sigma_x}{\bar{X}}\right)^2 = \left(\frac{2\sigma_u}{U}\right)^2\int_0^\infty |H(f)|^2\cdot|X_a(f)|^2\frac{S_u(f)}{\sigma_u^2}df \tag{4.83}$$

Approximate evaluation of σ_x:

$$\sigma_x^2 = k^{-2}\cdot\int_0^\infty |H(f)|^2\cdot S_p(f)df \approx A_R + A_B \tag{4.84}$$

where $A_R = k^{-2}\cdot\int_0^\infty |H(f)|^2 df$ $A_B = k^{-2}\cdot\int_0^\infty S_p(f)df$ $S_p(f_0) = \frac{\pi f_0}{4\varsigma_T k^2}S_p(f_0)$

4.4.3 Horizontal Line-Like Structures under Drag Excitation

Referring to Fig. 4.10, the equation of motion is

$$m\ddot{x} + (EIx'')'' = p(y,t) \tag{4.85}$$

The response can be expanded as a series of eigen modes:

$$x(y,t) = \sum_r \phi_r(y)\cdot q_r(t) \tag{4.86}$$

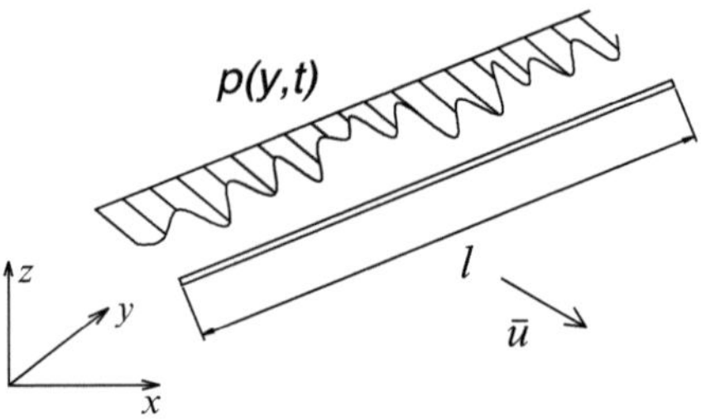

Fig. 4.10 Horizontal line-like structure under drag excitation

Introducing the modal quantities

$$M_r = \int_o^L m(y)\phi_r^2(y)dy$$
$$K_r = (2\pi f_r)^2 \cdot M_r$$
$$C_r = 2\varsigma_r(M_r K_r)^{1/2}$$
$$Q_r(t) = \int_0^L p(y,t)\phi_r(y)dy \tag{4.87}$$

The equation of motion is reduced to a modal equation for each mode as follows:

$$\ddot{q}_r + 2\varsigma_r\omega_r\dot{q}_r + \omega_r^2 q_r = Q_r(t)/M_r \tag{4.88}$$

The mean-square response is then given by

$$\sigma_x^2(y) = \sum_i \sum_j \overline{q_i q_j}\phi_i(y)\phi_j(y) \approx \sum_r \overline{q_r^2}\phi_r^2(y) \tag{4.89}$$

and

$$\overline{q_r^2} = (K_r)^{-2}\int_0^\infty |H_r(f)|^2 \cdot S_{Qr}(f)df \tag{4.90}$$

where

$$S_{Qr}(f) = \int_0^L\int_0^L S_p(y_1, y_2, f) \cdot \phi_r(y_1)\phi_r(y_2)dy_1 dy_2$$
$$S_p(y_1, y_2, f) = S_p(f) \cdot \tilde{R}_p(y_1, y_2, f) \tag{4.91}$$

The force coherence $\tilde{R}_p(y_1, y_2, f)$ is not known but in conventional buffeting analyses it has been often assumed to be the same as the velocity coherence $\tilde{R}_u(y_1, y_2, f)$ as it was originally introduced by Davenport. Then

$$
\begin{aligned}
\tilde{R}_p(y_1, y_2, f) &= \tilde{R}_u(y_1, y_2, f) = \exp(-fD/U) \text{ where } D = |y_1 - y_2| \\
S_{Qr}(f) &= S_p(f) \cdot |J_r(f)|^2 \\
|J_r(f)|^2 &= \int_0^L \int_0^L \tilde{R}_p(y_1, y_2, f) \cdot \phi_r(y_1)\phi_r(y_2) dy_1 dy_2
\end{aligned} \tag{4.92}
$$

which is called the joint acceptance function. Substituting these expressions, the mean-square modal response becomes as follows:

$$
\begin{aligned}
\overline{q_r^2} &= (K_r)^{-2} \int_0^\infty |H_r(f)|^2 \cdot S_{Qr}(f) df = (K_r)^{-2} \cdot \int_0^\infty |H_r(f)|^2 |J_r(f)|^2 \cdot S_p(f) df \\
|H_r(f)|^2 &= \left[\left(1 - \Omega_r^2\right)^2 + \left(2\varsigma_{T,r}\Omega_r\right)^2 \right]^{-1} \\
\Omega_r &= f/f_r \\
S_p(f) &= (2\bar{P}/U)^2 |X_a(f)|^2 \cdot S_u(f)
\end{aligned} \tag{4.93}
$$

in which $\varsigma_{T,s}$ is the total effective damping given by the sum of structural ($\varsigma_{s,r}$) and aerodynamic damping. The aerodynamic damping for the r-th mode is approximately given by

$$
\begin{aligned}
\varsigma_{T,r} &= \rho A C_D U / (4\pi f_r m_r) \\
m_r &= \int_0^L m(y)\phi_r^2(y) dy \Big/ \int_0^L \phi_r^2(y) dy
\end{aligned} \tag{4.94}
$$

The aerodynamic admittance function $|X_a(f)|^2$ needs to be defined in one way or another.

It should be remembered that there are a few major assumptions introduced to this analysis as follows:

1. Cross-coupling terms ignored: not much influence.
2. Span-wise force coherence: should be probably higher.
3. Aerodynamic admittance: scapegoat.
4. Unsteady forces: better to use measured derivatives (Davenport 1961, 1962).

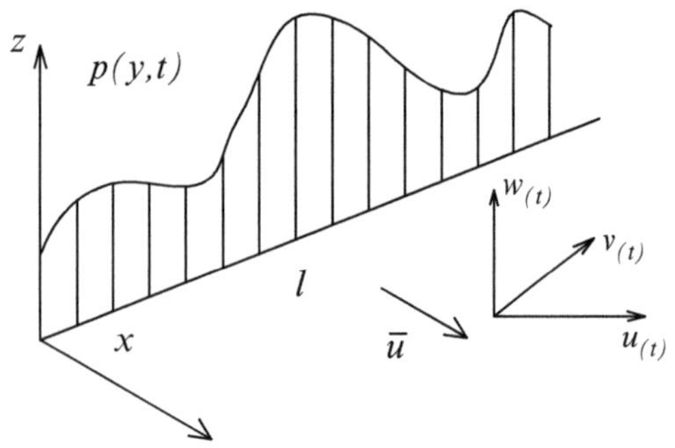

Fig. 4.11 Horizontal line-like structure under lift excitation

4.4.4 Horizontal Line-Like Structures under Lift Excitation

Referring to Fig. 4.11 and expressing the response as (Davenport 1961, 1962)

$$z(y,t) = \sum_r q_r(t) \cdot \phi_r(y)B \tag{4.95}$$

the modal equation of motion is

$$\ddot{q}_r + 2\varsigma_{T,r}\omega_r\dot{q}_r + \omega_r^2 q_r = Q_r(t)/M_r \tag{4.96}$$

in which

$$M_r = \int_0^L m(y)\phi_r^2(y)dy \tag{4.97}$$

$$\varsigma_{T,r} = \varsigma_{s,r} + \varsigma_{a,r} \tag{4.98}$$

The aerodynamic damping can be expressed, by using the Scanlan type aerodynamic derivatives, as

$$\varsigma_{a,r} = -\frac{H_1^*}{2\mu_r L} \cdot \int_0^L \phi_r^2(y)dy \tag{4.99}$$

in which $\mu_r = M_r/(L\rho B^2)$ is the modal mass ratio and the derivative H_1^* is a function of reduced frequency $K = B\omega_r/U$. If the quasi-steady approach is taken as before,

$$\varsigma_{a,r} = \frac{1}{4\mu_r L} \cdot \frac{1}{K} \cdot \frac{dC_L}{d\alpha} \int_0^L \phi_r^2(y)dy \tag{4.100}$$

the generalized buffeting force can be formulated as

$$Q_r(t) = \frac{\rho U}{2} \cdot \frac{dC_L}{d\alpha} \cdot \int_0^L w(y,t)\phi_r(y)dy \tag{4.101}$$

Auto-correlation of $Q_r(t)$ is

$$R_{Qr}(\tau) = \left(\frac{\rho U}{2}\frac{dC_L}{d\alpha}\right)^2 \cdot \int_0^L\int_0^L \overline{w(y_1,t)w(y_2,t+\tau)}\phi_r(y_1)\phi_r(y_2)dy_1dy_2 \tag{4.102}$$

By applying Fourier transform, the spectral density function of $Q_r(t)$ can be obtained. Considering the concept of aerodynamic admittance again,

$$S_{Qr}(f) = \left(\frac{\rho U}{2}\frac{dC_L}{d\alpha}\right)^2 \cdot |X_a(f)|^2 \cdot \int_0^L\int_0^L C_w(y_1,y_2;f)\phi_r(y_1)\phi_r(y_2)dy_1dy_2 \tag{4.103}$$

in which $C_w(y_1,y_2;f) = S_w(f) \cdot R_w(y_1,y_2;f)$ is the co-spectrum of velocity component $w(t)$. Again, the use of velocity correlation instead of force correlation is usually the practice here. It is an approximation which is probably not right and will, most likely, lead to the underestimation of the structural response. The modal response spectrum is

$$\begin{aligned} S_{qr}(f) &= \frac{|H_r(f)|^2}{K_r^2} S_{Qr}(f) \\ |H_r(f)|^2 &= \left[\left\{1-(f/f_r)^2\right\}^2 + \left\{2\varsigma_{T,r}f/f_r\right\}^2\right]^{-1} \\ \sigma_{qr}^2 &= \int_0^\infty S_{qr}(f)df = \left(\frac{\rho BU}{2K_r}\cdot\frac{dC_L}{d\alpha}\right)^2 \int_0^\infty S_w(f)|J_r(f)|^2|H_r(f)|^2df \end{aligned} \tag{4.104}$$

where the joint acceptance function is defined as

$$|J_r(f)|^2 = \int_0^L\int_0^L \tilde{R}_w(y_1,y_2;f)\cdot\phi_r(y_1)\phi_r(y_2)dy_1dy_2 \tag{4.105}$$

and the mean-square response is given by

$$\sigma_z^2(y) \approx \sum_r \sigma_{qr}^2\phi_r^2(y) \tag{4.106}$$

4.5 Vortex Induced Vibration

4.5.1 Concept of Vortex Shedding

Considering the resonance of structural frequency (f_s) with the frequency (f_v) of wake vortex formation, characterized by the Strouhal number $S_t = f_v D/U$

$$\begin{aligned} m\ddot{y} + c_s\dot{y} + ky &= F(t) \\ F(t) &= \left(\rho U^2/2\right)DC_L \cdot \sin(\omega_v t) \end{aligned} \tag{4.107}$$

Bishop and Hassan (1964) indicated a strong self-excited and self-limited nature of the phenomena, which led to the proposal of oscillator models. Also, the measurement of motion-dependent aerodynamic damping made a progress

$$\begin{aligned} K_a &= -m_a/\rho B^2 = K_a(U_r, \dot{y}, \sigma_u/U, \mathrm{Re}) \\ \mathrm{Re} &= UD/\nu \end{aligned} \tag{4.108}$$

The span-wise correlation of the vortex-induced loads is basically decided by the distance between the sections. It improves with the response amplitude and becomes less in higher turbulence. The aspect ratio of the structure also affects on it.

4.5.2 Mathematical Formulation

Assume that the total response can be expressed by modal functions as follows:

$$y(z,t) = \sum_{i=1}^{\infty} a_i(t)\Psi_i(z) \tag{4.109}$$

where the mode functions are normalized, for example, by

$$\int_0^h \Psi_i(z)dz = h \tag{4.110}$$

Equation of motion for the i-th mode of vibration is given by

$$M_i\ddot{a}_i + C_i\dot{a}_i + K_i a_i = Q_i(t) \tag{4.111}$$

where

$$M_i = \int_0^h m(z)\Psi_i^2(z)dz = \text{generalized mass}$$
$$K_i = (2\pi f_i)^2 M_i = \text{generalized stiffness}$$
$$C_i = 2(M_i K_i)^{1/2}\varsigma_i = 4\pi M_i f_i \varsigma_i = \text{modal damping coefficient} \tag{4.112}$$

and

$$Q_i(t) = \int_0^h w(z,t)\Psi_i(z)dz \tag{4.113}$$

is the generalized for the i-th mode when $w(z, t)$ is given as the vortex-induced force per unit length of the body. Buffeting due to lateral components of the free stream turbulence is not included in the present calculation.

$$\sigma_{ai}^2 = \int_0^\infty S_{ai}(f)df = \frac{1}{K_i^2} \cdot \int_0^\infty |H_i(f)|^2 \cdot S_{Qi}(f)df \tag{4.114}$$

Considering the structural admittance function

$$|H_i(f)|^2 = \left[\left\{1 - (f/f_i)^2\right\}^2 + \{2\varsigma_{Ti} f/f_i\}^2\right]^{-1} \tag{4.115}$$

Equation (4.114) can be approximated by

$$\sigma_{ai}^2 = \frac{\pi f_i}{4\varsigma_{Ti}} \cdot \frac{S_{Qi}(f_i)}{K_i^2} \tag{4.116}$$

where τ_{Ti} is the total, structural plus aerodynamic, damping corresponding to the i-th mode of vibration.

4.5.3 *Vortex Induced Motion of Bridge Decks*

4.5.3.1 Prediction by Wind Tunnel Tests

Vortex induced oscillation of bridge decks has been so frequently observed that it usually consists the first concern when the wind stability of a bridge is considered. It is often regarded as a relatively easy matter that can be determined by carrying out simple section model wind tunnel tests. It is chiefly because the design criteria applied to this phenomenon have been simply to say a section is "good" or "bad", meaning whether or not any appreciable vortex induced response could be observed in possible wind speed range and with anticipated structural damping. However, a question still remains: how "bad" is bad and how "good" is good. For more exact prediction, it becomes necessary to indicate actual response level rather than simply saying if "there could be a vibration".

Obviously the most reliable method at this moment for this prediction is to use an aeroelastic model of the whole bridge with properly simulated natural wind conditions. A possible problem for this case is with the wind speed. Often the available wind speed range of the wind tunnel is too low for the measurement. On the other hand, if the section model tests are projected, question remains if it is possible to make a reliable response prediction by using any available information from these tests?

4.5.3.2 Response Prediction by Section Model Tests

Usual practice of dynamic tests by the use of a section model is to construct a rigid model of the bridge girder with proper scaling of the following factors:

Mass ratio: $m/(\rho B^2)\&I/(\rho B^4)$
Structural damping: $\varsigma_V\&\varsigma_T$
Frequency ratio: f_T/f_V

By giving the measured dynamic response in dimensionless form as functions of reduced wind speed, $U_r = U/(fB)$ the universal response curves are obtained and the response of fall-scale structure can be read directly from these curves. This is the basic idea.

Even these parameters are chosen properly, it does not mean that the predicted response is the predicted bridge response itself since the vibration mode shapes, the distribution of aerodynamic forces and aerodynamic damping effects are not properly simulated in this test.

In order to make the response prediction of the prototype structure from the test results properly, it is necessary to have knowledge of magnitude of the aerodynamic lift force, frequency spectrum of the lift force, span-wise coherence of the lift force, aerodynamic damping and how amplitude dependent these quantities are.

The measured dynamic response could be converted to the magnitude of lift force coefficient. Past experience suggests that the frequency spectral form of the lift force would be probably Gaussian. However, since the test can provide no knowledge of the frequency bandwidth and span-wise coherence, these have to be assumed based on some theoretical considerations. The determination of span-wise coherence of the lift force is usually not in the scope of the section model tests. The measurement of lift coherence could be earned out by measuring the sectional pressure distribution of the girder at longitudinally different locations with the existing technique.

4.5.3.3 Mathematical Formulation

The lift force becomes

$$L(y,t) = \left(\rho U^2/2\right)BC_L \tag{4.117}$$

under the quasi-steady conditions and the root-mean-square (rms) of it is therefore given by

$$\sigma_L(y) = \left(\rho U^2/2\right)B\sigma_{CL} \tag{4.118}$$

Again even when the flow is smooth the fluctuating lift force is not strictly periodic and has a narrow-band spectrum over frequencies adjacent to the Strouhal frequency f_s, where f_s is again given by S_tU/D. The force spectrum is approximately given by

$$\frac{fS_L(f)}{\sigma_L^2} = \frac{f/f_s}{b\sqrt{\pi}} \cdot \exp\left\{-\left(\frac{1-(f/f_s)}{b}\right)^2\right\} \tag{4.119}$$

Where σ_L is rms lift force defined by Eq. (4.118), and b is a bandwidth parameter.

The bandwidth parameter b primarily depends on the large scale turbulence. The above spectrum is expected to fit over the frequency range of $f_s(1 \pm b)$. The spectrum of $Q_i(t)$ can be given in terms of the spectrum of $L(y, t)$ and its normalized co-spectrum as follows:

$$S_{Qi}(f) = S_L(f) \cdot \int_0^L \int_0^L R(y_1, y_2) \cdot \Psi_i(y_1)\Psi_i(y_2)dy_1dy_2 \tag{4.120}$$

in which $S_L(f)$ is given by Eq. (4.119). $R(y_1, y_2)$ is the span-wise correlation of the fluctuating lift forces acting on a stationary structure.

A suggested functional form of $R(y_1, y_2)$ based on some experimental results is

$$R(y_1, y_2) = \exp(-|y_1 - y_2|/nD) \tag{4.121}$$

which gives

$$\int_0^L \int_0^L R(y_1, y_2)dy_1dy_2 = 2(nD)^2[L/(nD) - 1 + \exp\{-L/(nD)\}] \tag{4.122}$$

if the mode shape is unity.

The motion dependent forces can be presented by

$$L_z(y,t) = \left(\rho U^2/2\right)(\omega D/U)^2\left[H_4^* \cdot z + H_1^* \cdot (\dot{z}/\omega)\right] \tag{4.123}$$

where generally speaking aerodynamic coefficients or derivatives H_1^* and H_4^* are amplitude dependent. Of particular importance for the response calculation is the magnitude of the aerodynamic damping, which can be modeled by

$$H_1^* = H_{1o}^* \cdot \left[1 - (\sigma_z/\sigma_{zL})^2\right] \tag{4.124}$$

where σ_z is rms amplitude andσ_{zL} is the rms limiting amplitude reached as $H_1^* \to 0$. H_{1o}^* is the value of H_1^* when $z \to 0$ and its magnitude depends largely upon the reduced velocity U_r, Reynolds number *Re* and the aspect ratio *L*/*D*. Ordinary measurement of aerodynamic derivatives intended for flutter analysis does not cover this range of reduced velocity but it could be done and needs to be done for the present purposes.

The aerodynamic damping previously referred is related to the derivative by

$$\begin{aligned} \varsigma_{aero} &= -H_1^*/(2\mu) \\ \mu &= m/\left(\rho B^2\right) \end{aligned} \tag{4.125}$$

$$\varsigma_{aero}(U_r, \sigma_z) = \frac{H_{1o}^*(U_r)}{2\mu} \cdot \left[\left\{\frac{\sigma_z}{\sigma_{zL}(U_r)}\right\}^2 - 1\right] \tag{4.126}$$

which can be obtained from the free vibration tests in wind.

Thus the section model dynamic tests could give the force coefficient σ_{cL} and aerodynamic damping ratio $\varsigma\tau\Theta_{aero}$. Possible source of errors in these measurements is the lack of information regarding the lateral force correlations unless they are measured separately. The prediction of bridge response becomes possible once these parameters are decided.

References

Aynsley RM, Melbourne WH, Vickery BJ (1977) Architectural aerodynamics, Applied Science Publishers

Biggs JM (1961) Wind forces on structures: final report. ASCE Task committee on wind forces

Bishop RED, Hassan AY (1964) The lift and drag forces on a circular cylinder oscillating in a flowing fluid. Proc Roy Soc Lond 277:51–75

Bisplinghoff RL, Ashley H, Halfman RL (1955) Aeroelasticity. Dover Publications, Inc., New York

Bleich F (1949a) Dynamic instability of truss-stiffened suspension bridges under wind action. Trans ASCE 114:1177–1222

Bleich F (1949b) Dynamic instability of truss-stiffened suspension brigdes under wind action. Trans ASCE 114:1177–1232

Blevins RD (1990) Flow-induced vibration. Van Nostrand Reinhold, New York

Davenport AG (1961) The application of statistical concepts to the wind loading of structures. Proc ICE, London, 19:449–472

Davenport AG (1962) Buffeting of a suspension bridge by storm wind. Proc ASCE 88:ST3

Davenport AG (1977) Wind engineering—ancient and modern—the relationship of wind engineering research to design. In: Proceedings of the 6th CANCAM, Vancouver

Dowell EH, Curtiss HC Jr., Scanlan RH, Sisto F (1989) A modem course in aeroelasticity, 2nd edn. Kluwer Academic Publishers, Dordrecht, Netherlands

Forsching H (1968) Aeroelastic stability investigations on prismatic beams. In: Proceedings of the symposium on wind effects on buildings and structures, Loughborough, p 22

Fung YC (1969) An introduction to the theory of aeroelasticity. Dover Publications, Inc., New York

Ge YJ (1999) Probability-based assessment and full-mode flutter analysis of cable-supported bridges against aerodynamic forces. Post-Doctoral research report, University of Ottawa

Ge YJ, Tanaka H (2000) Aerodynamic flutter analysis of cable-supported bridges by multi-mode and full-mode approaches. J Wind Eng Ind Aerodyn 86:123–153

Hirai A (1942) Aerodynamic torsional instability of suspension bridges. J JSCE 28

Hnston DR (1987) Flutter derivatives from 14 generic deck sections. In: Proceedings of the ASCE structures congress, Orlando

Jakobsen JB, Hjorth-Hansen E (1993) Determination of the aerodynamic derivatives by a system identification method. In: Proceedings of the 1st IAWE European & African regional conference, Guernsey

Jensen AG, Nielsen FG (1995) A forced motion mechanism for the investigation of aerodynamic loads on structures moving in an air flow. In: Proceedings of the 9th ICWE, New Delhi

Karpel M (1981) Design of active and passive flutter suppression and gust alleviation. NASA contractor report No. 3482

Kloppel K (1963) Teilmodellversuche zur Beurteilung des Aerodynamischen Verhaltens von Brücken. Der Stahlbau 32, 75–79 and 113–121

Modi VJ, Slater JE (1972) Quasi-steady analysis of torsional aeroelastic oscillators. In: Proceedings of IUTAMIAHR symposium on flow-induced structural vibrations, Karlsruhe, pp 355–372

Moisseiff L, Lienhardt F (1933) Suspension bridge under the action of lateral forces. Trans ASCE 98:1080–1095

Morimitsu Y, lto M, Miyala T (1973) Aeroelastic oscillation of structures due to nonlinear aerodynamic forces. Proc JSCE 219:27–36

Nakamura Y, Mizota T (1975) Torsional flutter of rectangular prisms. Proc ASCE 101:EM2

Novak M (1969) Aeroelastic galloping of prismatic bodies. Proc ASCE 95 EMI:115–142

Parkinson GV, Brooks NPH (1961) On the aeroelastic instability of bluff cylinders. J Appl Mech 28:252–258

Poulsen NK (1990) System identification in determination of the flutter coefficients. Skibsteknisk Laboratorium

Roger KL (1977) Airplane math modeling methods for active control design. AGARD-CP-228

Sarkar PP, Jones NP, Scanlan RH (1991) System identification for estimation of flutter derivatives. In: Proceedings of the 8th ICWE, London (Ont.)

Sarkar PP, Jones NP, Scanlan RH (1993) A comparative study of the aeroelastic behavior of three flexible bridges and a thin airfoil. In: Proceedings of the 7th US national wind engineering conference, UCLA

Scanlan RH (1978) The action of flexible bridges under wind: I. flutter theory. J Sound Vib 60:201–211

Scanlan RH, Lin WH (1978) Effects of turbulence on bridge flutter derivatives. Proc ASCE 104 EM4:719–733

Scanlan RH, Sabzevari A (1967) Suspension bridge flutter revisited. In: ASCE structural engineering conference, Seattle
Scanlan RH, Tomko JJ (1970) Ailfoil and bridge deck flutter derivatives. Proc ASCE 97 EM6
Selberg A (1957) Aerodynamic stability of suspension bridges. IABSE 17:209–216
Selberg A (1963) Oscillation and aerodynamic stability of suspension bridges. Acta Polytechnica Scandinavica 6:Ci 13
Tanaka H, Davenport AG (1979) Effect of turbulence on aeroelastic instability of suspension bridges. In: Proceedings of the 7th CANCAM, Sherbrooke
Tanaka H, lto M (1969) The characteristics of the aerodynamic forces in self-excited oscillation of bluff structures. Trans JSCE 1:209–236
Ukeguchi M, Sakata H, Nishitani H (1966) An investigation of aeroelastic instability of suspension bridges. In: Proceedings of the international symposium on suspension bridges, Lisbon
Yamada H, Miyata T, Ichikawa H (1991) Measurement of aerodynamic coefficients by system identification methods. In: Proceedings of the 8th ICWE, London (Ont.)

Chapter 5
Wind-Induced Vibrations of Structures: With Special Reference to Tall Building Aerodynamics

Kenny C.S. Kwok

Abstract This chapter examines the wind-induced vibrations of buildings and structures. In particular, the alongwind and the crosswind excitation mechanisms and response processes are discussed. Turbulence buffeting is the dominant excitation mechanism in the alongwind direction. In the crosswind direction, excitations associated with incident turbulence, vortex shedding and motion dependent excitations are the prominent mechanisms causing crosswind response. Many advanced wind codes and standards have adopted statistical and spectral analysis processes to predict wind loads and responses caused by these alongwind and crosswind excitations. The potential impact of interference excitations of tall buildings and beneficial effects of aerodynamic modifications of building shape are also reviewed in this chapter. Interference effects on the alongwind, crosswind and torsional responses of tall buildings from either an upstream or a downstream building are outlined. The beneficial effects and the economic perspectives of aerodynamic modifications of building shape in reducing the wind induced load and response of tall buildings are discussed.

Keywords Aerodynamic treatments • Excitation mechanisms • Interference • Tall buildings • Wind-induced vibrations

5.1 Introduction

The tremendous growth in research and testing activities in the field of wind engineering during the past 30 years has resulted in significant advances in the understanding of wind effects on buildings and structures. Much of the research

---

K.C.S. Kwok (✉)
Institute for Infrastructure Engineering, University of Western Sydney, Locked Bag 1797, Penrith, NSW 2751, Australia
e-mail: k.kwok@uws.edu.au

Y. Tamura and A. Kareem (eds.), *Advanced Structural Wind Engineering*,
DOI 10.1007/978-4-431-54337-4_5, © Springer Japan 2013

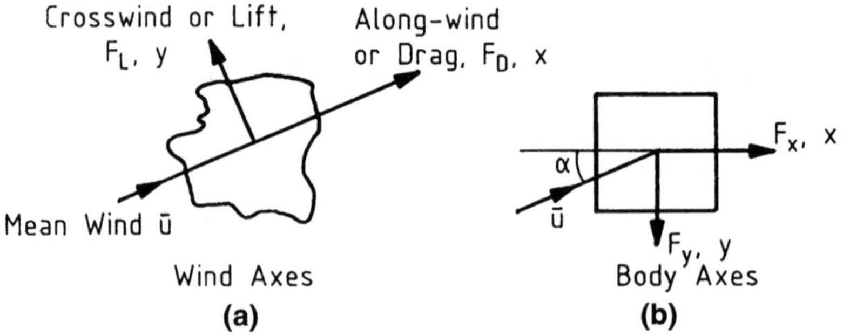

Fig. 5.1 Components of wind-induced forces and response. (**a**) Wind axes; (**b**) body axes

findings have been incorporated in current wind codes and standards around the world. However, due to the complexity of wind/structure interaction, evolution of design concepts, introduction of new material and significance of meteorological input, it is expected that experimental approaches such as wind tunnel model tests will continue to play an important role in wind engineering research and in wind resistance design of modern tall buildings and structures.

In the design of a building or structure against wind actions, a question which often comes to mind for the designer is how to determine if the building or structure is wind sensitive. A quick examination of the shape, size and special features of the building or structure will usually indicate whether a code approach is adequate for the estimation of wind load. However, for wind sensitive structures, amplification of wind load and structural response due to wind/structure interaction are dependent on the mass distribution, stiffness (both translational and torsional) and structural damping characteristic of the structure. Complex analytical and experimental procedures are required to assist in the design of these structures.

The excitation mechanisms causing alongwind, crosswind and torsional responses and the characteristics of these responses have been studied and reported extensively. The excitation mechanisms studied include turbulence buffeting, vortex shedding, and those associated with aeroelastic effects such as galloping, amplitude dependent crosswind and torsional lock-in excitations, and the separation/reattachment mechanism for torsion. The presence of eccentricity and coupling, both structural and aerodynamic, has also been found to have a significant effect on the excitation and response processes. Furthermore, these excitation mechanisms do not always occur in isolation and often more than one mechanism may contribute to the response of a body.

The wind-induced forces and responses are conveniently resolved into two orthogonal directions as shown in Fig. 5.1. These may be parallel (alongwind or drag) and perpendicular (crosswind or lift) to the mean wind direction, or parallel and perpendicular to a direction related to the geometry of the body. The first part of this chapter is concerned primarily with the identification of alongwind and crosswind excitation mechanisms, and the effects of the fluctuating loads on buildings and structures.

The new generation of tall buildings is taller and more slender, has futuristic shapes and sophisticated structural systems. The reduced wind velocities these buildings are required to operate in are steadily increasing and approaching, even surpassing, the threshold above which aeroelastic effects become important strength and serviceability design considerations.

Interference excitations caused by the presence of adjacent tall buildings, either upstream or downstream, have been shown to readily increase the alongwind, crosswind and torsional re-sponses by 30–40 % at low- to mid-range of reduced wind velocities, and by much more at the high range of reduced velocities and at critical building separations and positions. Potential increases in response of such a magnitude are important considerations in the design of modern tall buildings.

The roles of mass, stiffness, damping and aerodynamic shape in the reduction of wind-induced vibrations of tall buildings is dependent on the type of loading and are rather complex except that an increase in the damping capacity is always beneficial. Wind-induced displacement response generally is not very sensitive to changes in either the mass or stiffness although an increase in the mass or stiffness of a building can be advantageous in reducing acceleration response. Therefore, not surprisingly, a considerable amount of resources and effort has been devoted in recent years to the research on passive and active vibration controls. Similarly, the potential benefits of aerodynamic modifications of building shape have increasingly been adopted in modern tall building design. For example, relatively minor modifications to the aerodynamic shape of a building while retaining the basic plan shape, such as the introduction of chamfered corners, vertical and horizontal through building openings or scultured building top, have been shown to significantly reduce wind induced responses, particularly crosswind response.

The second part of this chapter reviews recent advances in the understanding of interference excitations and the associated increase in response of tall buildings, and the potential beneficial effects of aerodynamic modifications of building shape in reducing the wind induced responses of tall buildings.

5.2 Alongwind Excitation and Response of Buildings and Structures

The turbulence or gustiness in the natural wind causes buffeting which directly produces fluctuating forces on a structure in the flow. The fluctuating wind velocities, wind induced forces and the consequent responses are usually random in nature. The magnitude of the fluctuating load is a function of the intensity of turbulence and its length scale. The turbulence intensity determines the local magnitude of fluctuation forces while the turbulence length scale, in relation to the size of the structure, determines how well the fluctuations are correlated over the structure.

Fig. 5.2 Components of a time dependent variable

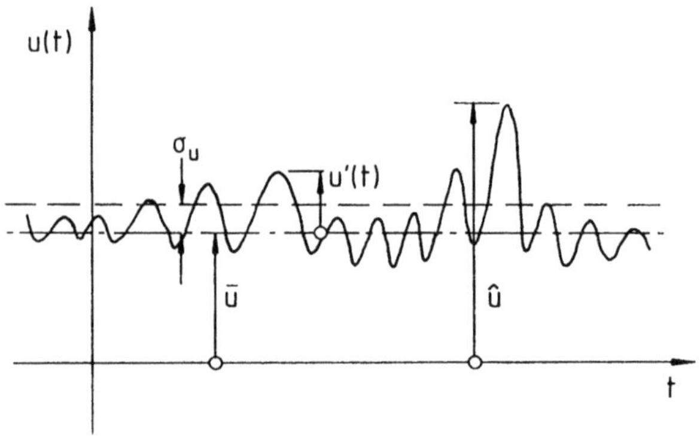

In order to determine the relationship between the turbulent wind and the wind load on a structure, it is more convenient to think both in terms of wavelengths and frequencies. For example, if the mean wind speed at some height z is $\bar{u}(z)$, then we can associate a wavelength $\lambda = \bar{u}(z)/n$ with a frequency component n. λ is then a linear measure of the size of gusts or eddies producing load fluctuations at a frequency n. The effectiveness of a gust in terms of producing a load on a large structure depends largely on the gust size in relation to the size of the structure, say b; that is, the ratio λ/b or $\bar{u}$/nb. In the case of high frequency components, $\lambda/b \ll 1$, the pressures or forces produced are well correlated or organised only on quite small areas of the structure. The total effect produced by these high frequency components is small because in some areas they will tend to produce increased loads while simultaneously at other parts of the structure there may be a decrease in load. On the other hand, the very low frequency components, where $\lambda/b \gg 1$, are felt over the whole or at least large areas of the structure simultaneously.

5.2.1 Fluctuating Drag Forces on a Bluff Body

The longitudinal wind velocity in an atmospheric boundary layer flow is shown in Fig. 5.2 and is given by

$$u(t) = \bar{u} + u'(t) \tag{5.1}$$

in which $\bar{u}$ is mean longitudinal wind speed and $u'(t)$ is fluctuating longitudinal wind speed. For a small structure of area A placed in such a flow, the drag force experience by the structure is

$$\begin{aligned} F_D(t) &= C_D \frac{1}{2} \rho A u^2(t) \\ &= C_D \frac{1}{2} \rho A \left[\bar{u}^2 + 2\bar{u}u'(t) + u'^2(t) \right] \end{aligned} \tag{5.2}$$

in which C_D is the drag coefficient. Ignoring the second order term which is small,

$$\begin{aligned} F_D(t) &\approx C_D \frac{1}{2} \rho \bar{u}^2 A + C_D \rho A \bar{u} u'(t) \\ &= \bar{F}_D + F'_D(t) \end{aligned} \tag{5.3}$$

in which $\bar{F}_D$ is mean drag and $F'_D(t)$ is fluctuating drag. Therefore the spectral representation in a frequency domain of this fluctuating drag force is

$$\begin{aligned} S_D(n) &= (C_D A \rho \bar{u})^2 S_u(n) \\ &= 4 \frac{\bar{F}_D^2}{\bar{u}^2} S_u(n) \end{aligned} \tag{5.4}$$

where $S_u(n)$ is the spectrum of the longitudinal turbulence. However, the force acting on the structure depends on the size of the disturbance or gust in relation to the size of the structure and this dependence can be expressed in terms of an aerodynamic admittance function

$$\chi\left(\frac{n\sqrt{A}}{\bar{u}}\right)$$

Therefore

$$S_D(n) = 4 \frac{F_D^2}{\bar{u}^2} \chi^2 \left(\frac{n\sqrt{A}}{\bar{u}}\right) S_u(n) \tag{5.5}$$

The following empirical relationship, which is shown in Fig. 5.3, can be fitted to χ values for flat plates, discs and prisms

$$\chi(n) = \frac{1}{1 + \left(\frac{2n\sqrt{A}}{\bar{u}}\right)^{\frac{4}{3}}} \tag{5.6}$$

Some properties of χ are fairly obvious; as $n\sqrt{A}/\bar{u} \to 0$, that is the low frequency components, $\chi \to 1$, and as $n\sqrt{A}/\bar{u} \to \infty$, that is the high frequency components, $\chi \to 0$.

5.2.2 Gust Response Factors

If we consider a structure which is assumed to be a linear single degree-of-freedom system with a natural frequency n_0, a structural damping ζ_s and a stiffness k, the alongwind response of the structure due to loading by wind gusts is

$$x(t) = \bar{x} + x'(t) \tag{5.7}$$

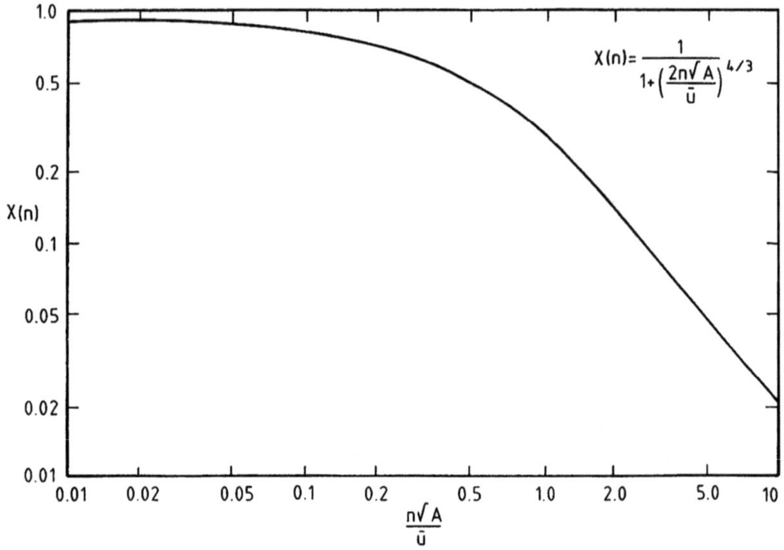

Fig. 5.3 Aerodynamic admittance for flat plates, discs and prisms

which (as shown in Fig. 5.2) is the sum of a mean and a fluctuating component. The mean or static component of the response is

$$\bar{x} = \frac{\bar{F}_D}{k} \tag{5.8}$$

The spectrum of the displacement response can be expressed as

$$\begin{aligned} S_x(n) &= \frac{1}{k^2} \frac{1}{\left[1 - \left(\frac{n}{n_0}\right)^2\right]^2 + 4\zeta_S^2 \left(\frac{n}{n_0}\right)^2} S_D(n) \\ &= \frac{1}{k^2} |H(n)|^2 S_D(n) \\ &= 4 \frac{\bar{F}_D^2}{\bar{u}^2} \frac{1}{k^2} \chi^2 \left(\frac{n\sqrt{A}}{\bar{u}}\right) |H(n)|^2 S_u(n) \end{aligned} \tag{5.9}$$

in which $H(n)$ is the mechanical admittance function. The variance of the fluctuating component of the displacement is

$$\sigma_x^2 = 4 \frac{\bar{F}_D^2}{\bar{u}^2} \frac{1}{k^2} \int_0^\infty \chi^2 \left(\frac{n\sqrt{A}}{\bar{u}}\right) |H(n)|^2 S_u(n) dn \tag{5.10}$$

If we substitute

$$\bar{x} = \frac{\bar{F}_D}{k}$$

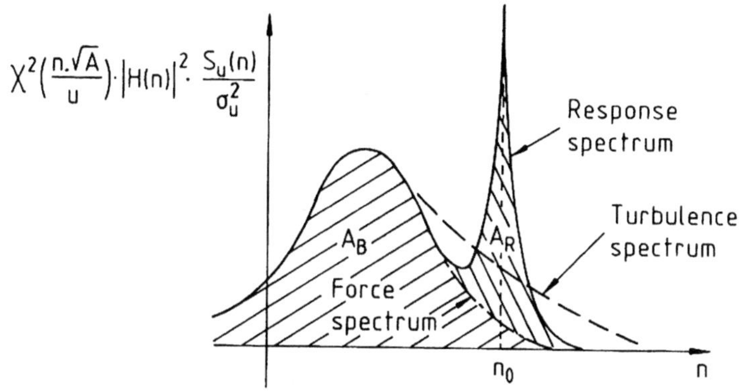

Fig. 5.4 Spectral representation of a single degree-of-freedom system responding to turbulence buffeting

and express the velocity spectrum in the more conventional normalized form

$$\frac{S_u(n)}{\sigma_u^2}$$

we may write

$$\begin{aligned}\frac{\sigma_x^2}{\bar{x}^2} &= 4\frac{\sigma_u^2}{\bar{u}^2}\frac{1}{k^2}\int_0^\infty \chi^2\left(\frac{n\sqrt{A}}{\bar{u}}\right)|H(n)|^2\frac{S_u(n)}{\sigma_u^2}dn \\ &= 4\frac{\sigma_u^2}{\bar{u}^2}[A_B + A_R]\end{aligned} \tag{5.11}$$

The background excitation due to low frequency gusts is represented by

$$\begin{aligned}A_B &= \int_0^{n_0}\chi^2\left(\frac{n\sqrt{A}}{\bar{u}}\right)\frac{S_u(n)}{\sigma_u^2}dn \\ &= B\end{aligned} \tag{5.12}$$

The resonant component is given by

$$\begin{aligned}A_R &= \chi^2\left(\frac{n_0\sqrt{A}}{\bar{u}}\right)\frac{S_u(n_0)}{\sigma_u^2}\int_0^\infty |H(n)|^2 dn \\ &= \chi^2\left(\frac{n_0\sqrt{A}}{\bar{u}}\right)\frac{S_u(n_0)}{\sigma_u^2}\frac{\pi n_0}{4\zeta_S}\end{aligned} \tag{5.13}$$

These components are shown in Fig. 5.4.
Let S be a size reduction factor

$$S = \chi^2\left(\frac{n_0\sqrt{A}}{\bar{u}}\right) \tag{5.14}$$

and E be a gust energy factor

$$E = \frac{\pi n_0}{4} \frac{S_u(n_0)}{\sigma_u^2} \tag{5.15}$$

so that

$$A_R = \frac{SE}{\zeta_s} \tag{5.16}$$

Then

$$\frac{\sigma_x^2}{\bar{x}^2} = 4\frac{\sigma_u^2}{\bar{u}^2}\left(B + \frac{SE}{\zeta_s}\right) \tag{5.17}$$

and the ratio of the standard deviation deflection to the mean deflection is

$$\frac{\sigma_x}{\bar{x}} = 2\frac{\sigma_u}{\bar{u}}\left(B + \frac{SE}{\zeta_s}\right)^{\frac{1}{2}} \tag{5.18}$$

The expected peak value of x is

$$\begin{aligned} x_{\max} &= \hat{x} \\ &= \bar{x} + x'_{\max}(t) \\ &= \bar{x} + g_f \sigma_x \end{aligned} \tag{5.19}$$

The factor g_f is an average peak factor which depends on the time interval and the form of the spectrum of x. In most cases, g_f is between 3 and 4 and a value of 3.5 is a fair approximation. Therefore

$$\begin{aligned} \hat{x} &= \bar{x} + g_f \bar{x} 2\frac{\sigma_u}{\bar{u}}\left(B + \frac{SE}{\zeta_s}\right)^{\frac{1}{2}} \\ &= \bar{x}\left[1 + 2g_f\frac{\sigma_u}{\bar{u}}\left(B + \frac{SE}{\zeta_s}\right)^{\frac{1}{2}}\right] \end{aligned}$$

and

$$\begin{aligned} G &= \frac{\hat{x}}{\bar{x}} \\ &= 1 + 2g_f\frac{\sigma_u}{\bar{u}}\left(B + \frac{SE}{\zeta_S}\right)^{\frac{1}{2}} \end{aligned} \tag{5.20}$$

where G is a gust response factor.

The foregoing derivation represents a simplified version of derivations first proposed by Vickery (1966) and Davenport (1967). The excitation and response

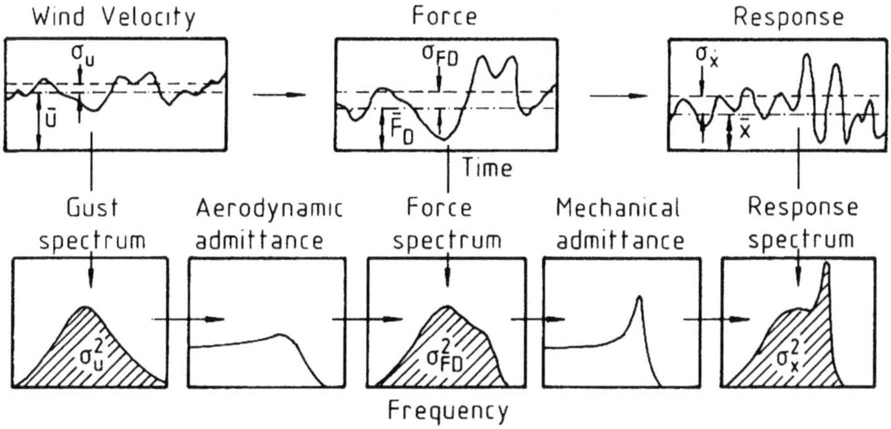

Fig. 5.5 Alongwind excitation and response process (after Davenport 1967)

processes are illustrated in Fig. 5.5. In general, the approach is adequate for the estimation of alongwind response of buildings, slender towers, suspension cables and some bridge structures, although some modifications may be necessary to account for the velocity gradient in a boundary layer flow and for the mode shape of the structure under consideration. More detailed descriptions of the application of the gust response factor approach are given in a companion chapter of this book.

5.2.3 Aerodynamic Damping

Fluctuating forces induced by turbulence buffeting exist irrespective of the movements of the structure. However, aerodynamic damping forces are induced only by the movement of the structure. Although the magnitude of these forces is small compared with the inertia and stiffness forces of a structure, they are appreciable in comparison with the forces due to structural damping and can markedly influence the behaviour of a structure in wind.

Assuming the structure is a linear system and the amplitude of vibration is small, the drag force acting in the x-direction, as shown in Fig. 5.1a, is

$$F_D(t) = C_D \tfrac{1}{2} \rho u_r^2(t) A \tag{5.21}$$

The relative wind speed is

$$u_r(t) = \bar{u} - \dot{x} \tag{5.22}$$

Hence

$$F_D(t) = C_D \tfrac{1}{2} \rho \bar{u}^2 A - C_D \rho \bar{u} \dot{x}\, A + C_D \tfrac{1}{2} \rho \dot{x}^2 A \tag{5.23}$$

In general, $\dot{x} \ll \bar{u}$ so that the last term in Eq. 5.23 may be ignored. The first term is the mean drag force and the second term is an aerodynamic damping force. The negative sign of the damping force means that it opposes the motion and energy is extracted from the structure. That is, the aerodynamic damping ζ_a is positive and adds to any structural damping.

For vertical structures, assuming a linear deflection mode shape, and if the mass distribution, cross-sectional shape and C_D are constant, the expression for aerodynamic damping is

$$\zeta_a = \frac{3\rho\bar{u}(h)C_D}{4\pi\rho_s(3+\gamma)n_0 d} \tag{5.24}$$

in which ρ is the density of the air, ρ_s is the density of the structure, d is the depth of the structure (in the direction of wind), and γ is the exponent of the mean wind velocity profile. For typical values:

$\rho = 1.2$ kg/m^3
$\rho_s = 200$ kg/m^3
$C_D = 1.2$
$d = 30$ m
$\bar{u}(h) = 50$ m/s
$\gamma = 0.25$

we have

$$\zeta_a \approx \frac{10^{-3}}{n_0} \tag{5.25}$$

For a typical tall building where $n_0 = 0.2$ Hz, ζ_a is about 0.5 % of critical damping. This is a significant addition to the structural damping which is typically 1–2 % of critical damping. As a result, the alongwind response of the building will be reduced. Although the aerodynamic damping in this case is positive, it should be noted that aerodynamic damping associated with some crosswind excitations may be negative, as will be discussed in a later section. In that case, the overall damping of a structure is reduced, which may result in a significant increase in crosswind response.

5.3 Crosswind Excitation and Response of Buildings and Structures

The crosswind excitation mechanisms according to Melbourne (1977) can be divided into three categories associated with:

1. Turbulence in the free stream flow;
2. The wake, in particular vortex shedding; and
3. Crosswind motion of the structure.

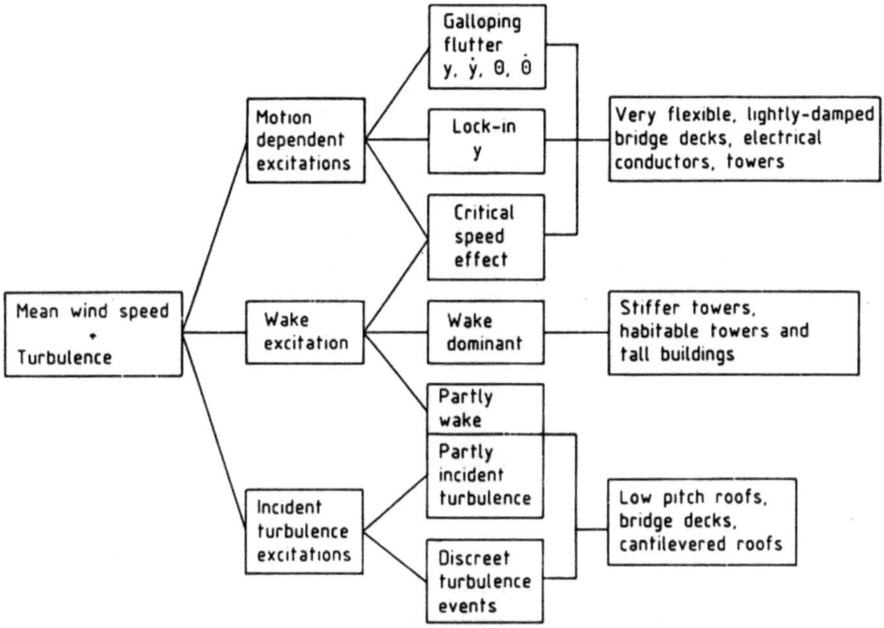

Fig. 5.6 Main variables and mechanisms associated with the crosswind response of structures (after Melbourne 1975)

The main variables and mechanisms associated with the crosswind response of a range of structures are shown in Fig. 5.6. Although these excitations are separately identifiable, they often superimpose in causing response. Despite significant advances in recent years, there is as yet no generalised analytical method available to calculate accurately the crosswind response of modern tall buildings and structures.

5.3.1 Excitation Due to Incident Turbulence

Turbulence in the natural wind gives rise to changes in the speed and direction of the incident wind, which directly induces fluctuating drag, crosswind forces and moments on buildings and structures. The significance of incident turbulence in causing a crosswind response depends on the ability of the incident turbulence to generate a crosswind force on the structure as a function of the longitudinal wind velocity and the angle of incidence of the mean wind. Incident turbulence may excite one or more natural modes of vibration or the structure may respond to a well-correlated discrete gust. A quasi-steady spectral approach, similar to that used to determine the alongwind response of structures, is considered to be applicable to the computation of crosswind response of structures due to incident turbulence.

The fluctuating crosswind force per unit length of a structure is

$$F'_L(t) = C_L \rho \bar{u} b u'(t) + \left(C_D + \frac{\partial C_L}{\partial \alpha} \right) \rho \bar{u} b v'(t) \tag{5.26}$$

in which $v'(t)$ is the lateral or crosswind component of turbulence, and $\partial C_L/\partial \alpha$ is the rate of change of lift coefficient with respect to the angle of wind incidence α. In general, crosswind excitation due to incident turbulence is significant only for structures with a long afterbody, such as some bridge decks, flat deck roofs and aircraft wings. For example, C_L and $\partial C_L/\partial \alpha$ are zero for a circular section. Therefore, the crosswind force is

$$F'_L(t) = C_D \rho \bar{u} b v'(t) \tag{5.27}$$

This compares with the alongwind force

$$F'_D(t) = C_D \rho \bar{u} b u'(t) \tag{5.28}$$

Since the crosswind (lateral) intensity of turbulence is usually about one half or less of the alongwind (longitudinal) intensity, the crosswind excitation on a circular cross-section will not normally be of significance.

For a streamlined bridge deck, C_D and C_L are both small, but $\partial C_L/\partial \alpha$ can be close to the theoretical value for a flat plate of 2π. The crosswind force then becomes

$$F'_L(t) = 2\pi \rho \bar{u} b w'(t) \tag{5.29}$$

in which $w'(t)$ is the vertical component of turbulence. A quasi-steady spectral approach was proposed by Davenport (1962) and many others have since adopted the methodology to determine the response of bridges due to turbulence buffeting.

One of the less well known effects of incident turbulence is its ability to control the development of flow in the near wake and hence change the crosswind force characteristic of a structure by altering the shedding mechanism of the shear layers. This effect is outlined in a later section.

5.3.2 Wake Excitation

For buildings and structures under wind action, the most common source of crosswind excitation is associated with vortex shedding; hence the commonly used term, "vortex excitation". For most structures, the shed vortices have a dominant shedding frequency n_s which is defined by the Strouhal Number

$$St = \frac{n_s b}{\bar{u}} \tag{5.30}$$

The Strouhal Number is typically of the order of 0.1–0.4 and is a function of the Reynolds Number *Re*, cross-sectional shape, surface roughness and free stream turbulence. During the vortex shedding process, a structure is subjected to a periodic pressure loading which results in an alternating crosswind force. If the natural frequency of the structure coincides with the vortex shedding frequency, resonant response may occur and this is commonly referred to as the critical speed effect. The oscillation of a lightly-damped chimney in its fundamental mode at a critical velocity is one of the best known examples of resonant vibrations caused by critical speed effect. It has been shown by Scruton and Walshe (1957) that such resonant vibrations can be effectively suppressed by the addition of a helical strakes. Perforated shrouds and splitter plates have also been shown to be effective in reducing the oscillating forces in wind tunnel tests. Helical strakes have been used in many full-scale structures but a major disadvantage is the increased drag force.

In most real life situations, vertical structures are exposed to a turbulent wind in which both the turbulence intensity level and the mean velocity vary with height so that excitation due to vortex shedding is effectively broad-band. Therefore, it is more appropriate to use the term “wake excitation” to include all forms of excitation associated with the wake and not just those associated with critical wind speeds.

Wake-induced crosswind forces have been studied extensively for various structures and incident turbulence. Despite some significant differences in experimental approach, the distribution of these forces can be summarised as in Fig. 5.7, and a number of important and fundamental conclusions reported by Saunders and Melbourne (1977) are summarised as follows:

1. Freestream turbulence has a profound effect on the crosswind forces associated with wake excitation. An increase in turbulence often leads to some reduction in the strength of vortex shedding, which results in a broadening of the crosswind force spectrum. Therefore, for the same structure in a more turbulent flow, there will be a significant increase in crosswind response at reduced velocities significantly lower than that associated with the vortex peak, and a significant reduction in response at reduced velocities near the vortex peak.
2. Vortex shedding from structures with a low height to breadth ratio is less organised, which results in a broadening of the crosswind force spectrum. The Strouhal Number, which defines the vortex shedding frequency, tends to decrease with a decreasing height to breadth ratio and hence the critical speed tends to increase.
3. Structures with a short afterbody produce a narrower bandwidth of the crosswind force spectrum than for a square cross-section, while structures with a long afterbody produce a wider force spectrum. Consequently, these effects result in a lower crosswind response for structures with a long afterbody than those with a short afterbody under the same operating conditions.

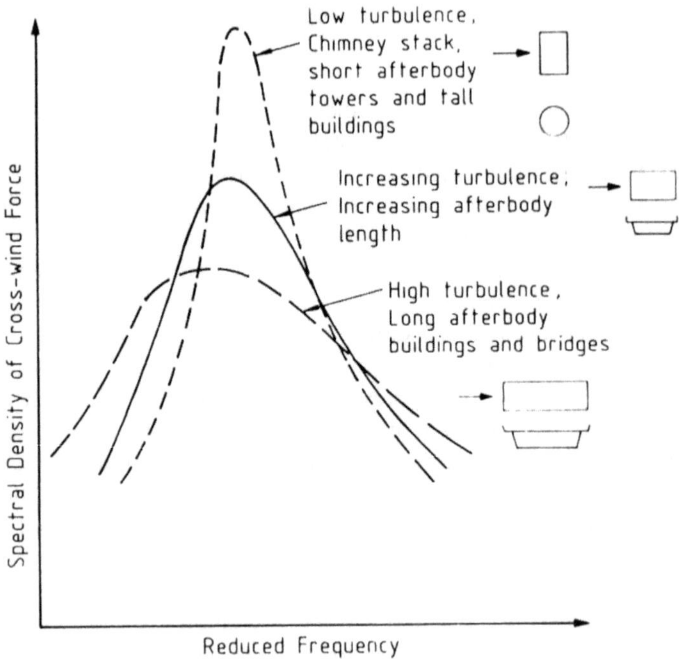

Fig. 5.7 Frequency distribution of wake-induced forces for various structures and free stream turbulence (after Melbourne 1975)

4. For most tall buildings and structures, crosswind forces are often significantly greater than alongwind forces as wind velocity is increased. Therefore in terms of design, crosswind response tends to be more important than alongwind response.

A Gaussian type crosswind force spectrum (per (unit length)^2) of the form shown below was also suggested by Vickery and Clark (1972)

$$S_L(n) = \left(C_L \tfrac{1}{2}\rho \bar{u}^2 b\right)^2 \frac{1}{B^* n_s \sqrt{\pi}} \exp\left[-\left(\frac{1-\frac{n}{n_s}}{B^*}\right)^2\right] \tag{5.31}$$

in which C_L is the standard deviation crosswind force coefficient, and B^* is the effective bandwidth of the spectrum, measured as a fraction of the centre frequency. By assuming that the wake excitation force is random, the crosswind response of buildings and structures can be evaluated by modal analysis, as done by Vickery and Clark (1972), Kwok (1982) and others. Considerable simplification of the evaluation process can be achieved if the response is assumed to be predominantly in the first (fundamental) mode of vibration. Then the variance of crosswind response at the top of a structure is given by

$$\sigma_y^2 = \int_0^\infty S_y(n)dn$$
$$= \frac{1}{(2\pi n_0)^4 m^2} \int_0^\infty S_L(n)|H(n)|^2 dn \qquad (5.32)$$

in which $S_y(n)$ is the crosswind displacement spectrum and m is the modal mass of the structure.

Eq. 5.32 can be further simplified:

$$\sigma_y^2 \approx \frac{1}{(2\pi n_0)^4 m^2}\left[\int_0^\infty S_L(n)dn + S_L(n_0)\int_0^\infty |H(n)|^2 dn\right]$$
$$\approx \frac{1}{(2\pi n_0)^4 m^2}\left[\int_0^\infty S_L(n)dn + \frac{\pi n_0 S_L(n_0)}{4\zeta_s}\right] \qquad (5.33)$$

If low frequencies background excitation is small and the structural damping is low (less than 10 % of critical damping), then the excitation bandwidth is wide-band compared with the resonant bandwidth, and the first term in Eq. 5.33 can be neglected. Hence

$$\sigma_y^2 \approx \frac{1}{(2\pi n_0)^4 m^2}\left[\frac{\pi n_0 S_L(n_0)}{4\zeta_s}\right] \qquad (5.34)$$

and the standard deviation response is proportional to $\zeta_s^{-\frac{1}{2}}$.

In general, the application of Eqs. 5.32 or 5.34 to predict crosswind response of tall buildings and structures is satisfactory when wake excitation is the dominant mechanism and when the operating reduced velocity is low. However, at high reduced velocities (greater than 10) at which other forms of excitation such as galloping excitation are prevalent, and particularly at low values of structural damping, such a prediction procedure may grossly underestimate the crosswind response.

For practical application, databases of wind force spectra for tall buildings and structures of various shapes, aspect ratios, height to breadth ratios, and wind exposures have been made available, such as the NatHaz Aerodynamic Loads Database published by NatHaz (Natural Hazards) Modeling Laboratory at the University of Notre Dame. For codification purposes, simplified calculation procedures, similar to the spectral based formulation outlined in Eqs. 5.32, 5.33, and 5.34, have also been adopted by wind codes and standards such as the Australian and New Zealand Standard AS/NZS 1170.2:2011, in which crosswind force spectrum coefficients for simple buildings are given both in graphical form and as empirical formulae.

5.3.3 *Excitations due to Crosswind Motions*

There are three commonly recognised crosswind excitations under this heading; namely, galloping, lock-in and flutter. These are frequently referred to as wind-induced instabilities or aerodynamic instabilities which occur because of either the flow pattern generated by the shape of the structure or the change in aerodynamic force due to the motion of the structure. Much research into aspects of these aerodynamic instabilities has been carried out since the collapse of the Tacoma Narrows Suspension Bridge and the occurrence of large amplitude oscillations of transmission lines and steel chimney stacks.

5.3.3.1 Galloping Excitation

Galloping is the term used to describe the large amplitude, single degree of-freedom motion of a structure associated with a sectional aerodynamic force characteristic which causes an increase in crosswind force in the direction of and, in phase with, the crosswind motion. A classical example is the violent oscillation of ice-laden transmission lines which was first investigated by Den Hartog in 1930. Galloping occurs readily in both smooth and turbulent flows and at high reduced velocities for bluff cross-sections which have fixed separation points, such as square and rectangular bodies, D-shaped and cruciform sections.

Consider a square cross-section in the presence of a wind normal to a face and the plane of motion, during galloping, any crosswind motion of the section will create a surface pressure distribution which results in a force F_y in the direction of the motion so that the displacement y is further increased. An oscillating motion will develop at the natural frequency of the body and the amplitude of oscillation will increase until the energy dissipated per cycle by the structural damping balances the energy input per cycle from the wind.

The fundamental assumption used in the analysis of galloping is the quasi-steady approximation which, because galloping oscillations are generally slow, assumes that for every instant during the oscillation, the aerodynamic force acting on the body is the same as for a static test on the same rigid body at the same angle of incidence of the mean wind.

The aerodynamic force is, according to Fig. 5.1

$$F_y = -(F_L \cos \alpha + F_D \sin \alpha) \tag{5.35}$$

and the transverse force coefficient is

$$C_{Fy} = -(C_L \cos \alpha + C_D \sin \alpha) \tag{5.36}$$

Now

$$\frac{\partial C_{Fy}}{\partial \alpha} = -\left(\frac{\partial C_L}{\partial \alpha}\cos\alpha - C_L \sin\alpha + \frac{\partial C_D}{\partial \alpha}\sin\alpha + C_D \cos\alpha\right) \tag{5.37}$$

and for small α

$$\begin{aligned}\alpha &= \tan^{-1}\frac{\dot{y}}{\bar{u}} \\ &\approx \frac{\dot{y}}{\bar{u}}\end{aligned} \tag{5.38}$$

Therefore

$$\frac{\partial C_{Fy}}{\partial \alpha} \approx -\left(\frac{\partial C_L}{\partial \alpha} + C_D\right) \tag{5.39}$$

or

$$C_{Fy} \approx -\left(\frac{\partial C_L}{\partial \alpha} + C_D\right)\frac{\dot{y}}{u} \tag{5.40}$$

The criterion for galloping instability is the slope of the transverse force coefficient and theoretically galloping will occur if $\partial C_{Fy}/\partial \alpha > 0$. The aerodynamic force associated with galloping excitation may be expressed in terms of an equivalent aerodynamic damping. For vertical structures exposed to a turbulent boundary layer flow, and by assuming a linear first mode shape and constant $\partial C_{Fy}/\partial \alpha$ along the structure, the equivalent aerodynamic damping is, according to Kwok and Melbourne (1980),

$$\zeta_a = -\frac{3\rho}{8\pi\rho_s(3+\gamma)} \cdot \frac{\bar{u}(h)}{n_0 b} \cdot \frac{\partial C_{Fy}}{\partial \alpha} \tag{5.41}$$

If the equivalent aerodynamic damping is negative so that the resultant damping, that is aerodynamic plus structural damping, is zero or negative, the crosswind response amplitude will grow until it reaches a steady magnitude governed by the non-linearity of the aerodynamic damping. Even if the wind velocity and structural damping are such that the resultant damping is neither zero nor negative, the presence of a negative aerodynamic damping effectively reduces the total available damping of the structure. These results in an increase in response due to other excitations such as those associated with turbulence and the wake.

More refined non-linear theories of galloping have been developed by Parkinson (1971), Novak (1972) and others. Although these are useful in the detailed analysis of galloping, it has generally been concluded that these theories are only valid at very high reduced velocities and hence much larger amplitudes of displacement. Therefore, these theories are applicable to a very limited range of structures such as suspended cables, and are not applicable to modern buildings and structures at

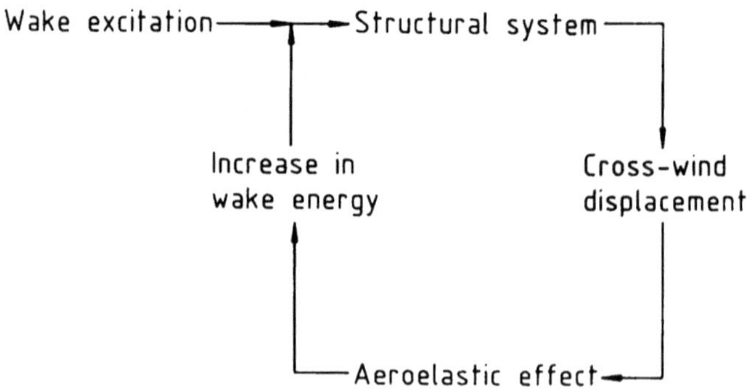

Fig. 5.8 Mechanism of lock-in excitation

normal wind speeds. Simple linear approximation generally leads to conservative estimations of galloping response but is often accepted as a fair indication of possible galloping instability for many structures.

Galloping type excitation and responses have also been observed for inclined slender square-sectioned structures such as bridge pylons, architectural features and sculptures (e.g. Shum et al. 2009).

5.3.3.2 Lock-in Excitation

"Lock-in" is a term commonly used to describe the phenomenon in which the large amplitude crosswind response of a structure occurs at a fairly wide range of wind velocities at which the vortex shedding frequency, as defined by the Strouhal Number, is close to the natural frequency of the structure. When the amplitude of oscillation reaches a certain critical magnitude, the frequency of oscillation and the vortex shedding frequency tend to become synchronised and the resultant cross-wind force increases significantly. Once lock-in becomes well established, the large amplitude response can generally be maintained over a fairly wide range of reduced velocities and is independent of the natural vortex shedding frequency.

When the velocity fluctuations of flow past a nearly stationary and a vibrating square prism are measured, a much more significant peak can be clearly identified in the wake spectrum of the vibrating prism, indicating an increased regularity in the vortex shedding process. It is well known that fluctuating pressures on a stationary structure are poorly correlated along the length of the structure. Cross-wind motion of sufficiently large amplitude can cause a significant increase in the span-wise correlation of pressure fluctuations and hence result in a large lock-in type crosswind force. The mechanism of lock-in excitation is shown diagrammatically in Fig. 5.8. A wide range of structures, including square, rectangular, circular and D cross-sections, is susceptible to lock-in excitation in both smooth and turbulent flows. The critical crosswind response amplitude, which represents the standard deviation crosswind displacement above which lock-in excitation is significant, can be determined experimentally. For tall buildings and structures in a turbulent boundary layer flow, the critical amplitude was reported by Kwok and Melbourne (1981) to be as low as 0.6 % of the diameter for circular cylindrical structures and about 2.5 % for prismatic structures.

When lock-in becomes established, the crosswind forces acting on the structure are well organised and well correlated along the span of the structure. The fluctuating crosswind force at any time *t* has a discrete frequency close to the natural frequency of the structure and the vortex shedding frequency, and may be assumed to be sinusoidal and takes the form

$$F'_s(t) = \sqrt{2}C_{FS}\frac{1}{2}\rho\bar{u}^2 b\sin(2\pi n_s t) \tag{5.42}$$

in which C_{FS} is a sinusoidal lock-in force coefficient. This force coefficient is non-linear and is dependent on parameters such as cross-sectional shape, turbulence, crosswind response amplitude and operating reduced velocity. As a first approximation, the non-linearity may be ignored in evaluating the response of structures due to lock-in excitation. The crosswind response due to lock-in excitation of a structure with a linear first mode shape is then

$$\frac{\sigma_y}{b} = \frac{C_{FS}\rho b^2 h}{2(1+\gamma)\zeta_s m\left[4\pi\frac{n_0 b}{\bar{u}(h)}\right]^2} \tag{5.43}$$

The standard deviation response is proportional to ζ^{-1} compared with $\zeta^{-\frac{1}{2}}$ when wake excitation is the dominant mechanism. For tall, slender and flexible circular-sectioned structures such as chimney stacks, a more accurate prediction of lock-in response can be achieved by incorporating the non-linear characteristic of the aerodynamic force in the mathematical model such as that proposed by Basu and Vickery (1983).

As far as the significance of lock-in excitation in modern tall buildings and structures is concerned, it depends very much on the ability of other forms of crosswind excitation, such as wake excitation, to generate crosswind response of significant magnitude which enables lock-in to become established. Therefore, it is reasonable to assume that lock-in is more likely to occur for tall buildings and structures which have relatively low stiffness, are lightly damped and are operating near the peak of the wake energy spectrum.

5.3.3.3 Flutter

Flutter is more complex than galloping and lock-in in that motions in two or more degrees-of-freedom are involved, commonly as combined bending and torsional motions. It is similar to galloping excitation in that the fluctuating forces are controlled by the motion of the body, but differs in an important respect that either or both degrees-of-freedom may have stable motions individually and it is the cross-coupling which extracts energy from the wind to produce the combined unstable motions.

Consider a bluff body which is able to move in vertical motion z and rotation θ, that is in bending and torsional motions respectively, the equations of motion can be written as follows (e.g. Scanlan and Tomko 1971):

$$\ddot{z} + 2\zeta_z\omega_z\dot{z} + \omega_z^2 z = \frac{F_z(t)}{m} + H_1^*\dot{z} + H_2^*\dot{\theta} + H_3^*\dot{\theta} \tag{5.44a}$$

$$\ddot{\theta} + 2\zeta_\theta\omega_\theta\dot{\theta} + \omega_\theta^2\theta = \frac{F_\theta(t)}{I} + A_1^*\dot{z} + A_2^*\dot{\theta} + A_3^*\dot{\theta} \tag{5.44b}$$

in which ζ_z, ζ_θ are structural damping ratios in translational and rotational modes, ω_z, ω_θ are the circular natural frequencies in translational and rotational modes, H_i^*, A_i^* $(i = 1, 2, 3)$ are aerodynamic coefficients, or commonly referred to as flutter derivatives, in translational and rotational modes which describe aerodynamic forces due to the motion of the body, and $F_z(t)$, $F_\theta(t)$ are aerodynamic excitations in translational and rotational modes due to other wind excitation mechanisms such as turbulence buffeting and vortex shedding.

Equation (5.44a) and (5.44b) are coupled second order linear differential equations, the solution of which can be readily determined mathematically if A_i^* and H_i^* are assumed to be constant. For certain values of A_i^* and H_i^*, unstable flutter type oscillations may result. The bending and twisting motions of aeroplane wings is a well known example of flutter and is usually called classical flutter. The occurrence of flutter of modern buildings and structures remains relatively rare, and is generally restricted to wing-like structures such as suspension bridge decks, cantilever roofs and cable-stayed roofs. The H-section used in the Tacoma Narrows Suspension Bridge is prone to flutter and galloping, and its failure is probably a combination of both phenomena. Readers are referred to the large volume of information available on the subject of bridge aerodynamics.

The oscillation of an airfoil about its stalling angle is called stall flutter. Strictly speaking, this form of oscillation is not flutter, as it involves only one degree-of-freedom motion. This type of excitation involves a hysteresis in the separation and reattachment of the air flow as the structure oscillates about the stalling angle, and bowl-shaped structures such as radio telescopes and radar reflectors are susceptible to this form of excitation.

5.4 Other Fluctuation Forces and Response Phenomena

5.4.1 Torsional Vibrations

Non-uniform wind pressure distributions, particularly for asymmetrical or complex building shapes, generate mean and dynamic torques (e.g. Isyumov and Poole 1983; Lythe and Surry 1990; Cheung and Melbourne 1992). Dynamic torque and resultant torsional vibrations of tall buildings can also be induced by turbulence buffeting, either associated with the free stream flow or caused by neighbouring tall buildings, vortex shedding, or re-attaching flow onto a long afterbody (e.g. Xu et al. 1992; Beneke and Kwok 1993; Zhang et al. 1993). The types of building most susceptible are tall slab-like buildings and those with complex shapes, particularly those with an eccentricity between the elastic and mass centres.

The structural systems commonly employed in modern tall buildings usually result in a natural frequency in the torsional mode that is significantly greater than the fundamental natural frequency in the translational mode. Hence severe translational vibrations would have been induced well before any significant torsional response is manifested. Provided that the aerodynamics of the building is not unusually unfavourable and the wind resistance structural system is properly designed, the torsional loads can usually be accommodated. However, the torsional vibrations may exaggerate occupant perception of the building motion, particularly at the extremity of the building.

Modern tall buildings and structures increasingly adopt complex shapes and structural systems that tend to present an eccentricity between the elastic and mass centres and introduce a significant unbalanced wind load. Hence, torsional loads and response have assumed a greater importance in the design process. Analytical procedures for the estimation of torsional loads and vibrations have not yet reached a sufficiently advanced stage to be routinely included in current wind codes and standards, and wind tunnel tests remain a useful technique for predicting such a response.

5.4.2 Alongwind and Ovalling Vibrations Induced by Vortex Shedding

The vortex shedding process can induce fluctuating drag forces which cause alongwind vibrations, and fluctuating pressures which cause ovalling vibrations of open-ended thin-walled structures. Alongwind vibrations due to vortex shedding in air flow are rare and the resultant motions are usually small in comparison to vibrations induced by turbulence buffeting.

Open-ended thin-walled structures such as large diameter steel chimneys and storage bins and silos, particularly those unstiffened circumferentially, are most susceptible to ovalling vibrations but the occurrence is rare. During ovalling vibrations, a circular cross-section changes to an oval shape which elongates periodically in the wind direction and then normal to it, or at an angle of $\pm 45^{\circ}$ to the wind direction. If the natural frequency of the structure in the ovalling mode n_o is known, ovalling vibrations can theoretically occur when the vortex shedding frequency is equal to n_o, $n_o/2$, $n_o/3$ and so on. Thus ovalling vibrations are possible at a range of wind speeds which corresponds to the critical wind velocities of

$$u_o = \frac{n_o D}{St}, \frac{n_o D}{2St}, \frac{n_o D}{3St}, \text{ etc.} \tag{5.45}$$

in which D is diameter and St is the Strouhal Number of the structure.

5.5 Wind-Induced Interference Excitation of Tall Buildings

When several structures are grouped closely together, the proximity of one structure to another can influence the flow field around, and the wind forces on, individual members of the group. Such an interference effect is particularly marked for groups of identical or similar structures such as a line of chimneys. Since a group of identical structures have the same natural frequency of vibration and other structural characteristics, they are very susceptible to interference caused by each other. When the vortex shedding frequency of the upstream structure coincides with the natural frequency of the downstream structure, large amplitude vibrations may develop. Such an occurrence is well known for in-line steel chimneys and is referred to as resonant buffeting.

Wake galloping and wake flutter can also occur when a structure is submerged in the wake of another structure. For structures with a cross-sectional shape known to be prone to galloping, the velocity fluctuations in the wake of another structure can enhance the galloping type force. For very flexible structures such as transmission lines and exposed lift cables, coupled alongwind and crosswind motions may develop and these are referred to as wake flutter.

In a typical city environment where many tall buildings and structures are grouped closely together, the proximity of one building to another can influence the flow field around and the wind forces acting on the buildings. The interactions of neighbouring tall buildings, commonly referred to as interference or proximity effects, can cause a significant increase in wind loads and exacerbate the problem of occupant perception of motions due to enhanced dynamic vibrations. Potential increases in building response due to such interference effects have become an important consideration in the design of modern tall buildings. Evidently, the owner of a new building may be held responsible for any adverse wind effect on existing buildings, after what is believed to be the first lawsuit in which Metropolitan Life Insurance Co. alleged the World Trade Centre to have caused "unusual, increased and unnatural wind pressures" on its building nearby. This adds a new dimension in litigation and professional liability.

The alongwind, crosswind and torsional responses of both upstream and downstream buildings in a group may be affected for wind from all directions. Consequently, the effects of interference from adjacent buildings, existing or proposed, are now routinely studied during wind tunnel model testing of tall buildings and structures. A wealth of useful information has been collected through these tests but rarely is this information widely available because of the proprietary nature of those tests. Nevertheless, significant advances in the understanding of the various interference excitation mechanisms and their potentials in causing significant increases in the response of tall buildings have been achieved.

There are many factors affecting the interference excitations of tall buildings. Some of the major factors are: size, shape, and dynamic properties of the buildings, wind velocity and direction, type of approach terrain, and the location and form of the neighbouring buildings. In the past decades, attention has increasingly been given to identify and quantify the effects of these factors on the overall wind-induced loadings and responses of square and rectangular buildings, e.g. Saunders and Melbourne (1980), Ruscheweyh (1980), Sykes (1983), Blessmann (1985), Bailey and Kwok (1985), Kareem (1987), Kwok (1989), Taniike (1991), Yahyai et al. (1992), Zhang et al. (1994, 1995), Khanduri et al. (2000), Thepmongkorn et al. (2002), Tang and Kwok (2004) and others. A detailed review of these investigations can be found in review papers by Kwok (1994) and Khanduri et al. (1998). This section summarises some of the important findings of these investigations.

5.5.1 Alongwind and Crosswind Responses of Tall Buildings Under Interference Excitation

The effects of interference on the alongwind and crosswind responses of tall buildings have been studied by Saunders and Melbourne (1980), Ruscheweyh (1980), Sykes (1983), Bailey and Kwok (1985), Blessmann (1985), Kareem (1987), Kwok (1989), Taniike (1991), Yahyai et al. (1992), Thepmongkorn et al. (2002), Tang and Kwok (2004) and others using aeroelastic techniques. These tests were undertaken in simulated turbulent boundary layer flows ranging from that over flat open terrain to highly turbulent urban terrain. Most of the test buildings were of either square or rectangular cross-section, with a height to breadth ratio ranging from 2 to 9. These tests covered a full range of reduced wind velocities ranging from 2 to 14, which spans design wind speeds typical for serviceability requirement to that well in excess of ultimate limit state strength design.

Interference effects are often expressed as contours of buffeting factor (BF) defined as:

$$BF = \frac{Overturning\ moment\ (neighbouring\ structure\ present)}{Overturning\ moment\ (isolated\ structure)} \tag{5.46}$$

An example buffeting contour plot is presented in Fig. 5.9 for the alongwind and crosswind responses of a building under interference excitation by an identical building in an open country terrain and at a reduced wind velocity of 6. The contours show large increases in alongwind and crosswind responses of up to 80 % for an upstream interference situation and more than 400 % in the alongwind response when the interfering building was in a critical downwind position.

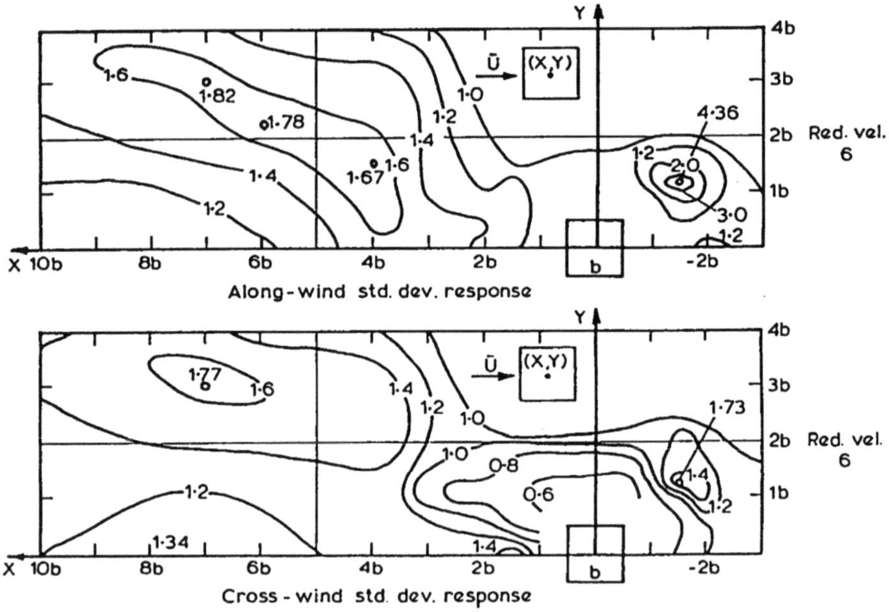

Fig. 5.9 Buffeting factors caused by an identical interfering building (after Bailey and Kwok 1985)

Interference excitation mechanisms were studied in detail by Bailey and Kwok (1985) and Kwok (1989) who examined the force and wake spectra of the principal building. The major excitation mechanism is enhanced turbulence buffeting or resonant buffeting, with high values of both alongwind and crosswind buffeting factors fanning out diagonally from the principal building, as shown in Fig. 5.9, suggesting the excitation forces intensify near the edge of the wake of the upstream building.

A downstream interfering building generally has very little effect on the principal building except for a small critical region, as shown in Fig. 5.9. When the spacing between the two buildings was at a critical width which produces a channel wide enough for the wind to converge and stream between the two buildings, the significantly altered wake characteristics of the principal building induced resonant type response for the principal building upstream. The principal building underwent large elongated elliptical motions, as shown in Fig. 5.10, along the diagonal line between the two buildings.

When the vortex shedding frequency of the upstream interfering building coincides with the natural frequency of the principal building, resonant buffeting occurs with large increases in excitation forces and responses. The critical reduced wind velocity at which resonant buffeting occurs depends on the vortex shedding characteristics. Very large increases in both alongwind and crosswind responses have been observed in wind tunnel studies using an upstream building of different

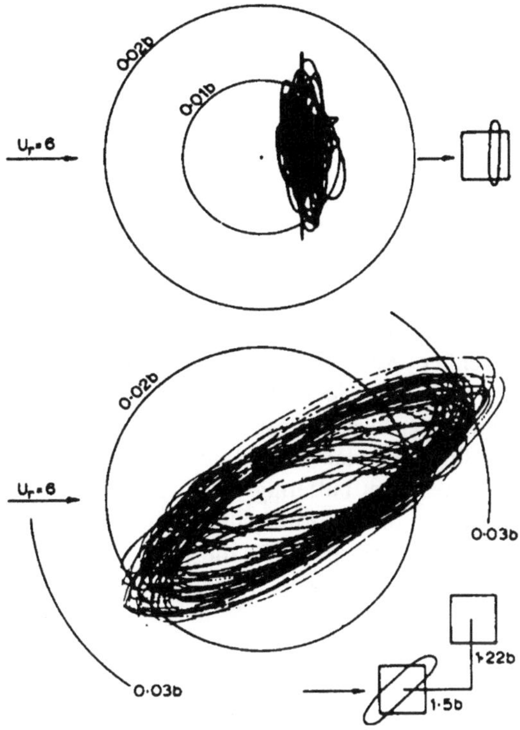

Fig. 5.10 Interference caused by a downstream building (after Bailey and Kwok 1985)

shape and smaller windward dimension to that of the principal building, e.g. Bailey and Kwok (1985), and Taniike (1991). Furthermore, such resonant buffeting has been found to occur at lower and more frequently occurring wind speeds than that for an isolated building.

5.5.2 *Torsional Response of Tall Buildings Under Interference Excitation*

Zhang et al. (1994) studied the torsional response of tall buildings under interference excitation by an interfering building of different shapes and sizes. The results show that interference caused by adjacent tall buildings significantly increases the torsional response of the principal building, depending on the reduced wind velocity, building separations and positions, and the shape and size of the interfering building. The torsional interference excitation mechanisms were found to be similar to those for the alongwind and crosswind responses and are dominated by resonant buffeting which occurs when the frequency of vortex shedding from the upstream building coincides with the natural torsional frequency of the principal building.

Zhang et al. (1995) and Thepmongkorn et al. (2002) further showed that a 10 % eccentricity between the mass and elastic centres of a building substantially enhanced the effects of interference.

5.5.3 Design Considerations

The potentials of interference excitation in causing large increases in alongwind, crosswind and torsional responses of tall buildings have clearly been demonstrated for either upstream or downstream interfering building. Upstream interference can be felt at a considerable distance away and is most prominent when the principal building is located near the edge of the wake shed by the upstream interfering building. For plan shape buildings, vortices shed by the interfering building can cause resonant buffeting when the shedding frequency coincides with the natural frequency of the principal building, potentially at a lower and more frequently occurring wind speed. On the other hand, downstream interference is confined to a relatively small area which produces a critical separation between adjacent buildings in close proximity.

The magnitude of increase of wind load and response due to interference is strongly dependent on the turbulence characteristic of the approach terrain. Tall buildings situated near the waterfront or near the edge of a city centre are the most vulnerable. For tall buildings located well within a city centre with its highly turbulent wind, the effects of interference are significantly reduced. However, when only a few tall buildings or other tall structures protrude well above the surrounding building canopy, the potential for mutual interference is not diminished and should be assessed in accordance with the wind flow condition above the canopy.

The range of reduced wind velocities within which interference effects have been found to be significant encompasses wind speed relevant for serviceability consideration, particularly in relation to occupant comfort, to that well in excess of ultimate limit state design requirement.

A set of generalised percentage increase contour diagrams for interference effects in open, suburban and urban terrain has been proposed by Kwok (1989) and is presented in Fig. 5.11. This can be used for a preliminary assessment of the potential effects of interference of an existing building on a proposed building, or vice versa, for buildings of similar shape and height. As a general rule, an increase in wind load of between 10 % and 30 % may be expected to be within the normal safety factor in design. However, an increase of between 30 % and 50 %, while not necessarily catastrophic, would be of considerable concern, particularly in satisfying the serviceability requirement in terms of occupant comfort. An increase of greater than 50 %, let alone the more than doubling in wind load which has been observed in some of the wind tunnel experiments, will be well in excess of the safety margin normally expected in modern design practice.

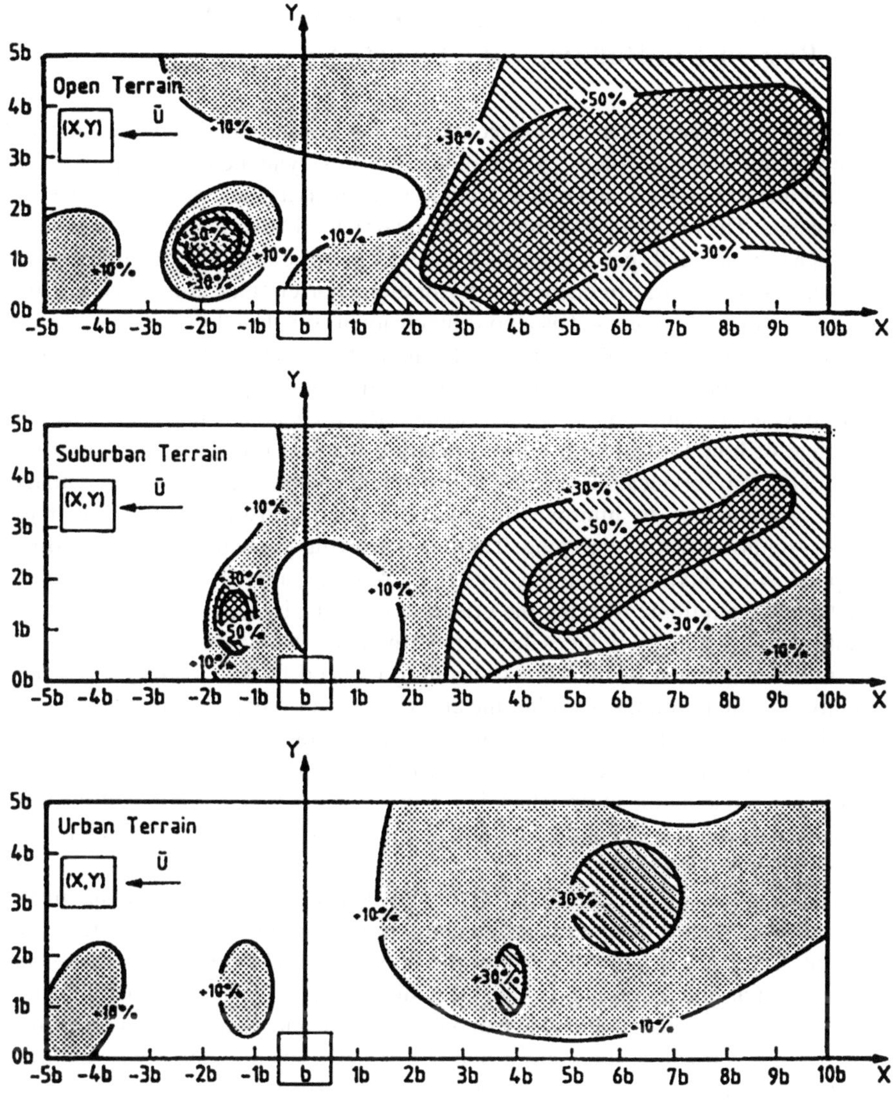

Fig. 5.11 Increase in dynamic load due to interference (after Kwok 1989)

Although the interference excitation mechanisms and resultant increases in wind load and responses are well understood, the complex nature of the interference effects and the influence of widely varying parameters, such as building shape and height, and approach terrain, make any codification difficult. Consequently, detailed investigation of the effects of interference from adjacent buildings, both existing and proposed, is now routinely part of the comprehensive wind tunnel model test program for the design of modern tall buildings and structures.

5.6 Aerodynamic Measures to Reduce Wind-Induced Response of Buildings and Structures

Aerodynamic measures can offer effective alternatives for reducing both static and dynamic wind-induced forces. A common source of dynamic excitation is due to the shedding of von Karman vortices from a structure as a whole or from its external components. The resulting dynamic crosswind or lateral "lift" response can become large at a critical wind speed when the frequency of vortex shedding is in resonance with one of the natural frequencies of vibration. For circular cross-sections, this critical wind speed, expressed as a non-dimensional or reduced velocity, is approximately 5. For buildings and members of square cross-section, this critical reduced velocity is about 10. Very tall slender and flexible buildings tend to operate in a reduced velocity range of 5–10 and sometimes higher, and designers face the challenge of how to deal with vortex shedding induced effects. These can affect strength design, serviceability issues and can create a potential for fatigue damage.

One approach is to improve the dynamic characteristics by altering mass, stiffness and/or effective damping. There is a growing population of tall buildings equipped with auxiliary damping systems. Another approach is to search for aerodynamic solutions. Altering the effective width D, by tapering the building or by introducing setbacks, reduces the spatial organization of the vortex shedding and lowers the magnitude of the vortex shedding induced forces. A procedure found effective for tall buildings is to alter the cross-section or "shape" along the height of a building or along the span of an exposed member. Various aerodynamic treatments, including strakes, spoilers, shrouds and others, have also been found effective in reducing the vortex shedding induced forces for chimneys and stacks of circular cross-section. A detailed review of such aerodynamic treatments has been made by Zdrakovich (1981).

The designers of tall buildings often find it difficult to use aerodynamically advantageous shapes or to add aerodynamic mitigating measures or devices. The planned occupancy usually determines the floor plan, and therefore the shape, and concerns for aesthetics often preclude the addition of passive or active aerodynamic devices and appendages, which have to be of a significant relative size in order to be effective. Nevertheless, there are a number of aerodynamic measures which can be readily adopted by designers of tall buildings in order to reduce the wind-induced responses. Some of these are discussed below.

5.6.1 Effects of Aerodynamic Modifications of Building Shapes on Wind Induced Response of Tall Buildings

Wind tunnel experiments were conducted by Kwok and Bailey (1987), Kwok et al. (1988), and Kwok (1988) to investigate the effects of aerodynamic devices, building edge configurations, and through building opening on the wind induced vibrations of tall buildings. Aeroelastic models of buildings with square and rectangular cross-section and a height to breadth ratio of 9 have been tested to examine the effects of fins and vented fins at the corners, slotted corners, chamfered corners, and horizontal slots. The modifications to the building corners ranged from around 9–16 % of the building breadth. Evidently aerodynamic modifications which increase the projected area or the effective width of a building, such as fins at corners, are generally not beneficial.

Horizontal slots, slotted corners and chamfered corners were found to be effective in causing significant reductions in the dynamic alongwind and crosswind responses of the rectangular cross-section CAARC Standard Tall Building. With chamfered corners, there was up to a 40 % reduction in the alongwind response within a reduced wind velocity range of 3–20 compared with that for the plain building.

The effects of building plan shape on the crosswind response of tall buildings have also been investigated by Melbourne and Cheung (1988) and Melbourne (1989). It was concluded from wind tunnel aeroelastic model studies and sample calculations for practical applications that for tall buildings with nominally symmetrical or square plan shapes, modest rounding or chamfering of corners up to 10 % does not significantly reduce the crosswind response at low values of reduced wind velocities relevant to the serviceability acceleration criterion for occupant comfort. However, significant reductions in the ultimate limit state design loads can be achieved at higher reduced velocities. For more substantial rounding or chamfering of corners such that the building plan shape approaches that of a rough circular shape, for example octagonal or hexagonal shape, the crosswind force spectrum was found to reduce at both the lower and higher ranges of reduced velocities. A similar characteristic was also observed for tall buildings with reduced upper level plan areas, achieved by tapering, cutting or dropping off corners progressively with increasing increases. However, for building plan shapes which are elliptical or elongated octagonal with a major to minor axis ratio of about 3:2, the critical reduced wind velocities were found to be significantly lower. The resultant high crosswind response and acceleration level make it more difficult to meet occupancy comfort criteria. Furthermore, these plan shapes were found to exhibit significant torsional response about the vertical axis.

5.6.2 Through Openings and Advantages of Base Bleed

One promising aerodynamic treatment is the "base-bleed" technique, where air is injected into the wake cavity or the "base-region" in order to increase the mean base pressure. This stabilises the wake and, depending on the change in base pressure, either disrupts or totally eliminates organized or "narrow-band" vortex shedding. This can be achieved by introducing "through openings" which allow the positive pressure on the windward face to be communicated or "bled" into the base region. Experimental data to illustrate the effectiveness of this technique are presented by Isyumov et al. (1989) and Dutton and Isyumov (1990) in which wind tunnel model tests were carried out to examine the effect of through openings on the wind forces acting on a tall building with a square cross-section in turbulent boundary layer flow. The model was mounted on a force balance which measured the mean and dynamic drag and crosswind base overturning moments. Measurements were also made with all gaps closed in order to provide a comparison.

The introduction of the "gaps" had a minimal effect on the drag forces. However, there was a dramatic reduction in the crosswind dynamic forces, which occur for winds approximately normal to one of the building faces. Alongwind through building openings and in particular combined alongwind and crosswind openings were found to cause a pronounced reduction of the vortex shedding induced forces and hence the crosswind dynamic deflection of the building. The critical reduced wind velocity shifted to a slightly higher value which implied that resonant vibrations of the building would be postponed to a higher wind speed with a longer return period. However, there was a second spectral peak and hence a corresponding increase in crosswind response at a reduced velocity of around 6 which may make it more difficult to meet occupancy comfort criteria. The level of disruption to the vortex shedding process varied with the width of the opening and large reductions were observed for openings as small as 4 % of the building width. Results obtained from a pressure model study suggested the mean base pressure coefficient is a good indicator of the overall aerodynamic, including the vortex shedding process, of a tall building.

5.6.3 Influence of Building Top, Tapering, Twisting, Setback and Other Composite Building Shapes

The potential beneficial effects of a "sculptured" building top of a 390 m high office tower project were studied in a wind tunnel using both force balance and aeroelastic model techniques and reported by Isyumov et al. (1992). The results show that a more sculptured building top, in this case an open triangular pyramid shape for the top 15 % of the building, can moderate both the alongwind and crosswind responses. A venting or "bleeding" of the building wake provided by a through building opening near the top can result in additional reductions in these responses.

More recently, Kim et al. (2011) and Tanaka et al. (2012) conducted comprehensive wind tunnel experiments to further understand the aerodynamic characteristic of tall buildings with various shape configurations including models that were tilted, winding, tapered, helical, opening and various composite models.

5.6.4 Design Considerations

Aerodynamic modifications of building shape, including slotted and chamfered corners, horizontal and vertical through building openings, sculptured building top, tapering, and dropping off corners have been shown to significantly reduce the wind induced loads and responses of tall buildings. Reductions of up to 40 % in the alongwind response can be achieved by aerodynamic modifications for a wide range of reduced wind velocities which encompass serviceability and ultimate limit state design wind speeds for most modern tall buildings. Crosswind response at serviceability design wind speed can be reduced by around 30 % while very substantial reductions can be achieved at the ultimate limit state design wind speed. In general, modifications to the building corners such as slotted or chamfered corners need to be applied to the corner region greater than about 10 % of the building breadth to be beneficial. However, modifications which increase the projected area or the effective breadth of the building would in general not be beneficial.

The vortex shedding characteristics of a building are often altered by aerodynamic modifications. If the Strouhal Number is lowered by the modification, the critical reduced wind velocity at which the vortex shedding frequency coincides with the natural frequency of the building will be raised to a higher value with a longer return period. On the other hand, if the modification causes an increase in the Strouhal Number, resonant response will occur at a lower reduced velocity with a shorter return period which may increase the ultimate limit state design load or make it more difficult to meet serviceability requirements such as occupant comfort criteria. Occasionally secondary spectral peak may appear and the corresponding increase in crosswind response may adversely affect the performance of the building.

Evidently, the wind load and wind induced response characteristics of a tall building are dependent on the basic plan shape of the building. The potential benefit of a plan shape which has a lower Strouhal Number has already been mentioned and is a parameter which can offer significant benefit when correctly selected. In the design of modern tall buildings, particular for those with an unusual plan shape and for super tall buildings, a comprehensive program of wind tunnel model test is always beneficial and provides useful information which can be used to assess the potential benefits of altering the plan shape or employing other forms of aerodynamic modifications to the building. However, an aerodynamically efficient plan shape may come at the cost of usable floor area that, in turn, may require additional compensatory storeys, which obviously increase the wind loads and construction cost again. On one hand, a less-efficient planform in terms of interior space-planning usually attracts lower rental or sales returns; on the other hand, additional upper level storeys have the potential to increase revenue. Therefore, it is not straightforward for engineers to decide on aerodynamic modifications, such as the size and type of corner modifications, with regard to both the structural responses and the effects on economic rewards.

Tse et al. (2009) conducted wind tunnel test to investigate variations of a 1:400 scale reference building, having a basic square shape with dimensions equivalent to 48 m $\times$ 48 m $\times$ 240 m ($B \times D \times H$) in full-scale, where different sizes of corner modifications were sequentially introduced to the reference building. To maintain the same usable floor area for the entire building, additional storeys are required to compensate for any losses of the floor area due to the corner modifications, resulting in changes to the building heights and hence overall aspect ratio (H/B) and natural frequencies for each modified-corner building configuration. A simplified cost model that took into consideration construction cost and return on investment, including floor-dependent rental return, annual rental increase, vacancy rate and bad debt losses, was adopted to determine the cost-benefit of aerodynamic modifications and to demonstrate the complexity of decision making based on commercial and financial factors. More realistic cost models that include inflation, operating cost and building depreciation can provide more accurate cost-benefit analysis for practical applications.

5.7 Conclusion

A detailed description of the alongwind and crosswind excitation mechanisms and response processes for buildings and structures has been presented. Turbulence buffeting is the dominant excitation mechanism in the alongwind direction. A gust factor approach, which is based on statistical and spectral analyses, has been outlined and is widely used in the prediction of alongwind response for a wide range of buildings and structures.

The crosswind excitation mechanisms and response processes of buildings and structures are considerably more complex than those in the alongwind direction. Excitations associated with incident turbulence, vortex shedding and motion dependent excitations are the prominent mechanisms causing crosswind response. Simplified methods for determining crosswind response of simple tall buildings and structures have been presented.

Recent advances in the understanding of interference excitations and the associated increases in the response of tall buildings, and the potential beneficial effects of aerodynamic modifications of building shape in reducing the wind induced response of tall buildings have been reviewed in this paper. Interference excitation from either an upstream or a downstream building can cause significant increases in the alongwind, crosswind and torsional responses of tall buildings. Effects of interference from adjacent buildings, both existing and proposed, should be carefully assessed in the design of tall buildings.

Aerodynamic modifications of building shape, including slotted and chamfered corners, horizontal and vertical through building openings, sculptured building top, tapering, and dropping off corners have been shown to significantly reduce the wind induced loads and responses of tall buildings. For modern tall buildings, particularly the new generation of super tall buildings, these aerodynamic modifications may be employed to meet serviceability requirements such as occupant comfort criteria and to reduce the ultimate limit state design wind load. However, aerodynamic modifications may come at the cost of reducing both the size and value of saleable floor area. In practice, a proper evaluation of the construction cost and investment return may be necessary to assess the impacts of aerodynamic modifications on the financial return of tall building projects.

Acknowledgements The author acknowledges the contributions to the research findings presented in this chapter by his research collaborators and research students at the University of Sydney and the Hong Kong University of Science and Technology for the past 30 years. Their contributions have been invaluable in advancing our understanding of wind-structure interactions and tall building aerodynamics.

References

AS/NZS 1170.2:2011 (2011) Structural design actions Part 2: Wind actions. Standards Australia/ Standards New Zealand

Bailey PA, Kwok KCS (1985) Interference excitation of twin tall buildings. J Wind Eng Ind Aerodyn 21:323–338

Basu RI, Vickery BJ (1983) Acrosswind vibrations of structures of circular cross-section, part II: development of a mathematical model for full-scale application. J Wind Eng Ind Aerodyn 12:75–97

Beneke DL, Kwok KCS (1993) Aerodynamic effect of wind induced torsion on tall buildings. J Wind Eng Ind Aerodyn 50:271–280

Blessmann J (1985) Buffeting effects on neighbouring tall buildings. J Wind Eng Ind Aerodyn 18:105–110

Cheung JCK, Melbourne WH (1992) Torsional moments of tall buildings. J Wind Eng Ind Aerodyn 41–44:1125–1136

Davenport AG (1962) Buffeting of a suspension bridge by storm winds. ASCE J Struct Div 88 (ST2):233–268

Davenport AG (1967) Gust loading factors. ASCE J Struct Div 93(ST2):11–34

Dutton R, Isyumov N (1990) Reduction of tall building motion by aerodynamic treatments. J Wind Eng Ind Aerodyn 36:739–747

Isyumov N, Poole M (1983) Wind induced torque on square and rectangular building shapes. J Wind Eng Ind Aerodyn 13:183–196

Isyumov N, Dutton R, Davenport AG (1989) Aerodynamic methods for mitigating wind-induced building motions. In: Proceedings of ASCE structures congress, San Francisco

Isyumov N, Fediw AA, Colaco J, Banavalkar PV (1992) Performance of a tall building under wind action. J Wind Eng Ind Aerodyn 41–44:1053–1064

Kareem A (1987) The effect of aerodynamic interference on the dynamic response of prismatic structures. J Wind Eng Ind Aerodyn 25:365–372

Khanduri AC, Stathopoulos T, Bédard C (1998) Wind-induced interference effects on buildings—a review of the state-of-the-art. Eng Struct 20(7):617–630

Khanduri AC, Stathopoulos T, Bedard C (2000) Generalization of wind-induced interference effects for two buildings. Wind Struct 3(4):255–266

Kim YC, Kanda J, Tamura Y (2011) Wind-induced coupled motion of tall buildings with varying square plan with height. J Wind Eng Ind Aerodyn 99:638–650

Kwok KCS (1982) Crosswind response of tall buildings. Eng Struct 4(10):256–262

Kwok KCS (1988) Effect of building shape on wind-induced response of tall buildings. J Wind Eng Ind Aerodyn 28:381–390

Kwok KCS (1989) Interference effects on tall buildings. In: Sun TF (ed) Proceedings of 2nd Asia-Pacific symposium on wind engineering, vol 1. Beijing, pp 446–453

Kwok KCS (1994) Aerodynamics of tall buildings. In: A state of the art volume on wind engineering, Wiley Eastern Limited, Delhi, 1994, pp 180–204

Kwok KCS, Bailey PA (1987) Aerodynamic devices for tall buildings and structures. ASCE J Eng Mech 113(3):349–365

Kwok KCS, Melbourne WH (1980) Freestream turbulence effects on galloping. ASCE J Eng Mech Div 106((EM2)):273–288
Kwok KCS, Melbourne WH (1981) Wind-induced lock-in excitation of tall structures. ASCE J Struct Div 107(STl):57–72
Kwok KCS, Wilhelm PA, Wilkie BG (1988) Effect of edge configuration on wind-induced response of tall buildings. Eng Struct 10(4):135–140
Lythe GR, Surry D (1990) Wind induced torsional loads on tall buildings. J Wind Eng Ind Aerodyn 36:225–234
Melbourne WH (1977) Crosswind response of structures to wind action. In: Eaton KJ (ed) Proceedings 4th international conference on wind effects on buildings and structures, Heathrow, 1977, Cambridge University Press, Cambridge, pp 343–358
Melbourne WH (1989) Wind loading and environment of tall building. In: Proceedings of international conference on tall buildings and city development, Brisbane, pp 159–164
Melbourne WH, Cheung JCK (1988) Designing for serviceable accelerations in tall buildings. In: Proceedings of 4th international conference on tall buildings, vol 1, Hong Kong and Shanghai, pp 148–155
Novak M (1972) Galloping oscillations of prismatic structure. ASCE J Eng Mech Div 98((EM 1)):27–46
Parkinson GV (1971) Wind-induced instability of structures. Philos Trans R Soc Lond A 269:395–409
Ruscheweyh H (1979) Dynamic response of high rise buildings under wind action with interference effects from surrounding buildings of similar size. In: Cermak JE (ed) Proceedings of 5th international conference on wind engineering, Fort Collins, Pergamon Press, Oxford, pp 725–734
Saunders JW, Melbourne WH (1977) Tall rectangular building response to crosswind excitation. In: Eaton KJ (ed) Proceedings 4th international conference on wind effects on buildings and structures, Heathrow, 1975, Cambridge University Press, Cambridge, pp 369–80
Saunders JW, Melbourne WH (1980) Buffeting effects of upstream buildings. In: Cermak JE (ed) Proceedings of 5th international conference on wind engineering, Fort Collins, Pergamon, Oxford, pp 593–606
Scanlan RH, Tomko JJ (1971) Airfoils and bridge decks flutter derivatives. ASCE J Eng Mech Div 97:1717–1737
Scruton C, Walshe DEJ (1957) A means of avoiding wind excited oscillations of structures of circular or near circular cross-section. National Physical Laboratory Aero Report 335
Shum KM, Kwok KCS, Hitchcock PA (2009) Galloping effects on inclined square cylinder in turbulent flow. In: Proceedings of 7th Asia-Pacific conference on wind engineering, Taipei (Full paper on CD-ROM)
Sykes DM (1983) Interference effects on the response of a tall building model. J Wind Eng Ind Aerodyn 11:365–380
Tanaka H, Tamura Y, Ohtake K, Nakia M, Kim YC (2012) Experimental investigation of aerodynamic forces and wind pressures acting on tall buildings with various unconventional configurations. J Wind Eng Ind Aerodyn 107–108:179–191
Tang UF, Kwok KCS (2004) Interference excitation mechanisms on a 3-DOF Aeroelastic CAARC Building Model. J Wind Eng Ind Aerodyn 92:1299–1314
Taniike Y (1991) Turbulence effect on mutual interference of tall buildings. ASCE J Eng Mech 117(3):443–456
Thepmongkorn S, Wood GS, Kwok KCS (2002) Interference effects on wind-induced coupled motion of a tall building. J Wind Eng Ind Aerodyn 90:1807–1815
Tse KT, Hitchcock PA, Kwok KCS, Thepmongkorn S, Chan CM (2009) Economic perspectives of aerodynamic treatments of square tall buildings. J Wind Eng Ind Aerodyn 97:455–467
Vickery BJ (1966) On the assessment of wind effects on elastic structures. Civil Engineering Transaction, Institution of Engineers, Australia, pp 183–192

Vickery BJ, Clark AW (1972) Lift and acrosswind response of tapered stacks. ASCE J Struct Div 98((ST1)):1–20
Xu YL, Kwok KCS, Samali B (1992) Torsional response and vibration suppression of wind-excited buildings. J Wind Eng Ind Aerodyn 43:1997–2008
Yahyai M, Kumar K, Krishna P, Pande PK (1992) Aerodynamic interference in tall rectangular buildings. J Wind Eng Ind Aerodyn 41–44:859–866
Zhang WJ, Kwok KCS, Xu YL (1994) Aeroelastic torsional behaviour of tall buildings in wakes. J Wind Eng Ind Aerodyn 51:229–248
Zhang WJ, Xu YL, Kwok KCS (1993) Torsional vibration and stability of wind-excited tall building with eccentricity. J Wind Eng Ind Aerodyn 50:299–308
Zhang WJ, Xu YL, Kwok KCS (1995) Interference effects on aeroelastic torsion response of structurally asymmetric tall buildings. J Wind Eng Ind Aerodyn 57:41–61
Zdrakovich MM (1981) Review and classification of various aerodynamic and hydrodynamic means for suppressing vortex shedding. J Wind Eng Ind Aerodyn 7:145–189

Chapter 6
The Gust Factor Approach to Evaluate the Along-Wind Response of Structures to Wind Excitation

Richard G.J. Flay

Abstract This chapter describes a relatively well developed method for determining the along-wind response of tall buildings to wind excitation using a theoretical approach. It has been found to give good predictions, and is now the basis of along-wind response predictions in some wind loading standards and codes. It is based on the premise that the along-wind motion of a relatively slender building is driven primarily by the onset turbulence, which excites the building as a single degree of freedom system with low damping.

The gust factor approach described in this chapter relies heavily on the previous work of researchers such as Vickery, Davenport and others, who developed this theory around 50 years ago.

Keywords Along-wind excitation • Along-wind response • Gust factor • Tall building motion

6.1 Wind Action on a Point-Like Structure Considered in the Time Domain

Consider the action of the wind on a relatively small (point-like) object, as shown in Fig. 6.1, where the bulk of the energy of the turbulence is at wavelengths much greater than a typical dimension of the structure.

In this analysis, it is assumed that the drag coefficient, frontal area and fluid density are all constant. Since wind speed u is a function of time, so is $F(t)$, the force on the object.

R.G.J. Flay (✉)
University of Auckland, Private Bag 92019 Auckland 1142, New Zealand
e-mail: r.flay@auckland.ac.nz

Y. Tamura and A. Kareem (eds.), *Advanced Structural Wind Engineering*,
DOI 10.1007/978-4-431-54337-4_6, © Springer Japan 2013

Fig. 6.1 Wind acting on a small point-like object

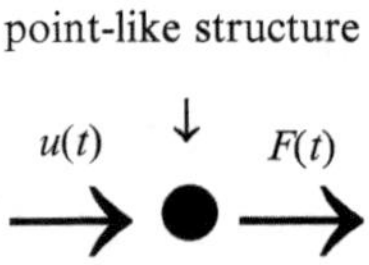

C_D - drag coefficient

A - frontal area

ρ - density of fluid

$$F(t) = C_D A \frac{1}{2} \rho u(t)^2 \tag{6.1}$$

Since the wind speed varies with time, it can be decomposed into a mean component and a fluctuating component. Here the mean is denoted by an overbar, and the fluctuating component is denoted with a prime subscript, following the Reynolds decomposition approach.

$$u(t) = \bar{u} + u'(t) \tag{6.2}$$

Now substitute Eq. (6.2) into (6.1).

$F(t) = C_D A \frac{1}{2} \rho (\bar{u} + u'(t))^2$ and expanding the squared term gives

$$F(t) = C_D A \frac{1}{2} \rho \left(\bar{u}^2 + 2\bar{u}u'(t) + u'(t)^2 \right) \tag{6.3}$$

The mean value of $F(t)$ is given by Eq. (6.4).

$$\bar{F} = \overline{F(t)} = C_D A \frac{1}{2} \rho \overline{\left(\bar{u}^2 + 2\bar{u}u'(t) + u'(t)^2 \right)} \tag{6.4}$$

Which can be simplified to give Eq. (6.5).

$$\bar{F} = C_D A \frac{1}{2} \rho \left(\bar{u}^2 + \underbrace{\overline{u'(t)^2}}_{\text{small}} \right) \tag{6.5}$$

The fluctuating part of the wind force can be found by subtracting the mean force off the instantaneous force which after simplification results in Eq. (6.6).

$$\begin{aligned} F'(t) &= F(t) - \bar{F} \\ &= C_D A \frac{1}{2} \rho \left(\bar{u}^2 + 2\bar{u}u'(t) + u'(t)^2 - \bar{u}^2 - \overline{u'(t)^2} \right) \end{aligned}$$

$$F'(t) = C_D A \rho \bar{u} u'(t) + C_D A \frac{1}{2} \rho \left(\underbrace{u'(t)^2 - \overline{u'(t)^2}}_{\text{normally small and is ignored}} \right) \tag{6.6}$$

If the turbulence intensity is say 10 %, then the second term on the right hand side of Eq. (6.6) will be only approximately 10 % of the size of the first term, and so can be ignored to keep the analysis straightforward. Subsequently a correction is added back in to account for the case when the turbulence intensity is high (say larger than 20 %). With the second term on the right hand side of Eq. (6.6) ignored, the following linear approximation is obtained.

$$F'(t) = (C_D A \rho \bar{u}) u'(t) \tag{6.7}$$

Now the instantaneous fluctuations $u'(t)$ above are part of a whole spectrum of fluctuations in velocity at different frequencies. Hence the action of the wind and its resulting forces in the frequency domain is now considered. Information on random data processing and transferring from the time to frequency domains and vice versa can be found in Bendat and Piersol (2011).

6.2 Wind Action on a Point-Like Object Considered in the Frequency Domain

Since $F'(t)$ and $u'(t)$ are functions of time related by a linear equation, the Fourier transform of both sides can be taken in order to obtain an expression in the frequency domain. Then at each frequency, the magnitude squared can be determined from the real and imaginary components to give the power spectrum. This can be expressed as:

$$\text{spectrum of } F'(t) = \text{spectrum of } u'(t)^* \text{ constant.}$$

The "constant" comes from Eq. (6.7) and is $(C_D A \rho \bar{u})^2$, and so the frequency domain expression becomes

$$S_F(f) = (C_D A \rho \bar{u})^2 S_u(f) \tag{6.8}$$

Since $F'(t)$ has the dimensions of [N], then $S_F(f)$ has the dimensions [N^2/Hz] and is proportional to *amplitude*^{2} of force fluctuations with frequency f over the differential frequency band *df*.

Equation (6.5) can be simplified to yield

$$\bar{F} = C_D A \frac{1}{2} \rho \bar{u}^2 \tag{6.9}$$

which can be rearranged to

$$C_D A\rho\bar{u} = \frac{2\bar{F}}{\bar{u}}$$

and then substituted back into Eq. (6.8) to give Eq. (6.10), which has terms which only result from force and wind speed.

$$S_F(f) = \frac{4\bar{F}^2}{\bar{u}^2} S_u(f) \tag{6.10}$$

This equation is quite satisfactory for gusts which completely envelope the structure, i.e. for the small point-like object considered in this analysis so far. However, now consider the situation where the object grows in size to become large, larger than some of the gusts. In such a situation, Eq. (6.10) would obviously overestimate the force for small gusts which are not completely correlated over the entire surface of the large object. It can be argued that this "lack of correlation" is related to the size of a gust with respect to the size of the structure. This thinking leads to the concept of the "aerodynamic admittance", as introduced by Vickery, Davenport and others (Davenport 1967; Vickery 1965, 1968).

6.3 The Aerodynamic Admittance

A gust wavelength can be defined by

$$\lambda = \frac{\bar{u}}{f} \tag{6.11}$$

where $f =$ frequency in Hz, $\bar{u} =$ mean speed, and if A represents the area of the object, then a characteristic width dimension for the structure is $\sqrt{A}$. Therefore how much gust force is transferred to a structure should be dependent on the ratio $\frac{\lambda}{\sqrt{A}} = \left(\frac{f\sqrt{A}}{\bar{u}}\right)^{-1}$. The effectiveness of a structure in capturing the wind is determined by its aerodynamic admittance, χ, which is a function of $\frac{\lambda}{\sqrt{A}}$, i.e. $\chi^2\left(\frac{f\sqrt{A}}{\bar{u}}\right)$.

The research that was carried out by Davenport, Vickery and others and referenced above, used both theoretical and experimental approaches to determine the shape of the function $\chi^2\left(\frac{f\sqrt{A}}{\bar{u}}\right)$. It was found that the empirical relationship given in Eq. (6.12) fitted the available information, and it is commonly used today.

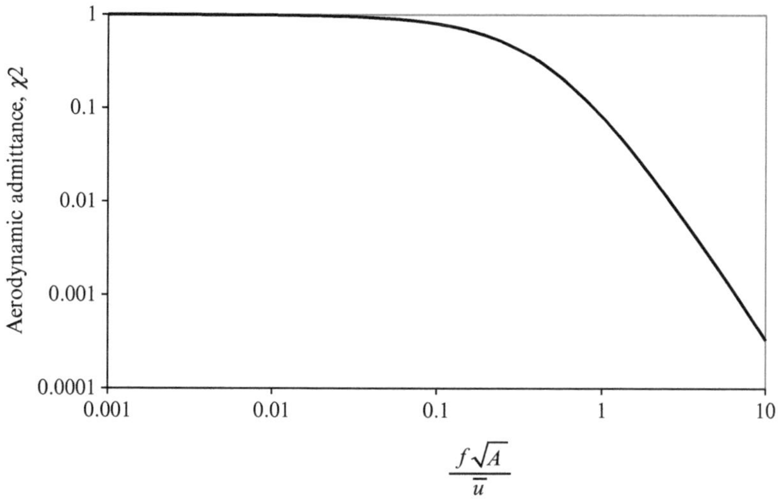

Fig. 6.2 Aerodynamic admittance

$$\chi^2\left(\frac{f\sqrt{A}}{\bar{u}}\right) \approx \left[\frac{1}{1+\left(\frac{2f\sqrt{A}}{\bar{u}}\right)^{\frac{4}{3}}}\right]^2 \tag{6.12}$$

Equation (6.12) has the shape shown in Fig. 6.2.

The spectrum of force actually transmitted to a "large" structure thus becomes the product of Eqs. (6.10) and (6.12).

$$S_F(f) = 4\frac{\overline{F}^2}{\bar{u}^2}S_u(f)\chi^2\left(\frac{f\sqrt{A}}{\bar{u}}\right) \tag{6.13}$$

This force spectrum will cause displacements in the structure. Obviously the effectiveness of the wind forces at producing a response of the structure will depend on the range of wind excitation frequencies in relation to the natural frequency of the structure. Hence the dynamic behaviour of the structure has to be considered. This is done for a simple single degree of freedom structure in the following section.

6.4 Review of the Dynamic Response of Single Degree of Freedom (SDF) Systems

Consider a single degree of freedom system consisting of a mass m connected to the ground through a spring of stiffness k, and a damper with a damping constant c, and subjected to an excitation force $F(t)$, as shown schematically in Fig. 6.3.

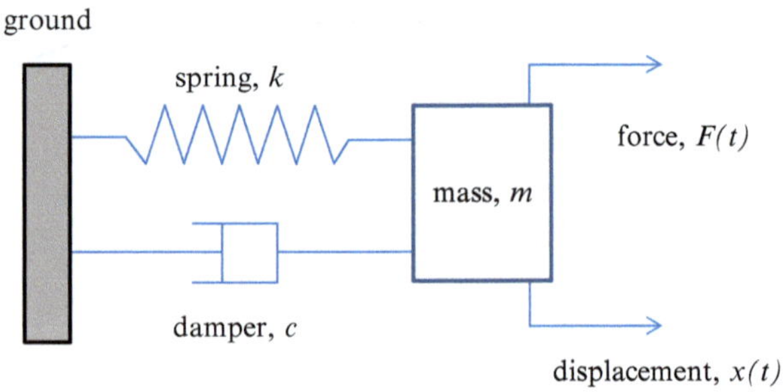

Fig. 6.3 Schematic diagram of a single degree of freedom system

From a free body diagram (FBD) it can be shown that

$$m\ddot{x} + c\dot{x} + kx = F(t) \tag{6.14}$$

The transient response is obtained by setting $F(t) = 0$. The solution is called the complementary function. Thus

$$m\ddot{x} + c\dot{x} + kx = 0 \tag{6.15}$$

$$\text{or } \left(mD^2 + cD + k\right)x = 0, \text{ where } D = d/dt.$$

to solve let $x = Ae^{qt}$, so that $\dot{x} = Aqe^{qt}$, $\ddot{x} = Aq^2e^{qt}$.

Now substitute for x, $\dot{x}$, and $\ddot{x}$ and form the auxiliary equation

$$\left(mq^2 + cq + k\right)Ae^{qt} = 0. \tag{6.16}$$

Equation (6.16) is satisfied if $A = 0$ (no motion – trivial solution), or if q is a solution to the quadratic, i.e.

$$q = \frac{-c \pm \sqrt{c^2 - 4mk}}{2m} \tag{6.17a}$$

$$= -\frac{c}{2m} \pm \sqrt{\frac{c^2}{4m^2} - \frac{k}{m}} \tag{6.17b}$$

$$= -\alpha \pm \beta \tag{6.17c}$$

where $\alpha = \frac{c}{2m}$ and $\beta = \sqrt{\frac{c^2}{4m^2} - \frac{k}{m}}$.

There are three distinct solution types for Eq. (6.17) which basically depend on how much damping there is in the system and which give distinctly different kinds of motion.

1. If β is imaginary, i.e. $\frac{k}{m} > \frac{c^2}{4m^2}$ then the motion is periodic, and the solution of (6.17) has the form (writing $\beta = i\omega_d$).

$$x = e^{-\alpha t}\left\{A_1 e^{i\omega_d t} + A_2 e^{-i\omega_d t}\right\} \text{ which is equivalent to writing}$$

$$x = e^{-\alpha t}\{B_1 \cos \omega_d t + B_2 \sin \omega_d t\}. \tag{6.18}$$

Here $\omega_d = \left(\frac{k}{m} - \frac{c^2}{4m^2}\right)^{\frac{1}{2}}$ and is called the damped natural angular frequency. An imaginary solution is the usual case for buildings.

2. If $\beta = 0$ i.e. $\left(\frac{k}{m} = \frac{c^2}{4m^2}\right)$, the motion is said to be critically damped and the solution is

$$x = e^{-\alpha t}(A_1 + A_2 t). \tag{6.19}$$

The critical damping coefficient is defined by $c_c = (4mk)^{1/2}$, and is an important value for assessing the relative amount of damping in systems. This solution is the boundary between periodic and aperiodic motion.

3. If β is real, i.e. $\frac{k}{m} < \frac{c^2}{4m^2}$, then the motion is again aperiodic, and the solution has the form

$$x = e^{-\alpha t}(B_1 cosh\beta t + B_2 sinh\beta t)$$

This solution is not important for wind engineering, but is included here for completeness.

If there is no damping present, then the undamped natural angular frequency is given by $\omega_n = \left(\frac{k}{m}\right)^{\frac{1}{2}}$. It is often useful to also write the damping as a fraction of critical, i.e. $\zeta = \frac{c}{c_c}$ (for buildings ζ is generally very small, i.e. of the order of 1 %.)

Using the symbols defined above, Eq. (6.15) can be written in terms of the undamped natural frequency and the damping ratio as follows.

$$\ddot{x} + \frac{c}{m}\dot{x} + \frac{k}{m}x = 0$$

Substituting for c and k/m enables further simplifications to be made.

$$\ddot{x} + \zeta\frac{(4mk)^{\frac{1}{2}}}{m}\dot{x} + \omega_n{}^2 = 0 \quad \text{or}$$

$$\ddot{x} + 2\zeta\omega_n\dot{x} + \omega_n{}^2 x = 0. \tag{6.20}$$

Consider now forced motion where $F(t)$ is periodic and is given by $F = F_0 \cos \omega t$. Then Eq. (6.13) becomes:

$$m\ddot{x} + c\dot{x} + kx = F_0 \cos \omega t \quad \text{or}$$

$$\ddot{x} + 2\zeta\omega_n\dot{x} + {\omega_n}^2 x = \frac{F_0}{m} \cos \omega t \tag{6.21}$$

Using D to denote differentiation with respect to time, Eq. (6.21) can be written:

$$\left(D^2 + 2\zeta\omega_n D + {\omega_n}^2\right)x = \frac{F_0}{m} \cos \omega t \quad \text{or}$$

$$x = \frac{\frac{F_0}{m} \cos \omega t}{D^2 + 2\zeta\omega_n D + {\omega_n}^2} \tag{6.22}$$

Now use the rule that since $D^2 \cos\omega t = -\omega^2 \cos\omega t$, we can write $D^2 = -\omega^2$ (Hannah and Stephens 1986), then Eq. (6.22) becomes

$$x = \frac{\frac{F_0}{m} \cos \omega t}{({\omega_n}^2 - \omega^2) + 2\zeta\omega_n D} \tag{6.23}$$

D in the denominator of Eq. (6.23) can be eliminated by multiplying the numerator and denominator by $({\omega_n}^2 - \omega^2) - (2\zeta\omega_n D)$ and results in

$$x = \frac{\left(({\omega_n}^2 - \omega^2) - 2\zeta\omega_n D\right)\frac{F_0}{m} \cos \omega t}{({\omega_n}^2 - \omega^2)^2 - 4\zeta^2{\omega_n}^2 D^2}$$

where again D^2 may be replaced by $-\omega^2$ and carrying out the differentiation in the numerator gives a solution in terms of sines and cosines.

$$x = \frac{F_0}{m}\left\{\frac{({\omega_n}^2 - \omega^2)\cos \omega t + 2\zeta\omega_n\omega \sin \omega t}{({\omega_n}^2 - \omega^2)^2 + 4\zeta^2{\omega_n}^2\omega^2}\right\} \tag{6.24}$$

From Eq. (6.24), it is useful to note the following behaviour.

1. When $\omega \to 0$

$$x \to \frac{F_0}{m}\frac{{\omega_n}^2 \cos \omega t}{{\omega_n}^4} = \frac{F_0}{m}\frac{m}{k} \cos \omega t \to \frac{F_0}{k} = \text{static displacement } X_{st}$$

i.e. the "slow" displacement is in phase with the excitation force.

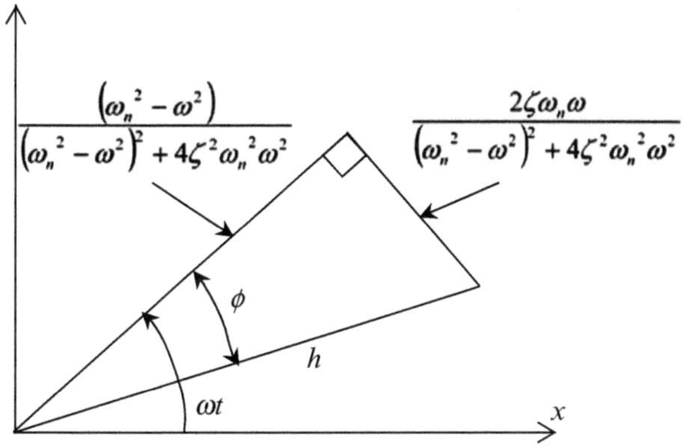

Fig. 6.4 Geometrical interpretation of Eq. (6.24)

2. When $\omega \to \infty$

$$x \to \frac{F_0}{m}\frac{(-\omega^2)\cos\omega t}{(-\omega^2)^2} = \frac{-F_0}{m}\frac{\cos\omega t}{\omega^2} \to 0$$

   and the "fast" displacement lags the excitation force by 180°, and tends towards zero.

3. When $\omega = \omega_n$ the excitation frequency is the same as the undamped natural frequency, and the displacement response is given by

$$x = \frac{F_0}{m}\frac{2\zeta\omega_n\omega_n\sin\omega t}{4\zeta^2\omega_n{}^2\omega_n{}^2} = \frac{F_0}{m}\frac{1}{\omega_n{}^2}\frac{\sin\omega t}{2\zeta} = \frac{F_0}{m}\frac{m}{k}\frac{\sin\omega t}{2\zeta} = \frac{F_0}{k}\frac{\sin\omega t}{2\zeta}.$$

   Here it can be seen that the displacement lags the excitation force by 90°, and it can obviously have large values for $\zeta \to 0$.

Equation (6.4) can be interpreted geometrically as shown in Fig. 6.4, where the distance "x" is measured from the origin along the x-axis, and the length of the lines of the two orthogonal arms of the triangle in Fig. 6.4 are the multipliers on the sine and cosine terms.

By expanding the denominator, Eq. (6.24) can be rewritten as follows.

$$x = \frac{F_0}{m}\frac{\left\{\frac{(\omega_n{}^2-\omega^2)\cos\omega t}{\left((\omega_n{}^2-\omega^2)^2+(2\zeta\omega_n\omega)^2\right)^{\frac{1}{2}}} + \frac{2\zeta\omega_n\omega\sin\omega t}{\left((\omega_n{}^2-\omega^2)^2+(2\zeta\omega_n\omega)^2\right)^{\frac{1}{2}}}\right\}}{\left((\omega_n{}^2-\omega^2)^2+(2\zeta\omega_n\omega)^2\right)^{\frac{1}{2}}} \tag{6.25}$$

Furthermore the hypotenuse, h, in Fig. 6.4 can be found by using the Pythagoras theorem.

$$h = \sqrt{\frac{(\omega_n{}^2 - \omega^2)^2 + (2\zeta\omega_n\omega)^2}{\left((\omega_n{}^2 - \omega^2)^2 + (2\zeta\omega_n\omega)^2\right)^2}} = \frac{1}{\left((\omega_n{}^2 - \omega^2)^2 + (2\zeta\omega_n\omega)^2\right)^{\frac{1}{2}}} \quad (6.26)$$

Which enables terms in Eq. (6.25) to be written as sines and cosines of the angle ϕ, as shown in Eq. (6.27).

$$x = \frac{F_0}{m}\frac{(\cos\varphi\cos\omega t + \sin\varphi\sin\omega t)}{\left((\omega_n{}^2 - \omega^2)^2 + (2\zeta\omega_n\omega)^2\right)^{\frac{1}{2}}} \quad (6.27)$$

The numerator in Eq. (6.27) can be simplified by using the cosine double-angle formula.

$$= \frac{F_0}{m}\frac{\cos(\omega t - \varphi)}{\left((\omega_n{}^2 - \omega^2)^2 + (2\zeta\omega_n\omega)^2\right)^{\frac{1}{2}}} \quad (6.28)$$

Equation (6.28) may now be simplified by substituting $\frac{1}{m} = \frac{\omega_n{}^2}{k}$ and then by dividing through by $\omega_n{}^2$.

$$x = \frac{F_0}{k}\frac{\cos(\omega t - \varphi)}{\left(\left(1 - \frac{\omega^2}{\omega_n{}^2}\right)^2 + \left(2\zeta\frac{\omega}{\omega_n}\right)^2\right)^{\frac{1}{2}}}; \quad \varphi = \tan^{-1}\frac{2\zeta\omega_n\omega}{(\omega_n{}^2 - \omega^2)} \quad (6.29)$$

Noting that $\frac{F_0}{k} = X_{st}$, the static displacement if the force F_0 were applied at zero frequency, and writing $r = \frac{\omega}{\omega_n}$, then the equation for the ratio of the amplitude of the forced motion to the static deflection, called the dynamic magnification or dynamic amplification can be found, and is given by Eq. (6.30). This is a standard equation, which is derived in many books on the mechanics of vibrations.

$$\left|\frac{x}{X_{st}}\right| = \frac{1}{\left(\left(1 - \left(\frac{\omega}{\omega_n}\right)^2\right)^2 + \left(2\zeta\frac{\omega}{\omega_n}\right)^2\right)^{\frac{1}{2}}} = H\left(\frac{\omega}{\omega_n}\right) = H(r) \quad (6.30)$$

Equation (6.30) has the characteristic shape shown in Fig. 6.5.

Thus the response of a SDF system depends on the frequency of the forcing function with respect to the natural frequency, i.e. of the wind force excitation, as stated previously, but now proved. The curve described by Eq. (6.30) and shown

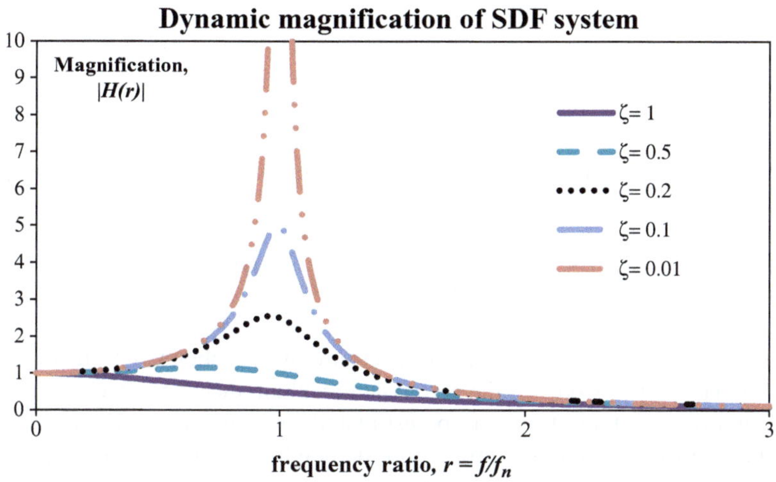

Fig. 6.5 Dynamic magnification of a single degree of freedom system

in Fig. 6.5 is also called the Mechanical Admittance in Wind Engineering terminology, and it must be integrated with the force spectrum to give the response as a function of frequency.

The analysis in Sect. 6.4 has used the circular frequency ω in radians per second. However, since $\omega = 2\pi f$, where f is frequency in Hz, we can interchange them, and for convenience with the prior sections, the argument f will be used in the mechanical admittance function henceforth.

From Eq. (6.30), the displacement spectrum can now be obtained from the force spectrum

$$\left|\frac{x}{X_{st}}\right| = H(f) \quad \text{so} \quad |x(f)| = \frac{Fo}{k}|H(f)|. \tag{6.31}$$

The effect of the dynamic behaviour of the structure, i.e. its mechanical admittance, can now be used to obtain an expression for the response of the structure, i.e. how much it moves under the action of an excitation force.

6.5 The Displacement Spectrum

Equation (6.13) is an expression for the spectrum of the excitation force acting on an object. The resulting behaviour, i.e. displacement, is obtained by forming the product of the force spectrum and the mechanical admittance function that was derived in Sect. 6.4.

This is given by Eq. (6.32).

$$S_x(f) = \frac{1}{k^2} |H(f)|^2 S_F(f) \tag{6.32}$$

and the following relationship for the displacement of a SDF system is obtained.

$$S_x(f) = \frac{4\overline{F}^2}{k^2 \overline{u}^2} S_u(f) \chi^2 \left(\frac{f\sqrt{A}}{\overline{u}} \right) |H(f)|^2 \tag{6.33}$$

Hence in principle the displacement spectrum can be obtained from Eq. (6.33), since we have expressions for the aerodynamic admittance, and mechanical admittance, and the wind spectrum can be estimated for a given location based on the roughness of the surrounding terrain and the height. The area of the displacement spectrum can be found by integration, for example simply by developing an appropriate spreadsheet, and thus the variance of the displacement. The peak displacement can be estimated by asserting that the peak displacement is the mean plus a certain number of standard deviations. However, this gust factor approach was developed long before users had access to spreadsheets, and so some clever analytical techniques were developed to enable the integration to be carried out. The power of these techniques is that they illustrate the importance of various parameters and their influence in controlling the mean and peak displacements of structures when subjected to wind excitation.

6.6 Integrating the Displacement Spectrum

The variance of the fluctuating component of displacement can be found by integrating the displacement spectrum, as expressed in Eq. (6.34)

$$\overline{x'^2} = \int_0^\infty S_x(f) df \tag{6.34}$$

Using Eq. (6.33), this becomes,

$$\overline{x'^2} = 4 \frac{\overline{F}^2}{k^2 \overline{u}^2} \int_0^\infty \chi^2 \left(\frac{f\sqrt{A}}{\overline{u}} \right) |H(f)|^2 S_u(f) df. \tag{6.35}$$

Now introduce the mean or static displacement,

$$\overline{x} = \frac{\overline{F}}{k}, \tag{6.36}$$

and a wind spectrum which has been normalised by the variance of the velocity fluctuations, which then has an area which is equal to 1.0, i.e.

$$S_u(f) = \overline{u'^2} * \underbrace{S'_u(f)}_{\substack{\text{normalised} \\ \text{spectrum} \\ \text{whose} \\ \text{area}=1}} \quad . \tag{6.37}$$

Substituting Eqs. (6.36) and (6.37) into (6.35) enables a dimensionless relationship to be obtained.

$$\frac{\overline{x'^2}}{\overline{x}^2} = 4\frac{\overline{u'^2}}{\overline{u}^2} \int_0^\infty \chi^2\left(\frac{f\sqrt{A}}{\overline{u}}\right)|H(f)|^2 S'_u(f)df \tag{6.38}$$

It was in carrying out this integration that the developers of gust factor approach introduced a clever and novel technique. They considered the integration to be made up of two parts, a background part A_B, which ignores any dynamic behaviour of the structure, and a resonant part A_R, which assumes a narrow band response at the natural frequency. This division of the integration results in the formulation expressed in Eq. (6.39).

$$\frac{\overline{x'^2}}{\overline{x}^2} = 4\frac{\overline{u'^2}}{\overline{u}^2}[A_B + A_R] \tag{6.39}$$

where

$$A_B \approx \int_0^\infty \chi^2\left(\frac{f\sqrt{A}}{\overline{u}}\right)S'_u(f)df = B \tag{6.40}$$

which assumes that $H(f) = 1$ over the whole spectrum, and

$$A_R \approx \chi^2\left(\frac{f_n\sqrt{A}}{\overline{u}}\right)S'_u(f_n)\int_0^\infty |H(f)|^2 df \tag{6.41}$$

where f_n is the natural frequency. The integral $\int_0^\infty |H(f)|^2 df$ has an analytical solution, namely $\int_0^\infty |H(f)|^2 df = \frac{\pi}{4}\frac{f_n}{\zeta}$

And hence Eq. (6.41) can be written as given in Eq. (6.42), which introduces the symbols S and E, which have a physical meaning.

$$A_R = \underbrace{\underbrace{\chi^2\left(f_n\frac{\sqrt{A}}{\overline{u}}\right)}\underbrace{S'_u(f_n)}}_{\substack{\text{evaluated at structure's}\\ \text{resonant frequency}\\ \text{and assumed constant}}}\frac{\pi}{4}\frac{f_n}{\zeta} = \frac{SE}{\zeta} \tag{6.42}$$

$S = \chi^2\left(f_n\frac{\sqrt{A}}{\overline{u}}\right)$ is a "size reduction factor", and accounts for the aerodynamic admittance reducing the loading on large objects due to the lack of wind speed correlation over the exposed surface.

$E = \frac{\pi}{4}f_n S'_u(f_n)$ is a "gust energy factor" and accounts for the spectral density, or the excitation energy of the wind at the natural frequency of the structure.

ζ = damping as a fraction of critical

$B = \int_0^\infty \chi^2\left(\frac{f\sqrt{A}}{\overline{u}}\right)S'_u(f)df$ is a "background" (low frequency) excitation factor, and accounts for the "quasi-steady" excitation of an object, ignoring any resonant response.

These rather complicated equations can be interpreted graphically as shown in Fig. 6.6.

Substituting Eqs. (6.40) and (6.42) into (6.39), then taking the square root results in Eq. (6.43), where the symbol σ has been used to denote the standard deviation. Equation (6.43) illustrates the contributions to the standard deviation of the displacement fluctuations. There is a background component, and a resonant component that can be reduced by increasing the damping.

$$\frac{\sigma_x}{\overline{x}} = 2\frac{\sigma_u}{\overline{u}}\left[B + \frac{SE}{\zeta}\right]^{\frac{1}{2}} \tag{6.43}$$

6.7 Estimating the Peak Response

The peak response may be estimated by assuming that it consists of the mean displacement plus several standard deviations of the response, as expressed in Eq. (6.44) where g is a peak factor with a value around 3–4.

$$\hat{x} = \overline{x} + g\sigma_x \tag{6.44}$$

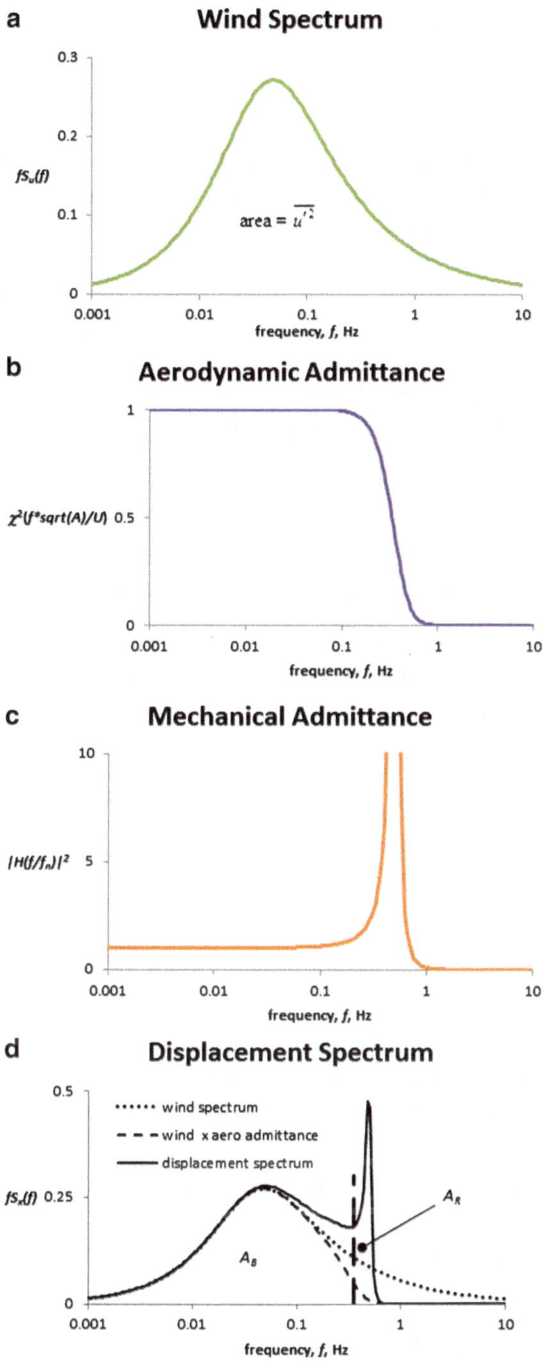

Fig. 6.6 Graphical representation of Eq. (6.35) to evaluate the mean square displacement response

If one is interested in the peak over say one hour, then g can be estimated using Eq. (6.45).

$$g \cong (2\ln(3600 f_0))^{\frac{1}{2}} \tag{6.45}$$

Combining Eqs. (6.43) and (6.44) gives Eq. (6.45) where the peak displacement is a function of the peak factor, the turbulence intensity of the wind, the background factor B, and the term $\frac{SE}{\zeta}$.

$$\hat{x} = \overline{x} + g\overline{x}2\frac{\sigma_u}{\overline{u}}\left[B + \frac{SE}{\zeta}\right]^{\frac{1}{2}}$$

The present derivation has been termed the gust factor approach, and the gust factor G is given by Eq. (6.46)

$$\frac{\hat{x}}{\overline{x}} = G = 1 + g\frac{\sigma_x}{\overline{x}} = 1 + gr\left[B + \frac{SE}{\zeta}\right]^{\frac{1}{2}} \tag{6.46}$$

Where the variable r is twice the turbulence intensity, i.e. $r = 2\frac{\sigma_u}{\overline{u}}$.

In high intensity turbulence it is not legitimate to ignore the term $u'(t)^2 - \overline{u'(t)^2}$ [see Eq. (6.6)] so a correction due to Vickery (1972) can be made to Eq. (6.46) by writing

$$G = 1 + gr\left[B(1+k)^2 + \frac{SE}{\zeta}\right]^{\frac{1}{2}} \tag{6.47}$$

where

$$k = \frac{gr\sqrt{B}}{4}. \tag{6.48}$$

This contribution is significant in high intensity turbulence, particularly when the resonant contribution is small. For a small structure, $B \to 1$ and if the resonant contribution is small, then Eq. (6.47) becomes

$$G = 1 + gr\left[(1+k)^2\right]^{\frac{1}{2}} = 1 + gr\left(1 + \frac{gr}{4}\right) = 1 + gr + \frac{g^2r^2}{4} = \left(1 + \frac{gr}{2}\right)^2$$

$$= \left(1 + \frac{g2\frac{\sigma_u}{\overline{u}}}{2}\right)^2 = \left(1 + g\frac{\sigma_u}{\overline{u}}\right)^2 = \frac{(\overline{u} + g\sigma_u)^2}{\overline{u}^2} = \left(\frac{\hat{u}}{\overline{u}}\right)^2$$

$$G = \left(\frac{\hat{u}}{\overline{u}}\right)^2 \tag{6.49}$$

and so it can be seen that with this correction, the gust factor approaches the square of the ratio of the peak gust speed to the mean speed, as required.

The above derivation is the essence of the "Gust Factor Method", first developed by Davenport and Vickery in the 1960s (Davenport 1967; Vickery 1968; Vellozi and Cohen 1968). It is applied to buildings by evaluating the terms gr, B, S, E for the subject building. Prior to the 1989 version of the Australian wind standard AS1170.2, an addendum containing a version of the above derivation was used along with some graphs to obtain the terms gr, B, S, E. The 1989 release of AS1170.2 supplied equations to determine these parameters based on the surrounding terrain roughness the building height, width and so on. The present version of the Standard, now AS/NZS1170.2:2010 also uses equations to determine the required parameters to obtain the gust factor. Furthermore, since that version of the standard uses a "gust" speed as its basis, an additional calculation is included to account for the use of the gust speed, as the foregoing analysis is based on a "mean" wind speed.

6.8 Estimating the Along-Wind Acceleration

The typical motion of a tall building when subjected to a strong wind is sketched in Fig. 6.7. It can be seen that there is a high frequency "resonant" motion superimposed on a low frequency "background" contribution.

For sinusoidal oscillations at a single frequency (e.g. $x = A \sin \omega t$) we know from simple harmonic motion theory that $\ddot{x} = -\omega^2 x$, where $\omega = 2\pi f, f =$ frequency in Hz.

Now, building motion is made up of a range of frequencies, and so it follows from the definition of a power spectrum that the area under it is the variance, and so not only do we have

$$\sigma_x^2 = \int_0^\infty S_x(f)df \tag{6.50}$$

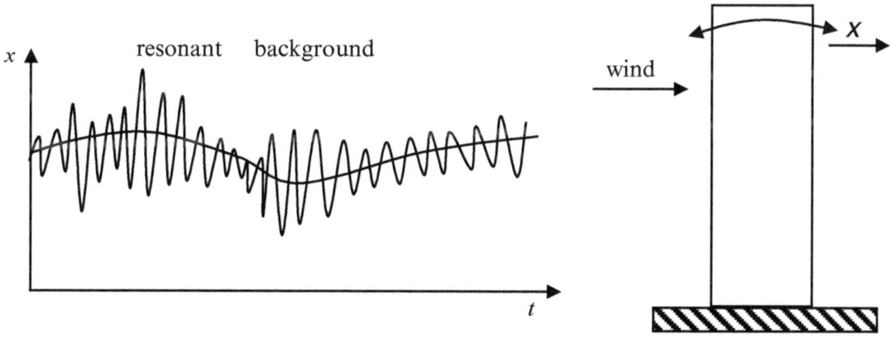

Fig. 6.7 Schematic diagram of building motion when subjected to wind excitation

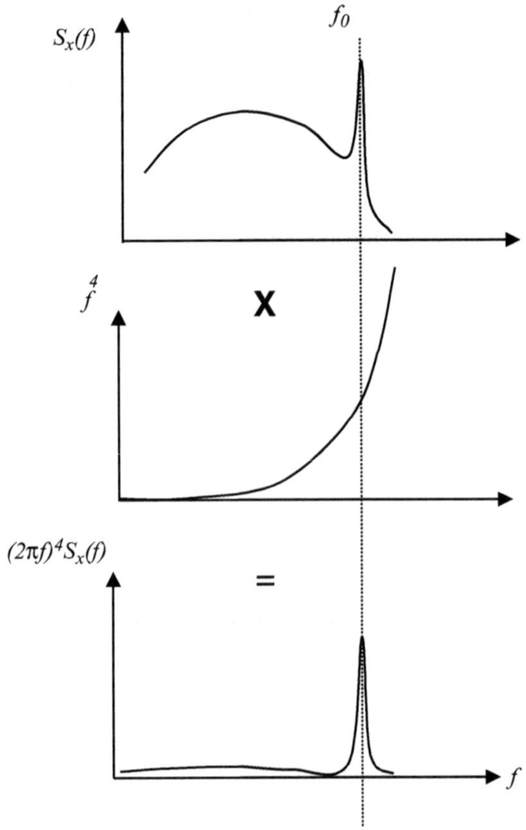

Fig. 6.8 Schematic diagram showing relationship between the displacement and acceleration spectra

but also there is a spectrum for the acceleration which can be related to the displacement spectrum approximately through Eq. (6.51), where the acceleration spectral estimate is the displacement spectral estimate multiplied by $2\pi f$ to the power of 4.

$$\sigma_{\ddot{x}}^2 = \int_0^\infty S_{\ddot{x}}(f)df \cong \int_0^\infty (2\pi f)^4 S_x(f)df. \tag{6.51}$$

Diagrammatically the integration contained in Eq. (6.51) looks like Fig. 6.8.

Hence the background part of the spectrum contributes very little to the acceleration, and so the acceleration calculation can be approximated by assuming that the acceleration is caused by motion at the resonant frequency only. This may be formulated in a similar manner as was done for the displacement spectrum in the foregoing.

$$\begin{aligned}\sigma_{\ddot{x}}^2 &= \int_0^\infty S_{\ddot{x}}(f)df \\ &\cong \int_0^\infty (2\pi f)^4 S_x(f)df \\ &\cong (2\pi f_n)^4 4\bar{x}^2 \frac{\sigma_u^2}{\bar{u}^2}\chi^2\left(\frac{f_n\sqrt{A}}{\bar{u}}\right)S_u{}'(f_n)\frac{\pi}{4}\frac{f_n}{\zeta} \\ &\cong (2\pi f_n)^4 4\bar{x}^2\left(\frac{\sigma_u}{\bar{u}}\right)^2\frac{SE}{\zeta}\end{aligned} \tag{6.52}$$

which can be simplified to Eq. (6.53).

$$\sigma_{\ddot{x}} = \bar{x}(2\pi f_n)^2 r\left(\frac{SE}{\zeta}\right)^{\frac{1}{2}} \tag{6.53}$$

The peak acceleration is found by applying an appropriate peak factor, Eq. (6.54), and there is no mean contribution, as the building oscillates about it mean deflected position.

$$\hat{\ddot{x}} = g\sigma_{\ddot{x}} \tag{6.54}$$

Which finally results in Eq. (6.55) where the symbols have the same meaning as used previously.

$$\hat{\ddot{x}} = g\bar{x}(2\pi f_n)^2 r\left(\frac{SE}{\zeta}\right)^{\frac{1}{2}} \tag{6.55}$$

6.9 Conclusions

The foregoing analysis has demonstrated an approach whereby the peak displacement and acceleration of a building can be found. It is assumed to behave as a single degree of freedom system, and is excited by the along-wind component of the wind turbulence. This derivation was published about 50 years ago, and is still used for prediction purposes today. It also forms the basis of some codified methods, for example in the Australian/New Zealand Standard AS/NZS1170.2:2011.

Acknowledgements The writer wishes to acknowledge the wind engineering he has learnt from such masters as Professors Alan Davenport and Barry Vickery, with whom he has worked at various times. The gust factor derivation was presented by Professor Vickery at several of the

Advanced Courses on Wind Engineering that were organised by Professor Bill Melbourne at Monash University, Australia. The writer was fortunate to attend two of these courses in the 1980s and 1990s.

References

Bendat JS, Piersol AG (2011) Random data analysis and measurement procedures, 4th edn. Wiley, New York

Davenport AG (1967) Gust loading factors. ASCE J Struct Div 93:11–34

Hannah J, Stephens RC (1986) Mechanics of machines, elementary theory & examples. Edward Arnold, London

Vellozi J, Cohen E (1968) Gust response factors. J Struct Div Proc ASCE 94(ST6)

Vickery BJ (1965) On the flow behind a coarse grid and its use as a model of atmospheric turbulence in studies related to wind loads on buildings. Aero Report 1143. National Physical Laboratory (UK)

Vickery BJ (1968) Load fluctuations in turbulent flow. ASCE J Eng Mech Div 94:31–46

Vickery BJ (1972) The provisions and limitations of the Australian Wind Loading Code, AS1170. In: Proceedings of the Australian conference on design & planning of tall buildings, Sydney

Chapter 7
Wind Loads on Building Components and Cladding

Gregory A. Kopp

Abstract Building components and cladding are important to the overall performance of buildings during extreme windstorms. Failure of even relatively small components can lead to internal pressurization, which increases the overall net loads on other components, or to significant rain water infiltration, which increases losses. In this chapter, we focus on the nature of wind loads on low-rise buildings and test methods used to define the ultimate capacity of building products and components. Wind loads are shown to have significant spatial and temporal variations, which are greatly simplified in building codes and standard test methods. Most building codes specify component loads as single peak values that have only a few different values over the various building surfaces for a small range of buildings. Many building components have complex and redundant load paths; the chapter discusses how load sharing can be handled in such systems. The design of some building component and cladding systems depend on the storm duration and the numbers of load cycles, which is also discussed.

Keywords Bluff body aerodynamics • Cladding • Low-cycle fatigue • Pressure equalization • Static fatigue • Wind loads

7.1 Introduction

The performance of building components and cladding systems in severe windstorms is important to ensuring life safety and mitigating losses. Failure or damage to a single panel can dramatically increase losses due to water ingress through openings created in the envelope. For example, when a failure occurs to a

---

G.A. Kopp (✉)
Boundary Layer Wind Tunnel Laboratory, University of Western Ontario,
London, ON, N6A 5B9, Canada
e-mail: gakopp@uwo.ca

Y. Tamura and A. Kareem (eds.), *Advanced Structural Wind Engineering*,
DOI 10.1007/978-4-431-54337-4_7,

component on a windward wall, the net wind loads on roofs, side walls, and leeward walls can be increased through internal pressurization. Such pressurization can cause subsequent failures, increasing the risk to life safety inside and outside of the building. The loss of a single roof panel can lead to substantial losses because of rain damage to the contents and interior finishings.

Wind loads for design are typically specified as

$$F_a = q \cdot C_e \cdot Cp_a \cdot C_g \cdot A \tag{7.1}$$

where F_a is the load, q is the dynamic wind pressure, C_e is the exposure factor, Cp_a is the effective pressure (or load) coefficient, and C_g is the gust factor. For the apparent simplicity of this formulation, there is actually a great deal of subtlety in it, since F_a is really an equivalent static load and Eq. (7.1) links the wind climate (through q), the wind loads, and the response. This is well known for the loads acting on the main structure, but it is also the case for building components. However, Cp_a is often determined solely from the aerodynamic pressures, as discussed below.

Building components and cladding systems are often statically indeterminate with complex and redundant load paths, using materials which are often non-structural and highly variable, but which may contribute substantially to the capacity. For many component and cladding systems, resonant effects are negligible, thus, the effective pressure or load coefficient, $Cp_a(\zeta,t)$, at time, t, for component location, ζ, which has a nominal tributary area, A, can be written as:

$$Cp_a(\zeta,t) = \int_{A_1} Cp_e(x,y,z,t) \cdot I(x,y,z) \frac{da}{A} \tag{7.2}$$

where $I(x,y,z)$ is the structural influence function,

$$Cp_e(x,y,z,t) = \frac{p_e(x,y,z,t) - p_o}{q} \tag{7.3}$$

is the external pressure coefficient at the location (x,y,z), p_e is the external pressure, and p_o is the static pressure. The integral is over area, A_I, in the region that $I(x,y,z)$ is non-zero, while the normalization is normally over the nominal tributary area, A, which is defined in Sect. 7.3. Thus, $Cp_a(\zeta,t)$ depends on both aerodynamic and structural factors. This is illustrated by Fig. 7.1.

Figure 7.1 gives an example of these concepts for a low-rise building with a gable roof. The spatial variations of the external pressure coefficients, Cp_e, can be seen by the colour contours in the main portion of the figure, which shows a snapshot of the coefficients at a single moment in time, with values from about 0 to about −3.0. In fact, this is the moment of the single largest magnitude point pressure, which occurs near the ridge of the roof. The time history of the pressure coefficient at this point is given in Inset A of Fig. 7.1, which shows that the “gust”

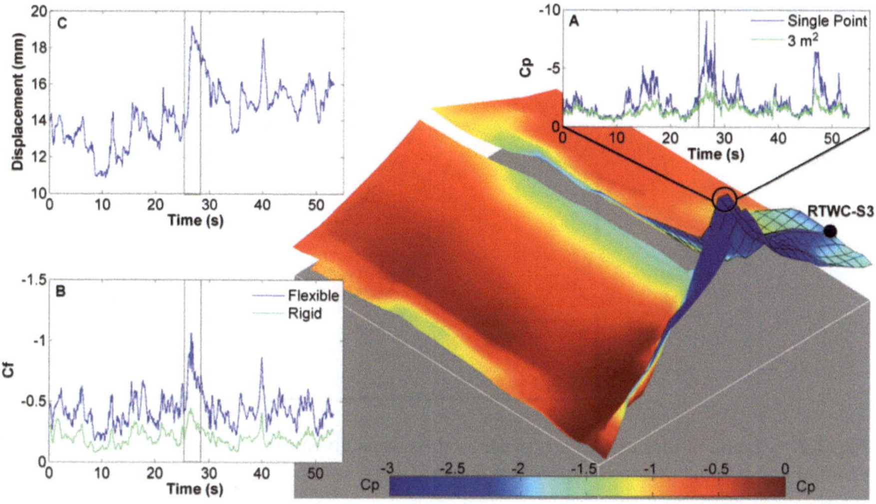

Fig. 7.1 Contours of the spatial gradients of external wind pressures on the roof of the test house. *Inset A*: wind tunnel pressure coefficients at a single point and averaged over a 3 m^2 area. *Inset B*: applied force coefficients for a single roof to wall connection (RTWC), considering no load sharing (flexible) and perfect load sharing (rigid). *Inset C*: the displacement of the RTWC-S3 under the loading shown in *Inset B* (from Morrison et al. 2012)

with increased suctions (i.e., negative pressures) occurs over about 3 s, with the actual peak being of very short duration (i.e., much less than a second). Considering Eq. (7.2) and the size of typical building components, this single point pressure is not of great relevance. However, if one considers the size of a typical sheathing panel on such a roof, say 3 m^2, the peak coefficient, Cp_a, can be seen to be reduced by a factor of about 3, for this particular case. The reason for this is that there are large spatial pressure gradients, and averaging over these larger areas yields significantly smaller aerodynamic force coefficients than those at a single point. In terms of the temporal pressure gradients, Inset A of Fig. 7.1 also indicates that, even for the spatially-averaged loads, the largest fluctuations occur intermittently.

Because of the uplift forces acting on the roof, there needs to be a vertical load path to resist these. For a wood roof, trusses are relatively closely spaced with connections between each roof trusses and a wall (Morrison et al. 2012). Inset B shows two possible time histories for the forces at a connection in such a roof, labeled as "RTWC-S3". One time history assumes the roof is rigid, while the other assumes that the roof is very flexible. Integration of Eq. (7.2), considering these assumptions, leads to different loads at each of the connections, with much higher loads at the worst loaded connections for the flexible roof. Of course, the flexible roof has minimal load sharing compared to the rigid roof since the values of the influence function, I, are non-zero over a much larger area of the roof with higher stiffness. As Inset B shows, how the structural system re-distributes the applied pressures has a significant effect on the net loads. Such re-distribution is important

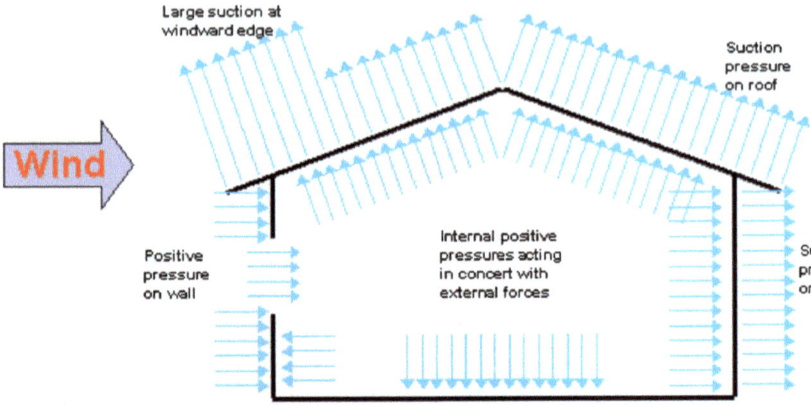

Fig. 7.2 Schematic illustration of typical wind loads as given in building codes, including both external and internal pressures (from Henderson 2010)

to define since the actual load paths have a substantial impact on the performance under extreme conditions. Inset C shows actual measured displacements for the connection; details can be found in Morrison et al. (2012).

Building codes, such as ASCE 7-10 (2010), account for these wind loads either explicitly, through the use of Eq. (7.1), or implicitly, through prescriptive design requirements or standardized tests. For component designs that depend directly on Eq. (7.1), building codes typically simplify the spatial pressure variations into a few zones, each with a constant pressure, while the temporal effects are accounted for by providing a single peak value (obtained using a statistical method). This is illustrated by Fig. 7.2. The effects of building geometry are handled by providing coefficients for a range of typical shapes. Further, typically, area-averaged pressure coefficients are provided under the assumption that $I = 1$ within the tributary area of the component or cladding element, so that it is left to the designer to establish the correct effective area for any particular component or system. In some cases, this is straightforward; in others, it is less so.

In this chapter, we will examine the nature of wind loads on (low-rise) buildings and how they are simplified for use in building codes and standardized test protocols. In particular, the assumptions required to link the aerodynamics with the response are examined. Wind loads are examined in Sect. 7.2, while the responses of some typical component or cladding systems, and how they may modify the design wind loads, are discussed in Sect. 7.3. In Sect. 7.4, we highlight some of the standardized test methods. Emerging, advanced test methods are also introduced, which reduce the need for many of the assumptions used in the standardized tests and in the interpretation of the building codes. All examples will be for low-rise buildings (which are the most common building type), but the concepts discussed apply equally to components and cladding on all buildings.

7.2 Wind Loads on Buildings

Wind loads depend on many parameters including the wind field, the wind direction, α, terrain roughness and surroundings, the Reynolds number, Re, and the building shape and size. The theoretical basis for all wind tunnel tests is the similitude of dimensionless groups containing all parameters relevant to the phenomena being measured. For pressures on a building, this leads to

$$Cp_e(x,y,z,t) = \frac{p_e - p_o}{\frac{1}{2}\rho V_H^2} = f\left(\frac{x}{H},\frac{y}{H},\frac{z}{H}, tV/H, Re, \frac{z_0}{H},\ \textit{building size and shape},\ \alpha\right) \quad (7.4)$$

for single isolated buildings in flat, uniform terrain, where the building dimensions are normalized, say, by the height, H, noting that $Re = VH/\nu$. Use of the Jensen number, z_o/H, implies that the atmospheric boundary layer can be defined by a single parameter, the aerodynamic roughness, z_o, which links the mean velocity profile and the details of the turbulence.

Since wind-induced pressures depend on a large number of parameters, it is not surprising that they have a great deal of variation. We can categorize these variations as being related to the location on the external surface of the building (spatial variations), temporal variations, geometric effects, pressures transmitted to the interior of the building, and wind direction. These are discussed in the following sections.

7.2.1 *Spatial Variations and Geometric Effects of Building Shape on Pressures*

Figures 7.3 and 7.4 show mean pressure coefficients for low-rise buildings with several roof shapes, roof slopes, eave heights, and wind directions. What is immediately apparent is that there are vastly different magnitudes of pressures and pressure distributions on these buildings. The wind separates from the building surfaces at the edges (i.e., the roof edge, the ridge, or at the wall edges), and then reattaches further downwind if the building is large enough. While the characteristics of the turbulence in the wind also affect this process, particularly altering reattachment points and the magnitudes of the pressures within the separated flow, the shape of the building plays a substantive role. Figure 7.3 shows that the roof slope alters the intensity of the two conical vortices formed at the leading corner of the roof when the wind is oblique to one of the walls, changing even the basic patterns of pressure as the slope is altered. Figure 7.4 shows that the size of the wall alters the intensity of the two conical vortices formed at the leading corner when the wind is oblique to one of the walls. Comparing the upper and lower

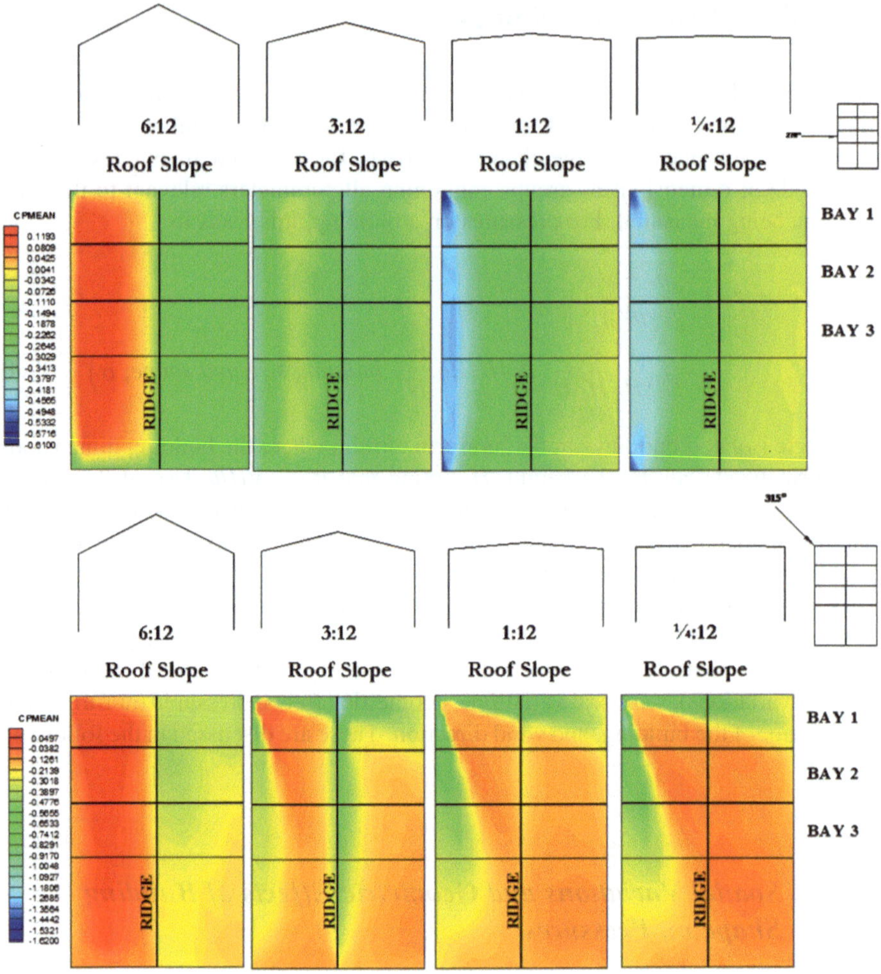

Fig. 7.3 Mean pressure coefficients on gable roofed buildings with different roof slopes for two different wind directions

parts of Fig. 7.3, it is evident that the wind direction is also a significant parameter affecting wind loads. So, again the spatial variation of the pressure across the building surfaces are evident, and these clearly depend on the building geometry. Changing the characteristics of the wind, say, by altering the terrain from open to suburban, also changes the magnitudes and some of the details pertaining to the spatial patterns. Changing the buildings in the immediate surroundings can have an even larger impact.

The spatial gradients of the external pressures are important for at least three reasons. First, they affect the area-averaged pressure coefficients, Cp_a, as illustrated by Figs. 7.1 and 7.5. Second, they indicate that components located at different

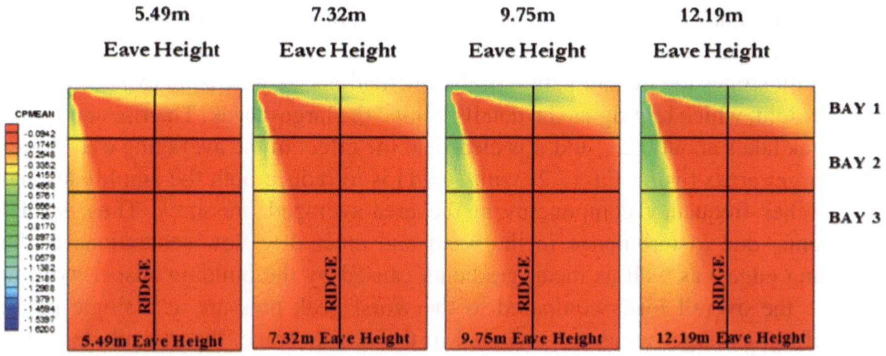

Fig. 7.4 Mean pressure coefficients on gable roofed buildings with different eave heights

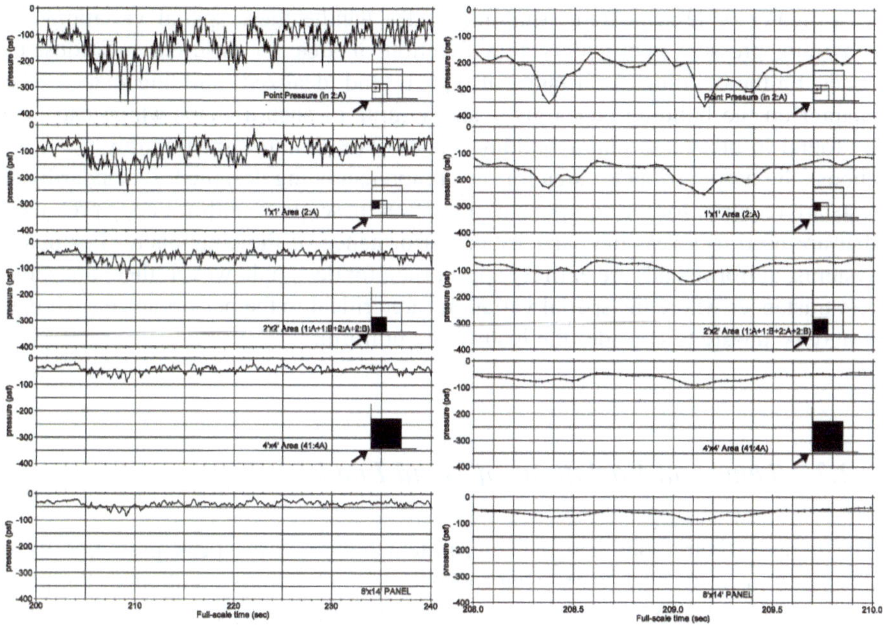

Fig. 7.5 Pressure variation on different tributary areas near the roof corner including the point with the largest minimum pressure (from Surry et al. 2007)

positions on the building surfaces experience different loads, and that the load path from the cladding, through the sub-structure, to the main structure, may be challenging to define analytically. Third, the spatial gradients affect pressure equalization for multi-layer roof and wall systems, as discussed in Sect. 7.2.3.

Figure 7.5 shows time histories of the external pressures acting on different sized areas of a roof, containing the instant when the largest peak local suction was observed on a flat roofed building for a cornering wind direction. As discussed by

Surry et al. (2007), the horizontal axis is full-scale time, using time scaling based on a severe tropical cyclone. Two different time axes are shown, the one on the left, which illustrates the overall nature of the fluctuations, as well as an expanded view on the right, which highlights the details around the largest peak. The rise and fall of the peak takes about 0.2 s, and it is clear that the effect of the averaging over larger and larger areas [using Eq. (7.2) with $I = 1$] is to reduce both the amplitudes and the higher frequency components in the area-averaged pressures. Thus, spatial gradients, due to turbulence in the wind and created at flow separations at the building edges, as well as mean gradients caused by the building shape, work to reduce the overall loads compared to the worst peak pressure at a single point. Therefore, as larger areas are included in the area-average, the magnitude of the peak pressure drops over that area. In Fig. 7.5, the difference between a point pressure and an area of 1.5 m^2 (16 ft^2) is a factor of 4, for this particular case.

These characteristics of pressure are repeated over all of the building's surfaces, to a greater or lesser extent. Building codes handle these spatial variations by providing pressure coefficients as a function of area for different locations, or zones, on the building, for a range of typical (and idealized) building shapes. The chapter on the building codes gives an example of such a chart, which can also be found in all of the major building codes. The designer then simply needs to choose the proper tributary area for the component being designed. If standardized tests are required for the component, it is often these quasi-uniform code pressures that are used to assess the capacity or performance of the component. For many systems, uniform loads are entirely appropriate for design; however, there are some circumstances where the spatial variations are critical to the performance such that it may be unsafe to consider the loads to be uniform.

7.2.2 Temporal Variations of Wind Loads

The temporal variations and the intermittent nature of the peak pressures were discussed in Sect. 7.2.1, pertaining to Figs. 7.1 and 7.5. As mentioned, the peak pressures are of very short duration due to the turbulence generated by the flow separation at the building edge and the turbulence in the wind. However, it is important to emphasize that there are a range of scales associated with these fluctuations, with energy up to roughly 10 Hz. Thus, it may be necessary to assess possible resonant effects when a particular component's natural frequencies are less than this. The temporal variations are also important for components which are sensitive to fatigue, low-cycle fatigue, or static fatigue. In these situations it is necessary to consider the duration of the storm (or storms during the design life), considering the numbers and sizes of load cycles.

Wind-induced pressures acting on a building during a windstorm depend on several additional parameters, beyond the building and wind parameters described by Eq. (7.4). These are primarily the size and translational speed of the storm, which then set the actual wind loading time history. To understand this, we introduce the idealized tropical cyclone given in Fig. 7.6, while details can be found in Jancauskas et al. (1994) and Kopp et al. (2010, 2012). Figure 7.6 depicts the variation of mean (15-min average) wind speed and direction for this idealized storm. Using measured pressure data from a wind tunnel experiment, the time history of Cp_e for a small segment of cladding on the roof of a building was converted to full-scale dimensional pressures, using Eq. (7.3), and is presented in the figure. The segment of cladding has an area of 0.15 m^2, which is typical of the tributary area associated with a single roof fastener. Such a fastener must withstand not only the peak load caused by the storm, it must withstand all of the loading cycles caused by the intermittent, fluctuating pressures.

Figure 7.6 shows that there is apparent "randomness" in the observed peak loads. This is due to the stochastic nature of wind loading. In this plot, actual peak values are shown, rather than statistical peak values obtained from fits to the original wind tunnel data. Because peak pressures are due to nonlinear interactions between the upstream (boundary layer) turbulence and turbulence generated by the building, they are relatively rare so that probability density functions are non-Gaussian. Peak factors above 15 at some locations on the roof have been observed, and peak factors of 8 are not uncommon (e.g., Kopp et al. 2005). Thus, design standards consider peak pressures with particular probabilities of exceedance (Cook 1990). It is worth emphasizing that the longer the duration of high winds, the higher probability of obtaining a high peak pressure and the greater the number of load cycles.

7.2.3 Internal Pressures and Pressure Equalization

When there is an opening in the building envelope, pressures and flow can enter the interior spaces. When such an opening is in a wall, like an open door or broken window, the interior volume can be pressurized. In this case, with the net load is altered such that

$$Cp_n(x, y, z, t) = Cp_e(x, y, z, t) - Cp_i(t) \tag{7.5}$$

where Cp_n is the net pressure across the wall or roof component and Cp_i is the internal pressure. Internal pressures are usually approximately uniform within the internal volume, so the spatial dependence is not given in the equation as it is for the external and net pressures. If the wall or roof component is a single layer, then Cp_n is the net pressure loading the cladding and components, according to Eq. (7.2). If the wall or roof component is an assembly made up of multiple layers, then Cp_n is the net load across the entire assembly. The theory of internal pressures is well established (Holmes 1979) and sufficiently accurate to predict the results of wind tunnel studies (Oh et al. 2007).

For multi-layer wall or roof assemblies, when the opening is a made by gap in the exterior surface of the building envelope, then the interior space is the relatively

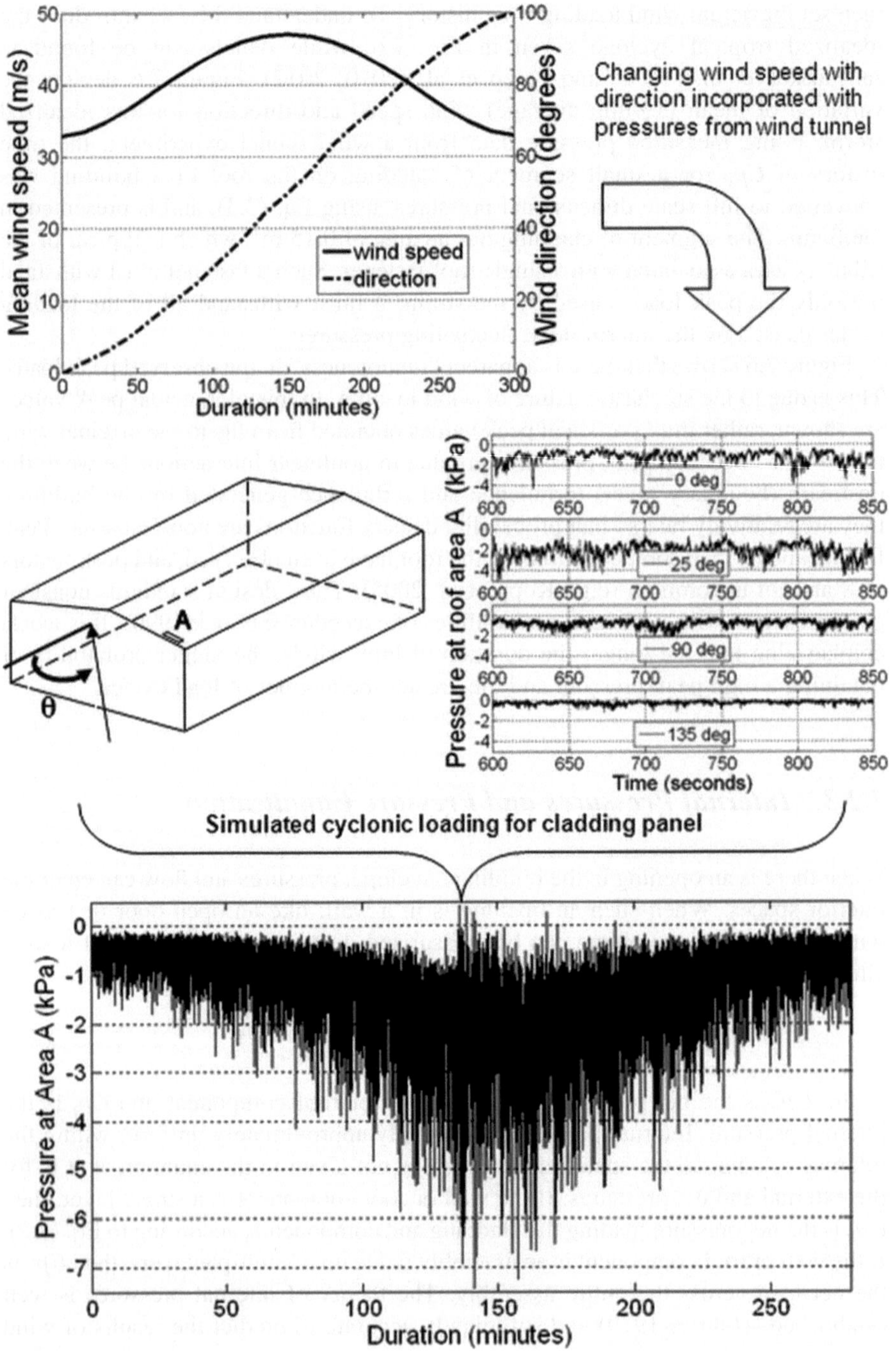

Fig. 7.6 Schematic diagram of the development of a pressure loading time history (*bottom*), indicating the combination of a storm time history (*top*) plus aerodynamic data from a wind tunnel (*middle*) (from Kopp et al. 2012)

small volume between the layers in a wall or built-up roof. Such openings can be either intentional, such as for rain screen walls or siding, or unintentional, such as for a crack or a gap where two cladding elements are butted together. So, in this case, the external pressure is also transmitted to the inner layer, but it is altered by the flow through the openings and the shape and size of openings and gaps, such that there is a different pressure between the layers, $p_l(x,y,z,t)$. Such pressures between layers will occur unless the outer layer is sealed.

Bienkiewicz and Sun (1992, 1997) have shown that the size of the gap or opening, relative to spacing between the layers, is a very important parameter for determining the pressure in the volume between the layers, p_l. When the gap to spacing between layers ratio is large, the inner layer pressure is much closer to that of the external pressure and has similar spatial variations. In this case, pressure equalization is said to occur such the p_l is similar to p_e, both in space and in time. In other words the pressure in the layer "tracks" the external pressure, although its magnitude will always be somewhat different. The similarity of the two pressures is given by the pressure equalization ratio, $R = p_l/p_e$ (Kopp and Gavanski 2012). When $R = 1$, there is "perfect" pressure equalization such that the net load on the layer is zero; when $R = 0$, there is none. When the ratio of the gap to the spacing between layers is small, the volume between layers becomes relatively larger so that p_l begins to behave like an internal pressure, p_i, and becomes more spatially uniform and R tends to be reduced. Additionally, the size of the panels or components will play a role, although this has not been well quantified in the literature. Reductions of loads for pressure equalizing systems could be up to 40–60 % in some situations, although a 90 % reduction in the loads (i.e., $R = 0.9$) may be possible. It is doubtful that greater reductions can be reliably achieved due to the sensitivities in the pressures to slight perturbations in the building dimensions, which are difficult to control with precision in typical building practice. The theory for pressure equalizing systems is not particularly advanced, so that in general $p_l(x,y,z,t)$ cannot be determined through analytical methods, as it can for $p_i(t)$.

7.2.4 Determination of Design Wind Loads

Design wind loads can be obtained from building codes or from a detailed wind tunnel study. Such wind tunnel tests need to be conducted by modeling both the wind and the building, as discussed in various standards [such as ASCE 49-12 (2012)]. For products used on many building shapes and sizes, such as roof-mounted solar arrays, one needs to consider the range of parameters where the product will be used (Kopp and Banks 2012). Typically, for components and cladding, one measures the external pressures, but internal pressures can be measured (Holmes 1979) or computed analytically (Oh et al. 2007). However, in many circumstances, inner layer pressures in multi-layer pressure equalizing systems cannot be directly measured in wind tunnels. When consideration of this is necessary for design, special wind tunnel approaches may be required and expert opinions should be sought.

Finally, it should also be noted that in design windstorms, there may be wind-borne debris. When such debris impacts a structure, it may breach the building envelope, breaking windows or doors or piercing the cladding, as discussed elsewhere in this book. This can lead to internal pressurization, as well as substantial water penetration. Various building codes and test standards exist to address these issues in design (e.g., Standards Australia 2011; BCA 2006), but they are not examined in detail here.

7.3 Response of Component and Cladding Systems to Wind Loads

In order to determine the response of building components or cladding, standardized tests are usually required. In some cases finite element analysis can be sufficiently accurate (e.g., Ali and Senseny 2003). There are many resources describing finite element methods, so these are not examined herein. Rather, the objective here is to discuss how the response of the components interacts with or modifies the design wind loads. For systems with complex or redundant load paths, load sharing and the choice of effective tributary areas are discussed. In these systems, the spatial variation of the wind loads over building surfaces is an important parameter determining the performance of such systems. Some components are dependent on the temporal variations; for these the numbers and magnitude of load cycles, and, therefore, the storm duration, are important.

7.3.1 *Tributary Areas and Load Sharing*

For a cladding fastener, such as nail for a wood panel or a clip for a standing seam metal roof, the nominal tributary area, A, for the fastener is normally assumed to be the area extending half way to the adjacent fasteners. For example, for a fastener on a panel with a spacing in one direction of 0.8 m and 0.2 m in the other, the nominal tributary area for the fasteners in the interior would be 0.8 m $\times$ 0.2 m $=$ 0.16 m^2 (for the fastener in the corner of the panel, the tributary area would be 0.4 m $\times$ 0.1 m $=$ 0.04 m^2). However, there is load sharing amongst the fasteners due to the bending stiffness of the panels. For a uniform, steady pressure applied to a panel, Eq. (7.2) can be modified as:

$$F_n = \int_{A_1} p_n(x,y,z) \cdot I(x,y,z)da = P_n \int_{A_1} I(x,y,z)da = P_n \cdot A_e \qquad (7.6)$$

Thus, the effective tributary area, A_e, which is given by the ratio of the net force applied to the fastener, F_n, divided by the net applied pressure, P_n, can be evaluated without precise knowledge of the influence functions. Uniform load tests could yield $A_e > A$ at the critical fastener because of load sharing (e.g., Farquhar et al. 2005), so that the actual critical load is larger than what one might assume using the nominal tributary area. For code-based design, then, P_c/P_n should be greater than A_e/A for the capacity to be greater than the load, where P_c is the pressure from the building code (notwithstanding the load factors). However, if the critical panel is not in the worst aerodynamic location on the building surface, the tributary method should be conservative (Morrison and Kopp 2010).

Since real pressures vary substantially over the surfaces, Eq. (7.6) is approximate. Farquhar et al. (2005) and Henderson (2010) have demonstrated that linear, elastic beam theory can be used to estimate influence functions with reasonable accuracy for certain cladding systems so that when wind tunnel, or other appropriate, data are available, responses can be accurately estimated. It should be noted that relatively stiffer panels will share loads more effectively, allowing weaker fasteners to shed their load to neighbouring fasteners (Khan 2012) so that they fail at a higher pressure (Rosowsky and Schiff 1996), allowing the panel to work more as a unit. This could be viewed as a reduction in A_e for the critical fastener, based on Eq. (7.6), at the cost of increased A_e for the surrounding fasteners. Beyond the increased capacity, the fact that the panel works more as a unit is beneficial since Cp_a drops with increasing area, as shown in Fig. 7.5, so that the load to be resisted is also reduced because of the spatial gradients.

In systems with non-linear behaviour, influence functions may be more difficult to determine analytically and results need to be based on, or validated by, testing. One such example is toe-nailed, roof-to-wall connections in wood truss roofs (Morrison et al. 2012). In this case, the nailed connections exhibit substantial yielding behaviour (i.e., nail slips) prior to failure (Morrison and Kopp 2011), which leads to continuously changing influence functions, caused by the nail slips at each connection, prior to the failure of the first connection (Morrison et al. 2012).

7.3.2 Low-Cycle Fatigue

Wind-induced pressures fluctuate significantly during wind storms. Some cladding systems and building components can fail due to low cycle fatigue caused by these fluctuations. One example is screw-fastened, metal roof cladding (Henderson 2010). During Cyclone Tracy in Darwin, Australia, large numbers of roof failures were observed when the metal roof decking pulled over the screw heads due to low cycle fatigue (Walker 1975a, b; Morgan and Beck 1977). Thus, this failure mechanism needs to be considered in design of such components and cladding systems. Usually, a cyclic-load test, such as SIDGERS (Baskaran and Chen 1998) or the low-high-low Australian standard (BCA 2006), must be performed to assess capacity.

Low-cycle fatigue is typically defined as failure occurring within 10,000 cycles. Examination of Figs. 7.1, 7.5 or 7.6 indicate that a "cycle" can be difficult to define, although the presence of cycles is clear. Henderson (2010) defines a load cycle with the sinusoidal feature that the cycle ends with the same value as it started with. During the cycle, the maximum load, S_{max}, and a minimum load, S_{min}, are considered, such that the load ratio, $R = S_{min}/S_{max}$, while the load amplitude, $S_{amp} = \text{abs}(S_{max} - S_{min})/2$. The rain-flow method is a common method for evaluating the number of cycles (e.g. Amzallag et al. 1994; Kumar 2000). Under wind loads, the vast majority of cycles are of low magnitude, with a few cycles with amplitudes that are large (i.e., comparable to the external or net pressures used in design). For example, Henderson (2010) reports that in one 15 min segment of a time history of roof pressures, that there were about 7,500 cycles in the range of 10 % of the peak design amplitude, while there was a single cycle with a magnitude similar to the peak pressure. Such load cycles can then be used to develop or evaluate a test standard, as was done by Baskaran and Chen (1998) for the SIDGERS method.

One of the limitations of the cycle counting approach is that the order of the cycles is lost, while actual performance may depend on order of the cycles in the sequence. However, in most storms, the peak loads tend to build gradually as the storm passes by the structure under consideration. Thus, the intensity and duration of the design-level storms also needs to be assessed for such building components, as discussed in Sect. 7.2.2; in some locations, structures may be exposed to many storms during their life time, and this also may need to be considered. Henderson (2010) provides a recent summary many of the details.

7.3.3 Static Fatigue and Glass

Window glass is often damaged in severe windstorms, the causes of which have been attributed to windborne debris, installation issues and wind-induced pressures (Minor and Beason 1976; Minor 1984; Kareem and Stevens 1984; Williams and Kareem 2003; Brewick et al. 2009). Glass windows are, of course, used in both residential structures and high-rise buildings. Glass is another example of component and cladding systems which require consideration of the storm intensity and duration effects in design standards.

Glass fails through a mechanism called static fatigue (Charles 1958); it does not fail at a particular stress level. This feature, the time-dependent reduction of strength due to the combination of duration and magnitude of loading, has been recognized in wind engineering for more than 30 years (Minor 1981). Static fatigue is well described by Brown's integral (Brown 1972, 1974),

$$DA = \int_0^t [\sigma_a(t)]^n dt \tag{7.7}$$

where DA is the damage accumulation, t is time, $\sigma_a(t)$ is the time-dependent applied stress at the critical crack (i.e., the particular crack leading to failure) on a glass surface, and n is a constant (generally accepted to be 16 for soda-lime silicate glass). A simplified form has been introduced using a power law relationship between stress and applied pressure (Brown 1972; Dalgliesh and Taylor 1990; Kawabata 1996; Calderone 1999) so that,

$$DA_{crit} = \int_0^{t_f} [p(t)]^s dt \tag{7.8}$$

where p is the applied load and s is a coefficient. A beneficial aspect of Eq. (7.8) is that it requires applied pressure instead of the stress induced at a crack and, hence, the critical crack location is not required for use. However, the values of s for specific glass panes differ due to the $p - \sigma_a$ relationship, which is non-linear and varies at each location on any given plate (but can be determined by analysis). Hence, it is often presented as a range for the various parameters which affect the stress (e.g., plate dimensions, glass type, etc.). Glass failure is defined as the moment when DA reaches its critical value, DA_{crit}, at time $t = t_f$.

Since the values of DA and DA_{crit} become large because of the high values of the exponent, n, the damage accumulation is typically converted into equivalent static loads. For example, if the reference time duration for this conversion is chosen to be t_{ref} sec, wind pressure time histories for a specific design situation, $p(t)$, would need to be converted into a t_{ref} equivalent static load, $p_{eq}(t_{ref})$. This $p_{eq}(t_{ref})$ corresponds to the design load in building codes. There are several conversion methods (e.g., Davenport 1975; Holmes 1985; Gavanski and Kopp 2011a, c), one of which is:

$$p_{eq}(t_{ref}) = \left(\int_0^T p^s(t)dt \Big/ t_{ref}. \right)^{1/s} \tag{7.9}$$

Assuming that the value of DA_{crit} is independent of the loading pattern, the calculated $p_{eq}(t_{ref})$ should be independent of the loading pattern as long as the rest of the parameters which affect DA_{crit} (such as glass type) are the same. Gavanski and Kopp (2011c) have examined how single peak pressures used in building codes can be interpreted for varying storm durations, using Eq. (7.9).

The failure pressure at a certain failure probability, P_f, under static load whose time duration is t_{ref} can be obtained using existing failure models and is expressed as $p_f(t_{ref}, P_f)$. Thus, $p_f(t_{ref}, P_f)$ becomes the load resistance in design standards. So, a glass type and geometry which satisfies

$$p_{eq}(t_{ref}) < p_f(t_{ref}, P_f) \tag{7.10}$$

is the appropriate design choice. Several models to predict the load resistance have been suggested by many researchers (Brown 1972; Beason and Morgan 1984). Gavanski and Kopp (2011b) provide a recent summary.

7.4 Test Methods for Building Components and Cladding

There are three methods to apply wind loads to a particular building component: (1) full-scale wind tunnel or field tests, (2) air-box pressure tests, and (3) applied force tests. Full-scale field measurements are reasonably common, but these tend to provide data at relatively low load levels, which are useful for evaluating analytical models. Wind tunnels, large enough to test building components in situ (i.e., on a building), have recently been built; however, at present, these are still research tools and not used to assess component capacities in practice. These full-scale tests are the most representative of reality, to the extent that the flow simulations are correct and the components are placed on a building typical for the component under consideration. For some current cladding systems, such as multi-layer, pressure equalizing systems, this is the only current method that is sufficiently accurate. However, in practice, air-box tests and force tests are much more common, and these are described below. For a recent discussion of full-scale test methods, see Kopp et al. (2012).

Applied force and air-box pressure tests typically only apply simplified loading patterns. Applied force methods apply point loads at single or multiple points, which require attachment to the cladding or component. These have been used to test full structures and also structural sub-components. However, the direct attachment implies that they are of limited use for determining cladding performance.

Air-box pressure tests have been used to test building products and components for a long time because of the relative simplicity of setting up a sealed test chamber with a fan used to pressurize or de-pressurized the volume. Early tests used a pressurized air-bag, like a bladder, to apply loads. Modern test rigs use rigid boxes, of which one side is the specimen. Such rigs are designed for relatively fast installation of the component so many samples can be tested, relatively inexpensively.

One of the significant limitations of air-box tests is that the pressure is uniform within the test volume. This tends to limit the use of such tests to single layer systems where pressure equalization is not playing a role, or where spatial gradients over the component area are negligible. Typically, the applied load with such methods is a step-and-hold or ramp pattern, however, it is straightforward to implement a simple control system to apply cyclic (usually sinusoidal) loads. Thus, many test standards require such methods and there are many commercial laboratories around the world that test building components and cladding using these methods.

Examples of such test methods can be found in Baskaran and Chen (1998) and Henderson (2010), and others; while test standards which require these methods include tests for roof uplift performance (ASTM 2005; Underwriters Laboratories 2006), cladding fatigue performance (BCA 2006), glass performance (ASTM 2007), and many others. These tests have the significant benefit of usually being relatively simple to perform, with uniform, static (or uniform, cyclic) loads being the norm. However, this simplicity comes at the cost of a severe relaxation of the nature of the true loading. As well, the effects of altered boundary conditions in standard tests may be important since both installation and the presence of the full structure may alter the performance.

As discussed above there have been significant recent developments to test full-scale structures in wind tunnels. There have also been significant advances in the pressure control systems used in air-box methods, such that complete structures can also be tested with realistic spatially and temporally-varying wind loads. Kopp et al. (2010, 2012) discuss these methods, as well as their relative advantages and disadvantages, in some detail. However, it should be noted that these methods now allow for the direct application of simulated (or directly measured) pressure time histories for a particular component on a particular building, such as that illustrated in Fig. 7.6.

One of the significant advantages of these advanced test methods is that the relationship between the pressure and velocity is clearly defined (via the definition of the pressure coefficient). In fact, the same assumptions required to convert wind tunnel data into a code formulation (e.g., St. Pierre et al. 2005) can be applied to the interpretation of the full-scale response tests, bringing greater consistency to the process. This is particularly relevant for systems where temporal or spatial variations play a significant role in the component performance. Such test methods may become more common in the future, as these advanced control systems become more widely available.

7.5 Summary

This chapter reviews the factors affecting the wind loads on building components and cladding. These can be summarized as follows:

- Wind-induced pressures vary continuously in time over all building surfaces. For simplicity of design, wind loads specified in building codes specify peak values in a handful of zones, each with uniform pressure. Most standard tests to evaluate component performance also apply uniform, static pressures.
- The spatial variations of the pressure, combined with a component's structural details, set the load paths and load sharing in statically indeterminate systems. Consideration of this can be important in design. However, the use of a simple tributary area approach will be conservative in many, though not all, situations.
- The performance of some building components and cladding systems depend on the fluctuating wind loads, the storm duration and the storm intensity, or, in other words, the number and magnitude of the load cycles.

References

Ali HM, Senseny PE (2003) Models for standing seam roofs. J Wind Eng Ind Aerodyn 91:1689–1702
Amzallag C, Gerey JP, Robert JL, Bahuaud J (1994) Standardization of the rainflow counting method for fatigue analysis. Int J Fatigue 16:287–293
ASCE 7-10 (2010) Minimum design loads for buildings and other structures. American Society of Civil Engineers, Reston
ASCE 49-12 (2012) Wind tunnel testing of buildings and structures. American Society of Civil Engineers, Reston
ASTM (2005) Standard test method for structural performance of sheet metal roof and siding systems by uniform static air pressure difference. ASTM E 1592-01, Philadelphia
ASTM (2007) Standard practice for determining load resistance of glass in buildings. ASTM E 1300-07, West Conshohocken
Baskaran B, Chen Y (1998) Wind load cycle development for evaluating mechanically attached single-ply roofs. J Wind Eng Ind Aerodyn 77–78:83–96
BCA (2006) Building Code of Australia. Australian Building Code Board, Canberra
Beason WL, Morgan JR (1984) Glass failure prediction model. J Struct Eng 110:197–212
Bienkiewicz B, Sun Y (1992) Wind-tunnel study of wind loading on loose-laid roofing systems. J Wind Eng Ind Aerodyn 43:1817–1828
Bienkiewicz B, Sun Y (1997) Wind loading and resistance of loose-laid roof paver systems. J Wind Eng Ind Aerodyn 72:401–410
Brewick P, Divel L, Butler K, Bashor R, Kareem A (2009) Consequence of urban aerodynamics and debris impact in extreme wind events. In: Proceedings of the 11th Americas conference on wind engineering, San Juan, Puerto Rico, 2009
Brown WG (1972) A load duration theory for glass design. Pub. No. NRC 12354, National Research Council of Canada, Ottawa
Brown WG (1974) A practicable formulation for the strength of glass and its special application to large plates. Pub. No. NRC 14372, National Research Council Canada, Ottawa
Calderone IJ (1999) The equivalent wind loading for window glass design. PhD Thesis, Monash University
Charles RJ (1958) Static fatigue of glass. J Appl Phys 29:1549–1553
Cook N (1990) The designer's guide to wind loading of building structures, Part 2. Static structures. BRE, Hertfordshire
Dalgliesh WA, Taylor DA (1990) The strength and testing of window glass. Can J Civ Eng 17:752–762
Davenport AG (1975) Discussion of 'Wind pressures on buildings-probability densities'. BLWT-4-1975, Boundary Layer Wind Tunnel Laboratory, University of Western Ontario, London
Farquhar S, Kopp GA, Surry D (2005) Wind tunnel and uniform pressure testing of a standing seam metal roof model. J Struct Eng 131:650–659
Gavanski E, Kopp GA (2011a) Glass breakage tests under fluctuating wind loads. J Archit Eng 17:34–41
Gavanski E, Kopp GA (2011b) Examination of load resistance in window glass design. J Archit Eng 17:42–50
Gavanski E, Kopp GA (2011c) Storm and gust duration effects on design wind loads for glass. J Struct Eng 137:1603–1610
Henderson D (2010) Response of pierced, fixed, metal roof cladding to fluctuating wind loads. PhD Thesis, James Cook University
Holmes JD (1979) Mean and fluctuating internal pressure induced by wind. In: Proceedings of the 5th international conference on wind engineering, Colorado State University, 435–450, 1979
Holmes JD (1985) Wind action on glass and Brown's integral. Eng Struct 7:226–230
Jancauskas ED, Mahendran M, Walker GR (1994) Computer simulation of the fatigue behaviour of roof cladding during the passage of a tropical cyclone. J Wind Eng Ind Aerodyn 51:215–227

Kareem A, Stevens JG (1984) Window glass performance and analysis in Hurricane Alicia. In: ASCE proceedings specialty conference on "Hurricane Alicia: One Year Later", New York, 178–186, 1985

Kawabata S (1996) Study on wind resistance design of glass plate for cladding. PhD Thesis, Nippon Sheet Glass Co Ltd (in Japanese)

Khan MAA (2012) Load sharing of toe-nailed, roof-to-wall connections under extreme wind loads in wood-frame houses. MESc Thesis, University of Western Ontario

Kopp GA, Banks D (2012) Use of the wind tunnel test method for obtaining design wind loads on roof-mounted solar arrays. J Struct Eng. doi:10.1061/(ASCE)ST.1943-541X.0000654

Kopp GA, Gavanski E (2012) Effects of pressure equalization on the performance of residential wall systems under extreme wind loads. J Struct Eng 138:526–538

Kopp GA, Surry D, Mans C (2005) Wind effects of parapets on low buildings: Part 4. Mitigation of corner loads with alternative geometries. J Wind Eng Ind Aerodyn 93:873–888

Kopp GA, Morrison MJ, Gavanski E, Henderson DJ, Hong HP (2010) The 'Three Little Pigs' project: hurricane risk mitigation by integrated wind tunnel and full-scale laboratory tests. Nat Hazards Rev 11:151–161

Kopp GA, Morrison MJ, Henderson DJ (2012) Full-scale testing of low-rise, residential buildings with realistic wind loads. J Wind Eng Ind Aerodyn 104–106:25–39

Kumar KS (2000) Prediction of wind-induced fatigue on claddings of low buildings. Comput Struct 75:31–44

Minor JE (1981) Window glass design practices: a review. J Struct Div 107:1–12

Minor JE (1984) Window glass performance and hurricane effects. In: ASCE proceedings specialty conference on "Hurricane Alicia: One Year Later", New York, 151–167, 1984

Minor JE, Beason WL (1976) Window glass failures in windstorms. J Struct Div 102:147–160

Morgan J, Beck V (1977) Failure of sheet metal roofing under repeated wind loading. Civ Eng Trans 19:1–5

Morrison MJ, Kopp GA (2010) Analysis of wind-induced clip loads on standing seam metal roofs. J Struct Eng 136:334–337

Morrison MJ, Kopp GA (2011) Performance of toe-nail connections under realistic wind loading. Eng Struct 33:69–76

Morrison MJ, Henderson DJ, Kopp GA (2012) The response of a wood-frame, gable roof to fluctuating wind loads. Eng Struct 41:498–509

Oh JH, Kopp GA, Inculet DR (2007) The UWO contribution to the NIST aerodynamic database for wind loads on low buildings: Part 3. Internal pressures. J Wind Eng Ind Aerodyn 95:755–779

Rosowsky DV, Schiff SD (1996) Probabilistic modeling of roof sheathing uplift capacity. In: Proceedings of the 7th speciality conference on probabilistic mechanics and structural reliability. Worchester, 334–337, 1996

St Pierre LM, Kopp GA, Surry D, Ho TCE (2005) The UWO contribution to the NIST aerodynamic database for wind loads on low buildings: Part 2. Comparison of data with wind load provisions. J Wind Eng Ind Aerodyn 93:31–59

Standards Australia (2011) AS/NZS1170.2: structural design actions, Part 2. Wind actions. Standards Australia, Sydney

Surry D, Sinno RR, Nail B, Ho TCE, Farquhar S, Kopp GA (2007) Structurally-effective static wind loads for roof panels. J Struct Eng 133:871–885

Underwriters Laboratories (2006) Standard UL 580. Test for uplift resistance of roof assemblies, Underwriters Laboratories

Walker GR (1975a) Report on cyclone Tracy – effect on buildings. Department of Housing and Construction, Australia

Walker GR (1975b) Estimation of wind field assessment of building damage. Aust J Struct Eng 7:209–224

Williams T, Kareem A (2003) Performance of building cladding in urban environments under extreme winds. In: Proceedings of the 11th international conference on wind engineering, Lubbock

Chapter 8
Windborne Debris in Horizontal Winds and Applications to Impact Testing

Chris Letchford, Ning Lin, and John Holmes

Abstract This chapter considers the trajectories of compact, rod-type and plate-type windborne debris in horizontal winds, using a combination of experimental and numerical studies. These studies indicate that the ratio of horizontal debris speed to wind gust speed is primarily a function of the horizontal distance traveled by the debris. Empirical expressions to approximate the horizontal speed of these debris as a function of travel distance and time, are developed, and may be used to establish rational debris impact criteria.

Keywords Aerodynamics • Debris • Horizontal wind • Impact criteria • Trajectories

8.1 Introduction

It is well recognized that windborne debris is a major cause of building damage and destruction in strong wind events such as hurricanes. Typical windborne debris are less well-fixed building components and damaged structural members, such as roof gravel, roof tiles, roof sheathing, and framing members. Fast-flying debris may penetrate building envelopes and threaten human life and property. Debris

C. Letchford (✉)
Department of Civil and Environmental Engineering, Rensselaer Polytechnic Institute, Troy, NY, USA
e-mail: letchc@rpi.edu

N. Lin
Department of Civil and Environmental Engineering, Princeton University, Princeton, NJ, USA

J. Holmes
JDH Consulting, Mentone, VIC, Australia

Y. Tamura and A. Kareem (eds.), *Advanced Structural Wind Engineering*,
DOI 10.1007/978-4-431-54337-4_8, © Springer Japan 2013

penetration also induces internal pressurization, approximately doubling the net loading on roofs, side walls, and leeward walls. Consequently, failed roofing structures, damaged wall cladding panels, and broken glass may become debris sources, starting a "chain reaction" of failures.

The debris problem may be summarized as one of generation, trajectory, and impact. Debris generation involves debris restraining conditions (for attached items), or structural damage (for framing members). Debris impact is a problem of material and structural resistance, while impact velocity (and energy and momentum) is determined by flight trajectory. Debris trajectories involve the aerodynamics and mechanics of flying objects, and are the main topic of this chapter.

8.2 Previous Published Studies

There have been far fewer published studies on windborne debris compared with those on wind loading, although there has been a renewed interest in the topic recently. This is well out-of proportion to the sources of damage produced in severe windstorms.

In the 1970s several trajectory models for missiles in tornados were developed based on assumptions about debris aerodynamics at different levels of complexity. A simple particle model considered only the drag force on the debris (e.g., McDonald et al. 1974). The concept of average drag coefficient was developed by Simiu and Cordes (1976) to account for random missile tumbling. Lee's model (1974) of tornado-generated missiles considered constant drag and lift during the flight. Twisdale et al. (1979) developed a three dimensional (3D) random orientation model which considered random drag, lift, and side force coefficients varying with the orientation of the missile with respect to the relative wind vector. Redmann et al. (1976) developed a full 6D trajectory model considering pitching, rolling and yawing moments, as well as drag, lift, and side forces. These investigations undertaken in the 1970s contributed insights into the aerodynamic characteristics of tornado-generated missiles; however, the numerical aerodynamic models were not validated by experimental data.

Tachikawa (1983) described wind tunnel experiments to support numerical simulation studies. He measured drag, lift, and moment force coefficients on rotating plates in uniform flow, developed experimental expressions for these coefficients as functions of rotational velocity, incorporated these expressions into numerical simulations, and compared the calculated trajectories with experimental trajectories. Tachikawa (1983) is primarily a summary of an extended study of windborne debris in Japan, a full description of which has, until recently, only been available in Japanese. Tachikawa (1983) also identified a dimensionless parameter, essentially the ratio of aerodynamic to gravity forces, as the principal determinant of debris trajectories.

Holmes (2004a) calculated the trajectories of "compact" debris and studied the influences of vertical air resistance and atmospheric turbulence. A dimensional analysis was carried out, and relevant non-dimensional parameters, including the Tachikawa Number, were discussed. The calculated trajectories of compact objects were subsequently validated against wind-tunnel simulations (Holmes et al. 2004; Holmes 2004b). The aerodynamic forces and moments, and the trajectories of flat plates, or "sheet-type" objects were studied by wind-tunnel and numerical simulations by Visscher and Kopp (2005), Lin et al. (2006), and Holmes et al. (2005, 2006a, b). Good agreement between the calculated and measured trajectories was obtained when lift forces proportional to the rate of rotation ("Magnus" effect) were incorporated into the calculations.

Baker (2004, 2007) proposed an alternative non-dimensional scheme to that of Tachikawa, and studied numerical solutions of the equations of motion. Asymptotic solutions at the start of the trajectories, and at long times after the start, were studied for both compact and sheet-type debris.

Clearly, the study of the aerodynamic characteristics of debris flight is the key to determining appropriate debris impact speeds. The values of aerodynamic force coefficients, which are required in numerical simulation of debris trajectories, can be established based on wind tunnel experiments. Wind tunnel simulation of debris trajectories not only provide comparison data for numerical simulation, but can also be used directly to study the aerodynamic characteristics of flying debris in an effort to reduce the complexity of the debris flight problem.

8.3 Debris Impacts and Impact Testing Criteria

Due to great direct and indirect damage and losses caused by windborne debris, building envelopes are required to be designed against windborne debris in hurricane and tornado zones. McDonald (1976) claimed that the biggest problem in tornado-resistant design of nuclear power plants and structures is protection from missiles. He also considered windborne debris a critical design factor for other structures: above-ground shelters, schools, and hospitals, where the protection of people is the primary concern.

In modern urban areas, windows and architectural glazing systems of tall buildings are among the structures which are most vulnerable to windborne debris. Minor (1994) illustrated the serious window damage caused by windborne debris in severe windstorms over a 20-year period, including Hurricane Alicia, in Houston, Texas in August, 1983, and Hurricane Andrew in Florida, in August, 1992. Lee and Wills (2002) reported on the significant glazing damage to one of Asia's tallest buildings, Central Plaza, Hong Kong, during Typhoon York, in September 1999.

Since the original impact testing criteria using timber roofing members were developed in the 1970s in Australia following Cyclone "Tracy" (Experimental

Building Station 1978), similar test protocols involving “2 by 4” impacts have emerged in the United States. This has resulted in a number of standard test specifications for debris impact (e.g., ASTM Standard E1886-05 2005; ASTM Standard E1996-05 2005; Texas Department of Insurance 1997; SFBC 1994). In particular, the horizontal impact of a 4.1 kg (9 lb) “2 by 4” roofing member at 15.2 m/s (50 ft per second) has become a requirement for the testing of wall claddings and window shutters in South Florida since Hurricane Andrew. A steel ball test for glazing above 9 m (30 ft) has also been introduced in Florida as a surrogate for the impact of roofing gravel. More severe impact criteria are proposed for Storm Shelters by the International Code Council (International Code Council 2005). The impact resistance of various building materials to debris impacts was discussed by McDonald (1990, 1999) and Minor (1994). Although the impact of debris impact criteria is evident in recent post-hurricane damage documentation most of these storms did not reach design wind strengths. Furthermore the missile impact speeds were chosen with apparently little consideration of the aerodynamics and knowledge of debris trajectories. More rational debris impact criteria will incorporate proper models of trajectories, particularly in the horizontal direction, and this undertaking is presented here.

The present work describes deterministic studies of debris trajectories in straight-line winds such as those in traditional boundary layers including hurricanes, based on collaborative studies at Texas Tech and Louisiana State Universities. Extensive simulations to investigate the aerodynamics of various flying debris have been carried out in the Texas Tech University wind tunnel as well as full-scale simulation, employing a C-130 aircraft to generate strong winds.

The experimental and numerical studies of plate-like debris have been presented in previous papers of Lin et al. (2006) and Holmes et al. (2005, 2006a, b). The present paper describes empirical solutions, compared with numerical solutions, of horizontal trajectories of debris of the compact and rod types, these types corresponding to missiles currently used in impact testing. The main question to be answered by these studies is: what is the debris speed at impact with a downwind building at a given distance from the source? Tachikawa’s non-dimensional scheme (Tachikawa 1983) is employed to analyze the simulation data. The results described herein are based on both wind-tunnel simulation data and numerical solutions of the equations of debris motion, and can be applied to improve debris impact criteria. Vertical displacements and velocities are of secondary importance, but need consideration for determining whether a debris object strikes the ground *before* impacting on a building wall.

Although debris trajectories are essentially probabilistic in nature it is necessary to fully study the trajectories in a deterministic way in order to identify the physical parameters affecting debris impact speeds and distances traveled. This information and the models developed can later be incorporated into risk and reliability studies, and into hazard models for windstorm damage modeling.

8.4 Debris Types and Equations of Motion

8.4.1 Generic Debris Types

Wills et al. (2002) classified diverse debris geometrically into three types: "compact-like", "plate-like", and "rod-like" (Fig. 8.1). This classification identifies various debris types by their shapes and characteristic dimensions. Examples are cube and sphere roof gravel for compact-like, plywood, roof tiles, and shingles for plate-like, and "2 by 4" timbers and bamboo poles for rod-like debris. The present research uses this classification to study flight trajectories of the compact and rod-type windborne debris in horizontal winds.

8.4.2 Equations of Motion

Based on Newton's second law, the following three equations describe the motion of a generic debris object moving in a 2D plane aligned with the wind direction in a uniform wind field, in the horizontal, vertical, and rotational directions, respectively (Tachikawa 1983).

$$m\frac{d^2x}{dt^2} = m\frac{du_m}{dt} = \frac{1}{2}\rho_a A[(U - u_m)^2 + {v_m}^2](C_D \cos\beta - C_L \sin\beta) \tag{8.1}$$

$$m\frac{d^2z}{dt^2} = m\frac{dv_m}{dt} = \frac{1}{2}\rho_a A[(U - u_m)^2 + {v_m}^2](C_D \sin\beta + C_L \cos\beta) - mg \tag{8.2}$$

$$I_m\frac{d^2\theta}{dt^2} = \frac{1}{2}\rho_a Al[(U - u_m)^2 + {v_m}^2]C_M \tag{8.3}$$

where, m is the debris mass; l is a reference length (along-wind dimension for a plate or rod); A is the reference debris area (usually taken as the largest face area—i.e. l^2 for a square plate, and $l \times d$, for a rod-type, where d is the largest cross-section dimension); I_m is the mass moment of inertia; x,z are horizontal and vertical displacement of debris respectively; θ is the angular rotation; u_m is horizontal debris velocity; v_m is vertical debris velocity; U is wind speed; ρ_a is the air density; C_D, C_L,

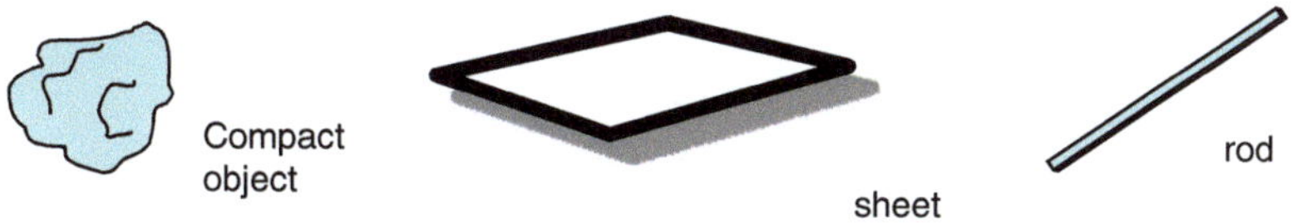

Fig. 8.1 Generic debris types (after Wills et al. 2002)

and C_M are drag, lift, and moment force coefficients, respectively; in general these are functions of the instantaneous angle of attack; β is the angle of the relative wind vector to the horizontal; g is the acceleration due to gravity; t is time.

In Eqs. (8.1), (8.2), and (8.3), the wind direction is assumed to be horizontal. These equations are coupled and must be solved simultaneously.

Compact objects are assumed to have only drag - no lift or rotational moment acting, and the drag coefficient is assumed to be independent of angle of attack. In that case, Eqs. (8.1) to (8.3) reduce to the following:

$$\frac{d^2x}{dt^2} = \frac{du_m}{dt} = \frac{\rho_a C_D A(U - u_m)\sqrt{(U - u_m)^2 + {v_m}^2}}{2m} \tag{8.4}$$

$$\frac{d^2z}{dt^2} = \frac{dv_m}{dt} = \frac{\rho_a C_D A(-v_m)\sqrt{(U - u_m)^2 + {v_m}^2}}{2m} - g \tag{8.5}$$

Equations (8.4) and (8.5) are considerably simpler to solve, but the coupling is retained. A further simplification for compact objects is possible by neglecting the air resistance to vertical motion in Eq. (8.4), and the horizontal components in Eq. (8.5). This results in the "decoupling" of Eqs. (8.4) and (8.5), and allows for "closed-form" solutions to be obtained (English and Holmes 2005). However, horizontal displacements and velocities are underestimated, and vertical displacements and velocities are over-estimated in these solutions.

Debris of the plate- and rod-types can also experience moments and rotations out of their flight plane and also out-of-plane side forces. However, the additional equations are not strongly coupled with the equation of horizontal motion (Eq. (8.1)), and have been ignored in the present work which focuses on the horizontal missile velocity at impact.

It is well known that wind fields in the atmosphere are non-uniform and turbulent. These effects can easily be incorporated into numerical calculations (e.g., Twisdale et al. 1996; Holmes 2004a). It is found that turbulence introduces variability into the trajectories. However, the effect of gusts can be treated approximately by taking the gust wind speed as a steady value for the duration of a missile flight (usually only 1–3 s). For missiles that impact at a similar height to their release point on a downwind building, the velocity variation with height over a trajectory may not be large. Hence, in the present work, in order to get usable results, a steady wind flow, uniform with height, has been assumed.

8.4.3 Non-Dimensional Forms

Tachikawa (1983) showed that Eqs. (8.1), (8.2), and (8.3) can be re-written in the following dimensionless forms.

$$\frac{d^2\bar{x}}{d\bar{t}^2} = \frac{d\bar{u}}{dt} = K[(1-\bar{u})^2 + \bar{v}^2](C_D \cos\beta - C_L \sin\beta) \tag{8.6}$$

$$\frac{d^2\bar{z}}{d\bar{t}^2} = \frac{d\bar{v}}{dt} = K[(1-\bar{u})^2 + \bar{v}^2](C_D \sin\beta + C_L \cos\beta) - 1 \tag{8.7}$$

$$\frac{d^2\theta}{d\bar{t}^2} = K\Delta Fr^2[(1-\bar{u})^2 + \bar{v}^2]C_M \tag{8.8}$$

where the dimensionless variables are:

$$\bar{x} = \frac{gx}{U^2} \text{ (dimensionless horizontal displacement)} \tag{8.9}$$

$$\bar{z} = \frac{gz}{U^2} \text{ (dimensionless vertical displacement)} \tag{8.10}$$

$$\bar{t} = \frac{gt}{U} \text{ (dimensionless time)} \tag{8.11}$$

$$\bar{u} = \frac{u_m}{U} \text{ (dimensionless horizontal debris velocity)} \tag{8.12}$$

$$\bar{v} = \frac{v_m}{U} \text{ (dimensionless vertical debris velocity)} \tag{8.13}$$

The dimensionless parameters, K, Δ, and Fr are defined as follows.

$$K = \frac{\rho_a U^2 A}{2mg} \tag{8.14}$$

K is a measure of the relationship between the aerodynamic force and the gravity force, and has been suggested to be named as Tachikawa Number (Holmes et al. 2006a).

$$\Delta = \frac{ml^2}{I_m} \tag{8.15}$$

Δ is a measure of the relationship between the mass and the rotational inertia.

$$Fr = \frac{U}{\sqrt{gl}} \tag{8.16}$$

Fr is a Froude Number.

8.4.4 Early Trajectory Solutions

For the early part of the flight, the equations can be decoupled and expressed for horizontal and vertical directions, respectively. Thus, when $\bar{u}$, $\bar{v} \ll 1$,

$$\cos\beta = \frac{1-\bar{u}}{\sqrt{(1-\bar{u})^2+\bar{v}^2}} \to 1, \quad \sin\beta = \frac{\bar{v}}{\sqrt{(1-\bar{u})^2+\bar{v}^2}} \to 0$$

neglecting small terms, in the horizontal direction, $\frac{d^2\bar{x}}{d\bar{t}^2} = \frac{d\bar{u}}{d\bar{t}} \to C_D K$. Integrating,

$$\bar{u} = C_D K\bar{t},\ \bar{x} = \frac{1}{2}C_D K\bar{t}^2,\ \text{and}\ \bar{u} = \sqrt{2C_D K\bar{x}} \tag{8.17}$$

Baker (2004, 2007) suggested similar limiting relationships, but only for compact (3D) objects. In fact, these relationships apply to all types of debris. Similarly, in the vertical direction, $\frac{d^2\bar{z}}{d\bar{t}^2} = \frac{d\bar{v}}{d\bar{t}} \to C_L K - 1$. Integrating,

$$\bar{v} = (C_L K - 1)\bar{t},\ \bar{z} = \frac{1}{2}(C_L K - 1)\bar{t}^2\ \text{and}\ \bar{v} = \sqrt{2(C_L K - 1)\bar{z}} \tag{8.18}$$

Equations (8.17) and (8.18) should be satisfied in the initial trajectory of any debris types by any empirical fit to experimental data, or numerical solutions.

8.4.5 Numerical Solutions

Although the initial trajectory solution as described above is valuable, accurate trajectories over the relevant ranges for real windborne debris must be obtained by numerical integration of the equations of motion. Thus, in solving the non-dimensional equations Eqs. (8.6) to (8.8) are used to determine the horizontal, vertical and angular accelerations at each time step. The velocities and displacements at the subsequent time step are estimated by extrapolation and interpolation. Thus the horizontal debris velocity at time step i + 1 may be obtained by:

$$\bar{u}_{i+1} = \bar{u}_i + \left(\frac{d\bar{u}}{d\bar{t}}\right)_i . \Delta\bar{t} \tag{8.19}$$

The horizontal displacement may then be estimated by:

$$\bar{x}_{i+1} = \bar{x}_i + 0.5(\bar{u}_i + \bar{u}_{i+1}).\Delta\bar{t} \tag{8.20}$$

Similar relationships are applied to the vertical and angular velocities and displacements. Note that for rod- (and plate-) type objects, the aerodynamic coefficients C_D, C_L, and C_M in Eqs. (8.6), (8.7), and (8.8) vary with the effective angle of attack of the relative wind at each time step. This angle consists of a component due to the angular rotation, θ, and an effective angle to the horizontal, β, due to the vertical motion of the object (Holmes et al. 2006a, b).

8.5 Experimental and Numerical Studies

Extensive simulations of trajectories of all types of debris have been carried out in the Texas Tech University wind tunnel. The cross section of the wind tunnel is 1.8 m wide by 1.2 m high. The wind tunnel was cleared of all roughness elements. The turbulence intensity over the flight path was 0.5 %. A circular electromagnet support (diameter $b = 18$ mm) was placed 0.6 m high and 6.65 m in front of a catch net. Small metal tabs glued to the models held them to the magnet. Wind velocities were measured by a 4-hole Cobra Probe (T.F.I. Series 100) located adjacent to the launch support. Switching off the current to the electromagnet allowed the debris to begin flight at desired wind speeds. An Olympus digital video camera (60 Hz, 0.0167 s per frame) captured the flight trajectories. The data of coordinates and the time intervals were obtained from the digital camera file and used to calculate flight speeds.

The models used in the wind-tunnel included 22 plates, 11 cubes and spheres, and eight rods, at wind speeds ranging from 4.5 to 26 m/s. The details of plate debris are given by Lin et al. (2006). The cubes ranged in weight from 1.3 to 21.1 g and in side length from 12 to 38 mm. The spheres ranged in weight from 3.6 to 47.5 g and in diameter from 19.1 to 65.9 mm. The K values of cubes and spheres ranged from 0.6 to 7.5. The "2 by 4" rods ranged in weight from 5.5 to 17.9 g and in lengths of 330 and 380 mm. The K values of rods ranged from 3.8 to 27.7. Rods were released with the major axis both perpendicular ($D/B = 0.03$ and 0.04) and parallel ($D/B = 26$ and 30) to the wind, and at initial angles of attack, α_0 of 0°, 15°, and 45°. The details of experimental procedures, data analysis methods, and all model and full-scale debris used have been reported by Lin (2005).

The trajectories of plate-like objects are relevant to debris such as roof tiles and sheathing, and are complicated by the generation of lift and angular momentum. Lin et al. 2006; Holmes et al. 2006a, b present a detailed analysis and only a summary of this work is presented here and in particular some comparisons with full scale tests conducted at Texas Tech University.

The trajectories of compact-like and rod-like objects are presented in this paper as these shapes are relevant to the debris types currently used in impact tests. Although compact objects (i.e., cubes and spheres), with only drag force acting, fell and traveled short distances in the experiments, their horizontal trajectories were mainly affected by K. With drag, lift, and pitching moments, akin to plates, rods were found to have more variation in the vertical displacement than plates; however, the horizontal component of the rod trajectories were comparable with

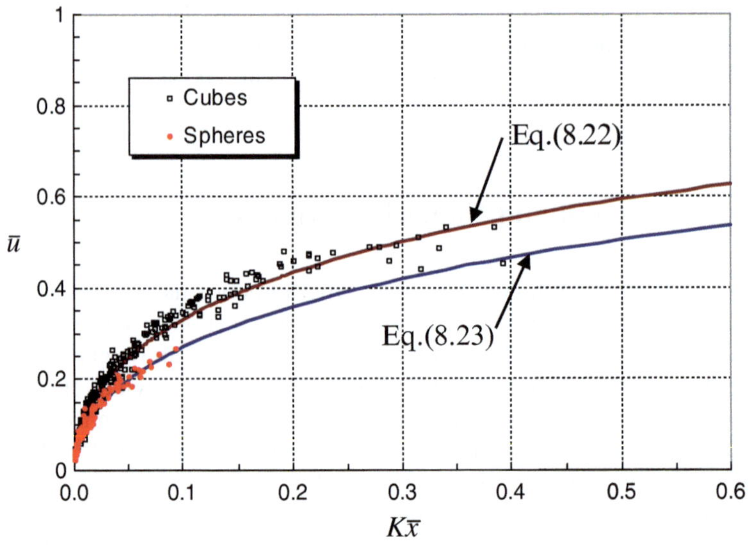

Fig. 8.2 Horizontal trajectories of compact debris as $\bar{u}$ versus $K\bar{x}$

those of plate and compact objects, and were mainly dependent on K. This characteristic of rod trajectories is discussed in detail by Lin et al. (2005a, b). Limited full scale studies were undertaken at the Reese Technology Center, Lubbock, Texas employing a C130 aircraft.

8.5.1 *Trajectories of Compact Objects*

Since the observed trajectories show that for a certain debris type, the horizontal trajectories are mainly influenced by K, and the early trajectory solution Eq. (8.17) indicates that $\bar{u}$ is basically a function of $K\bar{x}$, the experimental data for compact debris are presented in Fig. 8.2 as $\bar{u}$ versus $K\bar{x}$. The data collapse very well. Cubes fly farther and faster than spheres, due to a larger drag force.

The data show that for the entire observed trajectory, $\bar{u}$ may be approximated by an exponential function of $K\bar{x}$,

$$\bar{u} = 1 - e^{-\sqrt{2CK\bar{x}}} \tag{8.21}$$

Noting that for small values of $\sqrt{2CK\bar{x}}$, Eq. (8.21) approximates to $\bar{u} = \sqrt{2CK\bar{x}}$ (i.e., Eq. (8.17)).

Denoting C_c and C_s as the average aerodynamic coefficient for the horizontal trajectory of cubes and spheres, respectively, in Eq. (8.21), the least-square fit parameter is: $C_c = 0.81$ with a standard deviation, $\sigma = 0.020$, and $C_s = 0.50$

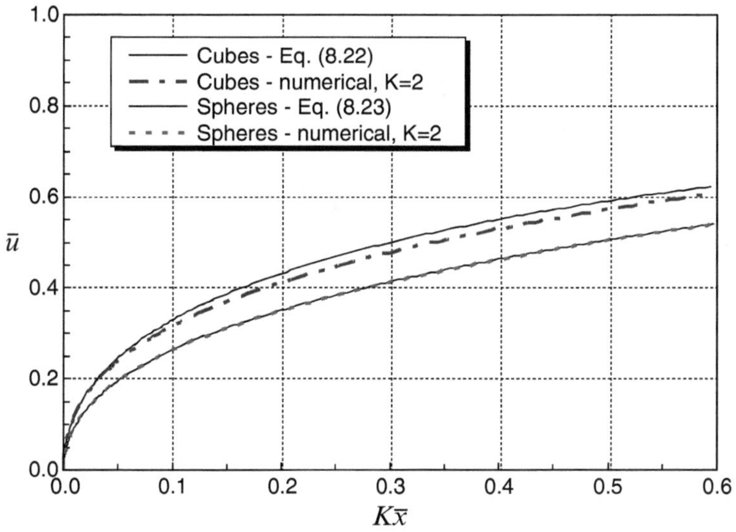

Fig. 8.3 Comparison of empirical and numerical speeds for compact debris

with $\sigma = 0.0087$. The empirically determined value of C_c and C_s approximates the drag coefficient for cubes ($C_D \approx 0.8$) and spheres ($C_D \approx 0.5$), respectively. Thus for cubes,

$$\bar{u} \approx 1 - e^{-\sqrt{1.6K\bar{x}}} \tag{8.22}$$

and for spheres,

$$\bar{u} \approx 1 - e^{-\sqrt{1.0K\bar{x}}} \tag{8.23}$$

Figure 8.3 compares these empirical expressions with numerical solutions for horizontal trajectories of compact objects. The numerical solutions are insensitive to the value of K when $\bar{u}$ is plotted against $K\bar{x}$ (English and Holmes 2005), and a single representative value of K equal to 2 has been used to generate the numerical results in Fig. 8.3. The numerical trajectory is slightly lower than the empirical trajectory for cubes, while the two are practically identical for spheres.

Equation (8.17) also indicates that $K\bar{x}$ is a function of $K\bar{t}$. Accordingly, the data for compact objects is re-presented in Fig. 8.4. The data shows that $K\bar{x}$ may be fitted by a polynomial function of $K\bar{t}$, expressed as:

$$K\bar{x} = \frac{1}{2}C(K\bar{t})^2 + a(K\bar{t})^3 + b(K\bar{t})^4 + c(K\bar{t})^5 + \ldots \tag{8.24}$$

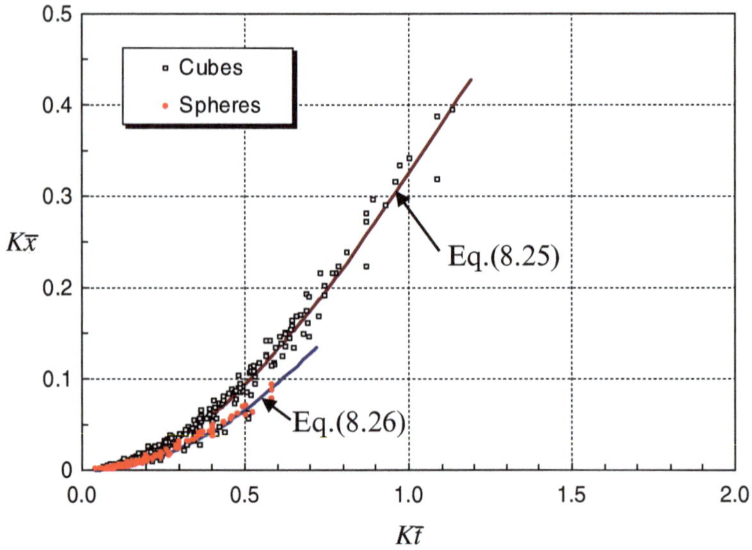

Fig. 8.4 Horizontal trajectories of compact debris as $K\bar{x}$ versus $K\bar{t}$

Note that, for small dimensionless times, Eq. (8.24) gives $\bar{x} = \frac{1}{2}CK\bar{t}^2$ (i.e., Eq. (8.17)).

With $C_c = 0.81$, the empirical fitted expression for cubes becomes,

$$K\bar{x} \approx 0.405(K\bar{t})^2 - 0.036(K\bar{t})^3 - 0.052(K\bar{t})^4 + 0.008(K\bar{t})^5 \qquad (8.25)$$

For spheres, $C_s = 0.50$,

$$K\bar{x} \approx 0.25(K\bar{t})^2 + 0.084(K\bar{t})^3 - 0.1(K\bar{t})^4 + 0.006(K\bar{t})^5 \qquad (8.26)$$

8.5.2 Trajectories of Rod-Type Objects

Experimental horizontal trajectories of rods are presented in Fig. 8.5 of $\bar{u}$ as a function of $K\bar{x}$. These data include trajectories for rods released both perpendicular and parallel to the wind, and at different initial angles of attack. Since the trend of horizontal trajectories of rods are similar to that of compact objects, Eq. (8.21) is fitted for rod data, obtaining the coefficient of rods, $C_r = 0.80$ with $\sigma = 0.062$. C_r may be regarded as the average aerodynamic coefficient for rod horizontal trajectory. Thus for rods,

$$\bar{u} \approx 1 - e^{-\sqrt{1.6K\bar{x}}} \qquad (8.27)$$

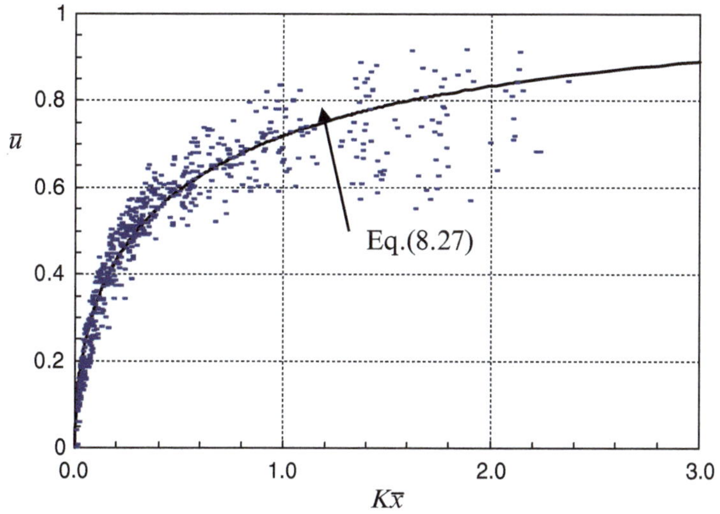

Fig. 8.5 Horizontal trajectory of rod-type debris as $\bar{u}$ versus $K\bar{x}$

Numerical trajectories for rod-type objects with the geometry of a "2 by 4" roofing member, both aligned perpendicular and parallel to the wind, were obtained using aerodynamic force coefficients generated by the model used by Twisdale et al. (1996), and assuming a center of pressure position varying between 0.25 times the length upwind from the geometric center at the angle of attack of 0^0, to the geometric center at the angle of attack of 90^0.

Figure 8.6 compares empirical and numerical results for the horizontal trajectory of rod-type objects. The numerical trajectories of $\bar{u}$ against $K\bar{x}$ are quite insensitive to the value of K for the rod-type debris, and a representative value of K equal to 10 has been used. The empirical trajectory is slightly higher than numerical trajectories. The fluctuations in the numerical trajectory are due to the angular rotations of the rod, especially for the rods aligned parallel to the wind. These fluctuations have been "smoothed" out in the empirical expression.

Figure 8.7 shows the rod data as $K\bar{x}$ versus $K\bar{t}$. The initial angle of attack (0°–45°) slightly influenced the horizontal trajectories, while values of $K\bar{x}$ of the rods released parallel to the wind are lower than those of the rods released perpendicular to the wind, and moved very slowly in early flight in the tests, resulting in very low values of $K\bar{x}$ at small $K\bar{t}$. With $C_r = 0.80$, fitting Eq. (8.22), for rods released perpendicular to the wind,

$$K\bar{x} \approx 0.40(K\bar{t})^2 - 0.16(K\bar{t})^3 + 0.036(K\bar{t})^4 - 0.0032(K\bar{t})^5 \qquad (8.28)$$

with $\sigma = 0.085$.

For rods released parallel to the wind,

$$K\bar{x} \approx 0.40(K\bar{t})^2 - 0.29(K\bar{t})^3 + 0.088(K\bar{t})^4 - 0.0082(K\bar{t})^5 \qquad (8.29)$$

with $\sigma = 0.15$.

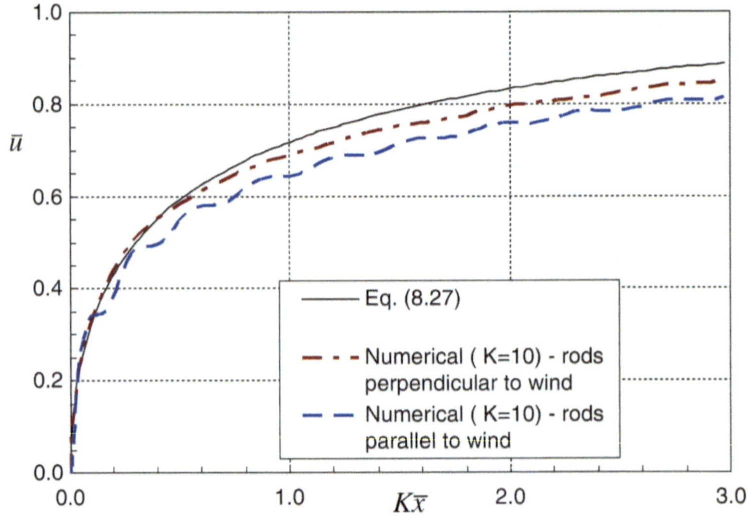

Fig. 8.6 Comparison of empirical and numerical speeds for rod-type debris

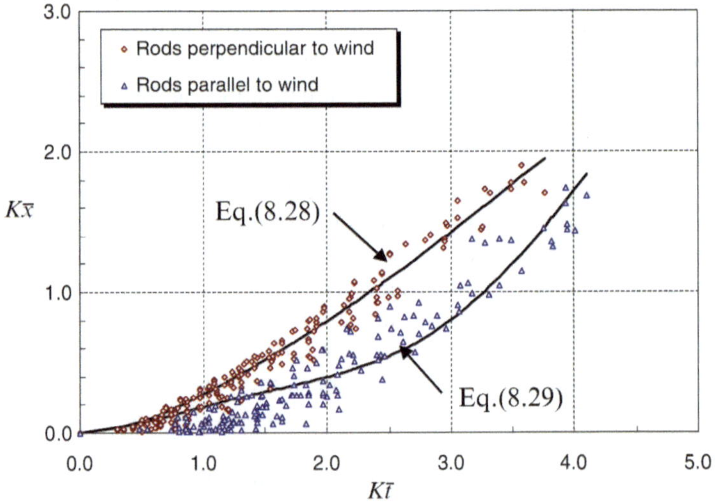

Fig. 8.7 Horizontal trajectories of rod-type debris as $K\bar{x}$ versus $K\bar{t}$

8.5.3 Trajectories of Plate-Type Objects

The flight of plate-type debris is much more complex because of lift generated by auto-rotation and experimental results indicate that, for a certain debris shape, the debris trajectory (T) is a function of at least nine parameters: wind speed (U), air density (ρ_a), plate dimensions (B, D, and h), plate density (ρ_m), support dimension (b), support position (s, e.g., center, corner, or edge), and initial angle of attack (α_0). Because debris impact criteria have concentrated on generating repeatable tests, to date they have focused on compact (ball bearings) and rod-type ("2 by 4"). Hence, plate type debris flight will not be discussed in detail here and the reader is referred to Lin (2005) for details.

8.6 Applications

The previous sections have shown that the ratio of the horizontal component of flight speed to the wind speed, for a given debris type, is a function of $K\bar{x}$. This relationship is apparent in the empirical relationship of Eq. (8.21). Since the wind speed cancels out of $K\bar{x}$, this means that the velocity ratio is simply a function of the horizontal distance traveled, for a given debris item. (Note: this is not the case for the vertical component of debris speed). The average aerodynamic coefficient C for debris horizontal trajectories is obtained from fitting Eq. (8.21) to simulation data, either experimental or numerical. In this way, the debris horizontal impact speed at a horizontal distance traveled may be estimated and used to establish rational debris impact criteria.

As application examples, Fig. 8.8 shows the ratio of horizontal missile speed to wind speed of a 2 g, 8 mm steel ball, and Fig. 8.9 of a 4.1 kg (9 lb) "2 by 4" missile, which are often specified as representative debris in impact testing standards. The numerically calculated trajectories in Figs. 8.8 and 8.9 are insensitive to the wind speed, but were obtained for a representative hurricane gust wind speed of 60 m/s.

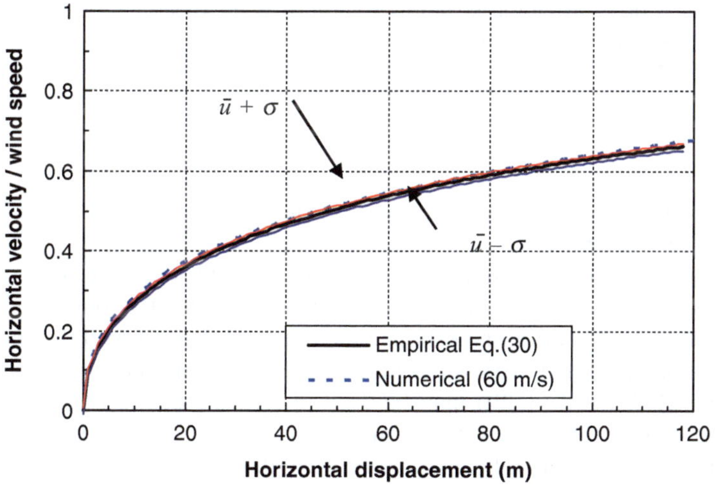

Fig. 8.8 Trajectory of a steel ball (8 mm, 2 g)

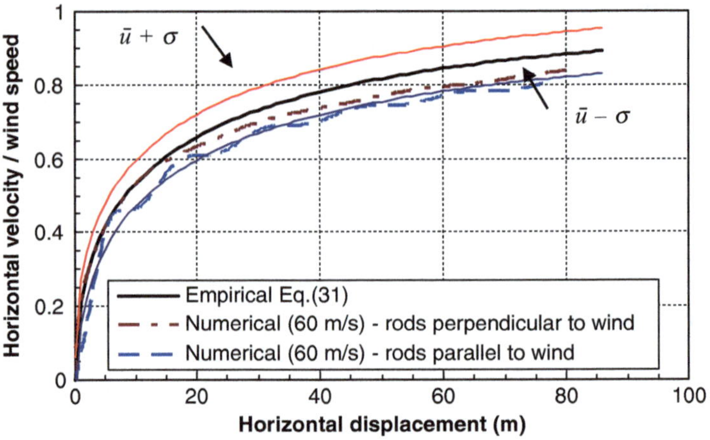

Fig. 8.9 Trajectory of a "2 by 4" missile (2.4 × 0.1 × 0.05 m, 4.1 kg, 0°)

ASTM Standards E1886-05 (2005) specified the same steel ball as the "small missile" representative with impact speed between 0.40 and 0.85 of the 3-s gust wind speed, and the "2 by 4" missile as a representative of the "large missile" with impact speed ratio between 0.1 and 0.55 of the 3-s gust wind speed. These specifications may overestimate or underestimate actual debris impact speeds, depending on how far the debris travels before impact.

Assuming $\rho_a = 1.2$ kg/m^3, the empirical expression for the 2 g, 8 mm diameter steel ball is obtained from Eq. (8.23) as:

$$\bar{u} \approx 1 - e^{-\sqrt{0.010x}}, \; \sigma = 0.0087 \tag{8.30}$$

where $\bar{u}$ is debris horizontal speed as a ratio of wind speed ($\bar{u} = u_m/U$), and x is debris horizontal displacement.

For the "2 by 4" missile (2.4 × 0.1 × 0.05 m, 4.1 kg) Eq. (8.27) gives:

$$\bar{u} \approx 1 - e^{-\sqrt{0.058x}}, \; \sigma = 0.0616 \tag{8.31}$$

Eqs. (8.30) and (8.31) may be directly used in specifying impact criteria for these particular debris items.

It should be noted that impact speed increases with flight distance; however, debris may have fallen to the ground before reaching the object of interest. Therefore, it is rational to first estimate the potential debris travel distance. Equation (8.24) and the fitted expressions for cubes (Eq. (8.25)), spheres (Eq.(8.26)), and rods (Eqs. (8.28) and (8.29)) may be used to estimate debris travel distance for given flight times. The debris flight time usually lasts for 1–3 s. Exact debris flight time depends on the initial height of debris and the vertical trajectory.

8.7 Discussion

This chapter has considered the trajectories of compact and rod-type windborne debris, using a combination of wind-tunnel studies, and numerical solutions of the equations of motion. These debris types are representative of the roof gravel and timber members currently used for testing building facades and storm shutters in Florida.

These studies have indicated that the ratio of horizontal debris speed to wind gust speed is primarily a function of the horizontal distance traveled by the debris as it accelerates towards the wind speed. Stating an "average" ratio of missile speed to wind speed over a trajectory (e.g., Wills et al. 2002) is not meaningful for a quantity that increases monotonically with distance traveled. The ratio of horizontal missile speed to wind speed approaches a value of 1.0 asymptotically. An intermediate maximum, followed by a falling value in that ratio with distance, cannot occur in a uniform wind field (as suggested or implied in previous studies). This would imply a negative horizontal acceleration which, if the wind speed exceeds the horizontal missile speed, could only be obtained with a negative drag coefficient—this is obviously not a physically realizable situation.

Simple empirical expressions for the relative horizontal missile speed with horizontal distance have been derived. These can be used to establish rational missile impact tests, both in terms of missile characteristics and speed for traveled distance (Lin et al. 2007). Previously, it appears these criteria were derived without any explicit consideration of building spacing.

Acknowledgements This work was carried out under the auspices of the U.S. Department of Commerce Texas Tech University /National Institute of Standards and Technology Windstorm Mitigation Initiative, the John P. Laborde Visiting Professor endowment, and Louisiana Sea Grant.

References

ASTM Standard E1886-05 (2005) Standard test method for performance of exterior windows, curtain walls, doors, and storm shutters impacted by missile(s) and exposed to cyclic pressure differentials. American Society for Testing and Materials, Inc., West Conshohocken

ASTM Standard E1996-05 (2005) Standard specification for performance of exterior windows, glazed curtain walls, doors and storm shutters impacted by windborne debris in hurricanes. American Society for Testing and Materials, Inc., West Conshohocken

Baker CJ (2004) Solutions of the debris equations. In: Proceedings of the 6th U.K. conference on wind engineering, Wind Engineering Society, Cranfield, 15–17 September 2004

Baker CJ (2007) The debris flight equations. J Wind Eng Ind Aerodyn 95:329–353

English EC, Holmes JD (2005) Non-dimensional solutions for trajectories of wind-driven compact objects. In: Proceedings of the 4th European and African conference on wind engineering (EACWE2005), Institute of Theoretical and Applied Mechanics, Prague, 11–15 July 2005

Experimental Building Station (1978) Guidelines for the testing and evaluation of products for cyclone prone areas, Technical Record 440. Department of Housing and Construction, Chatswood

Holmes JD (2004a) Trajectories of spheres in strong winds with application to wind-borne debris. J Wind Eng Ind Aerodyn 92:9–22
Holmes JD (2004b) The aerodynamics and mechanics of wind-borne debris. In: Proceedings of the 1st international symposium on wind effects and urban environment, Wind Engineering Research Center, Tokyo Polytechnic University, Tokyo, Japan, 8–9 Mar 2004
Holmes JD, English EC, Letchford CW (2004) Aerodynamic forces and moments on cubes and flat plates, with applications to wind-borne debris. In: Proceedings of the 5th international colloquium on bluff-body aerodynamics & applications, NRCC, Ottawa, 11–15 July 2004
Holmes JD, Letchford CW, Lin N (2005) Trajectories of windborne debris of the plate type. In: Proceedings of the 10th Americas conference on wind engineering, AAWE, Baton Rouge, 1–4 June 2005
Holmes JD, Baker CJ, Tamura Y (2006a) The Tachikawa number—a proposal. J Wind Eng Ind Aerodyn 94:41–47
Holmes JD, Letchford CW, Lin N (2006b) Investigations of plate-type windborne debris. II Computed trajectories. J Wind Eng Ind Aerodyn 94:21–39
International Code Council (2005) ICC draft standard on design, construction and performance of storm shelters. I.C.C., Falls Church. www.iccsafe.org
Lee AJH (1974) A general study of tornado-generated missiles. Nucl Eng Des 30:418–433
Lee BE, Wills JAB (2002) Vulnerability of fully glazed high-rise buildings in tropical cyclones. J Arch Engg (ASCE) 8:42–48
Lin N (2005) Simulation of windborne debris trajectories. M.S. thesis, Department of Civil Engineering, Texas Tech University
Lin N, Letchford CW, Gunn T (2005) Investigation of the flight mechanics of 1D (rod-like) debris. In: Proceedings of the fourth European & African conference on wind engineering (EACWE2005), Institute of Theoretical and Applied Mechanics, Prague, 11–15 July 2005
Lin N, Letchford CW, Holmes JD (2005) Experimental investigation of trajectory of windborne debris with applications to debris impact criteria. In: Proceedings of the 10th Americas conference on wind engineering, AAWE, Baton Rouge, 1–4 June 2005
Lin N, Letchford CW, Holmes JD (2006) Investigations of plate-type windborne debris. Part I. Experiments in wind tunnel and full scale. J Wind Eng Ind Aerodyn 94:51–76
Lin N, Holmes JD, Letchford CW (2007) Trajectories of windborne debris and applications to impact testing. ASCE J Struct Eng 133(2):274–282
McDonald JR (1990) Impact resistance of common building materials to tornado missiles. J Wind Eng Ind Aerodyn 36:717–742
McDonald JR (1999) Windborne debris impacts on metal wall panels and their implications. In: Larsen, Larose, Livesey (eds) Wind engineering into the 21st century. Balkema, Rotterdam. ISBN 90 5809 059 0, 1443-1449
McDonald JR, Mehta KC, Minor JE (1974) Tornado-resistant design of nuclear power-plant structures. Nucl Saf 15(4):432–439
Minor JE (1994) Windborne debris and the building envelope. J Wind Eng Ind Aerodyn 53:207–227
Redmann GH, Radbill JR, Marte JE, Degarabedian P, Fendell FE (1976) Wind field and trajectory models for tornado propelled objects. Electrical Power Research Institute, Technical report 1, Palo Alto
SFBC (1994). South Florida building code, 1994 Edition for Miami-Dade County
Simiu E, Cordes M (1976) Tornado borne missile speeds. NBSIR 76-1050, National Bureau of Standards, Gaithersburg
Tachikawa M (1983) Trajectories of flat plates in uniform flow with application to wind-generated missiles. J Wind Eng Ind Aerodyn 14:443–453
Texas Department of Insurance (1997) Building code for windstorm resistant construction. Texas Department of Insurance, Austin
Twisdale LA, Dunn WL, Davis TL (1979) Tornado missile transport analysis. Nucl Eng Des 51:295–308

Twisdale LA, Vickery PJ, Steckley AC (1996) Analysis of hurricane windborne debris risk for residential structures. Applied Research Associates, Inc., Raleigh

Visscher B, Kopp G (2005) Wind-tunnel measurements of trajectories of roof sheathing panels under high wind loads. In: Proceedings of the 10th Americas conference on wind engineering, AAWE, Baton Rouge, 1–4 June 2005

Wills JAB, Lee BE, Wyatt TA (2002) A model of wind-borne debris damage. J Wind Eng Ind Aerodyn 90:555–565

Chapter 9
Understanding Wind Codes and Standards: Fundamentals Behind Their Provisions

Theodore Stathopoulos

Abstract This chapter is a collection of notes describing the progress made in the last few decades on the assessment of wind loads on buildings through various codes of practice. This is necessary to understand the current code provisions for estimating wind-induced pressures on buildings, which in turn can help improve these provisions. Background studies carried out for the development of standard wind provisions almost exclusively used boundary layer wind tunnel experiments, with very limited attempts to capitalize on computational approaches. Topics presented in this chapter include: wind structure; wind speed models; wind pressure provisions for gable roof buildings with intermediate roof angles; pressure coefficients for hipped roofs; and internal pressures. It was generally observed that the current codes must incorporate the effects of upstream terrain roughness, different roof shapes and other architectural features. Advancements in Computational Wind Engineering (CFD and ANN approaches) have been discussed. Recent innovative approaches, such as wind load paths and Database-Assisted Design (DAD) using time histories of wind pressure coefficients for specific building geometries are expected to be used for the provisions of future codes of practice.

Keywords Buildings • Code provisions • Computational wind engineering • Design • Wind loads

---

T. Stathopoulos (✉)
Department of Building, Civil and Environmental Engineering, Concordia University, Montreal, QC, H3G 1M8, Canada
e-mail: statho@bcee.concordia.ca

Y. Tamura and A. Kareem (eds.), *Advanced Structural Wind Engineering*, DOI 10.1007/978-4-431-54337-4_9, © Springer Japan 2013

9.1 Introduction

This chapter includes lecture notes addressing some fundamentals in wind engineering useful to understand the development and evolution of wind standards and codes of practice with respect to the wind design of buildings. There has been a lot of progress during the last century in the development of wind engineering and the results originating from intensive research have been incorporated into upgraded editions of various national and international wind standards and codes of practice. Special emphasis is given to the low buildings, since tall buildings, particularly those of complex shapes and inhomogeneous surroundings are still tested in boundary layer wind tunnels, although computational attempts are always on the rise. Table 9.1 shows the pressures in pounds per square foot on a flat surface normal to the direction of the wind for different velocities in mph, as calculated by the formula $P = 0.004\ V^2$.

A one-page description of wind code specifications has now evolved hundreds of pages in books requiring detailed guides and commentaries to elaborate on these provisions and assist designers with proper interpretation and correct application. In fact, most national and international wind load standards have a common format in so far as they include their provisions in terms of descriptive instructions to the designer followed by a commentary or supplement or appendix with background information and explanations relaying to the origin of the provisions or rationale for their modifications from previous editions of the code/standard. Relevant references are provided for the interested reader for further reading and understanding, although some countries' standards are mute in this regard, presumably because code committees have utilized their collective wisdom significantly and altered or not considered parts of the content of the material of the original research.

Another common characteristic in the provisions of various wind loading standards is the basic equation for the evaluation of wind pressure loads acting on building surfaces, which takes a form such as:

$$P = q\,C_e\,C_p\,C_g \tag{9.1}$$

Table 9.1 Wind pressures specified by ARMCO steel buildings in 1944

Velocity (mph)	Pressure (lbs/ft^2)	Wind type
10	0.4	Fresh breeze
20	1.6	Mild wind
30	3.6	Strong wind
40	6.4	High wind
50	10.0	Storm
60	14.4	Violent storm
80	25.6	Hurricane
100	40.0	Violent hurricane

NB: For a rectangular building, the pressure may be taken as 80 % of that given in this table

This equation includes the dynamic velocity pressure (q) depending on wind climatology and terrain topography, an exposure factor (C_e) depending on height and upstream exposure characteristics, the pressure coefficient or shape factor (C_p) depending on the shape and geometric details of the building interacting with the wind and a gust effect factor (C_g), which expresses the dynamic character of the wind pressures acting on building surfaces, as well as the local fluctuations including the so-called "hot spots" influencing the design of cladding elements. Each of these basic factors will be discussed in his chapter.

Wind loading provisions included in the Commentaries of the National Building Code of Canada (NBCC 1995, 2005, 2010) are indeed pioneering whereas the American National Standard ASCE 7-10 (2010) is now considered one of the best in the world, since it incorporates the major research findings in the area of wind-building interaction. Upgrading the code provisions is of course an ongoing process and there are always issues to be settled and innovations to be implemented. Currently, critical issues include, among others, the evaluation of wind loads on buildings of different shapes, the stipulations for the evaluation of internal pressures, as well as the clear definition of exposure including cases of roughness transition. These are significant questions to dominate the research agenda in the years to come. In addition, the understanding and evaluation of load paths for various building configurations are real challenges, while work is expected to continue on various means to mitigate the effects of vortices generating very high suctions near roof corners, ridges and edges.

This chapter reviews the evolution of knowledge and its current state regarding the evaluation of wind loads on low buildings. A brief historical perspective is followed by some detailed description of the University of Western Ontario's research on wind loads on low buildings carried out in the 1970s. The influence of the research results on the formulation of design load provisions in contemporary wind standards and codes of practice will be shown. Finally, the status of computational wind engineering, as well as the so-called computer-aided wind engineering, in the evaluation of wind pressures on buildings will be discussed.

9.2 Wind Velocity

The dynamic velocity pressure $q = ½ \rho V^2$ of Eq. (9.1) depends heavily on the wind speed V.

It should be noted that:

- Ground obstructions retard the movement of air close to the ground surface, causing a reduction in wind speed (Fig. 9.1).
- At some height above ground, the movement of air is no longer affected by ground obstruction. This height is called gradient height, z_g, which is a function of ground roughness.

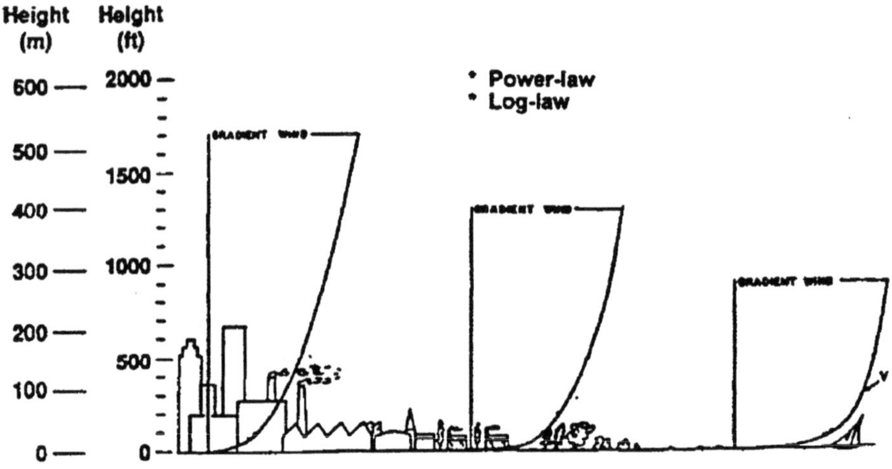

Fig. 9.1 Variation of wind speed with height (Davenport 1967)

Table 9.2 Typical values of velocity profile characteristics

Terrain category	Terrain description	Gradient height, z_g (m)	Roughness length, z_o (m)	Mean speed exponent (α)	Gust speed exponent
1	Open sea, ice, tundra, desert	250	0.001	0.11	0.07
2	Open country with low scrub or scattered trees	300	0.03	0.15	0.09
3	Suburban areas, small towns, well wooded areas	400	0.3	0.25	0.14
4	Numerous tall buildings, city centers, well developed industrial areas	500	3	0.36	0.20

The unobstructed wind speed is called gradient wind speed, V_g and it is constant above gradient height.

The power law is generally used by engineers to represent the variation of wind speed with height. This is mostly an empirical equation and is defined as

$$\frac{V_z}{V_g} = \left(\frac{z}{z_g}\right)^{\alpha} \tag{9.2}$$

where z_g and α are functions of ground roughness. Typical values of z_g and α are given in Table 9.2.

The logarithmic law is used more by meteorologists [Eq. (9.3)] and is based on physics of the boundary layer. It is valid in the bottom 20–30 % of the boundary layer and is defined as

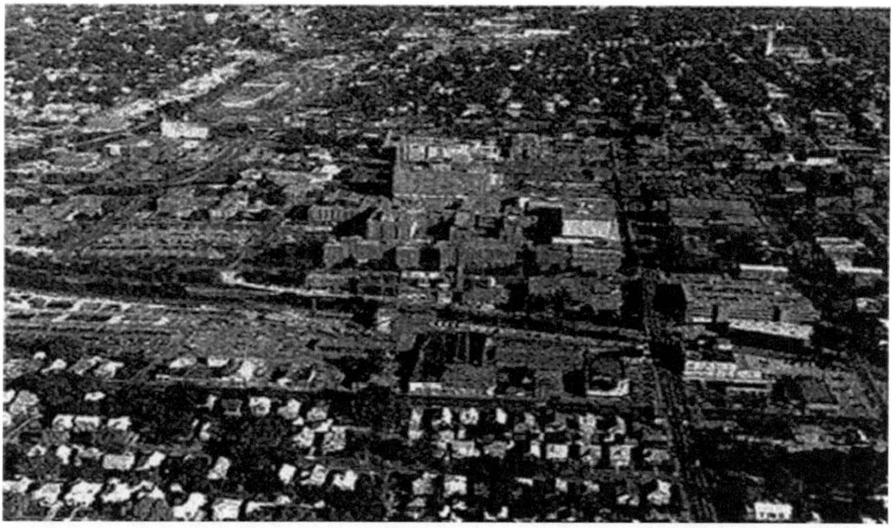

Fig. 9.2 Exposure B: urban area with numerous closely spaced obstructions having the size of single-family dwellings or larger. For all structures shown, terrain representative of surface roughness category B extends more than 20 times the height of the structure or 2,600 ft (792 m), whichever is greater, in the upwind direction (ASCE 2010)

$$V_z = 2.5u^* \ln\frac{z}{z_o} \tag{9.3}$$

where z_0 is the roughness length, $u*$ is the friction velocity $\left(\sqrt{\frac{\tau_o}{\rho}} \cong 1 - 2m/s - \text{for extreme winds}\right)$, τ_o is the shear stress at the ground surface and ρ is the air density. Table 9.2 also includes typical values of z_0 and the gust speed exponent calculated as 60 % of the mean speed exponent.

9.2.1 Exposures

A number of changes were made to the definitions of exposure categories in the most recent editions of North American wind standards. For instance, in the current ASCE 7-10 standard (ASCE 2010), Exposure A has been eliminated leaving the basic exposure types to three, namely B (suburban), C (open country) and D (shoreline areas). Exposure A was intended for very dense centers of city areas in close proximity to tall buildings. In these areas, the variability of the wind is indeed great to the point that only physical simulation in wind tunnels might predict local wind conditions.

Distinction made between surface roughness and exposure categories led to more precise definitions of exposures. Figures 9.2, 9.3, and 9.4 show typical

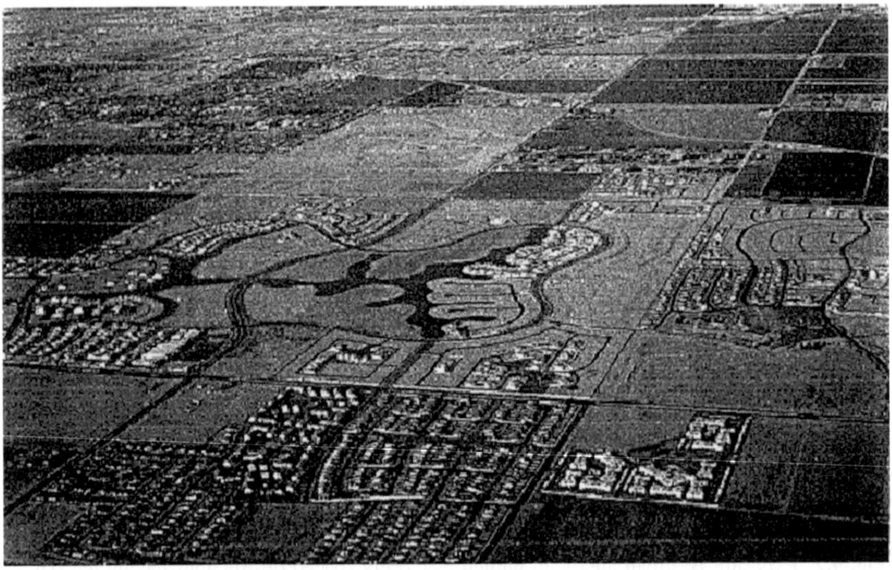

Fig. 9.3 Exposure C: open terrain with scattered obstructions having heights generally less than 30 ft for most wind directions. All 1-story structures with a mean roof height less than 30 ft in the photograph are less than 1,500 ft or ten times the height of the structure, whichever is greater from an open field that prevents the use of exposure B (ASCE 2010)

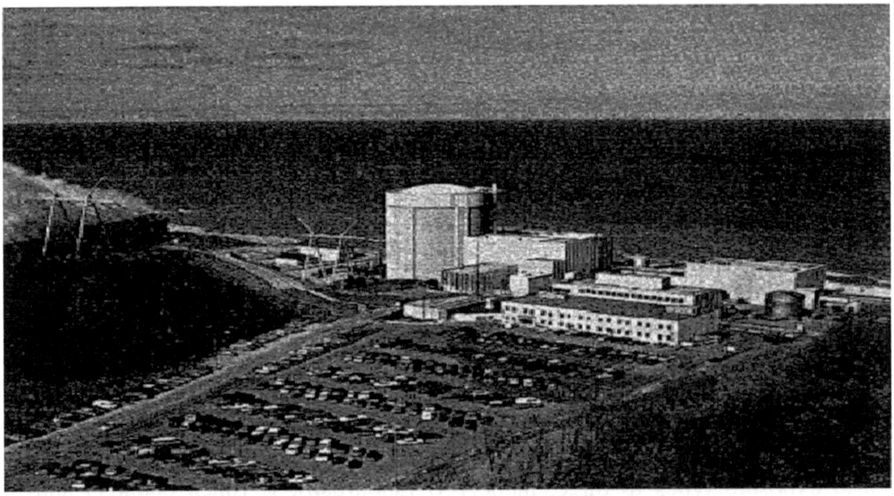

Fig. 9.4 Exposure D: a building at the shoreline (excluding shorelines in hurricane-prone regions) with wind flowing over open water for a distance of at least 1 mile. Shorelines in exposure D include inland waterways, the great lakes, and coastal areas of California, Oregon, Washington, and Alaska (ASCE 2010)

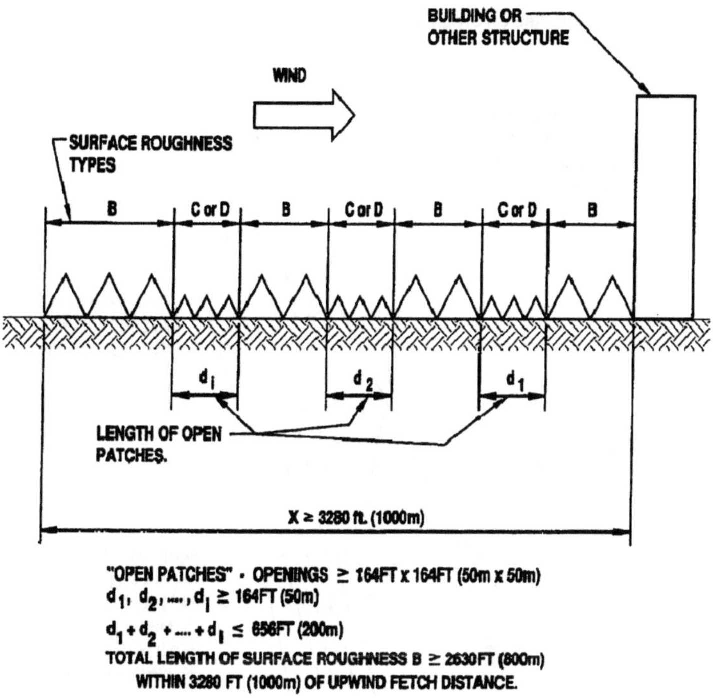

Fig. 9.5 Exposure B with upwind open patches (ASCE 2010)

cases of B, C and D exposure categories. It should be noted that the exposure for each wind direction was defined as the worst case of the two 45° sectors from either side of the direction under consideration.

The required fetch to characterize an exposure upwind of a tall building has been increased from 10 to 20 building heights. For exposure B, a surface roughness for at least 800 m upwind a building (with insufficient open patches) is required in order to name it such. An open patch is defined as an opening greater than or equal to about 50 m on each side. Figures 9.5 and 9.6 show the effect of open patches of surface roughness C or D on the use of exposure category B. Clearly the plan location of any open patch may have a different effect for different wind directions.

Good engineering judgment is a sine-qua-non for the proper assessment of exposure. It should also be kept in mind that the exposure for a particular building may change in several wind directions due to the future development and/or demolition of upstream buildings.

9.2.1.1 Upstream Exposure

A renewed interest in a more rigorous definition of upstream exposure and the roughness associated with it has recently surfaced. This is due to the perceived significant potential for reduction of design wind pressures on low buildings due to

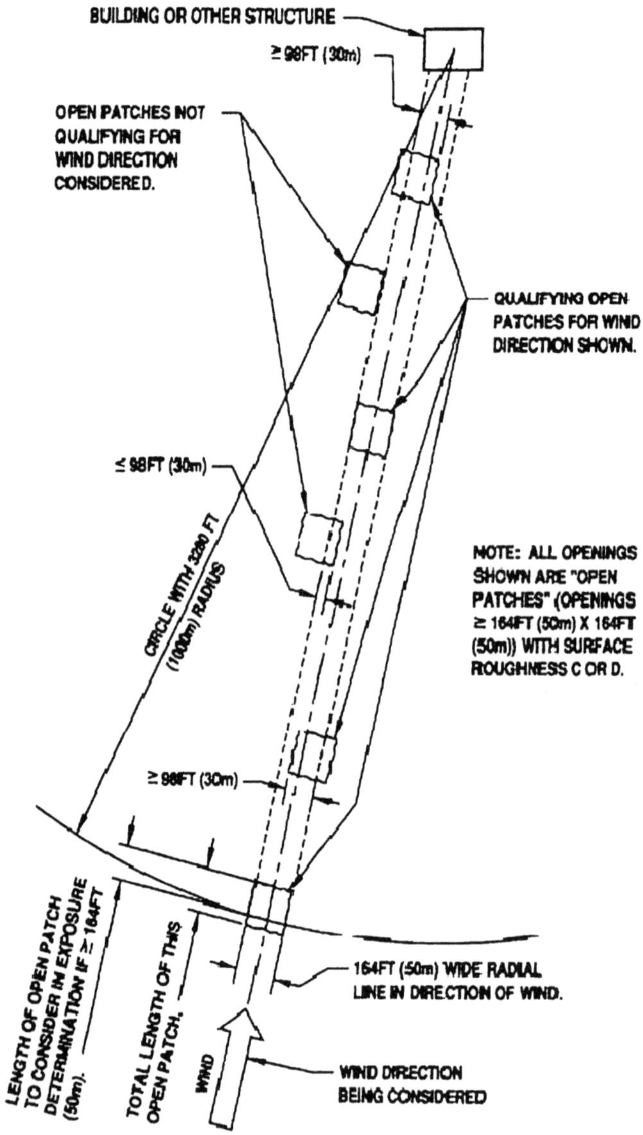

Fig. 9.6 Exposure B with open patches (ASCE 2010)

the actual retardation of the flow expected above rougher terrain conditions. The objective is to include in the design wind provisions more clear definitions of the exposure including the transition cases by specifying appropriate values for the exposure coefficient. A detailed elaboration of the exposure effects will increase the complexity of the design process but may indeed lead to more adequate and

Table 9.3 Values of z_0 and minimum fetch length of typical wind standards and codes of practice

Terrain class	–	NBCC (2010)	ASCE 7-10 (2010)	AS/NZS 1170.2 (2002)	BS 6399-2 (1997)
Coastal	z_0 (m)	In terms of α	0.005	0.002	0.003
	Min. fetch	–	Max (1,524 m, 20 H)	Variable[a], dependent on H	1,000 m
Open country	z_0 (m)	In terms of α	0.02	0.02	0.03
	Min. fetch	Max (1,000 m, 20 H)	–	Variable[a], dependent on H	–
Suburban/urban	z_0 (m)	In terms of α	0.3	0.2	0.3
	Min. fetch	Max (1,000 m, 20 H)	Max (792 m, 20 H); 457 m if $H_0 < 9.1\ m$	Variable[a], dependent on H	100 m (with $H_0 \geq 5$ m)
Large-city-centre	z_0 (m)	–	2	2	–
	Min. fetch	–	–	Variable[a], dependent on H	–

H Building height, H_0 Average roof level
[a]1,000 m (for $H < 50$ m); 2,000 m (for 50 m $\leq H < 100$ m); 3,000 m (for 100 m $\leq H < 200$ m); 4,000 m (for $H \geq 200$ m)

economical wind provisions. Regardless, the amount of research work dedicated to pressure coefficients in the last few decades is very disproportionate in comparison with the rather minimal effort made so far on the investigation of exposure effects. However, it would also be fair to mention that upstream exposure effects are much more difficult to study and indeed hard to codify for design purposes. Furthermore, there are currently significant discrepancies and inconsistencies on the definition of upstream terrain exposures in various wind design standards and codes of practice, as shown in some detail in Table 9.3. A typical example may be the minimum upstream fetch required to characterize the most frequently met suburban/urban exposure, which varies from a minimum of 100 m in the British Code (BSI 6399-2 1997) to a maximum of 1,000 m in the Canadian Code (NBCC 2010).

It should be noted that several current wind pressure coefficient provisions in codes and standards including those in the 2005 National Building Code of Canada (NBCC 2005) were based on a conservative open country exposure regardless of the actual situation and the prevailing wind direction. The American standard ASCE 7-98 (1998) has proceeded to allow the use of coefficients similar to those used in Canada—but with a 0.85 instead of a 0.8 directionality factor—with the velocity profile of a suburban exposure if the actual building is indeed exposed to a suburban corresponding roughness. This reflects the great majority of cases of low-rise buildings. St. Pierre (2002) has compared experimentally measured vertical uplift and horizontal thrust coefficients on an end bay of gabled roof low buildings in suburban terrain roughness with the Canadian-NBCC (1995), American-ASCE 7-98 (1998) and European-ENV (CEN 1995) corresponding provisions. It was

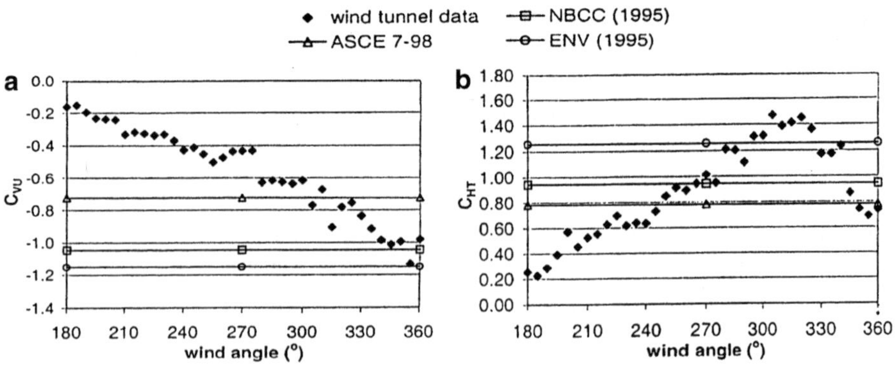

Fig. 9.7 Minimum vertical uplift and maximum horizontal thrust coefficients for the end bay of a gabled roof building in suburban terrain exposure (**a**) eave height: 9.75 m; roof slope: 3:12 (**b**) eave height: 7.32 m; roof slope: 1:12 (St. Pierre 2002)

shown that the American provisions underestimate the measured data distinctly more than the other specifications. Figure 9.7 shows typical results from this study. In fact it appears that the ASCE 7-98 coefficients may have been reduced more than warranted when the building design case under consideration is indeed in a suburban type of upstream terrain for all wind directions. Furthermore, the urban exposure (Exposure A) was eliminated in the 2002 edition of the ASCE 7 Standard, since it was deemed appropriate for the evaluation of design pressures to carry out a wind tunnel study to account for the proximity effects that are usually quite significant in such cases. Current efforts related to this area address the necessary roughness fetch requirements for the characterization of an exposure, and examine the transition equations in order to provide exposure coefficients appropriate for cases of upstream terrain roughness changes. The intent is to allow interpolation in exposure transition zones near changes in ground surface roughness and ESDU sources are the main origin of such provisions.

9.2.2 *Wind Speed and Turbulence Models*

Recently, a new wind speed model has been proposed for estimating design wind loads above fetch with roughness changes after an extensive wind tunnel study, as well as a numerical study, dealing with a variety of upstream exposure configurations for homogeneous and inhomogeneous terrain cases illustrated in detail in Fig. 9.8. The model uses *N* sections corresponding to *N* different roughness patches in the upstream terrain within a range no more than 4 km from the site, and the recognition of patch type and patch roughness information (z_o and α) can be referenced to a roughness classification after Davenport et al. (2000). The importance, relevance, and validation of the speed model have been discussed in detail in Wang and Stathopoulos (2005).

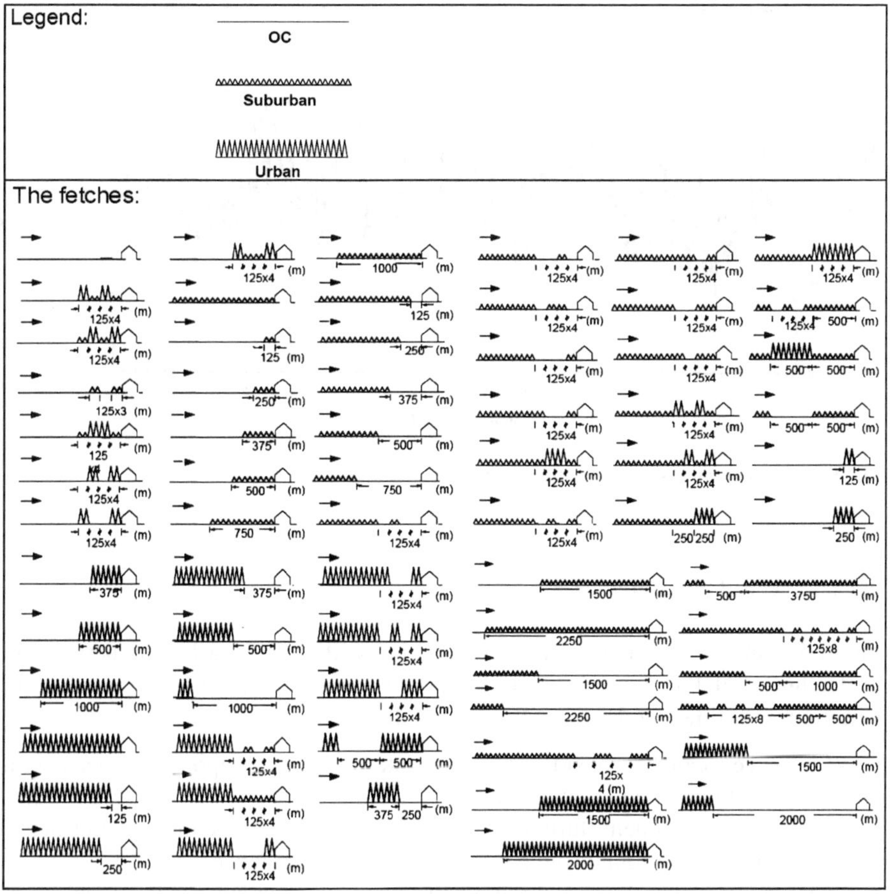

Fig. 9.8 The variety of fetches used in the experimental study (Wang and Stathopoulos 2005)

This wind-tunnel study has also provided an opportunity to establish a model estimating turbulence intensity (I_u) above various exposures, where I_u is defined as:

$$I_u = \sigma / V_z \tag{9.4}$$

where V_z = mean wind speed at elevation z and σ = standard deviation value of fluctuating velocity. It should be noted that, for homogeneous terrain, previous studies have provided measured I_u data, and a number of sources (e.g. ESDU 82026; Simiu and Scanlan 1996; AIJ 1996; ASCE 7-02 2003; Zhou and Kareem 2002) have proposed specific models; however, for inhomogeneous terrain, only few studies were found, among which Deaves (1981) furnished the numerical results that were formulated into datasets for typical inhomogeneous fetch examples in ESDU (84030).

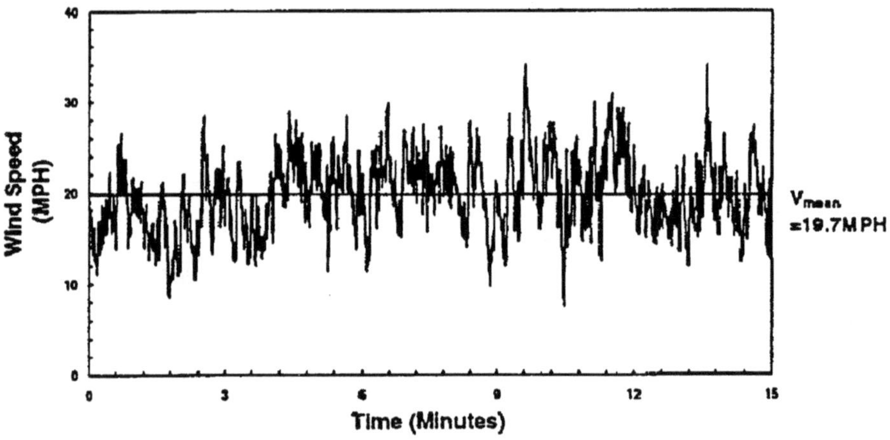

Fig. 9.9 Typical wind speed variation with time

9.2.2.1 Wind Turbulence

Figure 9.9 shows a typical wind speed variation with time for a 15-min period. The unsteadiness of the wind is clear. Important characteristics of this wind speed variation with time can be summarized as follows:

- Wind speed can be considered as consisted of two components: mean wind speed and fluctuating component.
- Fluctuating component (turbulence) is caused by convective movement (meteorological) and/or ground roughness (mechanical turbulence).
- In boundary layer flow at high wind speed, it is assumed that mechanical turbulence predominates; this is of interest to engineers.
- Turbulence (mechanical) is higher in rougher terrain than in smoother terrain; e.g. suburban vs. flat open terrain.
- Turbulence (mechanical) decreases with increasing height above ground.
- Turbulence (gust) parameters of importance are:
  - Turbulence intensity, as defined by Eq. (9.4),
  - Gust size (wind snapshot) referred to as integral scale,
  - Gust frequency.

9.2.2.2 Wind Structure

Figure 9.10 shows a typical record of wind velocity obtained from measurements at three different heights of a mast. The tendency of mean wind speed to stay relatively steady over periods of 10 min to an hour is significant (stationarity) and forms the basic idea for wind tunnel testing. Explanation of steadiness lies in the

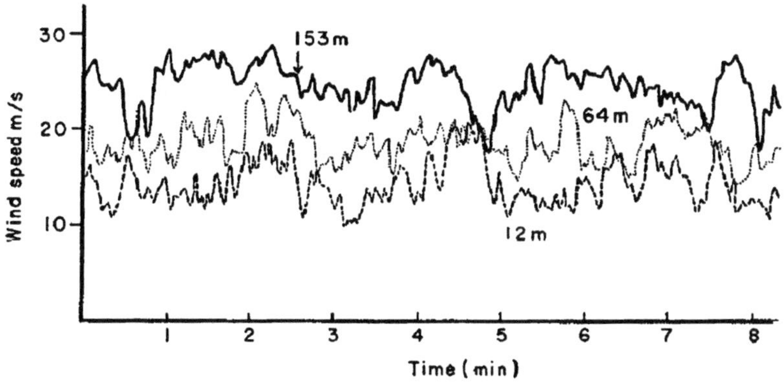

Fig. 9.10 Record of wind speed at three heights on a 153 m mast in open terrain

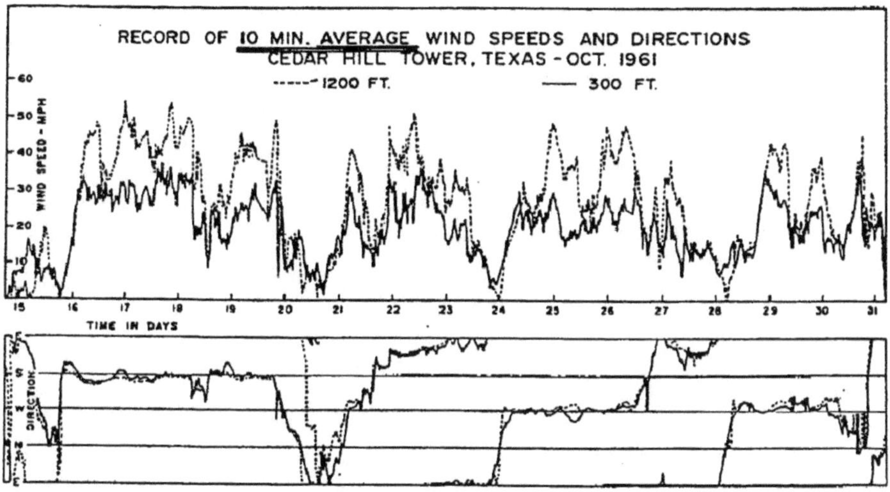

Fig. 9.11 Variation of mean velocity in the atmospheric boundary layer

fact that processes generating the mean flow have time scales larger than 1 h. However, mean speed does vary with time and large fluctuations cover a period of several days.

Some variation in the mean velocity does occur and this is illustrated in Fig. 9.11 by some wind records from a tall tower in Texas. Large fluctuations cover a period of several days while the speed variation with height is less apparent for some particular days. The variation of wind direction with height is almost nil.

Figure 9.12 shows the full wind spectrum, as proposed by Van der Hoven (1957). Distinctive features of the wind speed spectrum found from full-scale measurements are summarized here:

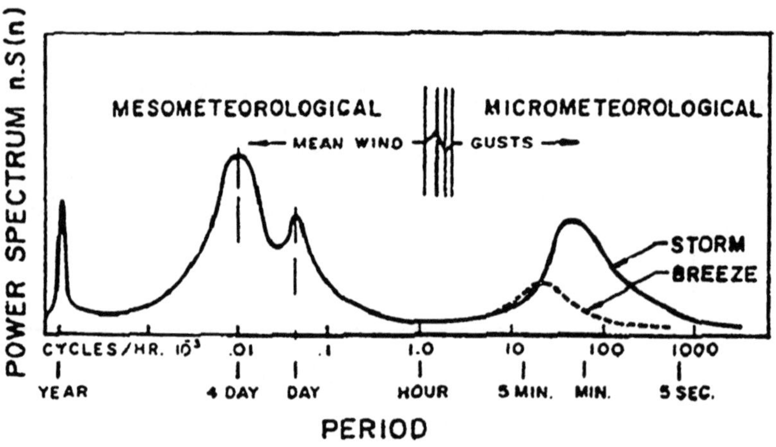

Fig. 9.12 Wind spectrum (Van der Hoven 1957)

- Energy appears into two major humps [$T = 4$ days $\rightarrow$ movement of large scale pressure systems ($T = 1$ day $\rightarrow$ diurnal frequency), $T = 1$ min $\rightarrow$ associated with turbulence] separated by a gap ($T = 10$ min to 1 h).
- The magnitude of high frequency hump depends on speed of mean flow.
- The gap enables a convenient distinction between "mean wind" and "gusts".
- The choice of an averaging period for mean wind of 1 h to 10 min provides fairly stable mean values (thunderstorms may be an exception).

In addition, it is apparent that wind speed fluctuates randomly, at all frequencies (Hz = cycles/second). Referring to Fig. 9.12,

- The vertical axis represents energy (relative number) at each frequency.
- The peak is around $n = 0.01$ (period of 100 s).
- The energy reduces drastically above 1 Hz.
- If the building or structure has fundamental frequency of greater than 1 Hz, gust induces negligible dynamic resonance (buffeting) response.

The micrometeorological part of the spectrum, which can be plotted in a smoothed form such as that of Eq. (9.5) (among others), fits well a series of full-scale data obtained in several locations, as Fig. 9.13 shows.

$$\frac{nS(n)}{k\bar{v}_{10}} = \frac{4.0\left(\frac{nL}{\bar{V}10}\right)^2}{\left[1+\left(\frac{nL}{\bar{V}10}\right)^2\right]^{4/3}} \tag{9.5}$$

where $L \cong 1{,}200$ m, k is the surface drag coefficient dependent on the terrain roughness; it has values of about 0.001, 0.005, 0.015 and 0.04 for terrain categories 1-4 or D-C-B-A respectively and $\overline{V}_{10}$ is the mean speed at height of 10 m.

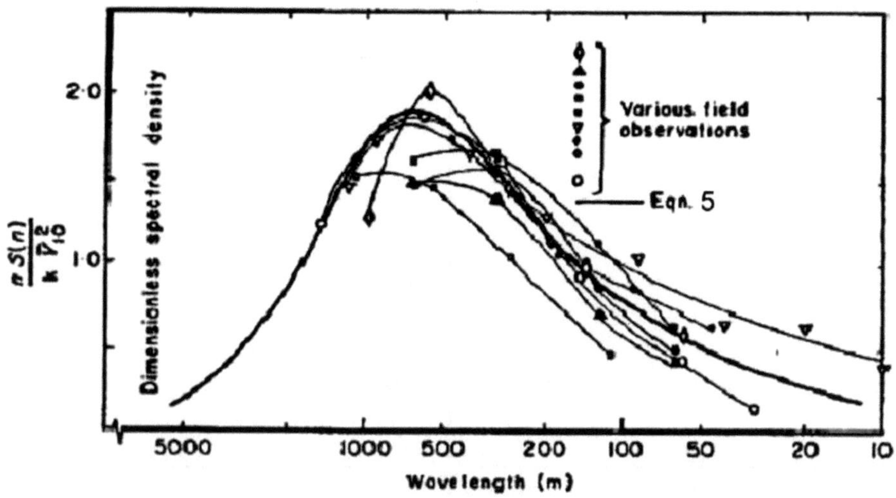

Fig. 9.13 Spectrum of turbulence in strong winds (Davenport 1967)

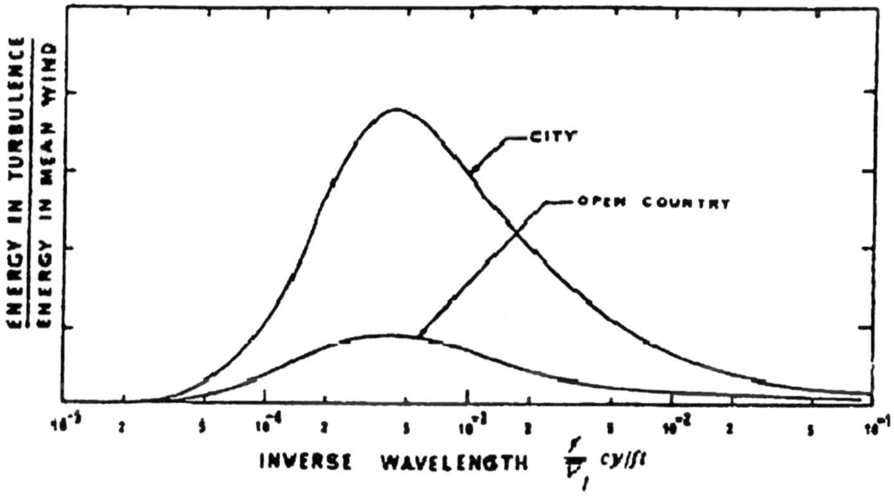

Fig. 9.14 Comparison of turbulence in the city and in open country

Even though mean wind speeds may be lower in a city, gustiness is much higher. In fact, a ratio between the surface drag coefficient in a rough urban area over that in an open terrain is about 10 to 1. Thus in city conditions, the percentage of turbulence in the air speed may be significantly greater. This is clearly indicated in Fig. 9.14.

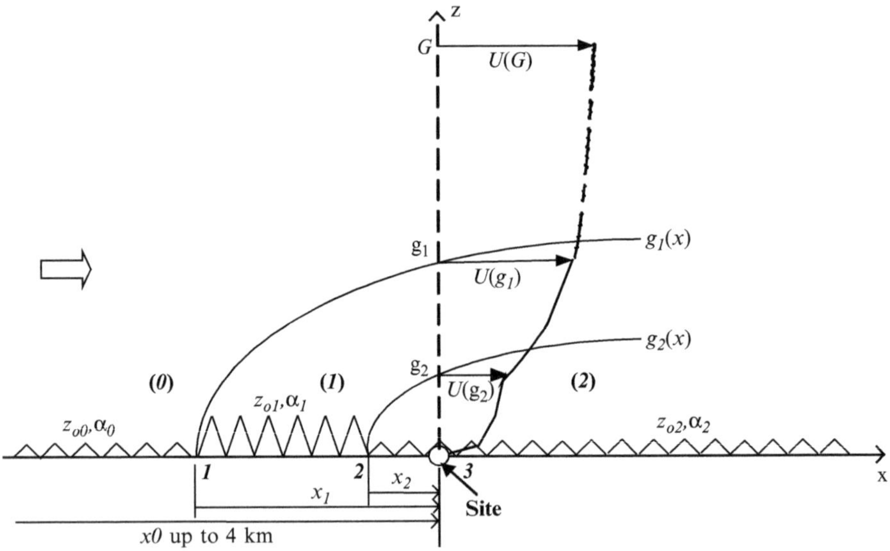

Fig. 9.15 Schematic boundary layer with two IBL's developing above a fetch with two roughness changes

9.2.2.3 Wind Speed Model

Recent wind-tunnel investigations confirmed that small-scale roughness changes play an important role on the speed variation above site (Schmid and Bunzli 1995; Zhang and Zhang 2001). Above a well-defined pertinent fetch, the patches may stratify the boundary layer regime into an outer sub-layer or a set of corresponding Internal Boundary Layers (IBL's). The IBL depth growth $g(x)$ may be modelled by a 0.8 power law, as mentioned in Garratt (1990), as follows:

$$g(x) \propto z_{o,r}^{0.2} x^{0.8} \tag{9.6}$$

where $z_{o,r}$ is the greatest of an upstream patch roughness length or its adjacent downstream patch roughness length; x is the distance from the change of roughness to the site. Each segment in the speed profile is dictated by the power law index of the corresponding patch. An illustrative example is shown in Fig. 9.15 in order to outline the general features of the model.

This example has a pertinent fetch of three patches: patch 0 from the 4 km upwind location to the first roughness change location 1; patch 1 from roughness change location 1 to 2; and patch 2 from roughness change location 2 to the site. The entire boundary layer is correspondingly stratified into three sub-layers: outer sub-layer (*0*) in the height range from G down to $g_1(x)$, IBL (*1*) from $g_1(x)$ down to $g_2(x)$, and IBL (2) from $g_2(x)$ down to the ground. Both $g_1(x)$ and $g_2(x)$ may be

Table 9.4 Roughness classification after Davenport et al. (2000) with power law α values after ASCE (1999)

Roughness class	z_o (m)	α	G (m)
1. Sea	0.0002	*0.09*	213
2. Smooth	0.005	*0.125*	213
3. Open country (OC)	0.03	0.15	274
4. Roughly open	0.1	*0.2*	274
5. Suburban (rough)	0.25	0.25	366
6. Very rough	0.5	*0.3*	366
7. Urban (closed)	1	0.33	366

NB: Number in *italics* is obtained by best fit of data

modelled with the 0.8 power law (2). The layer depths $g_1(x)$ and $g_2(x)$ break the speed profile at the site into three segments, each of which can be modelled by power-law equation with the patch-respective power-law indices α_o, α_1 and α_2.

The proposed speed model has the form:

Gradient height:

$$g_n(x) = G\ (n = 0) \tag{9.7}$$

IBL depth:

$$g_n(x) = 0.5z_{o,(n,n-1)}^{0.2}x_n^{0.8}\ (n = 1, 2, \ldots, N) \tag{9.8}$$

Speed profile segment:

$$U(z) = U(g_n(x))\left(\frac{z}{z_{g_n}}\right)^{\alpha_n}\ (g_{n+1} \le g_n; n = 0, 1, \ldots, N; \tag{9.9}$$

in which the subscript n denotes the patch number, i.e., patch N is the patch of the site; $g_o(x)$ denotes the gradient height; $g_n(x)$ denotes the depth of the nth IBL; x_n is the distance from the nth roughness change to the site; $z_{o,n}$ and α_n are roughness characteristic values of patch n; and $z_{o,(n,\ n-1)} = z_{o,n-1}$ or $z_{o,n}$, whichever is larger. For homogeneous terrain of no roughness change, $N = 0$, Eq. (9.8) does not apply and Eq. (9.9) reduces to the ordinary power law.

In order to use Eqs. (9.7), (9.8), and (9.9), the boundary conditions need to be specified for the pertinent fetch length, the patch information (x_n, $z_{o,n}$, α_n), the gradient height G, and the gradient wind speed $U(G)$. In this study, the pertinent fetch length is assumed as constant 4 km. Patch roughness information (x_n, $z_{o,n}$, α_n), and the gradient height G for the pertinent fetch can be found in Table 9.4 after Davenport et al. (2000) and ASCE (1999).

The highest roughness grade (Class 8, Chaotic, $z_o > 2$ m) of the original classification (Davenport et al. 2000) is combined within urban (closed) in compliance with ASCE 7-02. Values of $U(G)$ have already been available in wind standards and codes of practice.

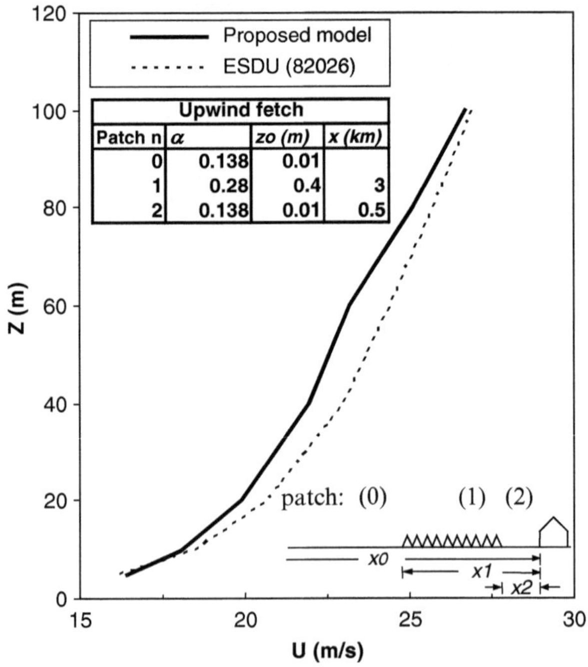

Fig. 9.16 The proposed speed model and the ESDU (82026) model for the ESDU (82026) example case (terrain condition shown) (Wang and Stathopoulos 2005)

Discussion

The application and validation of this new speed model has been discussed in depth by Wang and Stathopoulos (2005). This chapter intends to give an example demonstrating how the proposed model fits the actual ESDU (82026) values for a case of inhomogeneous upstream terrain. The example provided in the ESDU document was deemed the most appropriate case to check. The ESDU (82026) example requires to find the mean speed profile at a site downwind of two changes in surface roughness, given the reference speed $U(10) = 22$ m/s, and the fetch containing three patches, as shown in Fig. 9.16. ESDU (82026) appears to take a lot more steps than the proposed model in speed profile calculation. It is worth noting that the probability factor, which is taken into account by the original ESDU (82026) data, has been removed in the comparison of Fig. 9.16, which shows that the agreement between the proposed model and ESDU (82026) is reasonable, particularly below 20 m and above 80 m. For intermediate heights, ESDU tends to provide higher values than the proposed model, which thus appears less conservative than ESDU (82026). However, the proposed model agrees better with the full-scale investigation of Letchford et al. (2001) on a geometrically similar fetch configuration under hurricane conditions. It is of interest to note that Letchford et al. (2001) found that the ESDU (82026) transitional speed model may tend to overestimate, by as much as around 20 %, the increase in speed at 10 m height induced by an R-S (rough-to-smooth) upstream terrain change. It is also noteworthy that this particular case ensures conditions of neutral atmospheric stability considering the high wind speeds it refers to.

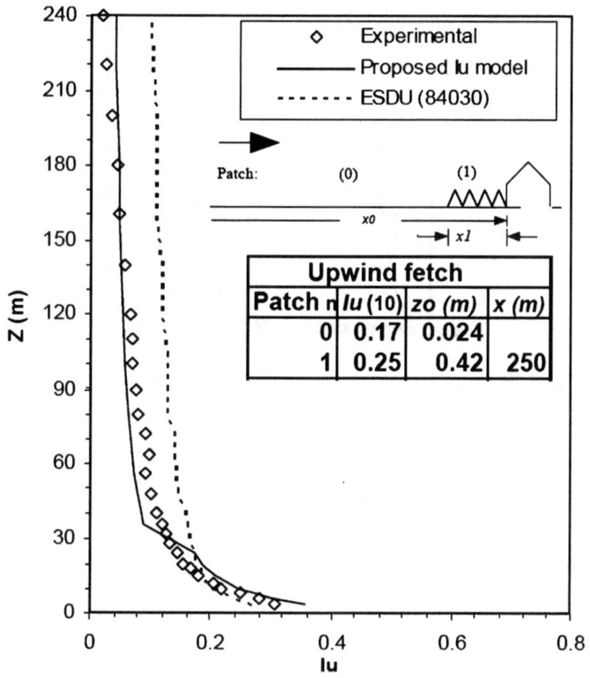

Fig. 9.17 The *Iu* model and wind-tunnel data for a fetch with a single roughness change, and the ESDU (84030) data for a very similar fetch (Wang and Stathopoulos 2005)

9.2.2.4 Turbulence Model

Although the detailed derivation of the turbulence model is not presented in this chapter, it can be found in Wang and Stathopoulos (2005). However, a comparison with the experimental data for a fetch with a single roughness change (from open country to suburban) is shown in Fig. 9.17. The proposed model agrees well with the experimental data, whereas the ESDU data overestimates the turbulence intensity at higher heights. However, there is excellent agreement among data, ESDU model and the present study results, as well as the ESDU data for heights less than 30 m, i.e. the height zone of most low buildings.

9.2.3 Topographic Effects

In the presence of an upstream topographic feature of height H within a distance equal to 100 H, designers should take into account the variation of wind speed caused by topography. In this regard, the following should be recalled:

- Local abrupt topography affects wind near ground.
- Topographic features create wind speed up.
- Wind speed-up depends on shape of hill, location of building, and height above ground.

Figure 9.18 shows a typical speed-up case for the case of an escarpment.

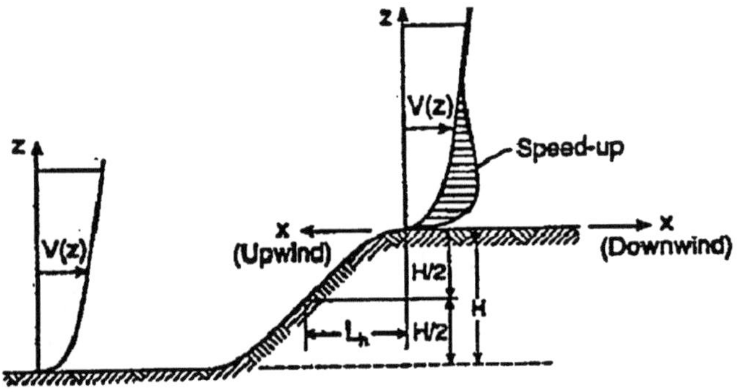

Fig. 9.18 Speed-up occurred on top of an escarpment

9.2.4 Wind Climatology

The most important utilization of climatological elements of wind is the establishment of a statistical basis to determine the prediction of strong winds for design purposes. A series of wind records would be necessary for the locality of the project in order to proceed with the analysis.

The following comes from an example:

- One hour mean is approximately 60 knots (1 knot = 1.15 miles/h)
- Ten minute mean is about 65 knots
- Two minute mean is about 75 knots
- One minute sustained value 80 knots
- Peak (3 s) value is 85 knots
- It is important to know the averaging time
- For fastest-mile wind speed (distance of 1 mile) averaging time $t = 3{,}600/V_{fm}$
- V_t = wind speed of averaging time t seconds
- $V_{3,600}$ = mean hourly wind speed
- Non-hurricane wind gust duration curve by Durst (1960)—see Fig. 9.19
- If sustained wind speed is 120 mph, what is the 3-s wind speed? Sustained wind is 1 min or 60 s average $V_3 = (120)\ (1.53/1.25) = 147$ mph

The Weibull distribution can also be used to describe the distribution of wind speed and direction. Climatological data can be gathered and then in each of, say, 16 directions a Weibull function fitted. This directional distribution can be expressed as:

$$P(>V,\theta) = A(\theta)e^{\left[-\left(\frac{V}{C(\theta)}\right)^{k(\theta)}\right]} \tag{9.10}$$

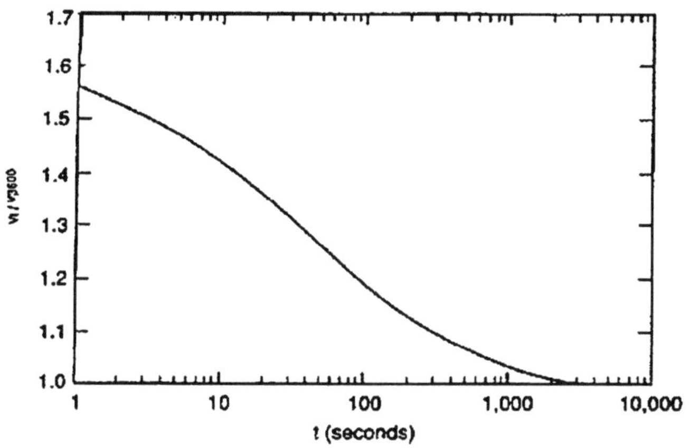

Fig. 9.19 Relationship between wind speeds averaged over an arbitrary time to the wind speed averaged over 1 h (Durst 1960)

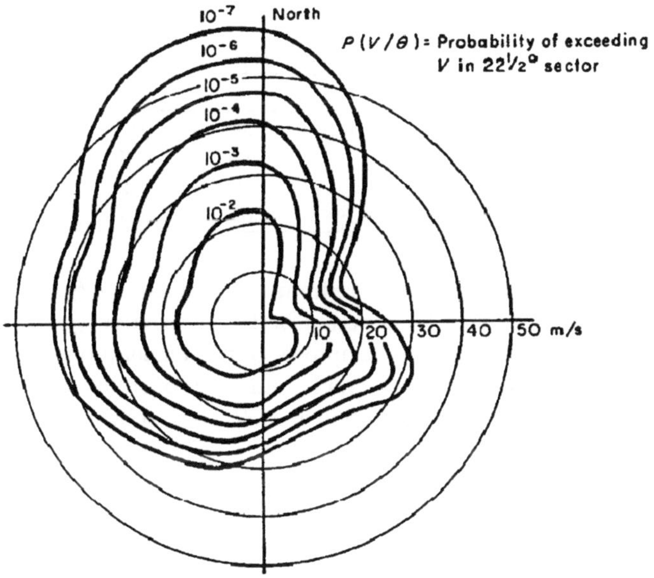

Fig. 9.20 Directional distribution of wind speed at 800 m height

Figure 9.20 shows one such distribution while Fig. 9.21 shows the Weibull fit for various locations on the basis of the most critical sets of wind speed data (regardless of direction) producing different constant (k and c) distribution values.

The annual probability of exceeding a specific wind speed is provided by using a Fisher–Tippett Type I Extreme Value Distribution. This is used to get the annual maximum reference wind speed for purposes of codification. A Fisher–Tippett

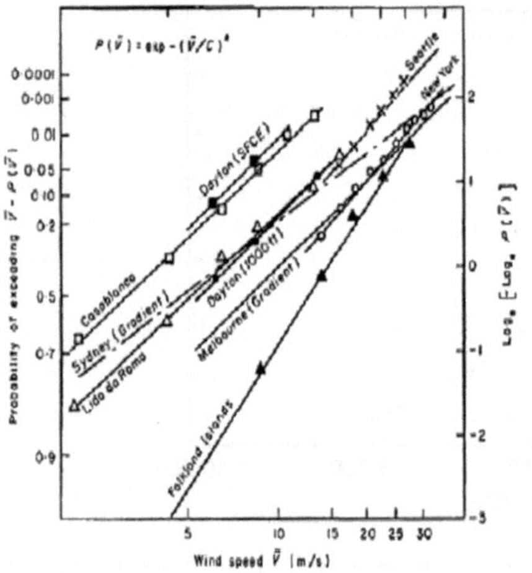

	C (m/s)	k
Dayton (Ohio) surface	3·9	1·7
Dayton (Ohio) (300 m)	7·1	1·7
Casablanca	4·2	1·5
Falkland Islands	14·8	2·5
Lido da Roma	6·8	1·5
New York (F.J.K. airport—500 m)	11·4	1·87
Seattle airport (300 m)	8·5	1·9
Sydney (gradient)	7·8	1·39
Melbourne (gradient)	10·5	1·64

Fig. 9.21 Weibull distribution of wind speed for different locations

Table 9.5 Probability of exceeding design wind P

Annual probability of exceedance P	Design life of structure (years) 1	5	10	25	50	100
0.04	0.04	0.18	0.34	0.64	0.87	0.98
0.02	0.02	0.10	0.18	0.40	0.64	0.87
0.01	0.01	0.05	0.10	0.22	0.40	0.64
0.005	0.005	0.02	0.05	0.10	0.22	0.39

Type I simplified procedure is given in Simiu and Scanlan (1996). Data required is annual maximum wind speed for a period of 10 consecutive years or more. The Mean Recurrence Interval (*MRI*) is reciprocal of the annual probability of exceeding a particular speed value. For example, annual probability $Pa = 0.02$ means *MRI* of 50 years. Wind speed of a particular *MRI* is given by

$$V_{MRI} \cong U + \frac{1}{\alpha}\ln(MRI) \tag{9.11}$$

where U and $1/a$ are the mode and dispersion respectively of the Type I distribution. Then the probability of exceeding reference wind speed during life of a structure (P) is evaluated by

$$P = 1 - (1 - P_a)^n \tag{9.12}$$

in which P_a is the annual probability of exceeding reference wind speed (reciprocal of *MRI*) and n is the design life or reference period in years.

Table 9.5 provides probabilities of exceeding design wind speed for various design lives of structures and *MRI*'s. For instance,

- If *MRI* is 50 years and design life of a building is 50 years, there is a 64 % probability that the reference wind speed will be exceeded during the life of the building
- If *MRI* is 50 years and life $n = 10$ years, there is an 18 % probability of exceedance

It can be shown that:

$$MRI = \frac{Life\ of\ structure}{Risk}\ \text{(for small risk)} \tag{9.13}$$

9.3 Wind Pressure Coefficients

Low buildings are usually defined in codes of practice and design standards as those with heights less than about 20 m. Including residential construction, the percentage of buildings characterized as low (or low-rise) probably exceeds the 90 % level. Since the early 1980s, the Canadian wind load provisions for low-rise buildings have gained an international reputation following the recognition of their innovative and pioneering character by researchers and practitioners across the globe. In this regard these provisions have been influential in the development and evolution of various national and international wind load standards, including the ASCE 7 in its current form and, to a lesser extent, in its previous versions.

In fact, the wind design provisions for low-rise buildings changed significantly in the National Building Code of Canada in 1980. The changes were based on a study carried out in the University of Western Ontario (Stathopoulos 1979), which clearly demonstrated that the tributary area is a key parameter in the evaluation of wind pressures on buildings. However, in spite of these revisions, wind codes of practice and design standards for low-rise buildings neglected to consider the variation of the dynamic (fluctuating) character of the pressures with the particular location of interest on the building envelope. Figure 9.22 shows the pressure coefficients for buildings with gable roofs as specified in the 1975 edition of the Commentaries of Part 4 of the National Building Code of Canada (NBCC 1975). These coefficients vary with the slope of the roof: suction dominates for roof angles lower than 22° and pressure for roof angles greater than about 39°. Roofs with angles between 22° and 39° have to be designed for both pressures and suctions. Although there are provisions for increased values of suctions acting on roof edges and corners, there is no recognition of the very high suctions occurring on ridges and ridge corners of intermediate or high roof slopes. It is interesting that these provisions are only somewhat different from those appearing in a 1948 document published by the US Department of Commerce (Whittemore et al. 1948). In contrast, typical provisions for the design of gable roofs of intermediate and high slopes included in the User's Guide (Structural Commentaries, Part 4) of the

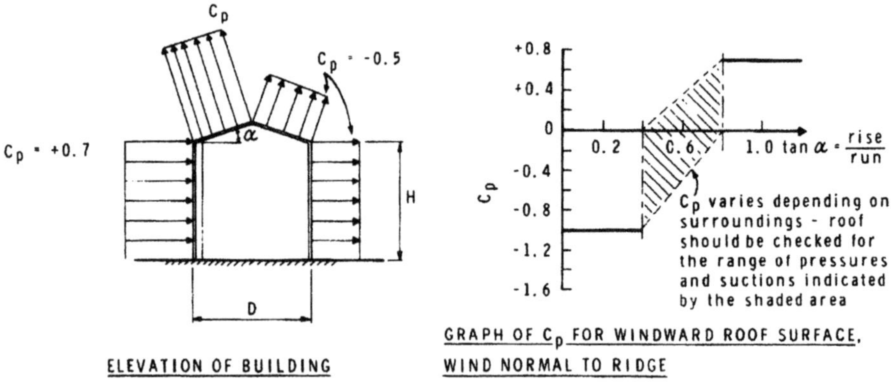

Fig. 9.22 Gable roof design pressure coefficients (NBCC 1975)

2010 National Building Code of Canada (NBCC 2010) show the dominant effect of the tributary area (particularly for roof corners, edges and ridges of intermediate slopes) and the critical suction values for specific roof areas and different ranges of roof slopes, in comparison with the 1975 data—see Fig. 9.23.

9.3.1 Area-Averaged Pressures

Historical reviews about the evaluation of wind loads on low buildings have been provided by Stathopoulos (1995) and Surry (1999). Although wind pressures on buildings reported prior to the postulation of Jensen's "Model Law for Phenomena in Natural Wind" (1958), may not be representative, some of the early ideas hold well up to date. As early as 1884, Sir Benjamin Baker, following wind load measurement experiments on boards of various sizes, stated that the mean wind pressure on a large area must be less than that on a small area because "*threads of the currents moving at the highest velocity will strike an obstruction successively rather than simultaneously*". Baker's experiments, discussed by Davenport (1975), led Stanton (1924) to invent and develop the wind pressure recorder, which is a mechanical mean pressure transducer to catch the effects of gustiness of the wind.

The apparatus shown on the left hand side of Fig. 9.24 consists of a series of chambers, each containing a flexible corrugated diaphragm, the displacements of which due to the difference of air pressure on its sides, is followed by a guide-rod carrying a light pulley. A thin platinum wire is fixed at one end and passes over the system of the pulleys shown, the free end carrying a pen in contact with a drum actuated by clock work. The motion of the pen integrates the separate motion of the diaphragms, so that a record of the average pressure differences in the chambers is obtained. It was this idea that led to the utilization of manifolds (see right hand side of Fig. 9.24) and the development of the well-known pneumatic averaging technique by Surry and Stathopoulos (1978).

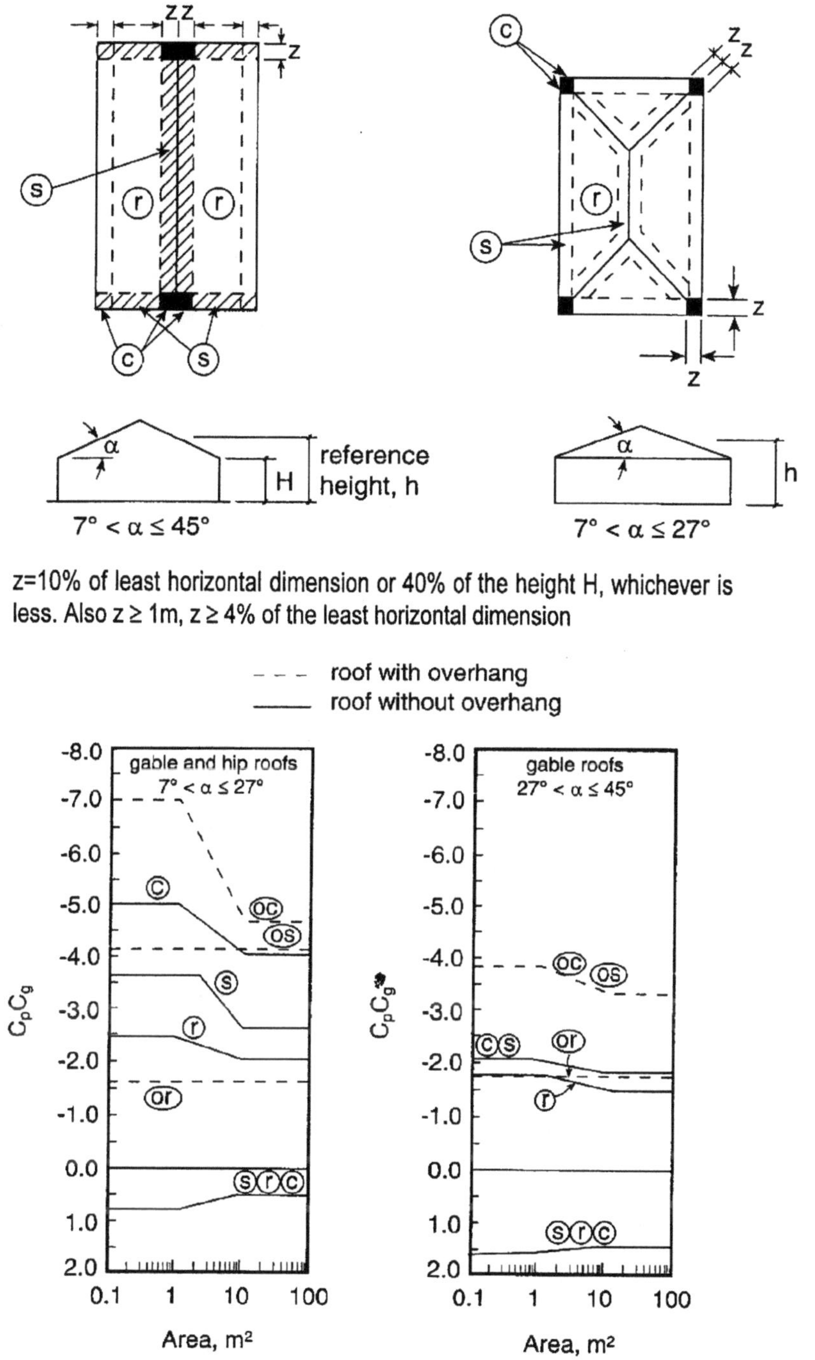

Fig. 9.23 Gable roof design pressure coefficients (NBCC 2010)

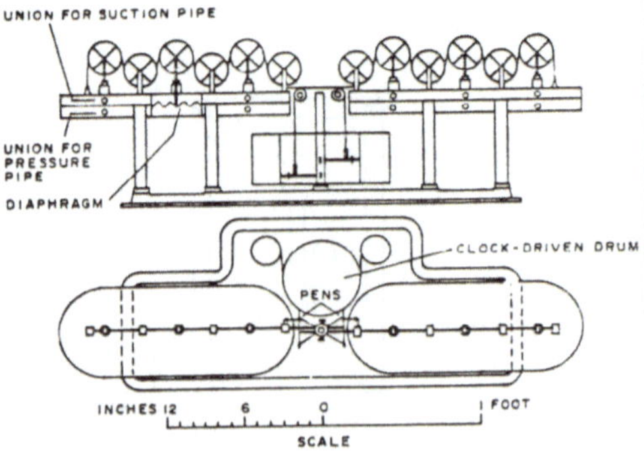

Wind pressure recorder (Stanton 1924)

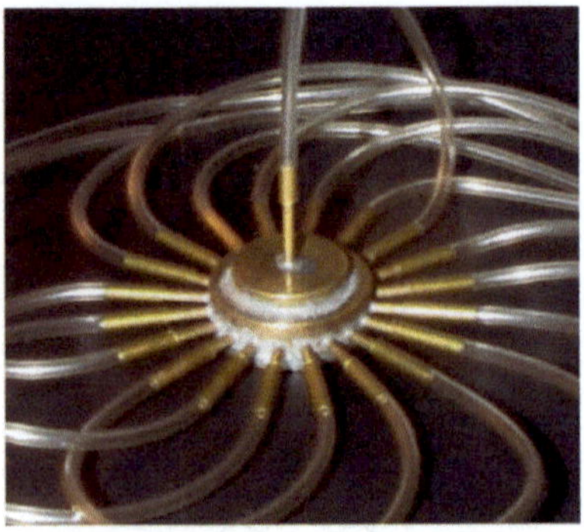

Pneumatic average (Surry and Stathopoulos 1978)

Fig. 9.24 Area-averaged pressure measurement devices

As Surry (1999) says, in the research on low buildings carried out under Alan Davenport in the 1970s, necessity was the mother of invention. In the absence of electronic pressure scanning, common manifold systems used for the measurement of static pressures at throats of wind tunnels, were explored for the measurement of area-averaged pressures. Although at the beginning, manifolds appeared to provide a non-linear pressure average, careful design and the realization that long tubes would force laminar flow in the system and hence linear pressure drops, led to a powerful tool to measure time-varying spatially-averaged pressures. It took a long time however, to convince the wind engineering community that this simple methodology was really working. Nevertheless, the pneumatic averaging technique has been used thereafter routinely in several laboratories.

However, the importance of the tributary area has an additional dimension when wind pressure loads acting on systems covering more than one surfaces of the building envelope are considered. A typical example is the frame of a low metal building. The lack of correlation of wind pressures acting on different building surfaces, in addition to the effects of tributary area, are significant factors to be considered in the evaluation of the actual wind load seen by the structure and, consequently, be used in its design. Notwithstanding the difficulties due to the experimental limitations at the time, this rather sophisticated part of the research determined values of structural actions, such as total uplift, horizontal shear and bending moments at various points of the indirectly-loaded primary structural systems with a variety of influence lines utilized on-line during the experiments. An optimization software routine was subsequently used to develop sets of *pseudo* pressure coefficients to generate loading conditions which would envelope the maximum induced force components to be resisted for the wind directions and exposures tested. The set of coefficients provided represents fictitious loading conditions, which conservatively envelope the maximum induced force components to be resisted independent of wind direction. Figure 9.25 taken also from the User's Guide—Structural Commentary I—of the National Building Code of Canada (2010) shows the end result.

Load case A: winds generally perpendicular to ridge

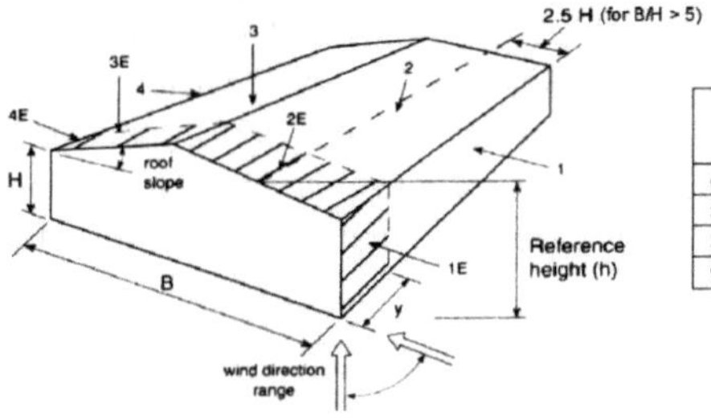

Roof slope	Building surfaces							
	1	1E	2	2E	3	3E	4	4E
0° to 5°	0.75	1.15	−1.3	−2.0	−0.7	−1.0	−0.55	−0.8
20°	1.0	1.5	−1.3	−2.0	−0.9	−1.3	−0.8	−1.2
30° to 45°	1.05	1.3	0.4	0.5	−0.8	−1.0	−0.7	−0.9
90°	1.05	1.3	1.05	1.3	−0.7	−0.9	−0.7	−0.9

Load case B: winds generally parallel to ridge

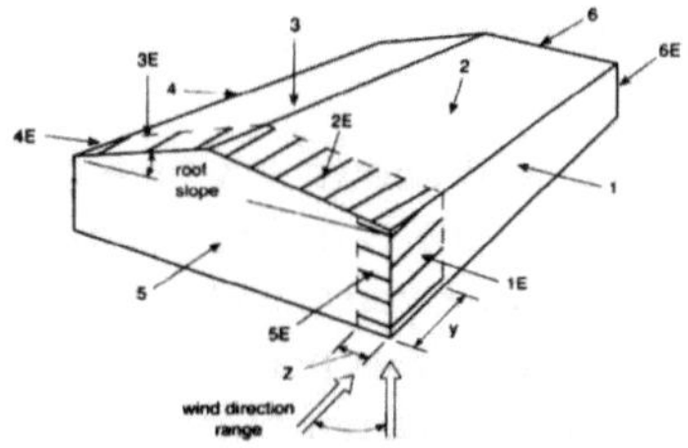

Roof slope	Building surfaces											
	1	1E	2	2E	3	3E	4	4E	5	5E	6	6E
0° to 90°	−0.85	−0.9	−1.3	−2.0	−0.7	−1.0	−0.85	−0.9	0.75	1.15	−0.55	−0.8

End zone width "y" should be greater of 6m or 2z, where "z" is the gable wall end zone defined for Case B. Alternatively, for buildings with frames, the end zone "y" may be the distance between the end and the first interior frame.

End zone width "z" is the lesser of 10% of the least horizontal dimension or 40% of height H, except that "z" must be at least 4% of the horizontal dimension and al least 1m

Fig. 9.25 External peak pressure coefficients, C_pC_g, for primary structural actions arising from wind load acting simultaneously on all surfaces (NBCC 2010)

The set of gust pressure coefficients provided represents fictitious loading conditions, which conservatively envelope the maximum induced force components to be resisted independent of wind direction. It is of interest that it took almost 20 years to persuade the ASCE-7 wind load committee to accept this approach and introduce a similar set of coefficients in the American wind standard.

Further to these advances, particularly the establishment of Jensen's number h/z_o, where h is the height of low building and z_o the roughness length of the upstream terrain, as a critical simulation parameter in order to duplicate the measured mean pressure coefficients in the wind tunnel, more recent experiments by Tieleman et al. (1999) have established additional criteria for the proper simulation of peak pressures acting on low buildings. In particular, it has been found that the duplication of the small-scale turbulence should take priority over the duplication of the large-scale turbulence, and the simulation of the small-scale turbulence and turbulence intensity near the leading edge from which the shear layers separate is of paramount importance. Peak suctions can be duplicated best by careful wind tunnel simulation of the horizontal turbulence intensities and their respective small-scale turbulence content.

The wind study on low buildings at UWO and some subsequent studies as well, examined the effect of parapets on the roof loading, the wind pressures on roof overhangs by integrating simultaneously the upper and lower surface loads, the probability distributions of wind-induced pressures and the very significant effect of internal pressure for various building envelope porosities and wall openings. In the course of the latter work, which supplemented Vickery's earlier studies (1976) and Holmes' parallel experimental and theoretical work (1979), the spatial uniformity of wind-induced pressures was clearly established. Finally, a more recent work by Zisis and Stathopoulos (2009) forms a first step to codify the wind loads applied on attached patio covers.

9.3.2 Reduction Factors in Codification Process

It was clearly Alan Davenport's idea to introduce the well-known **0.8 factor** in order to account for a variety of alleviating effects such as the variation in terrain roughness and wind direction in the codification process. Open country terrain exposure was giving generally the highest pressures but the great majority of low buildings were in suburban terrain exposure and since the design gust pressure coefficients (*CpCg*) were provided regardless of wind direction, it would be rather unlikely to expect the most critical wind speed to originate from the most critical wind direction for a particular building orientation. Therefore, codified wind pressure coefficients were formulated by considering the most critical measured values factored by 0.8. It is interesting to know that the introduction of this factor has generated one of the major controversies in the history of wind engineering. However, subsequent extensive studies such as that of Ho (1992) have justified Davenport's original approach not only regarding the selection of this factor but also its rationale and the reliability analysis he carried out—see Davenport et al. (1985)—requiring statistics of exposure factors, pressure coefficients and gust factors in addition to those of the wind dynamic velocity pressures.

In addition to the well-known 0.8 factor, there is another factor—also due to Davenport's ingenuity and intuition for wind related matters—that appears in the Canadian wind design specifications but has passed by rather quietly! It is well known that the wind applies to various structural and non-structural building components and elements either directly or indirectly. Roof cladding, for instance admits the wind loading directly, whereas the primary structural system admits the wind load indirectly through other secondary structural elements such as girders and purlins. Consequently, there is a degree of attenuation of the wind loading function depending on the wind load path necessary to reach a particular component of the building system. Clearly this concept applies the most to the foundation of the building. The Structural Commentary I "Wind Loads and Effects" of the National Building Code of Canada (2010) specifies in a note to the primary structural actions arising from wind load acting simultaneously on all surfaces, i.e. Fig. 9.25:

> For the design of foundations, exclusive of anchorages to the frame, only 70 % of the effective load is to be considered

There are only very few people knowing that this note is due exclusively to Davenport. A recent full-scale field study, after examining several real wind records has estimated that this reduction is at least 17 % (Zisis and Stathopoulos 2010; Zisis et al. 2011). Although this is the only experimental effort to verify this **0.7 factor**, there is no doubt in the author's mind that this value is sound and that subsequent research results would confirm its validity.

9.3.3 Additional Wind Load Considerations for Low Buildings

The UWO study under the direction of Alan Davenport in the 1970s has found that the effect of roof slope and building height are significant but the dependence of load coefficients on height reduces considerably by referencing them to the velocity pressure at eaves or mid-roof height; the influence of length on the wind loads is not so significant; the effect of tributary area was recognized as a key parameter in the wind load evaluation of various roof and wall components. The dynamic component of all loading effects is dominant over the mean component and this is particularly true of local pressures, less so for distributed load effects. The dynamic component also increases with rougher terrain due to the increased value of turbulence. However, the effect of exposure has not been fully investigated as of yet. The importance of internal pressures has been appreciated more than in the past. In the last few years, research efforts were made to assess the variability of wind loads on low buildings for different surroundings and to evaluate the wind loads on building roofs of different configurations. The effect of external projections (e.g. balconies, eaves, parapets) on the wind pressure distribution on low-rise buildings of different geometries has also been examined.

Questions have been raised regarding the appropriateness of having a single set of provisions for roofs with intermediate roof angles (10–30°). Over this roof angle range, the wind flow over the building roof may change drastically in comparison to the much more well-defined low and high roof slope ranges. For instance, depending upon the exact value of roof slope, a given roof region could be subjected to either positive or negative pressure. Furthermore, the current provisions for this intermediate roof range have originated from extensive wind tunnel tests on only one roof slope, namely 4:12, corresponding to a roof angle of 18.4° and therefore, they may not be appropriate for the entire range. Figure 9.26 taken from Stathopoulos et al. (2000), shows the minimum local pressure coefficients measured on five tappings near the ridge corner, which has been found particularly vulnerable to high suctions for intermediate roof slopes. Pressure coefficients on roofs of 10–25° show similar trends, with their highest peak occurring on the 20° roof. However, the behaviour of pressure coefficients on the 30° roof appears to be quite different from the other roofs in the intermediate slope range. The study found that the 10° roof-angle data would fit better in the intermediate roof slope provisions, whereas the 30° case would be much better represented by the 30–45° range. Very similar diagrams to those shown in Fig. 9.23 have been

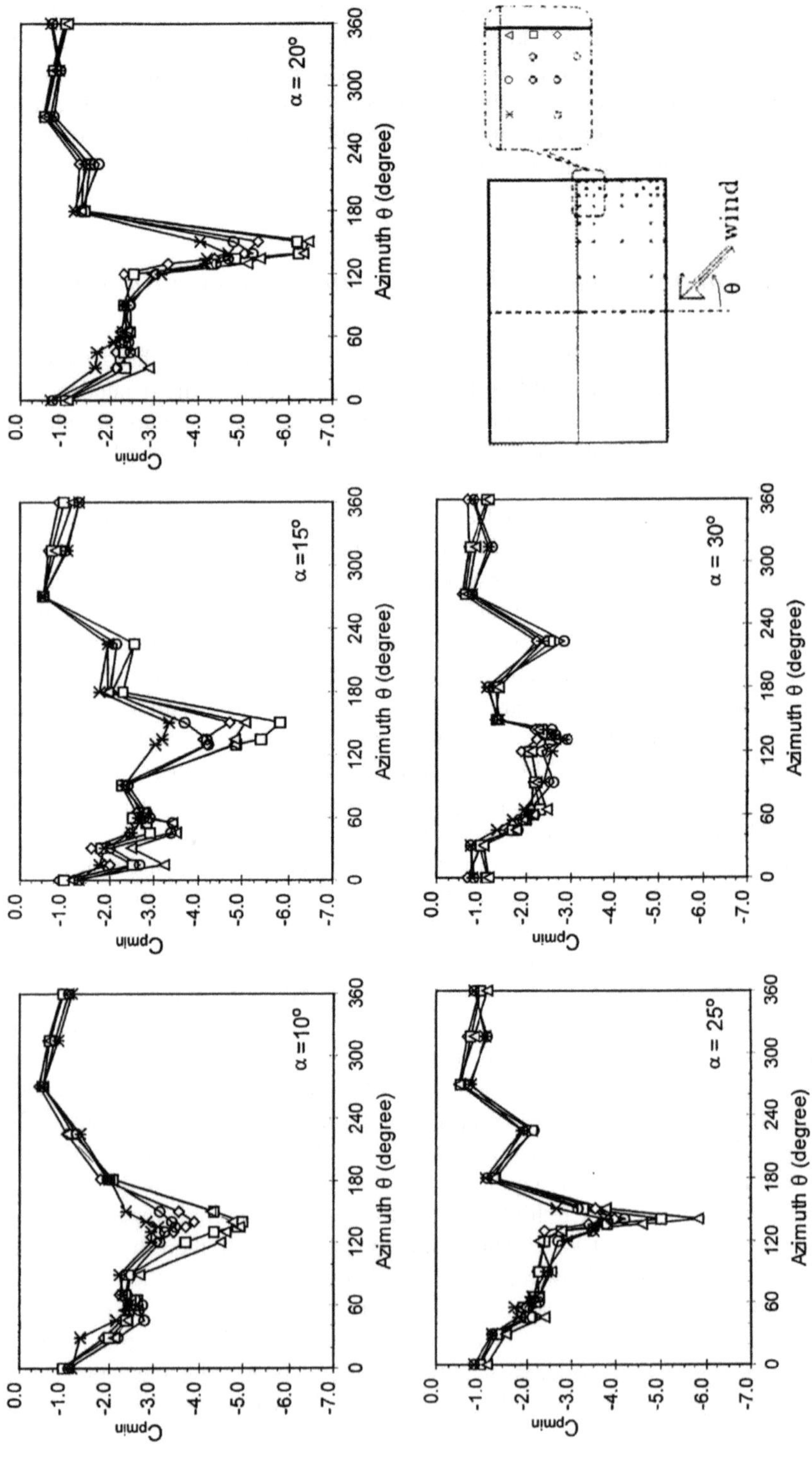

Fig. 9.26 Ridge corner pressures (Stathopoulos et al. 2000)

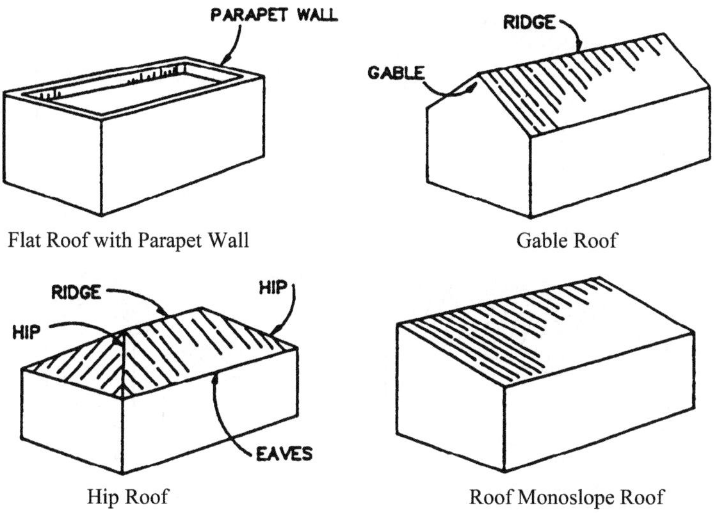

Fig. 9.27 Types of roofs addressed in wind codes and standards

included in ASCE 7-02 (2003)—as well as in subsequent editions—applying also on the same roof slope ranges, due again to the results of the above-mentioned research (Stathopoulos et al. 2001).

Contemporary wind codes and standards include provisions for the evaluation of local and area-averaged pressure coefficients appropriate for the design of flat, gabled, stepped, multi-gabled (folded), monoslope and saw-tooth roof shapes whereas guidelines are provided for hip roofs with roof angles varying from 10° to 30°. The latter are based on a study by Meecham et al. (1991) which found that the worst peak pressure on the hip roof was reduced by as much as 50 % from that on the gable roof but the study was based on a single roof pitch of 18.4°. Nevertheless, this finding justifies the better performance of hip roofs in comparison to gable roofs in severe hurricanes. Furthermore, a more recent wind tunnel study by Xu and Reardon (1998) found that the roof pitch does affect both the magnitude and distribution of hip roof pressures. For instance, for a roof pitch of 30° the largest suction coefficients are almost the same for gabled and hip roofs.

9.3.4 The Effect of Roof Shape and Geometric Consideration of the Low Building

Based on the previous comments and aerodynamic considerations, the roof shape is critical for the evaluation of wind pressures on roofs. Figure 9.27 shows the simple roof geometries for which extensive studies have been carried out and their results

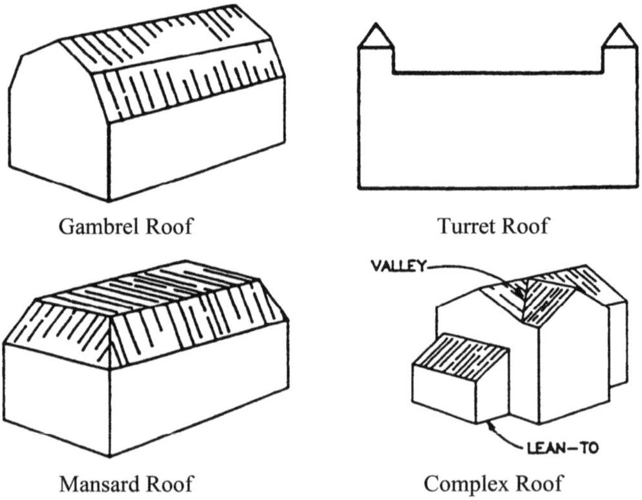

Fig. 9.28 Types of roofs not addressed in wind codes and standards

have been utilized for purposes of codification. On the other hand, there are many roofs for which no information is currently available to the designer. Figure 9.28 shows typical examples in the latter category including the complex but very realistic case of non-rectangular plan-form. The effect of the plan-view form of a low building on the values of design pressure coefficients has not been examined in detail, as yet. This generates several questions regarding the applicability of the pressure coefficients provided for rectangular roof and building shapes to buildings with non-rectangular plan-view forms. Similar questions are posed regarding the design of primary structural systems and other indirectly loaded surfaces of low buildings by the wind. Specific pressure coefficients for these cases are currently provided only for gabled-roof rectangular buildings and it is difficult to extend to cover other geometries. Although the effect of parapets on the wind loads on roofs has been studied quite extensively in the recent past, there was very little information available regarding the actual wind load affecting the parapet itself until very recently—see Stathopoulos et al. (2002) and Mans (2001).

In spite of the significant progress made in the course of the last 30 years or so in the design provisions of low buildings, there are still outstanding questions on several issues. These include but are not limited to the evaluation of design pressure coefficients for buildings of different shapes, the evaluation of internal pressures, the characterization of exposure effects and their evaluation on design pressures, the evaluation of load paths through the low building structural systems, as well as the development of successful mechanisms for the reduction of high suctions on roofs such as the utilization of special form parapets (Baskaran 1992) or special devices for roof vortex mitigation (Richardson and Surry 1994; Surry and Lin 1995).

9.4 Internal Pressures

Since buildings are very rarely airtight, the wind load assumed in design is taken as the difference between loadings acting on the external and the internal sides of the building envelope. Wind-induced internal building pressures can greatly affect the net loading of the building envelope. In enclosed buildings internal pressures are relatively small compared with the external pressures but their magnitude becomes comparable to that of external pressures for partially-enclosed buildings. In the past few years, there has been a renewed interest in wind-induced internal pressures, resulting in numerous publications and possible revisions of relevant provisions of building codes and standards.

Internal pressure loads depend on the distribution of external pressures and the location of building openings, if any, along with the magnitude and variability of the porosity of the building envelope. Figure 9.29, taken from Whittemore et al. (1948), shows internal pressure coefficients specified more than half a century ago. Clearly, for buildings with 30 % or more of the wall surface open, or subject to being opened or broken open, an internal pressure coefficient near +0.8 or –0.6 is specified for windward or leeward/parallel wall openings respectively.

Based on more recent research (Stathopoulos et al. 1979; Beste and Cermak 1996; Vickery and Bloxham 1992; Irwin and Dunn 1994), the current provisions for internal pressures, as they appear in ASCE 7-10 (2010), are much more detailed,

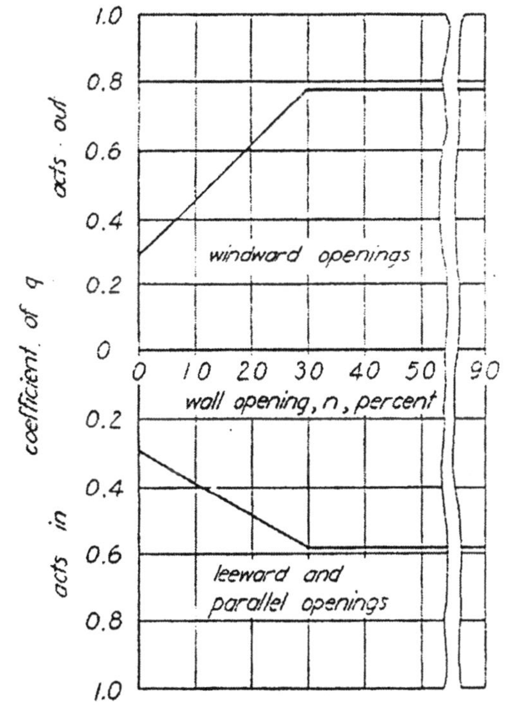

Fig. 9.29 Internal pressure coefficients, after US Department of Commerce (Whittemore et al. 1948)

Table 9.6 Internal pressure coefficients, after ASCE 7-10 (2010)

Enclosure classification	GC_{pi}
Open buildings	0.00
Partially enclosed buildings	+0.55
	−0.55
Enclosed buildings	+0.18
	−0.18

Notes

1. Plus and minus signs signify pressures acting toward and away from the internal surfaces, respectively
2. Two cases shall be considered to determine the critical load requirements for the appropriate condition:

(a) A positive value of GC_{pi} applied to all internal surfaces
(b) A negative value of GC_{pi} applied to all internal surfaces

although they preserve the physical reality of pressurization and de-pressurization phenomena. Table 9.6 (ASCE 2010) specifies three sets of internal pressure coefficients for three types of buildings defined as follows:

Open: A building having each wall at least 80 % open. This condition is expressed for each wall by the equation $A_o \geq 0.8A_g$ where:

A_o = total area of openings in a wall receiving positive external pressure
A_g = the gross area of that wall in which A_o is identified

Partially Enclosed: A building that complies with both of the following conditions:

1. The total area of openings in a wall receiving positive external pressure exceeds the sum of areas of openings in the balance of the building envelope (walls and roof) by more than 10 %, and
2. The total area of openings in a wall receiving positive external pressure exceeds 4 ft^2 (0.37 m^2) or 1 % of the area of that wall, whichever is smaller, and the percentage of openings in the balance of the building envelope does not exceed 20 %.

Enclosed: A building that does not comply with the requirements for open or partially enclosed.

The Canadian provisions for internal pressure coefficients also recognize that due to the changeability and uncertainty of the size and distribution of building openings, their values can be influential and wide ranging. However, the Canadian approach is somewhat different. There are also three design categories of buildings, defined as follows:

Category 1 (C_{pi} = −0.15 to 0; C_g = 1.0): Buildings without large openings but with small uniformly distributed porosity, less than 0.1 % of total surface area;

Category 2 (C_{pi} = −0.45 to 0.3; C_g = 1.0): Buildings with significant openings expected to be closed in storms and with non-uniformly-distributed porosity; design should cover the entire range of C_{pi}'s provided; and

Category 3 ($C_{pi} = -0.7$ to 0.7; $C_g = 2.0$): Buildings with large openings expected to transmit gusts to the interior; structures of post-disaster services are included in this category.

Notwithstanding the sound intent of these provisions, their interpretation and application in practice have caused and continue to cause problems to practicing engineers. For instance, when dealing with the ASCE-7 provisions, the internal pressure for a building goes from moderate (enclosed) to high (partially enclosed), back to moderate (partially open), to negligible (open). However, it is difficult to understand that the internal pressure goes back to a moderate level if the building does not meet the enclosed or partially enclosed requirements—users of these provisions tend to go directly to the open classification. The suggestion to add a fourth category of buildings (partially open) creates additional problems because part of the difficulty in transferring the knowledge to the practitioners is associated with the terminology used. For instance, the term "enclosed" may be used for a building with 70 % of each wall open! This is a critically sensitive area requiring additional work for its satisfactory solution. Consequently, there is an ongoing effort to clarify issues and better organize these provisions.

9.5 Database-Assisted Design (DAD)

The concept of database-assisted design was proposed by Simiu and Stathopoulos (1997) and Whalen et al. (1998) as a means of providing future code alternatives that would make direct use of stored pressure time series for the design of low buildings. The idea is not really new but it is only now possible because of the development of electronically-scanned pressure measurements and increased information storage and computational capacities. Pressure time series can thus be recorded simultaneously and stored for a large number of tap locations covering the entire envelope of a low building over a time interval representative of the duration of typical storms. It has already been shown by Simiu and co-workers that provisions which use aerodynamic databases containing pressure time histories and climatological databases containing wind speed data can lead to risk-consistent and therefore, more economical designs for low buildings subjected to wind action than those based on conventional standard provisions.

The first such database was generated by the Alan G. Davenport Wind Engineering Group at the University of Western Ontario. The database covers three building plan dimensions (40 ft $\times$ 62.5 ft, 80 ft $\times$ 125 ft and 160 ft $\times$ 250 ft) with eave heights ranging between 12 and 40 ft. Three roof slopes for the 80 ft $\times$ 125 ft building (1/4:12, 1:12 and 3:12) and a single roof slope (1:12) for the other two building plans have been tested. Thirty-seven wind angles in each of two terrains (open country and suburban) have been studied for each building configuration. Time series from the approximately 700 pressure taps covering each building envelope have been stored in a standard format and recorded on CD-ROMs, which will be available for public distribution.

In addition, since the data base generated consists of local pressures, the opportunity exists to calculate spatially-averaged loads over whatever structural tributary areas may be important, or to combine them with any structural influence lines of interest to provide loads "actually seen" by a variety of structural systems indirectly loaded by the wind. Thus, the designer can generate wind loads for specific types of buildings (geometries, configuration etc.) and for any wind storm pattern including directional effects, and proceed with a more economical design tailored to a specific rather than a generic type of building, as it happens with the application of current standards and codes of practice. More discussion about the potential development of this new generation of standards and their advantages can be found in Simiu and Stathopoulos (1997) and Whalen et al. (1998).

9.6 Computational Evaluation of Wind Loads

Computational Wind Engineering (CWE) deals with the application of Computational Fluid Dynamics (CFD) methodologies, typically numerical solutions of Navier–Stokes equations using appropriate turbulence models, into wind engineering problems. Methods for turbulence treatment can be classified into three groups, i.e. Reynolds-averaged Navier–Stokes equations with a turbulence model, large eddy simulation and direct numerical simulation. These methods were reviewed by Murakami (1997).

At the present time, the application of CFD into wind engineering is limited mainly due to difficulties related to the specific features of atmospheric boundary-layer flow and structures of wind engineering interest, such as high turbulence, high Reynolds number, 3D flow field, bluff bodies and associated flow separation and vortex shedding. Although it has tremendous potential, CWE is still in its infancy and has a long way to go to become truly useful to the design practitioner. However, for applications for which mean wind flows and pressures may be of design importance, CWE can be used, at least for preliminary design purposes, to take advantage of its reduced time and cost. Figure 9.30 compares experimental and computational results regarding mean pressure coefficients measured on a cube surface. A simple k-ε model has been used for the computation. The predictions on the windward face of the cube are in good agreement with the experimental data but numerical results on the roof appear problematic. The standard k-ε model fails to predict the suction at the upwind roof edge. Recent modifications of the standard k-ε model have improved the comparisons between experimental and computational data on the roof but such adjustments are of an ad hoc nature and improve the situation only for some particular cases. Overall however, the range of numerical pressure coefficients by using all its forms and improvements is very comparable to the variations of the experimental data obtained in previous studies and highlighted in Fig. 9.30. A comprehensive review on CWE has been carried out by Stathopoulos (2002).

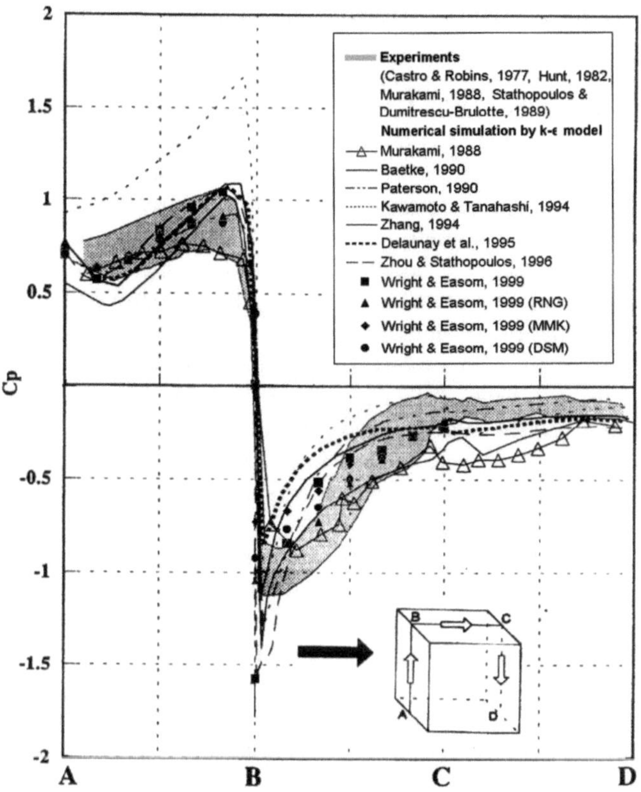

Fig. 9.30 Comparison of mean wind pressure coefficients: experimental measurements and numerical simulations by using several turbulence models (Stathopoulos 2002)

9.6.1 Computer-Aided Wind Engineering

9.6.1.1 Computer Simulation

Wind tunnel and field experimentation is the traditional approach for the investigation of wind-induced pressure fluctuations and time histories. However, the collection of long time histories of wind and pressure data might be time consuming and laborious, considering the inherent variability in such time histories affected by building geometry, measurement location, surroundings and other factors. This, on the other hand, can be efficiently handled by computer simulation using probabilistic/statistical models. Techniques for simulating Gaussian and non-Gaussian time histories can be broadly grouped into (1) FFT (Fast Fourier Transform) and (2) ARMA (Auto Regressive Moving Average). Seong and Peterka (1993) proposed a new technique by combining Fourier transform method, autoregressive model Gaussian and non-Gaussian time histories can be broadly grouped into (1) FFT

(Fast Fourier Transform) and (2) ARMA (Auto Regressive Moving Average). Seong and Peterka (1993) proposed a new technique by combining Fourier transform method, autoregressive model with non-Gaussian input process, and phase transformation of Fourier coefficients in Fourier representation of time series.

However, in order to apply this and other simulation methods for engineering purposes, more research efforts were necessary. The adequacy of the model has been shown for different building zones and flow characteristics; the simulation procedure as well as the parametric estimation procedure have been simplified; and the generalization in terms of codification or at least clear design procedure has been provided. Kumar and Stathopoulos (1997) have improved the simulation for both Gaussian and non-Gaussian wind pressures by addressing these issues. They have also applied their methodology for the analysis of building envelope fatigue. The complete details of the synthetic simulation approach can be found in Kumar and Stathopoulos (2001). Wavelets have been used for simulation purposes (Hajj and Tieleman 1996). For more general information on simulation techniques for stochastic wind effects, with mathematical background and wind engineering applications, see Kareem (1993).

The success of computer simulation in generating limitless amounts of data and the present capacity of storage and access to these data provides a new challenge and potential benefit to design professionals. Pressure coefficient time histories originating from either direct wind tunnel measurements or via computer simulation can be reconstituted as full-scale pressure time histories for any full-scale wind speed. Furthermore, this computer simulation can be used in combination with the wind tunnel generated pressure coefficient databases for purposes of interpolation between different building geometries and exposures. There is no doubt that the future in the codification approaches appears even more challenging and exciting than the past!

9.6.1.2 Artificial Intelligence

While wind-engineering research has been greatly benefited by the computer assistance, computers have played even more important roles in bringing research findings to engineering practice. In this regard, this paper discusses some effective computerized tools, such as knowledge-based expert systems, and neural networks, as well as their impact on wind engineering applications, related to the wind effects on low buildings.

Knowledge-Based Expert Systems (KBES) are interactive computer programs that incorporate human experience, judgment, rules of thumb and other expertise to provide knowledgeable advice on a wide range of topics. Because of the ill-structured nature of many wind-related design and analysis problems, it was predicted that the knowledge-based systems would play a major role in wind engineering in the 1990s (Reed 1990).

Code interpretation of wind loads on low buildings has been an area for KBES applications. Attempt has been made in Australia (Sharpe et al. 1990), Canada

(Stathopoulos and Wu 1994), USA (Schechter et al. 1995) and other countries for the correct and accurate interpretation of code specifications by KBES. Other applications can be found in Mehta et al. (1993) for evaluating wind damage to buildings (WIND-RITE™), in Minson et al. (1996) for estimating wind loads on building attachments using detailed local velocity data, and Cheng and Shang (1997) using GIS for surface wind field modelling. Work by the Alan G. Davenport Wind Engineering Group attempts to use KBES for the interpolation of measured pressure coefficient time series (see Sect. 9.5) to other geometries within the bounds of the tested buildings.

The advantages of KBES can be summarized as follows: (1) No special knowledge is required for running the system and complex domain knowledge becomes accessible to general users via a user-friendly interface, (2) a usually time-consuming consultation process can be executed consistently in a very short time period, (3) such a time saving feature also makes feasible the evaluation of different design alternatives at early design stages, (4) the system can be expanded readily as more information becomes available, since in a KBES, the domain knowledge is explicit and separate from the inference mechanism of the system.

While the use of KBES has been well demonstrated, applied and accepted, Neural Networks (NN) are relatively new to wind engineering. Similar to biological neural functioning, NN technology is capable of learning the complex relationships from training on data samples, which may be incomplete or inconsistent.

Sandri and Mehta (1995) used a back propagation neural network for predicting wind-induced damage to buildings. To demonstrate the use and adaptability of the artificial neural network technology to damage predictions, a database of buildings and their corresponding damage was simulated instead of being compiled from real damage documentation. The database was then split into two parts for training and testing. During the training process, the network repeatedly processed examples of input-output pairs and gradually adjusted a set of numerical weights between the neurons until the network error was reduced to a minimum. Finally, the testing database was used to evaluate the effectiveness of learning.

A similar approach has been adopted by English and Fricke (1999) to the quantification of shielding effects of two identical rectangular buildings as a function of reduced separation distance, aspect ratio and power law index; and by Khanduri et al. (1996) to the modelling of interference effects of buildings of different sizes on drag, lift and overturning moment. In both cases, wind-tunnel results obtained by the respective authors and from literature form the databases for training and testing. It should be pointed out that, in the latter case, a hybrid computer system has been developed with NN knowledge acquisition and KBES for knowledge presentation, since the NN lack of explanation capability can be compensated by the KBES' logic reasoning and inference. A generalization of this approach, which may be appropriate for codification purposes, has been described in Khanduri et al. (2000).

In all these problems, the interactions between the variables are too complex for an analytical solution. In general, there is no sufficient understanding in the problem domain. It is the pattern-recognition ability of NN, which is useful in the evaluation of wind effects on buildings. Compared to KBES, NN are capable of consistently predicting the outputs of new, previously unknown inputs, by using the significant features and complex relationships captured from a proper training. They are efficient in expanding existing knowledge and performing parametric studies, when wind-tunnel experiments are too laborious and time consuming.

9.7 Summary and Conclusions

This chapter reviews the fundamentals behind the provisions of contemporary wind standards and codes of practice, the latest developments and the progress made in the evaluation of wind loads on buildings. It refers to some of the most recent studies that have led or will lead to the updating of the North American and other national wind codes and standards. Such updates relate with the re-examination of wind-induced pressures on gable-roof low buildings with intermediate roof angles (10–30°), the introduction of pressure coefficient provisions for hipped roofs, the true magnitude of internal pressures by considering the alleviating lack of correlation between fluctuating external and internal pressures, to name a few. Areas of additional research such as that required for the incorporation of the effects of upstream terrain roughness, the evaluation of wind loads on primary structural systems of buildings with different roof shapes, the assessment of loads on parapets, carports and other architectural features for the formulation of new code provisions have been identified.

The advancements in the area of computational wind engineering and the effects of innovated techniques of artificial intelligence (knowledge-based expert systems, neural networks) in the evaluation of wind loads on buildings have been discussed. Finally, a new approach towards future codification by creating databases of time histories of pressures acting on building envelopes has been described. This approach will lead to more accurate and economical design of buildings since the loads specified will be tailored to each particular case of interest rather than to a generic type of building with enveloped loads as it is currently the case.

Acknowledgement The assistance of three ex-Ph.D. students, Dr. P. Karava, currently Assistant Professor at Purdue University and Dr. I. Zisis, currently Assistant Professor at Florida International University and, Dr. B. Hajra, currently postdoctoral fellow at Concordia University in putting these lecture notes, from which this chapter stems, together is gratefully acknowledged and highly appreciated.

References

American Society of Civil Engineers ASCE (1998) ASCE 7-98. Minimum design loads for buildings and other structures. ASCE, New York

American Society of Civil Engineers ASCE (1999) Wind tunnel studies of buildings and structures. ASCE manuals and reports on engineering practice no. 67, Aerospace Division of the American Society of Civil Engineers

American Society of Civil Engineers ASCE (2003) ASCE 7-02. Minimum design loads for buildings and other structures. ASCE, New York
American Society of Civil Engineers ASCE (2010) ASCE 7-10. Minimum design loads for buildings and other structures. ASCE, Reston
Architectural Institute of Japan (AIJ) (1996) Recommendation for loads on buildings. Architectural Institute of Japan, 26-30, Shiba 5-chome, Minato-ku, Tokyo 108-8414
Baskaran A (1992) Design of roofs with parapets. Construct Can 34:19–21
Beste F, Cermak JE (1996) Correlation of internal and area-averaged wind pressures on low-rise buildings. In: Third international colloquium on bluff body aerodynamics and applications, Blacksburg
British Standards Institution (BSI) (1997) BS 6399-2. Loading for buildings. Code of practice for wind loads
Cheng E, Shang J (1997) Complex terrain surface wind field modeling. J Wind Eng Ind Aerodyn 67:94–105
Davenport AD (1967) Gust loading factors. J Struct Div Proc ASCE 93(ST3):11–34
Davenport AG (1975) Perspectives on the full-scale measurement of wind effects. J Wind Eng Ind Aerodyn 41:15–22
Davenport AG, Stathopoulos T, Surry D (1985) Reliability of wind loading specifications for low buildings. In: Fourth international conference on structural safety and reliability (ICOSSAR '85), Kobe, 27–29 May 1985
Davenport AG, Grimmond S, Oke T, Wieringa J (2000) The revised Davenport roughness classification for cities and sheltered country. In: Third symposium on the urban environment. Davis, 14–18 August 2000, pp 7–8
Deaves DM (1981) Terrain-dependence of longitudinal r.m.s. velocities in the neutral atmosphere. J Wind Eng Ind Aerodyn 8:259–274
Durst CS (1960) Wind speeds over short periods of time. Meteorol Mag 89:181–186
Engineering Sciences Data Unit, ESDU. Data Item 82026. Strong winds in the atmospheric boundary layer. Part 1: hourly-mean wind speeds. Engineering Sciences Data Unit
Engineering Sciences Data Unit. ESDU. Data Item 84030. Longitudinal turbulence intensities over terrain with roughness changes. Engineering Sciences Data Unit
English EC, Fricke FR (1999) The interference index and its application using neural network analysis of wind-tunnel data. J Wind Eng Ind Aerodynamics 83(1–3):567–575
European Committee for Standardization, C.E.N (1995) (ENV). Eurocode 1: basis of design and actions of structures. Part 2–4: wind actions. ENV-1991-2-4, C.E.N., Brussels
Garratt JR (1990) The internal boundary layer – a review. Boundary Layer Meteorol 50:171–203
Hajj M, Tieleman H (1996) Application of wavelet analysis to incident wind in relevance to wind loads on low-rise structures. Trans ASME 118:874–876
Ho TCE (1992) Variability of low building wind loads. PhD Thesis, The University of Western Ontario, London
Holmes JD (1979) Mean and fluctuating internal pressures induced by wind. Wind engineering. In: Proceedings of the 5th ICWE, Fort Collins, July 1979, pp 435–450
Irwin PA, Dunn GE (1994) Review of internal pressures on low-rise buildings. RWDI Report 93-270 for Canadian Sheet Building Institute
Jensen M (1958) The model law for phenomena in the natural wind. Ingenioren (Internationl Edition) 2:121–128
Khanduri AC, Bedard C, Stathopoulos T (1996) Modelling wind-induced effects-integrating experimental and computerized approaches. In: Proceedings annual conference of Canadian Society of Civil Engineering, Edmonton
Khanduri AC, Stathopoulos T, Bedard C (2000) Generalization of wind-induced interference effects for two buildings. Wind Struct 3:255–266
Kareem A (1993) Simulation of stochastic wind effects. In: Proceedings of the 7th U.S. national conference on wind engineering, Los Angeles, 27–30 June 1993

Kumar SK, Stathopoulos T (1997) Computer simulation of fluctuating wind pressures on low building roofs. J Wind Eng Ind Aerodyn 69–71:485–495

Kumar SK, Stathopoulos T (2001) Generation of local wind pressure coefficients for the design of low building roofs. Wind Struct 4:455–468

Letchford C, Gardner A, Howard R, Schroeder J (2001) A comparison of wind prediction models for transitional flow regimes using full-scale hurricane data. J Wind Eng Ind Aerodyn 89:925–945

Mans C (2001) Wind loads on parapets. Master of Engineering Science Thesis, The University of Western Ontario, London

Meecham D, Surry D, Davenport AG (1991) The magnitude and distribution of wind-induced pressures on hip and gable roofs. J Wind Eng Ind Aerodyn 38:257–272

Mehta K, Smith D, Cheshier R (1993) A knowledge-based system for wind damage to buildings: Hurricanes of 1992. Andrew and Iniki One Year Later, ASCE, Miami

Minson A, Harries R, Wood C (1996) Towards an expert system for estimating wind loads on building attachments using detailed local velocity data. J Wind Eng Ind Aerodyn 62:11–36

Murakami S (1997) Overview of turbulence models applied in CWE-1997. Keynote Lecture. In: Proceedings of the 2nd European and African conference on wind engineering, Genova, 22–26 June 1997

NBCC (1975) National building code user's guide – commentaries on part 4 including requirements for plain and reinforced masonry. Canadian Commission on Building and Fire Codes, National Research Council of Canada, Ottawa

NBCC (1995) National building code user's guide – structural commentaries (part 4). Canadian Commission on Building and Fire Codes, National Research Council of Canada, Ottawa

NBCC (2005) National building code user's guide – structural commentaries (part 4). Canadian Commission on Building and Fire Codes, National Research Council of Canada, Ottawa

NBCC (2010) National building code user's guide – structural commentaries (part 4). Canadian Commission on Building and Fire Codes, National Research Council of Canada, Ottawa

Reed D (1990) Expert systems in wind engineering. J Wind Eng Ind Aerodyn 33:487–494

Richardson GM, Surry D (1994) The Silsoe structures building: comparison between full-scale and wind-tunnel data. J Wind Eng Ind Aerodyn 51:157–176

Schmid HP, Bunzli B (1995) The influence of surface texture on the effective roughness length. Q J R Meteorol Soc 121:1–21

Simiu E, Scanlan RH (1996) Wind effects on structures – fundamentals and applications to design, 3rd edn. Wiley, New York

Sandri P, Mehta KC (1995) Using back-propagation neural network for predicting wind-induced damage to buildings. In: Proceedings of the ninth international conference on wind engineering, New Delhi, 1989–1999

Schechter M, Schechter E, Simiu E (1995) Development of computer-based version of ASCE 7-95 standard provisions for wind loads. NIST technical note 1415, National Institute of Standards and Technology, Gaithersburg

Seong SH, Peterka JA (1993) Computer simulation of non-Gaussian wind pressure fluctuations. In: Proceedings of the 7th U.S. national conference on wind engineering, Los Angeles, 27–30 June 1993

Sharpe R, Marksjo B, Holmes J, Fitchett R, Ho F (1990) Wind loads on buildings: expert system-WINDLOADER. J Wind Eng Ind Aerodyn 36:1269–1277

Simiu E, Stathopoulos T (1997) Codification of wind loads on buildings using bluff body aerodynamics and climatological databases. J Wind Eng Ind Aerodyn 69–71:497–506

Standards Australia (2002) AS/NZS 1170.2:2002 – Australian/New Zealand standard, Structural design actions, part 2: wind actions. Standards, Australia

Stanton TE (1924) Report on the measurement of the pressure of wind on structures. In: Minutes of the proceedings of the institute of civil engineers, 16 December 1924

Stathopoulos T (1979) Turbulent wind action on low-rise buildings. PhD Thesis, The University of Western Ontario, London

Stathopoulos T, Surry D, Davenport AG (1979) Wind-induced internal pressures in low buildings. In: Proceedings of the 5th international conference on wind engineering, Fort Collins
Stathopoulos T, Wu H (1994) Knowledge-based wind loading for envelope design: beyond building codes. J Wind Eng Ind Aerodyn 53:177–188
Stathopoulos T (1995) Evaluation of wind loads on low buildings – a brief historical review, a state-of-the-art in wind engineering. International Association for Wind Engineering, Wiley Eastern Limited, New Delhi 110002
Stathopoulos T, Wang K, Wu H (2000) Proposed new Canadian wind provisions for the design of gable roofs. Can J Civil Eng 27:1059–1072
Stathopoulos T, Wang K, Wu H (2001) Wind pressure provisions for gable roofs of intermediate roof slope. Wind Struct 4(2):119–130
Stathopoulos T (2002) The numerical wind tunnel for industrial aerodynamics: real or virtual in the new millennium? Wind Struct 5(2–4):193–208
Stathopoulos T, Saathoff P, Bedair R (2002) Wind pressures on parapets of flat roofs. J Archit Eng ASCE 8(6):49–54
St. Pierre LM (2002) Evaluation of wind load provisions for low buildings. Master of Engineering Science Thesis, The University of Western Ontario, London
Surry D, Stathopoulos T (1978) An experimental approach to the economical measurement of spatially averaged wind loads. J Wind Eng Ind Aerodyn 2:22–25
Surry D (1999) Wind loads on low-rise buildings: past, present and future. In: Proceedings of the 10th international conference on wind engineering, Copenhagen, 20–24 June 1999
Surry D, Lin JX (1995) The effect of surroundings and roof corner geometric modifications on roof pressures on low-rise buildings. J Wind Eng Ind Aerodyn 58:113–138
Tieleman HW, Hajj MR, Rainhold TA (1999) Pressure characteristics for separated flows. In: Larsen, Larose, Livesey (eds) Wind engineering into the 21st century. Balkema, Rotterdam
Van der Hoven I (1957) Power spectrum of wind velocities fluctuations in the frequency range from 0.0007 to 900 cycles per hour. J Meteorol 14:1254–1255
Vickery BJ (1976) Wind loads on low-rise buildings. In: Proceedings of the D.R.C. Seminar, Darwin
Vickery BJ, Bloxham C (1992) Internal pressure dynamics with a dominant opening. J Wind Eng Ind Aerodyn 41–44:193–204
Whalen T, Simiu E, Harris G, Lin J, Surry D (1998) The use of aerodynamic databases for the effective estimation of wind effects in main wind-force resisting systems: application to low buildings. J Wind Eng Ind Aerodyn 77(78):685–693
Wang K, Stathopoulos T (2005) Exposure model for wind loading of buildings. In: Proceedings of the 4th European-African conference on wind engineering, Prague, 11–15 July 2005
Whittemore HL, Cotter JB, Stang AH, Phelan VB (1948) Strength of houses: application of engineering principles to structural design. Building materials and structures report BMS109, US Department of Commerce, National Bureau of Standards
Xu YL, Reardon GF (1998) Variation of wind pressure on hip roofs with roof pitch. J Wind Eng Ind Aerodyn 73:267–284
Zhang X, Zhang RR (2001) Actual ground-exposure determination and its influences in structural analysis and design. J Wind Eng Ind Aerodyn 89:973–985
Zhou YS, Kareem A (2002) Definition of wind profile in ASCE 7. J Struct Eng ASCE 128:1082–1086
Zisis I, Stathopoulos T (2009) Wind-induced cladding and structural loads on a low wooden building. J Struct Eng ASCE 135:437–447
Zisis I, Stathopoulos T (2010) Wind-induced pressures on patio covers. J Struct Eng ASCE 136 (9):1172–1181
Zisis I, Stathopoulos T, Smith I, Doudak G (2011) Cladding pressures and primary structural system forces of a wood building exposed to strong winds. Can J Civil Eng 38(9):963–973

Chapter 10
Numerical Simulation of Wind Effects

Ahsan Kareem and Megan C. McCullough

Abstract Modern structural systems are becoming increasingly complex and numerical simulation of the potential loads with which they will be affected is critical for analysis, design, and optimization of safe and reliable structures. Monte Carlo analysis approaches are often used, which involve the input of loads into a structural model and the output of responses. Besides being necessary for numerical analysis, digitally simulated data is also necessary to drive computer controlled test facilities. Both approaches necessitate an ensemble of input signals that accurately represents what the structure may expect to experience during its lifetime. Therefore, simulation of time histories of wind velocity, pressure, and force fluctuations are necessary, in addition to simulation of structural response, which allows assessment of attendant functionality and safety under service and design loads, respectively. Random processes simulated for analysis purposes are often assumed to be Gaussian and stationary for simplicity. Many wind events, however, are characterized by non-stationarity and non-Gaussianity. Therefore, simulation methodologies are necessary for univariate and multivariate processes, unidimensional and multi-dimensional fields, Gaussian and non-Gaussian data, stationary and non-stationary processes, and conditional and unconditional cases. In order to accomplish this task, methods based on the time, frequency, and time-frequency domains are employed. This paper summarizes a historical perspective, recent developments, and future challenges for simulation. Also included in the discussion are computational tools employed for data and response analysis. Examples are presented to illustrate some of the topics discussed.

Keywords Frequency • Gaussian • Monte Carlo • Non-Gaussian • Non-stationary • Simulation • Stationary • Time • Time-frequency domains • Transient events

---

A. Kareem (✉) • M.C. McCullough
NatHaz Modeling Laboratory, University of Notre Dame, Notre Dame, IN, USA
e-mail: kareem@nd.edu; mmccull3@nd.edu

Y. Tamura and A. Kareem (eds.), *Advanced Structural Wind Engineering*,
DOI 10.1007/978-4-431-54337-4_10, © Springer Japan 2013

10.1 Introduction

Simulation of wind related processes, such as wind velocities, pressures and force fluctuations, is essential for time-domain numerical analysis of structures subjected to wind loads, e.g., long span bridges, offshore structures, or tall buildings. Besides numerical analysis, digitally simulated data is also needed to drive computer controlled test facilities, e.g., wind tunnels with multiple fans or controlled damping systems for mitigation of motion. Typically simulation schemes are based on exact theoretical models, phenomenological or empirical models, or observed data of the process to be simulated. The schemes can be broadly categorized into two classes: (1) methods based on the summation of trigonometric functions (wave superposition) and (2) methods based on the convolution of white noise with a kernel function or integration of a differential equation driven by white noise (digital filtering). These techniques vary in their applicability, complexity, computer storage requirements and computing time.

Several of the key issues encountered in simulation include:

- The necessity of simulations at a very large number of points, e.g., along the span of a long bridge, which requires appropriate modeling and inclusion of imperfect correlation of the random processes/fields;
- Consideration of modulations in both time and frequency, e.g., non-stationarity present in transient wind fields of a downburst;
- Inclusion of non-Gaussian features, e.g., pressure fluctuations under separated flow regions; and
- The need for conditional simulation of data due to limited measurements or the loss of measured data.

In some cases, all of these issues may plague a single simulation. For example, hurricane winds are extreme and imperfectly correlated and may result in non-Gaussian loads on structures. In addition, limited observations are often available.

A simulation scheme may be appropriate for use with processes that are:

- Conditional or un-Conditional;
- Gaussian or non-Gaussian;
- Stationary/Homogenous or non-Stationary/non-Homogeneous; and
- Univariate/Uni-Dimensional or Multivariate/Multi-Dimensional.

Initially, the development of appropriate simulation techniques focused on the generation of univariate and one-dimensional processes. The more difficult task of simulating multivariate or multi-dimensional processes was first addressed by Borgman (1969), Shinozuka (1971), and Shinozuka and Jan (1972). In addition, conditional simulation, or stochastic interpolation, schemes have been researched, which can be used to simulate, or interpolate, data in between measured locations. These schemes will all be addressed in detail in the following sections.

10.2 Wave Superposition (Spectral) Based Schemes

10.2.1 Spectral Representation Approaches

One of the traditional, and most popular, approaches for simulation is to utilize a superposition of trigonometric functions with random phase angles (Shinozuka and Jan 1972). Using a large number of terms in the summation, the continuous target energy spectrum is approximated by its discrete form. The Gaussianity of the simulated data is established by virtue of the Central Limit Theorem. This process is applicable for multivariate and multi-dimensional processes/fields, but quickly becomes computationally inefficient as the number of variables increases. In an effort to improve computational efficiency, the summation of trigonometric functions may be carried out utilizing a fast Fourier transform (FFT) (Wittig and Sinha 1975). The use of FFT, however, increases demand on computer storage, which is a problem particularly for simulations that involve long-duration, multivariate or multi-dimensional processes (Chen and Kareem 2002; Li and Kareem 1993b; Samaras et al. 1985). In the last several decades, however, significant progress has been made improving the efficiency and extending the wave superposition schemes for the simulation of univariate and multivariate, uni-dimensional and multi-dimensional, and conditional and un-conditional processes, e.g., Shinozuka and Deodatis (1991), Shinozuka et al. (1990), Li and Kareem (1991; 1993b; 1997), Grigoriu (1993), Mann (1998), Chen and Kareem (2005), Carassale and Solari (2006).

Due to the limitations imposed by the computer memory using the FFT approach, Li and Kareem (1993b) introduced a numerical simulation scheme that combines the advantages of the wave superposition techniques and digital filtering to simulate continuous long-duration multivariate random processes. This approach offers the simple convenience of conventional FFT based schemes; it does not, however, suffer from the drawbacks of the required large computer memory that, in the past, has precluded the generation of long duration time series. This approach involves the simulation of a large number of time series segments utilizing the FFT algorithm. These segments are then synthesized by means of a digital filter to provide the desired duration of the simulated processes. The method offers computational efficiency, convenience, and robustness, and has immediate applications to the simulation of real-time processes. Other measures are possible to reduce the computational effort for a large number of simulation locations, e.g., decomposition at limited frequency locations and larger time steps. In addition, consideration of the decomposition method for the cross power spectral density matrix (XPSD) may yield additional means to reduce computational effort.

The Cholesky decomposition of the XPSD has been widely utilized in digital simulation of multivariate and multi-dimensional schemes (Shinozuka and Jan 1972). Li and Kareem (1990b, 1993b, 1995) introduced the concept of "stochastic decomposition" of the XPSD matrix for the simulation of stationary random processes, which was further extended to the simulation of non-stationary processes (Li and Kareem 1991, 1997). Central to stochastic decomposition is the decomposition of a correlated vector-valued random process into a set of vector-valued subprocesses. This decomposition must occur such that any two component processes from the same subprocess are statistically fully coherent, while any two component processes from different subprocesses are non-coherent. Therefore, the simulation of the parent process is simplified by the simulation of noncoherent subprocesses. The Schur decomposition is often more attractive than the Cholesky decomposition as the former permits a relatively small number of modes to be utilized in the simulation (Di Paola and Gullo 2001; Li and Kareem 1993b; Shinozuka et al. 1990). Similar gains are also possible with Cholesky decomposition through truncation of terms (Li and Kareem 1995).

In order to facilitate the use of stochastic simulation in the design office, the NatHaz Modeling Laboratory has developed a web-based simulation portal, which would allow simulation of multivariate Gaussian random processes with prescribed spectral characteristics. The NatHaz online wind simulator (NOWS), available at http://windsim.ce.nd.edu, enables users to simulate stationary random wind fields beyond temporal and geographical boundaries using an intuitive and user-friendly interface, shown in Fig. 10.1. NOWS supports both Metric (SI) and English units for inputs/outputs and provides an online unit converter. A similar concept has already been employed for the NatHaz Aerodynamic Load Database, http://aerodata.ce.nd.edu (Kwon et al. 2008). Three input options for multiple locations are provided for multivariate wind simulation: vertical coordinate location specification (e.g., high-rise building or chimney); horizontal coordinate location specification (e.g., long-span bridge deck); and arbitrary two-dimensional coordinate location specification (e.g., long-span bridge deck and tower). As for the mean wind speed, NOWS has employed an ASCE7-98-based concept, i.e., 3-s gust speed and four terrain exposure conditions (ASCE7-98), to calculate 1-h mean wind speeds at given heights. NOWS also offers the flexibility of selecting one of four different simulation schemes:

- Conventional spectral representation method (Shinozuka and Deodatis 1991);
- Discrete frequency function with Cholesky decomposition and FFT (Wittig and Sinha 1975);
- Ergodic spectral representation with Cholesky decomposition and FFT (Deodatis 1996; Ding et al. 2006); and
- Schur decomposition by AR model with polynomial approximation (Di Paola 1998; Di Paola and Gullo 2001).

For more details on these four schemes, please see the documentation available at http://windsim.ce.nd.edu and the accompanying citations. The results interface is shown in Fig. 10.2. The selected simulation scheme and the user's input parameters are shown in addition to three samples of the simulated wind speeds. NOWS allows the user to then download an ASCII file which contains the time and fluctuating wind speeds at all locations. This allows for the user to perform further off-line analysis, e.g., structural dynamic analysis under wind loads in the time domain.

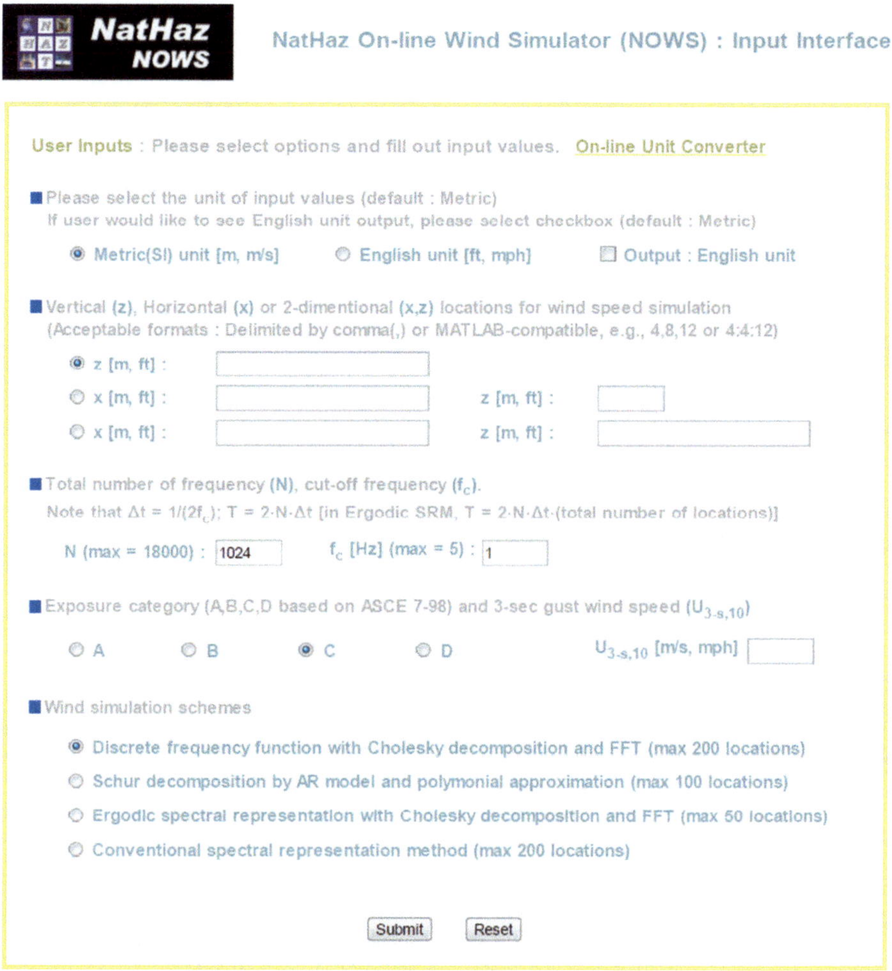

Fig. 10.1 NatHaz NOWS web-based simulation portal

10.3 Digital Filtering (Time-Series Approach) Based Schemes

10.3.1 Parametric Time-Series Methods

Digital filtering based schemes offer efficient methods for simulating random processes. Typical schemes are based on autoregressive (AR) models, moving average (MA) models, and their combination, autoregressive moving average (ARMA) models (Li and Kareem 1990a, 1991, 1993b; Mignolet and Spanos 1992; Samaras et al. 1985; Spanos and Mignolet 1992). The ARMA representation entails

NatHaz NOWS

NatHaz On-line Wind Simulator (NOWS) : Result interface

Message:

Ergodic spectral representation method with Cholesky decomposition and FFT was used for wind simulation

■ Input parameters by user

Number of Locations	Vertical 3 points	
Exposure category	Exposure C	
Total number of frequency	N	1024
Cut-off frequency [Hz]	f_c	1
3-sec gust wind speed [m/s]	$U_{3\text{-}s,10}$	40
Input unit	Metric(SI)	
Output unit	Metric(SI)	

■ Fluctuating wind speed at Location 1, 2 and 3

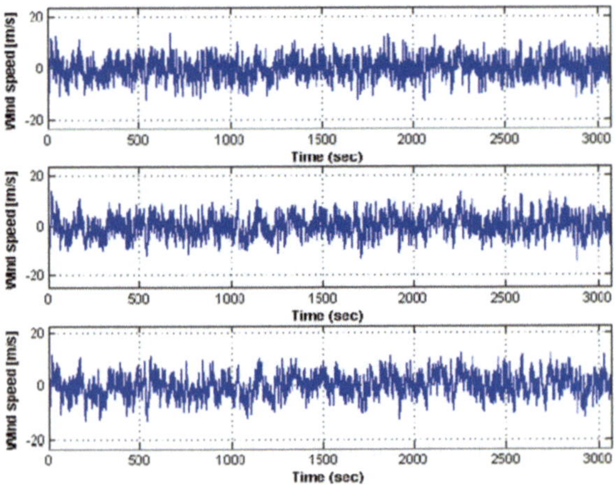

■ Download data file including all wind force components : w_out_all_ml00167.mat

(MATLAB variables: t_scale = time steps [sec]; mean_speed = mean wind speed(s) [m/s];
Y_speed = fluctuating wind speed(s) [m/s])

Back to the Input Interface

Fig. 10.2 NatHaz NOWS web-based simulation results

weighted recursive relations that connect the random quantity being simulated at successive time increments. Unlike the FFT-based approaches, this scheme does not require large computer memory; rather, only a limited amount of information (e.g., coefficient matrices) is stored and long duration time series may be simulated at successive time increments. Determination of an appropriate parametric model for certain applications may not be straightforward, especially for those processes with filtered white noise as a spectral description. Due to potential difficulties in selecting appropriate model orders, Aas-Jakobsen and Strommen (2001) developed a method that uses a spectral approach (Chen and Kareem 2002).

In addition to proper selection of model order, the estimation of the optimal model parameters is also sensitive to the time increment selected (Li and Kareem 1990a, 1993b; Mignolet and Spanos 1992). Based on the system's dynamic characteristics, the time-integration schemes require that the time increment not exceed a prescribed value. This concern has been noted in a study by Chay et al. (2006), which proposed an integrated model for simulating downburst wind fields. Turbulent fluctuations were simulated using ARMA models and linear interpolation was used to address instabilities associated with an insufficient time step. Interpolation, however, introduces its own problems. In order to efficiently simulate realizations of wind loads at prescribed time increments, Li and Kareem (1990a) offers two approaches. The first involves an ARMA algorithm based on a three stage matching method and the other is based on a scheme which combines ARMA with a digital interpolation filter. This topic has been revisited by Wang (2007), which introduces an improved interpolation filter and demonstrates the shortcoming of the linear interpolation often used in practice.

Digital filtering simulation approaches can either be directly implemented for multivariate and multi-dimensional processes/fields or they can be simulated as univariate processes in conjunction with a decomposition approach, such as Cholesky or Schur. In Li and Kareem (1990a), the dynamic response of structural systems to a variety of random excitations using recursive models was presented. The methodology permits analysis of stationary, non-stationary, or transient response of structures under stochastic multi-input correlated processes, e.g., wind loads. For non-stationary or transient excitation, the analysis involves a direct computation of the output covariance from the non-stationary correlation structure of the input. The ARMA representation of the response is obtained in terms of the ARMA description of the stationary excitation. The usefulness of the recursive approach was also highlighted for nonlinear systems.

Time series approaches allow for numerical evaluation of commonly used operations. For example, discrete convolution models can be employed for linear transformation of a given time history, discrete differentiation models can be used to obtain derivatives, and discrete interpolation models can be used for interpolation of intermediate time steps in time series. Hybrid combinations of these models can provide simulation schemes that are particularly efficient and effective (Li and Kareem 1993a). For example, a hybrid combination of discrete convolution and differentiation models results in a simulation scheme that is very effective in evaluating the response of structures with a displacement feedback, e.g., the

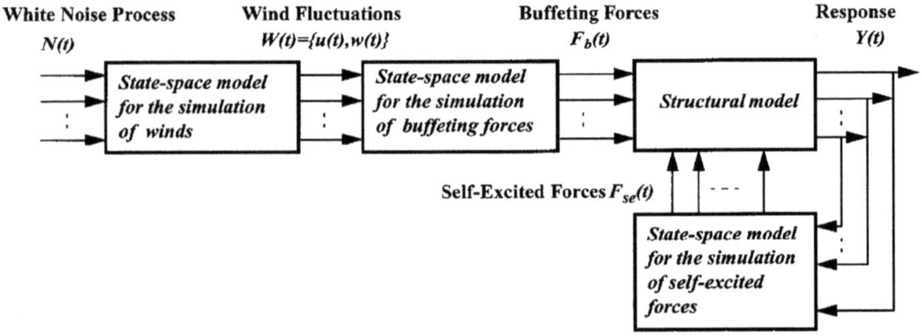

Fig. 10.3 Integrated modeling of dynamic response of wind-excited structures (Chen and Kareem 2001)

response of a large floating platform exposed to ocean waves and wind requires that loads be evaluated at the instantaneous displaced position of the platform (Li and Kareem 1993a).

10.3.2 State Space Modeling

In most modern control and dynamics studies, state space representations are often used for system modeling. In a state space approach, a model of the loading is first determined. This model can then be augmented to the structural state space representation for the overall dynamic response analysis. Unfortunately, due to a large number of correlated wind-related processes with prescribed spectral characteristics, researchers have had difficulty finding a state space model for wind (Goβmann and Waller 1983; Kareem 1999). Models for special applications, however, have been tailored to meet problem-focused needs. For example, Kareem and Mei (2000) use a combination of stochastic decomposition and AR models to formulate a state space representation of wind effects. The XPSD matrix of the wind-related process is decomposed utilizing Proper Orthogonal Decomposition (POD). Each component process is then modeled as an AR model, which is subsequently expressed in a state space format. By stacking these component models, the overall model for the entire system can be obtained in a straightforward manner.

One of the most elegant applications of the state space model, presented in Chen and Kareem (2001), addresses aeroelastic analysis of bridges under multi-correlated winds. An integrated state-space model of a multi-input and multi-output system with a vector-valued white noise input is used to model the dynamic response of bridges. Following Fig. 10.3, a vector of white noise is transformed first to multi-correlated wind fluctuations, then to unsteady buffeting and aerodynamic forces, and finally to structural motions via successive state-space models. Both the equations of motion at the full order in the physical coordinates or at a reduced-order in the generalized modal coordinates can be utilized in this approach. Use of the state-space format allows the use of tools based on linear system theory for response analysis and structural control design (Chen and Kareem 2001). The integrated linear time-invariant state space model of the response allows for direct evaluation of the covariance matrix of the response using the Lyapunov equation. This approach has much higher computational efficiency than the conventional spectral analysis approach. This approach also allows for time domain simulation of multi-correlated wind fluctuations, the associated aerodynamic forces, and the buffeting response using Monte Carlo simulation. Figure 10.4a shows an example simulation of the u and w components of wind fluctuations at the midpoint of the main span of a long-span bridge. Figure 10.4b–c show the drag, lift, and moment components of the buffeting and self-excited forces, and Figure 10.4d shows the buffeting response in the lateral, vertical, and torsional directions. It is important to note that the state space model more accurately represents the frequency dependent unsteady characteristics of the aerodynamic forces compared to time domain bridge aerodynamic studies, which often use a straightforward quasi-steady approach. In addition, this approach allows for structural and aerodynamic coupling effects among structural modes, which are critical for the multi-modal response of bridges, to be automatically included in the computation. The analysis framework can also be immediately and simply extended to the analysis of bridges/structures with structural and aerodynamic nonlinearities by using a time-variant system model. See Chen and Kareem (2001) for an additional example of the application of this approach to a long-span cable-stayed bridge.

10.4 Simulation of Inflow Conditions

Flow around sharp corners poses a substantial challenge for turbulence modeling in applications of computational fluid dynamics (CFD). Stagnation, separation, transition, curved shear layer, reattachment, and wake unsteadiness exhibit a chain effect and add to the complexity of effective turbulence modeling. These flow features are also more sensitive to inflow conditions than features associated with flows not characterized by sharp corner separation. Therefore, the inflow conditions are critical to proper turbulence modeling and estimation of wind loads using CFD (Murakami 1997; Spalart and Rumsey 2007). In early studies, smooth uniform inflow conditions were introduced and combined with periodic boundary conditions to avoid difficulties associated with inflow and outflow conditions. Utilization of a uniform inflow may be acceptable in some cases in which the turbulence intensity is small or the object is assumed to be Reynolds number insensitive. Most civil structures, however, are located within the atmospheric boundary layer, where turbulence levels are generally high. Therefore, uniform inflow is no longer an acceptable approximation, even for shapes with sharp corners that may be considered Reynolds number insensitive (Haan and Kareem 2009). For direct comparison with experimental results, it is highly desirable to perform CFD simulations with inflow conditions that are similar to those employed in experiments. One such approach is to generate inflow turbulence utilizing stochastic simulation methods such as those discussed in the previous sections. For example, Kondo et al. (1997) utilized the

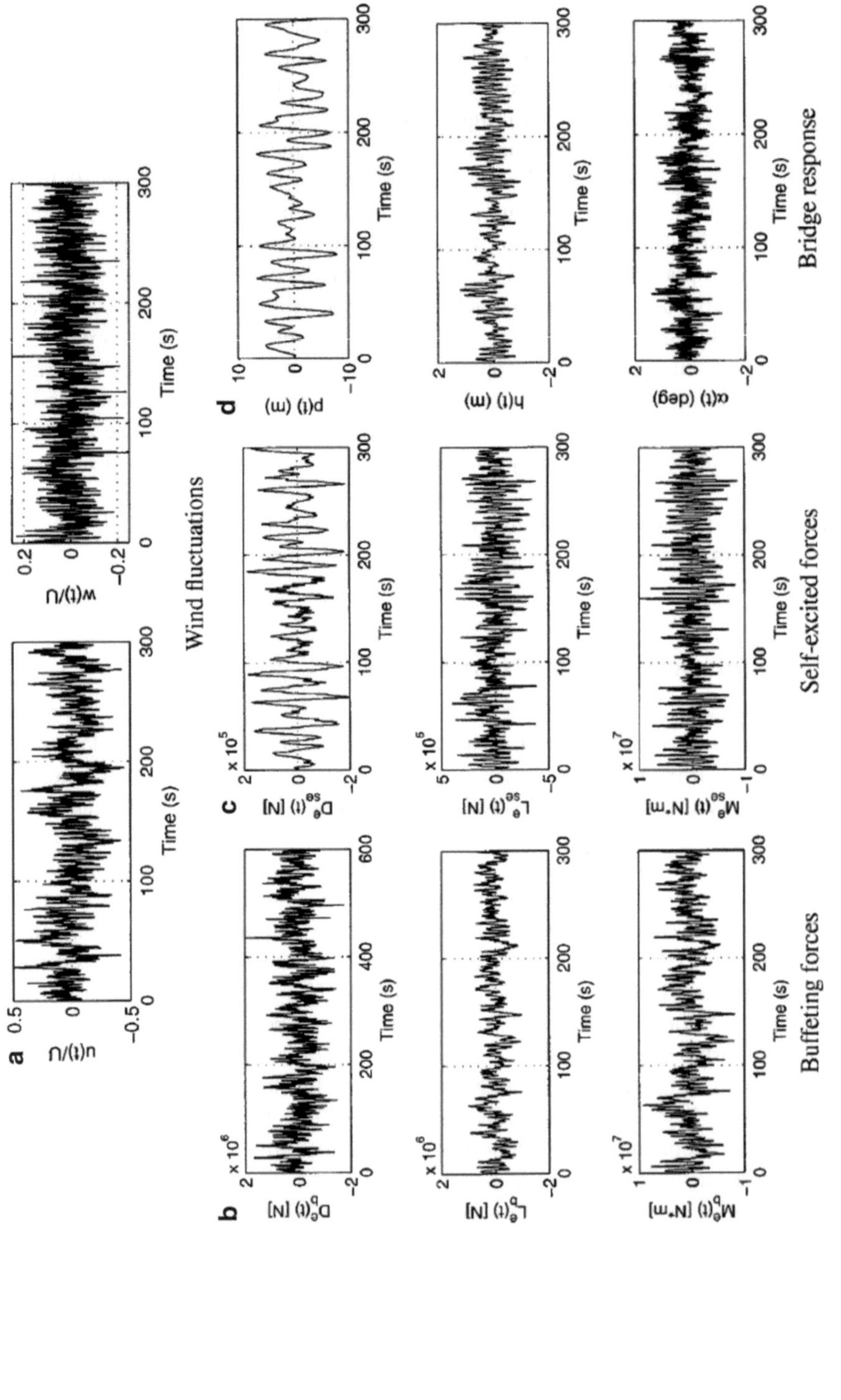

Fig. 10.4 Simulation of wind fluctuations (**a**), buffeting and self-excited forces (**b–c**), and bridge response (**d**) (U = 60 m/s, at midpoint of main span) (Chen and Kareem 2001)

wave superposition approach in conjunction with FFT to simulate velocity fluctuations for large eddy simulation (LES). These simulations were then used by Tamura and Ono (2003) to study the influence of turbulence on the aerodynamics of prisms. To improve the convergence of the LES computation, the generated inflow turbulence needs to be modified to satisfy the continuity equation by divergence-free operation. An iterative approach may be required and should also consider the decay of turbulence as the flow moves into the computational domain. Conditional simulation can also be utilized in this framework to simulate flow at many grid points. Measurements of inflow turbulence in the wind tunnel can be used as input to an appropriate conditional simulation scheme to simulate inflow turbulence at a number of locations to improve flow simulation in the computational domain. Conditional simulation will be discussed in an additional section.

Consideration of a boundary-layer profile in addition to turbulent fluctuations adds even more complexity to the simulation of inflow conditions. One approach is to generate more realistic inflow conditions in a wind tunnel by generating flow over rectangular blocks whose size is adjusted based on the built conditions. Basic parameters such as turbulence intensity, turbulence length scale, or turbulence viscosity can then be measured in the wind tunnel and imposed in the computational domain for the estimation of wind effects (Tamura et al. 2002). While this approach may be appropriate for use with LES simulations, it can lead to misleading and erratic behavior in Reynolds-averaged Navier Stokes (RANS) simulations (Spalart and Rumsey 2007). Most RANS models are not able to properly model turbulence properties for the inflow and require low turbulence levels. Therefore, RANS models must be used considering specific guidelines for inflow that facilitate convergence and ensure accuracy.

For CFD simulation to become a widely used and trusted tool for the study of flow characteristics around bluff bodies, a database of high-fidelity data-validated CFD setups for basic building categories is necessary. Most current CFD studies concerning turbulent flow simulation are problem-specific and lack thorough validation. A database of experimentally validated CFD setups and results for a number of geometric profiles would aid users in determining the most appropriate turbulence models, inflow characteristics, body fitted mesh, etc. for their particular purpose.

10.5 Non-Stationary/Non-Homogeneous Processes

In comparison with atmospheric boundary-layer winds, which are generally regarded as stationary, windstorms such as hurricanes and thunderstorms/downbursts are characterized by wind speeds and turbulence intensities that are in a constant flux as the storm passes over a fixed location. Recently, non-stationary analysis of wind-excited structures has gained interest in the aftermath of increasing damage to structures in extreme winds. The significance of transient wind events and their load effects can be readily surmised from an analysis of hurricane and thunderstorm databases both in the U.S. and around the world, e.g., the Florida Coastal

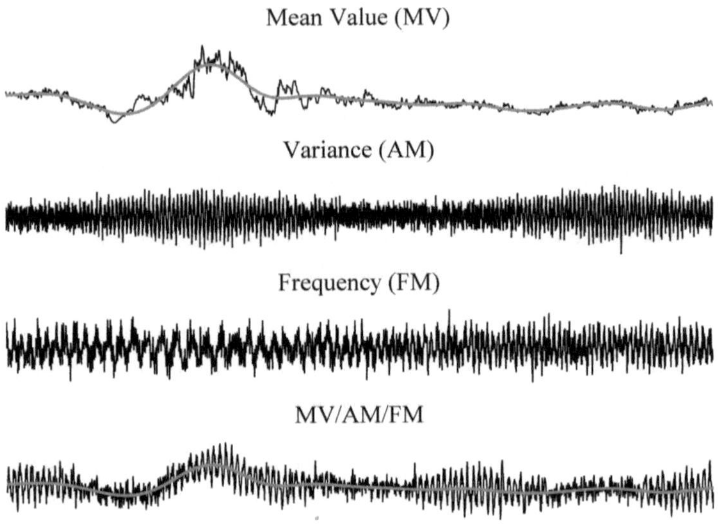

Fig. 10.5 Depiction of different types of non-stationarity (McCullough and Kareem 2013)

Monitoring Program (FCMP) (FCMP 2011), the Joint Airport Weather Studies (JAWS) project (Fujita 1981; McCarthy et al. 1982), and the Northern Illinois Meteorological Research on Downburst (NIMROD) project (Fujita 1985). Observations suggest that the mechanics of gusts associated with these transient events differ significantly from conventional turbulence both in kinematics and dynamics, e.g., their contrasting velocity profile and their transient nature. Extreme loads on structures may be sensitive to the influence of transient flows, i.e., the load coefficients may be enhanced based on the gust form and the resulting localized, rapid changes in the surrounding flow. In addition, the characteristics of these flow structures suggest that the resultant load effects are likely to be correlated over larger areas than in conventional flows. These aerodynamic consequences suggest higher loads on structures than would be predicted by present codes and standards. To accurately assess structural performance and safety during the passage of an extreme wind event, more appropriate modeling and simulation procedures for non-stationary wind events are necessary. Non-stationarity in wind events may reveal itself as changes in the mean value, amplitude modulations (AM) or changes in the variance, frequency modulations (FM), or any combination of these, as shown in Fig. 10.5. Therefore, it is essential to establish a model for extreme winds that reflects the evolutionary behavior of extreme wind events accurately. An appropriate model can then be used to develop a more effective simulation scheme for non-stationary/non-homogeneous processes.

Many of the methods for simulation of non-stationary processes have been developed by the seismic community for generation of earthquake ground motion records. Several of these non-stationary stochastic models utilize the concept of evolutionary power spectral density (EPSD) (Priestley 1965; Shinozuka and Jan 1972). These models include the modulated filtered Poisson process

(Amin and Ang 1968; Shinozuka and Sato 1967), the modulated stationary process (Narayana Iyengar and Sundara Raja Iyengar 1969), and the filtered modulated stationary process (Levy et al. 1971). Despite their theoretical soundness, most of these non-stationary models have practical difficulty simulating real natural phenomena. It is, in many cases, prohibitively challenging/impossible to model phenomena as complex as earthquake ground motions or thunderstorm downbursts, especially considering the importance of capturing transient amplitude and frequency features.

The development of autoregressive moving average (ARMA) models for stationary processes (Li and Kareem 1990a; Mignolet and Spanos 1992; Samaras et al. 1985; Spanos and Mignolet 1992) has prompted the application of the parametric time series model to the simulation of non-stationary random processes. There are several ways to introduce non-stationarity into ARMA models, including piecewise stationarity (Chang et al. 1979), variance-stabilizing transformations (Cakmak et al. 1985), and time-dependent ARMA models in an iterative Kalman filtering scheme (Conte 1992). Deodatis and Shinozuka (1988) proposed an efficient simulation procedure involving a time-dependent AR model. The computational efficiency of these parametric models has resulted in their frequent application to simulation of random processes; however, the application of an AR/ARMA model for evolutionary processes may be more difficult due to the elusive definition of the time-varying correlation functions and/or spectral functions. Additionally, the selection of an appropriate AR/ARMA model order is not always straightforward.

To improve computational efficiency compared to parametric models, Li and Kareem (1991, 1997) introduced an FFT-based simulation scheme for multivariate non-stationary processes. The utilization of FFT was made possible through the application of stochastic decomposition advanced by the author and discussed earlier in this paper. In this scheme, a decomposed spectral matrix is expanded into a weighted summation of basic functions and time-dependent weights that are simulated by the FFT-based algorithm. The effectiveness of the proposed technique was demonstrated using ground motion records that could be separated into time- and frequency-dependent components as well as for those in which both AM and FM modulations are present. This approach requires a description of the target evolutionary spectral characteristics, which may be prescribed in a closed form or a set of numerical values at discrete frequencies.

For hurricanes and thunderstorms/downbursts in addition to earthquake ground motions, a lack of sufficient data makes it difficult to establish a theoretical model for the physical process. Therefore, reliance must be made on constructing the model based on available data. In the case of extreme events, the data may be very limited and often only a single sample of the event is available. The seismic community has been seeking to estimate EPSDs based on a single realization of a non-stationary random process (Der Kiureghian and Crempien 1989; Kameda 1975; Saragoni and Hart 1974; Scherer et al. 1982; Spanos et al. 1987). This can be viewed as the beginning of time-frequency analysis and the modeling of transient signals. With the development of the short-time multi-window (STMW) spectrum estimator, Conte and Peng (1997) proposed a non-stationary analytical stochastic model simulating real earthquake accelerograms by fitting the analytical evolutionary power spectrum to the estimated power spectrum using a least-square fitting method. It is observed that this analytical model is able to capture the temporal variation of both the amplitude and the frequency components of actual earthquake ground motions. With these developments, a move toward simulation methods that incorporate time-frequency analysis and modeling can be seen. Due to the constraints of the Heisenberg Uncertainty Principle, however, optimal resolution in both time and frequency cannot be achieved.

Wavelet transform has emerged as a convenient tool to obtain multi-resolution time-frequency information of a non-stationary signal (Daubechies 1992; Morlet et al. 1982) and has been incorporated into non-stationary simulation schemes (Iyama and Kuwamura 1999; Kwon et al. 2007; Liang et al. 2007; Spanos and Failla 2004). For example, Gurley and Kareem (1999) used wavelet transform to model ground motion records as a sum of modulated stationary processes centered at narrow-banded frequencies, similar to those of Der Kiureghian and Crempien (1989). Modulated by energy present in the wavelet coefficients at each frequency band, the simulation preserves the non-stationary characteristics of the energy distribution of the target random process. Huang and Chen (2009) used a wavelet-based approach to estimate EPSDs of multivariate evolutionary processes including earthquake ground motions and downburst winds. The EPSDs can be used for frequency domain analysis of wind load effects on structures and time domain simulation of multivariate non-stationary processes. Empirical mode decomposition offers an alternative to wavelet based techniques, which has its advantages and disadvantages in comparison with wavelet transform based approaches (Kijewski-Correa and Kareem 2006).

Wang and Kareem (2004b, 2005a, b, 2006) examined the evolutionary behavior of measured earthquake ground motions, hurricane winds, and thunderstorm/downburst events using a time-frequency framework involving wavelet and Hilbert transforms. Stationary wavelet transform (SWT) first decomposes a sample of a multi-component non-stationary random process into a set of mono-component signals. These signals are subsequently transformed to analytic signals using the Hilbert transform (HT), which yields the instantaneous amplitudes and frequencies. The observation that the instantaneous frequency at each frequency band follows a Gaussian distribution has laid the groundwork for an efficient simulation method for non-stationary processes based on a measured record. This simulation approach does not depend on the customary assumption of piecewise stationarity or the assumption of parametric models. The method is extended to the simulation of multivariate random processes utilizing proper orthogonal decomposition. Analytical expressions for the statistical properties of the underlying random processes are formulated and can be found in Wang and Kareem (2004a, 2005a, b, 2006)

The effectiveness of the SWT/HT procedure is demonstrated using downburst data recorded simultaneously at three different towers located 263 m apart at a height of 10 m, as shown in Fig. 10.6a. This data was provided through the 2002 Thunderstorm Outflow Experiment conducted by the Department of Atmospheric Science of

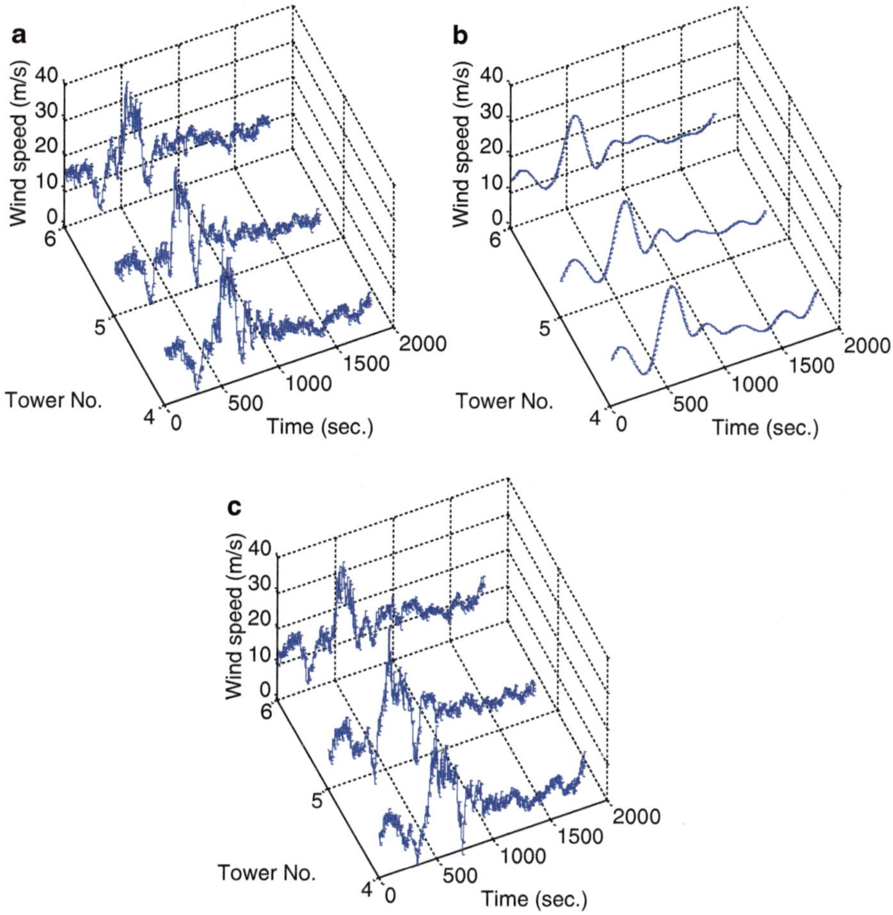

Fig. 10.6 Multivariate simulation of downburst data: (**a**) downburst data measured simultaneously at horizontally distributed towers; (**b**) time-varying means extracted using SWT; (**c**) simulation that results using multivariate simulation scheme (Wang and Kareem 2005a)

Texas Tech University (Orwig and Schroeder 2007). Figure 10.6b shows the time-varying means that are extracted from each time history using SWT and Fig. 10.6c shows the simulations that result using the multivariate simulation scheme. It is observed that the simulation exhibits non-stationary amplitude characteristics similar to the measured record. The Hilbert spectra for each measured and simulated signal are compared in Fig. 10.7. The simulation reasonably reproduces the evolutionary energy associated with the measurement as demonstrated by the apparent equivalence of the Hilbert spectra. For the simulation of multivariate processes, it is equally critical to capture the instantaneous correlation structure among components. Figure 10.8 presents the coscalograms between the different measurements from the three towers. The similarity between the coscalograms of the measurement and those of the simulation shows that the simulation satisfactorily captures the correlation between each component. Further examples concerning the

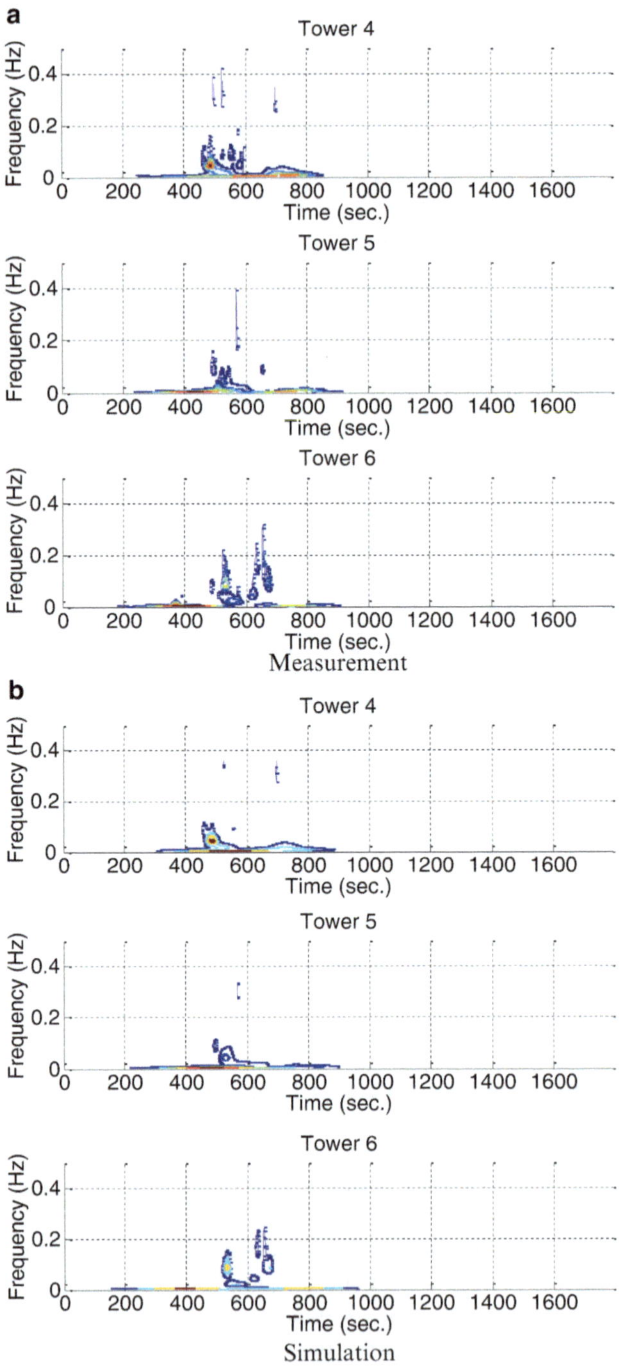

Fig. 10.7 Comparison of (**a**) measurement and (**b**) simulation of multivariate downburst data in the time-frequency domain using Hilbert contour spectra

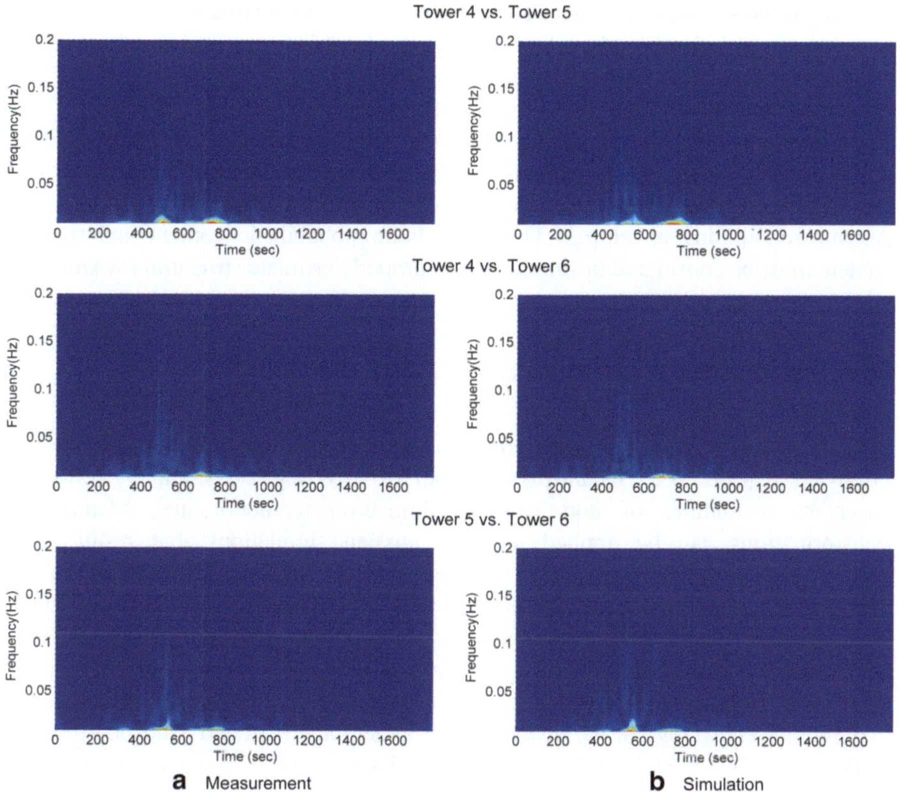

Fig. 10.8 Comparison of correlations between signals in the time frequency of the (**a**) measured signals and (**b**) simulated signals for the multivariate RFD data utilizing cross-scalograms

simulation of downburst wind data and earthquake ground motions are presented in Wang and Kareem (2004b, 2005a, b, 2006).

Alternative methods for modeling downburst type non-stationary winds have used commercially available CFD software packages, e.g., FLUENT (Chay et al. 2006; Kim et al. 2005). The time-varying means are captured using wavelet shrinkage and amplitude modulated fluctuations based on separable time and frequency features (Chen and Letchford 2005, 2007). Other methods have relied on the use of time-varying ARMA models to add turbulence to non-turbulent downburst winds generated using CFD (Chay et al. 2006).

10.6 Non-Gaussian Simulation

Significant progress has been made over the last several decades in the modeling and simulation of univariate and multivariate, conditional and unconditional Gaussian random processes as discussed in previous sections. Many processes associated with natural hazards, however, are characterized by non-Gaussianity. For example, velocities in turbulent winds, which may or may not be non-Gaussian themselves, often have a nonlinear relationship with the resulting pressures on a structure, leading to non-Gaussian features in the pressure distribution, especially in separated flow regions. These non-Gaussian loads may affect the structure in unexpected ways, especially if the design loads were assumed to be Gaussian. The extreme negative pressures that may be present in non-Gaussian pressure distributions have been known to cause localized damage to building envelopes. Therefore, both probabilistic content and spectral content must be considered in simulations to properly estimate structural responses.

10.6.1 Spectral Representation-Based Schemes

Spectral representation methods for the simulation of stationary, Gaussian processes, discussed in Sect. 10.2.1 (Shinozuka and Jan 1972; Wittig and Sinha 1975), form a basis for a number of non-Gaussian simulation methodologies. Memoryless transformations can be applied to the Gaussian simulations that result using spectral representation simulation methods to modify the probabilistic content to match a prescribed probability density function (PDF) (Deodatis and Micaletti 2001; Grigoriu 1984; Gurley et al. 1996). Several different memoryless transfromations may be used for this purpose, including analytical filtering or cumulative distribution function (CDF) mapping. Application of the memoryless transformations, however, may require knowledge of the inverse of the target CDF (Liu and Munson 1982). In response to this limitation, Filho and Yacoub (2008) proposed a method for coloring non-Gaussian sequences. A sequence is first generated that is a zero-mean unity-variance Gaussian process with the appropriate autocorrelation. Separately, an independent and identically distributed sequence is generated that has the target CDF. Finally, the samples of the sequence with the target CDF are rearranged to match the rank of the Gaussian process. In this procedure, the inverse of the target CDF may be used but it is not required. Instead, rejections methods may be used to generate values matching a target CDF (Papoulis and Pillai 2002). A similar method, known as the amplitude adjusted Fourier transform (AAFT) surrogate method, was also developed in the physics literature for the purpose of hypothesis testing (Schreiber and Schmitz 1996, 2000; Theiler et al. 1992). In this method, the original dataset is reordered to match the rank of the Gaussian process, rather than data being sampled from the target CDF. Unfortunately, the Gaussian to non-Gaussian transformations in each of these steps leads to whitening of the spectra. In some cases, the whitening may lead to significant deviations between the target spectrum and the spectrum of the simulation.

10.6.1.1 Correlation Distortion

Alternative simulation approaches that may be able to more faithfully represent the target spectrum are known as correlation distortion schemes. These methods begin with the transformation of the target spectrum of the non-Gaussian process to the underlying spectrum of the Gaussian process. The spectral representation method is then used to produce Gaussian simulations with the specified underlying spectrum. Finally, the Gaussian simulations are transformed back to the non-Gaussian processes. Methods of this type seek to find the underlying Gaussian spectrum that becomes the target spectrum of the non-Gaussian process due to the nonlinear transformation from Gaussian to non-Gaussian (Conner and Hammond 1979; Deutsch 1962; Johnson 1994). Yamazaki and Shinozuka (1988) use an iterative procedure to find the correct underlying spectrum of the transformed Gaussian process. In each iteration, the spectrum of the simulated and transformed non-Gaussian process is compared to the target non-Gaussian spectrum, if the discrepancy is unacceptable, the underlying Gaussian spectrum is updated and the iteration begins again with Gaussian simulation and non-Gaussian transformation (Gurley et al. 1996). A similar method developed by Grigoriu (1998) uses an analytical method to determine the underlying Gaussian spectrum that is not iterative. Deodatis and Micaletti (2001) introduced an algorithm that combines the previous two algorithms with additional small changes, such as the utilization of CDF mapping rather than a nonlinear Gaussian to non-Gaussian transformation and an updated Gaussian spectrum correction step. For highly non-Gaussian systems or narrow-banded spectra, these methodologies may not yield acceptable results, instead conforming directly to either the spectra or the marginal CDFs instead of both (Masters and Gurley 2003).

10.6.1.2 Direct Transformation

Direct transformation schemes begin with a sample of a non-Gaussian time history rather than its spectrum, which was used in the correlation distortion methods (Gurley et al. 1996, 1997). The non-Gaussian time history is first transformed into a Gaussian time history using direct transformation methods. Spectral representation-based methods are then used to simulate a new Gaussian time history. Finally, the resulting Gaussian simulation is transformed back to a non-Gaussian process again using nonlinear transformation methods. A direct transformation methodology that utilizes Hermite polynomials and iteratively corrects the parameters for the direct transformation is better able to match the non-Gaussian characteristics of the measured process than the correlation distortion schemes (Gurley et al. 1996). If a target spectrum and non-Gaussian characteristics are available, rather than the original signal, the direct transformation method may not be used.

10.6.1.3 Spectral Correction

In response to the limitations of both the correlation distortion and direct transformation schemes, a spectral correction scheme was introduced (Gurley and Kareem 1997; Gurley et al. 1996). This scheme requires the specification of the non-Gaussian statistical characteristics in the form of the first four moments, which include the target skewness and kurtosis, and the frequency content in the form of the spectrum. These target values may be determined using a sample process or they may be available via other means, such as an analytical PDF or spectra. The target spectrum is first used to produce a Gaussian process in the same manner as the spectral representation-based methods. The Gaussian process is then put through the moment correction part of the scheme in which a back transformation is utilized to yield a Gaussian process and then a forward transformation is utilized to yield a process that has the desired moments that were specified. The process that results from the moment correction portion of the scheme, however, no longer matches the target spectra. Therefore, spectral correction is utilized to produce a modified process that once again matches the target spectra. Unfortunately, the spectral correction process distorts the target moments, such that the resulting process may need to be sent back into the moment correction part of the scheme. This process continues until the moments of the process produced after spectral correction match the target moments within a set tolerance. Therefore, the resulting simulation data will match the target spectra exactly and the target moments approximately. Unfortunately, however, this method utilizes only the first four moments to define the desired PDF, always resulting in a third-order Hermite polynomial PDF. Therefore, there is a significant limitation as to the types of CDFs that may be used for the simulations (Masters and Gurley 2003).

Several methods have been developed in the engineering and physics literature that overcome this limitation, but maintain the ability to iteratively correct the probabilistic and spectral content. The CDF map-based spectral correction method proposed by Masters and Gurley (2003) uses CDF mapping for the moment correction part of the scheme rather than the backward and forward Hermite polynomial corrections. The CDFs can be determined empirically from the sample data or analytically if it is feasible. This algorithm allows significantly more flexibility for the CDF than the traditional spectral correction method by allowing the CDF to be defined in a number of ways. The iterative amplitude adjusted Fourier transform (IAAFT) surrogate method is an improvement upon the AAFT surrogate method that incorporates iterations in order to better approximate the spectrum while still matching the CDF (Schreiber and Schmitz 1996, 2000). This scheme uses rank matching as the moment correction step rather than CDF-mapping or Hermite polynomials.

The spectral correction schemes can be compared using wind velocity data taken from a gable roofed house in uniform, open terrain with a 4:12 roof slope (Fig. 10.9). This data was made available to us by the Alan G. Davenport Wind Engineering Group of the Boundary Layer Wind Tunnel Laboratory at the University of Western Ontario. Wind direction is perpendicular to the ridge and normal to one of the walls. The asterisk in Fig. 10.9 shows the location of the measurement and a portion of the measured data. A portion of a simulation using the CDF-map based spectral correction scheme is also shown. The mean PSDs and PDFs resulting from 100 simulations are compared with the PSD and PDF of the original dataset in Fig. 10.10. It is apparent that all the methods match the target PSD reasonably well, but only the CDF-map based spectral correction and the IAAFT methods match the PDF. The PDF for the spectral correction method is forced to match a Hermite polynomial PDF model.

All three of the spectral correction methods can be extended to multivariate processes in a straightforward manner. Multivariate simulations are important for simulating correlated loads with multiple components or at spatially distributed

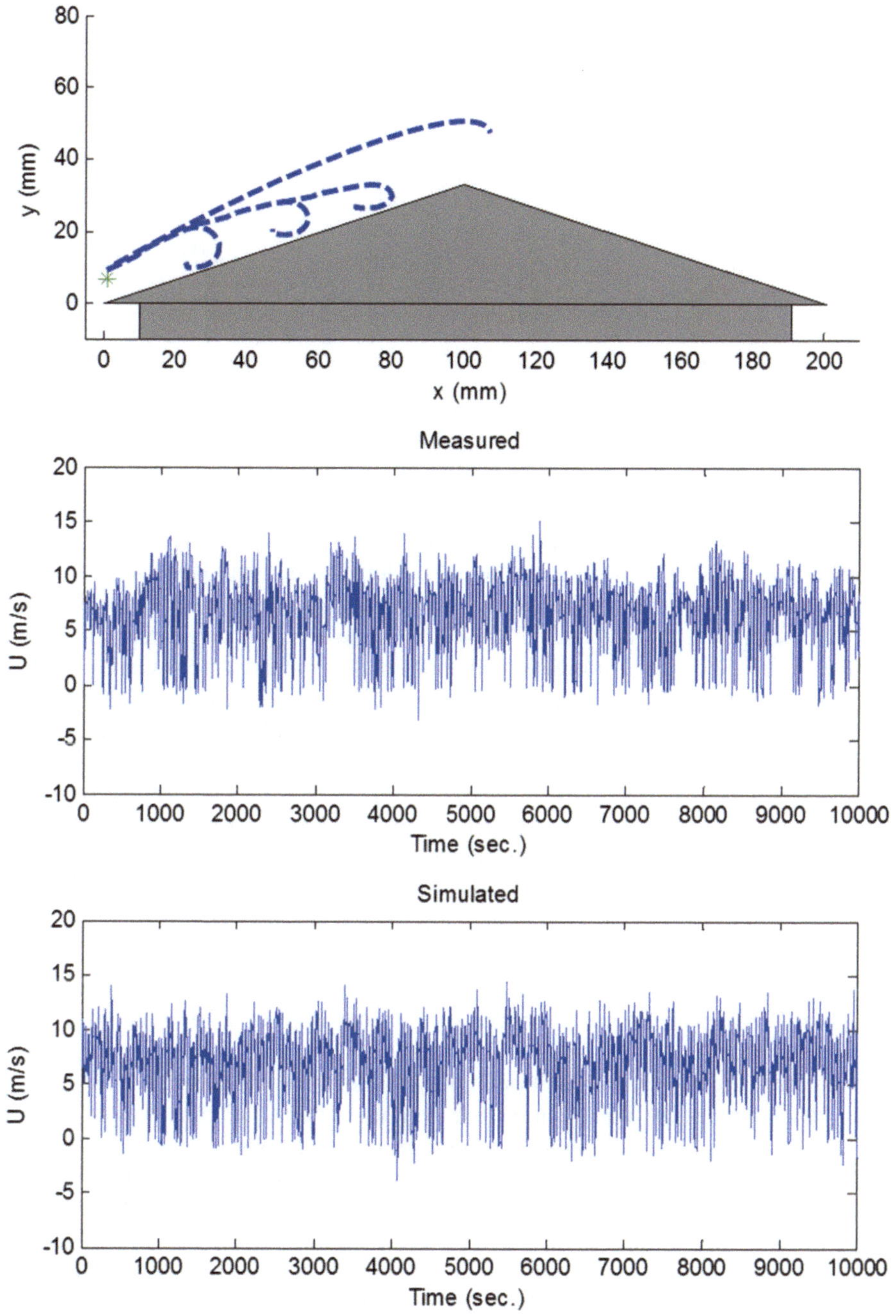

Fig. 10.9 Non-Gaussian wind velocity time series: location of measurement on roof (*top*); portion of measured signal (*middle*); portion of simulated signal using CDF-map based spectral correction (*bottom*)

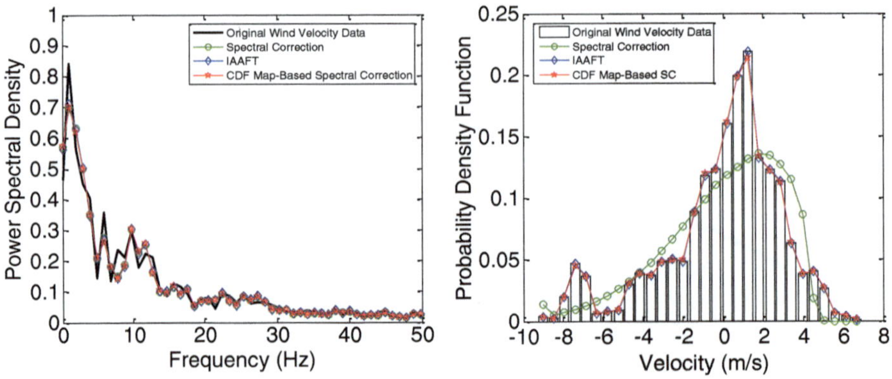

Fig. 10.10 Comparison of PSDs (*left*) and PDFs (*right*) of original data and spectral correction methods

locations in order to accurately determine the associated structural response. For example, correlated wind pressures that vary spatially are necessary to fully characterize a boundary layer pressure field. Multivariate non-Gaussian simulation schemes must be able to match the cross- and auto-spectra in addition to the non-Gaussian statistical characteristics. The multivariate spectral correction method and CDF-map based spectral correction method can be extended to include a cross-correlation correction step within each iteration to account for the correlations between measurements (Gurley and Kareem 1998b). Similarly, the IAAFT method is extended to include a step for updating the phases (Schreiber and Schmitz 2000). The IAAFT algorithm takes advantage of the fact that the cross-spectrum reflects the Fourier amplitudes of the two variables but the relative phases only. Therefore, to preserve the individual spectra as well as the cross-spectra, the Fourier amplitudes need to be fixed as well as the difference in the phases for each pair of variables. Therefore, multivariate IAAFT includes a step to correct the phase difference between each pair of variables.

Unfortunately, the multivariate spectral correction methods only match the marginal PDFs, with no consideration of the joint PDF (JPDF), which may also exhibit significant non-Gaussianity. The joint occurrence of extremes is important, necessitating accurate representation of the JPDF in the simulations. A new method is in development by the authors that combines the CDF map-based spectral correction method and the iterative amplitude adjusted Fourier transform methods and incorporates a JPDF correction step. In each iteration, the JPDF correction step involves registration of the simulated data with the original data. This can be accomplished most efficiently and effectively with non-rigid point set registration using coherent point drift (Myronenko and Song 2010). Coherent point drift treats registration as a maximum likelihood estimation problem with the coherence of the velocity field as the constraint. Therefore, this method finds the best transformation model between the simulated data and the original data, which is a realization of the target JPDF. Alternatively, a different target dataset could be simulated from the target JPDF. The iterative process then continues until the spectra and PDFs are matched to an acceptable degree.

10.6.2 Transformation Schemes with Memory

When the only available information is a sample of the final process to be simulated, the spectral representation-based schemes are efficient and effective. If more information is available, such as the upwind wind velocity, then it is possible to better simulate the desired process by establishing a system identification model. Transformation schemes with memory allow for all information available to be used (Gurley et al. 1996, 1997). A model that relates Gaussian input to non-Gaussian output may easily be applied to the simulation of the output process by passing simulations of the Gaussian input through the model. Important phase characteristics are then retained in the output through the memory transformation. Though there has been limited progress in this direction, several approaches with promise will be discussed.

Volterra series-based schemes lend themselves to the simulation of non-Gaussian wind-related processes for which the Volterra kernels are available or estimated (Gurley et al. 1996, 1997; Kareem et al. 1998). In the Volterra series formulation, the input–output relationship may be expressed in terms of a hierarchy of linear, quadratic, and higher-order transfer functions or impulse response functions (Kareem and Li 1988; Schetzen 1980; Spanos and Donley 1991). These transfer functions, or kernels, can be determined from experimental data or from theoretical considerations. The memory retained by use of the higher order transfer functions facilitates the simulation of processes that are able to match not only the spectrum and PDF, but the high-order spectra as well. For example, use of the quadratic transfer function (QTF) would enable the bispectrum to be matched in the simulations. Unfortunately, the determination of the higher-order kernels from measured data becomes prohibitively difficult due to the number of parameters needed to describe them (Gurley et al. 1996).

An alternative approach involves representation of the system using a discrete differential equation provided by a non-linear autoregressive moving average models with exogenous inputs (NARMAX) model (Gurley et al. 1996). A number of convenient algorithms are available that use input/output data to define a polynomial model. A polynomial order and maximum lag are specified by the user and the algorithm determines the parameters associated with the particular structure specified. NARMAX provides a more flexible representation of a non-linear system, and generally requires fewer parameters than a Volterra model. More information about NARMAX modeling can be found in Chen et al. (1989) and Billings and Tsang (1989).

Another recently developed approach to non-Gaussian simulation utilizes artificial neural networks, which consist of layers of interconnected nodes, or processing elements. A neural network is able to model the relationships between inputs and outputs by adjusting a weighting factor between each of the nodes and applying various nonlinear functions at each of the nodes (Kung 1993). A realization of the process to be simulated can be used to train and test the neural network. The neural network can then be used to simulate output, necessitating only the input. Gurley et al. (1996) utilized neural networks to simulate a non-Gaussian sea state. It is demonstrated that the neural network model is better able to match the shape and magnitude of the bispectrum of the original sample than the methods that do not incorporate memory.

10.6.3 Phase Tailoring

A class of single point non-Gaussian processes, e.g., pressure fluctuations at a location, can also be simulated through manipulation of the phase (Seong and Peterka 1998). The assumption in spectral representation based methods for Gaussian simulation is that the phase spectrum contains no relevant information. More recent analyses, however, indicate that phase coupling results in a non-Gaussian signal (Gurley et al. 1997). In Seong and Peterka (1998), the phases were acquired from a time series produced by a peak-generation model. Higher-order spectral analysis, such as bi- and tri-spectra, can be used to identify distinctive phase characteristics of non-Gaussian data (Gurley et al. 1997).

10.7 Conditional Simulation

Simulation of velocity and pressure time series at un-instrumented locations conditioned on measured records are often necessary for wind related studies. For example, malfunctioning instruments may leave a hole in a data set or information may be lacking due to limited number of sensors. The concept of conditional simulation is similar to conditional sampling in experiments or numerical simulations. Two main approaches have been developed; the first is simulation based on linear estimation, or kriging, and the second is simulation based on a conditional probability density function (Borgman 1990; Kameda and Morikawa 1992; Krige 1966). The approaches are equivalent for univariate, uni-dimensional processes (Shinozuka and Zhang 1996). In addition to facilitating the generation of time series at new locations, both methods can also be used to extend existing records beyond the sampling time for cases where conditioning time series is limited to small subintervals. Gurley et al. (1997) utilized the conditional probability density function approach to simulate one of the measured velocity records based on the availability of the other three on a full-scale tower. Figure 10.11 shows these records together with the conditionally simulated record (Gurley et al. 1997). Figure 10.12 shows a comparison of the target, measured, and conditionally simulated cospectrum and PSD (Gurley et al. 1997). The apparent jaggedness in the cospectrum is a result of the variance inherent in the individual realizations. Utilizing ensemble averages would yield a smooth cospectrum that lies on top of the target cospectrum. The PSDs also show good agreement. The other application of the conditional simulation methodology is shown in Fig. 10.13. A measured record up to 2,500 s shown in the top figure is assumed to be only available

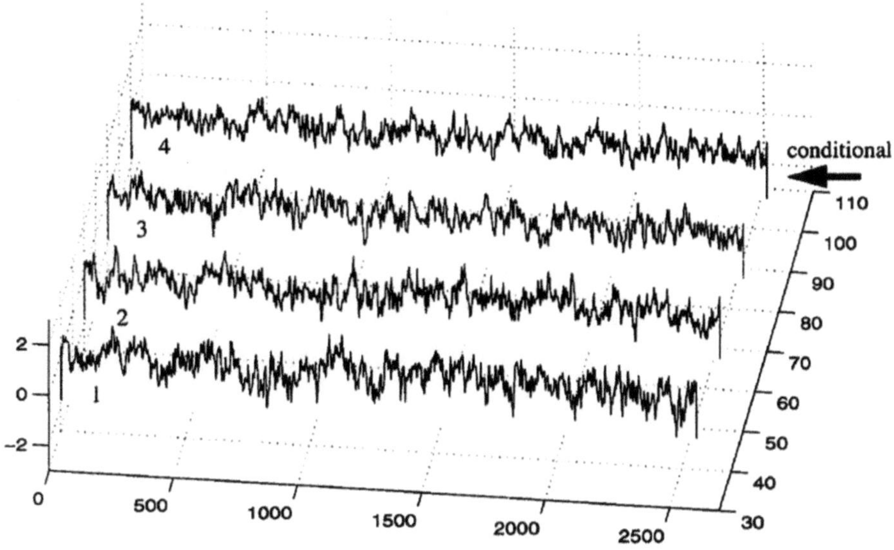

Fig. 10.11 Example of a conditionally simulated time history (Gurley et al. 1997)

up to 980 s, indicated by the darkened portion of the record. The remaining record from 980–2,500 s is simulated, as shown in lighter color in the bottom figure, based on the information derived from the first 980 s of data. Both examples demonstrate the effectiveness and utility of conditional simulation of Gaussian processes. An interesting application of this approach has been noted in developing inflow conditions for a CFD study in which measured turbulence at limited locations is expanded to the entire grid through conditional simulation (Maruyama and Morikawa 1995).

Similar schemes are possible using linear stochastic estimation (LSE) in which conditional information at reference points are used to estimate information over the entire flow domain (Adrian 1994). In this approach estimated values of the flow field at points conditioned on measurements at reference locations and time are expressed in terms of the Taylor series expansion. By minimizing the mean quadratic error of cross-correlation and retaining the first-order terms, desired estimates are obtained. Essentially, LSE estimates a random variable as a linear combination of some known reference variables. LSE has the ability to reconstruct the entire random field by using a few reference variables in conjunction with attendant correlations. This approach can also be used to couple experiments with CFD. In this application, the experimental time histories, measured at limited locations, are interfaced with the inlet section of a time-varying numerical simulation of a spatially developing flow (Deville et al. 2000). Applications in wind engineering have been explored in Chen et al. (2003) in which LSE was utilized to reconstruct wind-induced pressure time series from the covariance matrix for structural load analyses on a low building roof. It is a noted that this approach helps to reduce data storage for large scale databases, where one may reconstruct data based on stored covariance matrices and a few measured pressures for reference. It should be noted that the authors were unable to faithfully reproduce individual time series, but the integrated effect of several time histories resulting in a

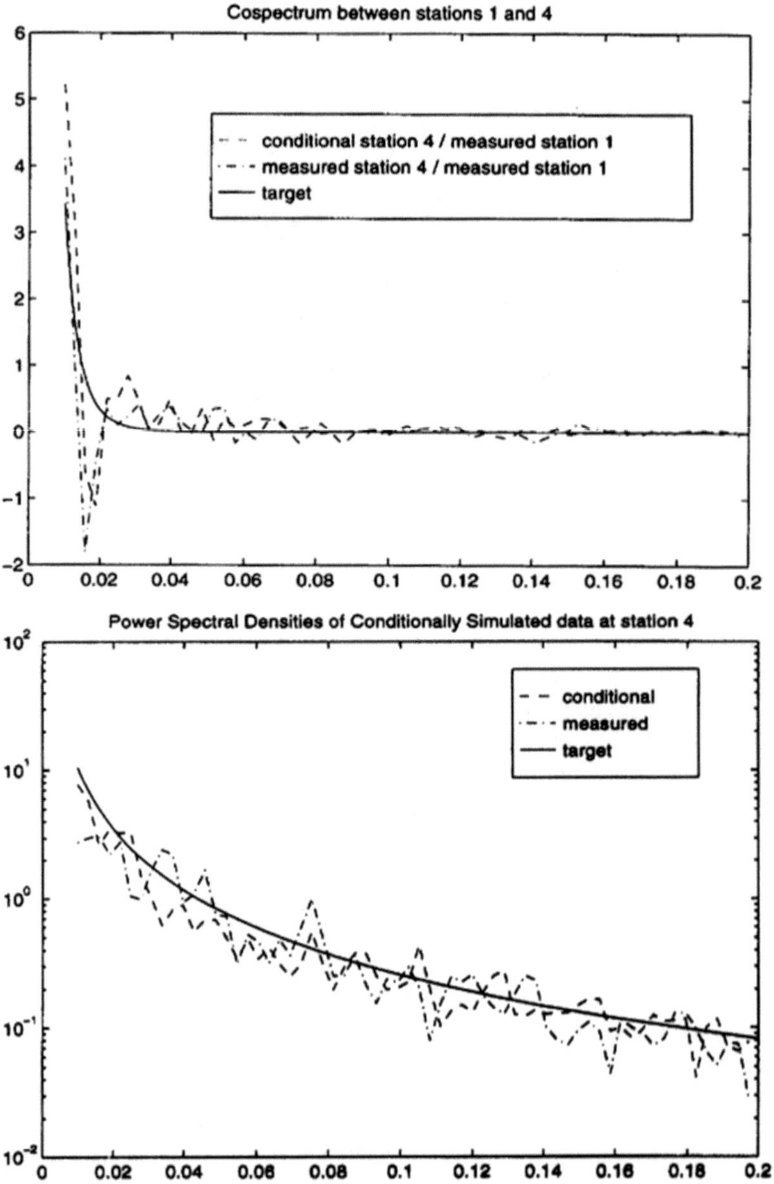

Fig. 10.12 Conditional, measured, and target cospectrum between stations 1 and 4 (*top*) and conditional, measured, and target PSD at station 4 (*bottom*) (Gurley et al. 1997)

panel load were successfully reproduced. This difficulty in matching individual time history may result from the non-Gaussian nature of the data. Therefore, conditional simulation methodologies compatible with non-Gaussian data are necessary and will be discussed in the next section.

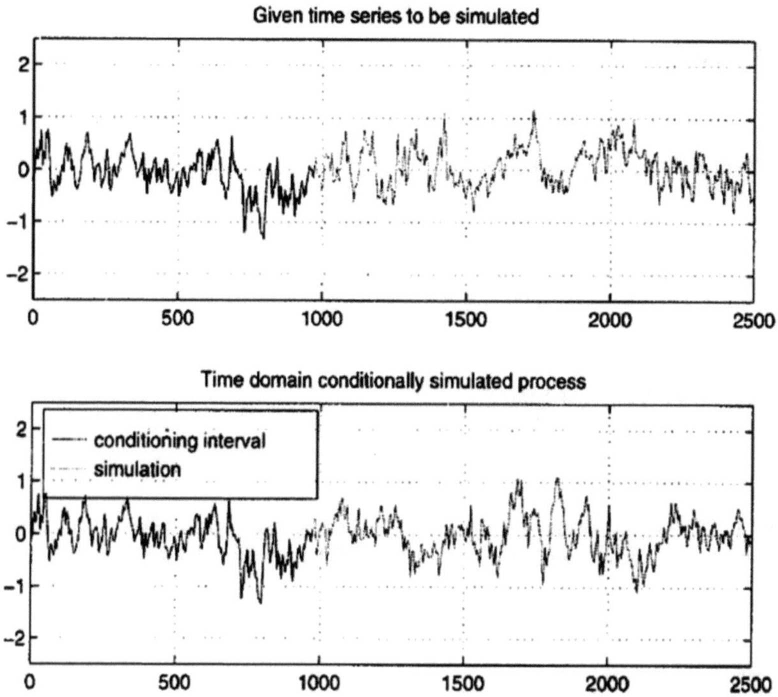

Fig. 10.13 Measured wind velocity (*top*) and time domain conditional simulation (*bottom*) (Gurley et al. 1997)

10.7.1 Non-Gaussian Processes

The concept for conditional simulation can be extended to non-Gaussian cases often experienced in wind engineering, such as pressure fluctuations in separated flow regions or velocity fluctuations in extreme atmospheric flow conditions. A non-Gaussian conditional simulation given in Elishakoff et al. (1994) addressed problems related to wide-band processes. In this scheme, the non-Gaussian random processes were first mapped into Gaussian processes, which allowed the use of traditional Gaussian conditional simulation. The simulations could then be mapped back into non-Gaussian processes after conditional simulation had been performed. In Gurley and Kareem (1998a), a multivariate non-Gaussian conditional simulation scheme is proposed that is an extension of the spectral correction non-Gaussian simulation scheme discussed in a preceding section. The forward and backward modified Hermite transform allows a convenient mapping between Gaussian and non-Gaussian processes. This method utilizes the conditional probability density function approach, which is applicable to both narrow- and wide-banded processes. Details of the simulation scheme are available in Gurley and Kareem (1998a). An example of the conditional simulation, presented in Gurley

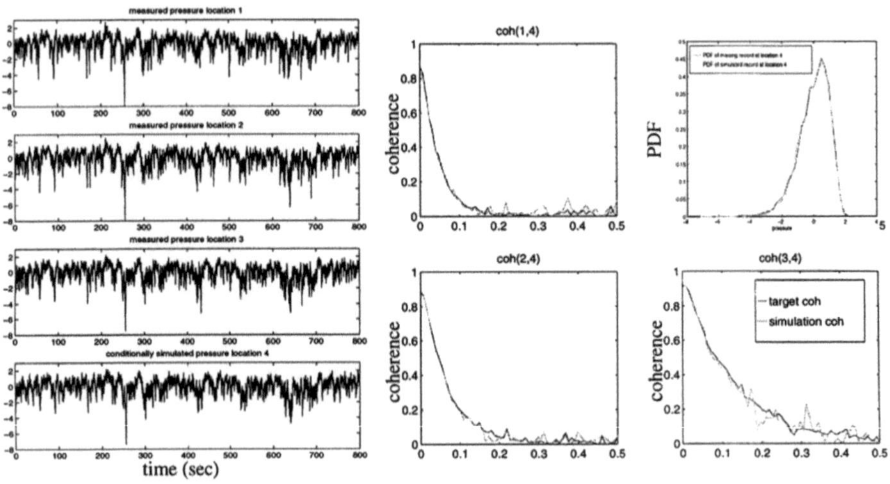

Fig. 10.14 Three measured pressure records and a conditionally simulated record (*left*) and comparison of target and simulated coherence functions and PDF (*right*) (Gurley and Kareem 1998a)

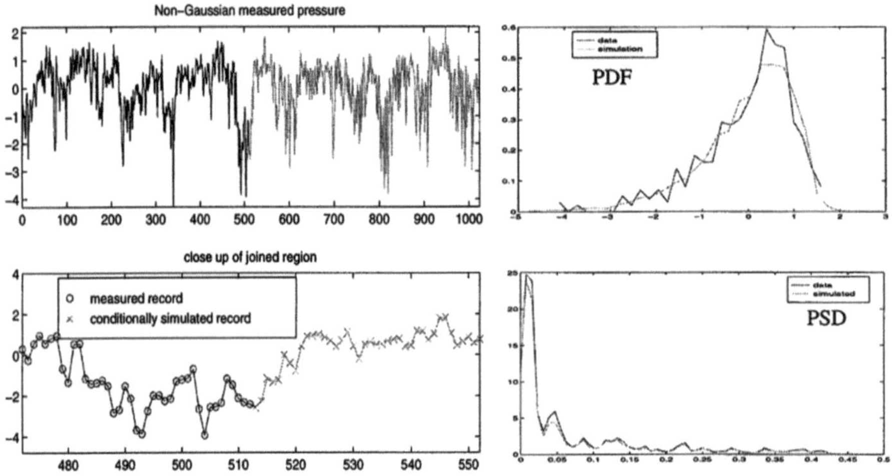

Fig. 10.15 Example of extended data record and close-up of extension region (*left*) and comparison of PDF and PSD of measured and extended pressure record (*right*) (Gurley and Kareem 1998a)

and Kareem (1998a), involves pressure fluctuations given at three locations and a conditional simulation of the fourth location. The left side of Fig. 10.14 shows the three records and the simulated fourth record. The right side of the figure shows the target and measured coherence between locations 1 and 4, 2 and 4, and 3 and 4. The far right top in Fig. 10.14 is a comparison of the PDF of the missing record and its replacement. Similar to the conditional simulation in Fig. 10.13, extension of a missing record is shown in Fig. 10.15, with a close-up of the transition from the measured to simulated data location, to illustrate the absence of discontinuity. These examples demonstrate that the conditional simulation scheme presented in Gurley and Kareem (1998a) ensures excellent agreement between the target and simulated marginal distribution and spectral descriptions.

10.7.2 Non-Stationary Processes

In the previous two sections, discussion was focused on the conditional simulation of Gaussian and non-Gaussian vector-valued processes. Conditional simulation of non-stationary random processes has been elusive and limited efforts are reported in the literature, especially for wind related studies. Application of the conditional simulation of ground motion records marked by non-stationary behavior are available in the literature, e.g., Vanmarcke and Fenton (1991), Hoshiya (1995), Shinozuka and Zhang (1996), Kameda and Morikawa (1992) and Heredia-Zavoni and Santa-Cruz (2000). These techniques are based on kriging, the conditional probability density function method, or their combination. Heredia-Zavoni and Santa-Cruz (2000) first mapped the non-stationary data to a domain where conditional simulation can be performed for stationary processes. The non-stationary characteristics are then applied to the stationary processes using amplitude and frequency modulation functions. Morikawa and Kameda (2001) utilized group delay time spectra to include the non-stationary properties in the simulations. Relationships between the time histories and their phase spectrums could be used to develop PDFs of the Fourier amplitudes and phases in the frequency domain in order to properly represent non-stationarity.

Wang (2007) developed a new method for non-stationary conditional simulation that utilizes time-frequency analysis tools. This scheme utilizes evolutionary cross-correlation in terms of cross wavelet coefficients, which allows the kriging method to be extended to the time-frequency domain. Two schemes are introduced: a mapping-based conditional simulation scheme that utilizes a modified Hermite transform and a non-mapping conditional simulation scheme that determines kriging coefficients for localized wavelet coefficients. The non-mapping scheme reduces the simulation error caused by the non-linearity of mapping. A numerical example of the conditional simulation of downburst wind velocities is shown in Fig. 10.16 and Fig. 10.17. The data is simultaneously recorded on June 15, 2002 at different heights in a field experiment by the Department of Atmospheric Science at Texas Tech University (Orwig-Gast and Schroeder 2005). The input to the scheme is the data measured at 4 and 15 m heights (Fig. 10.16) and data is simulated at 10 m (Fig. 10.17). The simulation preserves the temporal fluctuations of the real measurement, as compared in Fig. 10.17. The time-varying correlation structure, not shown here for the sake of brevity, shows that the simulated results maintain the correlation between multiple locations.

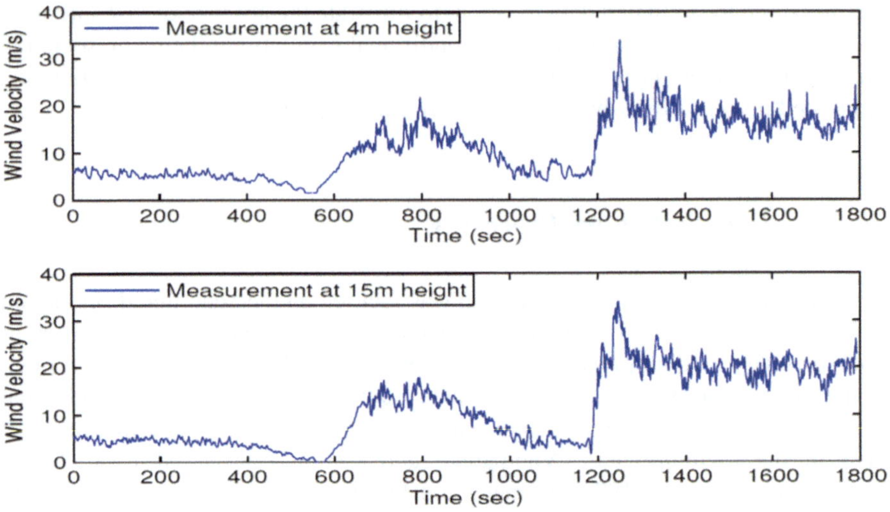

Fig. 10.16 Measurements at 4 and 15 m heights (Wang 2007)

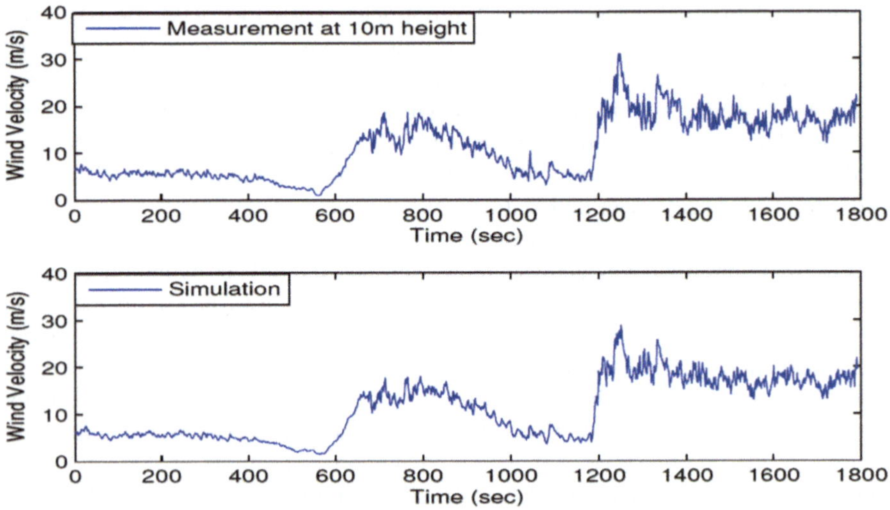

Fig. 10.17 Measurement and simulation at 10 m height (Wang 2007)

10.8 Computational Modeling Tools

10.8.1 Proper Orthogonal Decomposition

Multi-correlated stationary random processes/fields, such as those associated with wind velocity and pressure fluctuations, can be transformed into a set of subprocesses by diagonalizing their covariance or cross power spectral density

(XPSD) matrices using either Cholesky (lower or upper triangular) or Schur (eigenvector) decomposition. The eigenvector decomposition offers physically meaningful insight into the process as each eigenvector (eigenmode) may be characterized on the basis of its spatial distribution. It is also recognized that only a small number of eigenmodes corresponding to eigenvalues with larger magnitudes are dominant, such that one may ignore some of the eigenmodes associated with small eigenvalues in the description of a large-size random field. Accordingly, this technique provides a unique tool for data compression and facilitates a reduced-order modeling of large-size random fields. The eigenvector decomposition is theoretically based on the Karhunen–Loeve expansion, which is also known as the proper orthogonal decomposition (POD) (Loeve 1963).

The POD technique based on the covariance matrix has been widely used in many fields, such as fluid mechanics, image processing, signal analysis, data compression, and others. A comprehensive review on the subject is available in Solari et al. (2005). Lumley (1970) and Armitt (1968) introduced this technique to address turbulence and wind-related problems, respectively, and it was later used by many researchers in describing pressure fluctuations on buildings and structures and a host of wind-related problems, e.g., Lee (1975); Kareem and Cermak (1984); Holmes (1992); Kareem (1999); Tamura et al. (1999); Carassale et al. (2001). In stochastic structural mechanics, the POD technique based on the covariance matrix has been utilized for the simulation of spatially varying correlated random variables, e.g., Yamazaki and Shinozuka (1990), stochastic finite element analysis, e.g., Ghanem and Spanos (1991), and stochastic dynamics, e.g., Li and Kareem (1990b, 1995), Vasta and Schueller (2000).

The numerical advantage of the POD technique, akin to modal analysis in structural dynamics, relies on the reduced-order representation through truncation of the higher eigenmodes associated with small eigenvalues. This reduced-order representation must retain the important characteristics of the random field and ensure that related quantities either remain unchanged or that the modification that results is acceptable. Several studies on the covariance matrix-based POD technique have demonstrated that truncating higher wind loading modes helps to expedite computations of global wind loads and their effects, e.g., Tamura et al. (1999); Chen and Kareem (2005). Truncation of higher modes, however, may not work effectively in the case of local response, which may lead to an underestimation of the local wind loads and their effects (Chen and Kareem 2005; Rocha et al. 2000).

Chen and Kareem (2005) presented a theoretical framework of the reduced-order modeling of random vector-valued processes using POD based on the covariance and XPSD matrices. Utilizing this framework, a physically meaningful link between wind loads and the attendant background and resonant response of structures can be found. The XPSD matrix-based POD technique facilitates simulation of multi-correlated random processes by generating reduced order vector-valued subprocesses. The subprocesses can be simulated using spectral representation based methods, parametric time series models, or other techniques discussed in preceding sections.

10.8.2 Time-Frequency Representation and System Identification

Like many physical processes of interest, wind effects often manifest non-stationary and nonlinear features. If these features are present, the representation in the frequency domain may not accurately represent the data, since transient or time-dependent components become spread over the entire frequency range and thus are assumed to occur over the entire length of the time series. Time-frequency representations are able to more faithfully represent a non-stationary or nonlinear time series due to the localization of energy in both time and frequency. The multi-resolution capabilities of the wavelet transform are particularly attractive for civil engineering applications. Wavelet transform has the ability to identify transient components of non-stationary processes with applications in earthquake, wind, and ocean engineering (Gurley and Kareem 1999; Kijewski and Kareem 2003). Wavelet analysis has also been utilized in a number of situations where Fourier transforms (for frequency domain analysis) or Hilbert transforms (for time domain system identification) were traditionally used to define quantities of interest.

Wavelet transform involves the scaling and translating of an initial, mother wavelet over the duration of a signal. The scaling process involves comparison of the compressed or expanded mother wavelet to the signal to determine the scale and location that best matches the original signal. In this manner, the resolution both in time and in frequency can be adjusted based on the frequency being analyzed at a specific location of the signal. A map of the squared magnitudes of the wavelet coefficients is known as a scalogram, and is able to reveal the frequencies at which the signal energy content is strongest (Kijewski and Kareem 2003). The concept of the scalogram can also be used to identify correlation between signals. In coscalograms, the squared coefficients are replaced with the product of the wavelet coefficients of two different processes (Gurley and Kareem 1999; Gurley et al. 2003). Coscalograms reveal time-varying pockets of correlation over frequency.

The coherence concept in the frequency domain can also be extended to the time-frequency domain utilizing spectra. Wavelet-based coherent maps are determined by taking the ratio of the wavelet cross spectrum to the product of the wavelet autospectra of the two signals (Gurley et al. 2003). The resulting map provides a view of the localized correlation with respect to both time and frequency. To prove the utility of wavelet-based coherence maps, Gurley et al. (2003) simulated a Gaussian wind velocity signal and a pressure record that are correlated over selected frequency ranges only at certain times. They are correlated from 512 to 768 s between 0.0625 and 0.25 Hz and from 1,536 to 1,792 s between 0.19 and 0.37 Hz. The resulting coherence map is shown in the top left of Fig. 10.18. Several different methods to reduce spurious coherence in the bi-coherence maps are then discussed. The "smart" ridge extraction methodology is shown in Fig. 10.18 to effectively filter the wavelet coherence map, eliminating noise and further localizing where the signals are correlated (Gurley et al. 2003; Kareem and

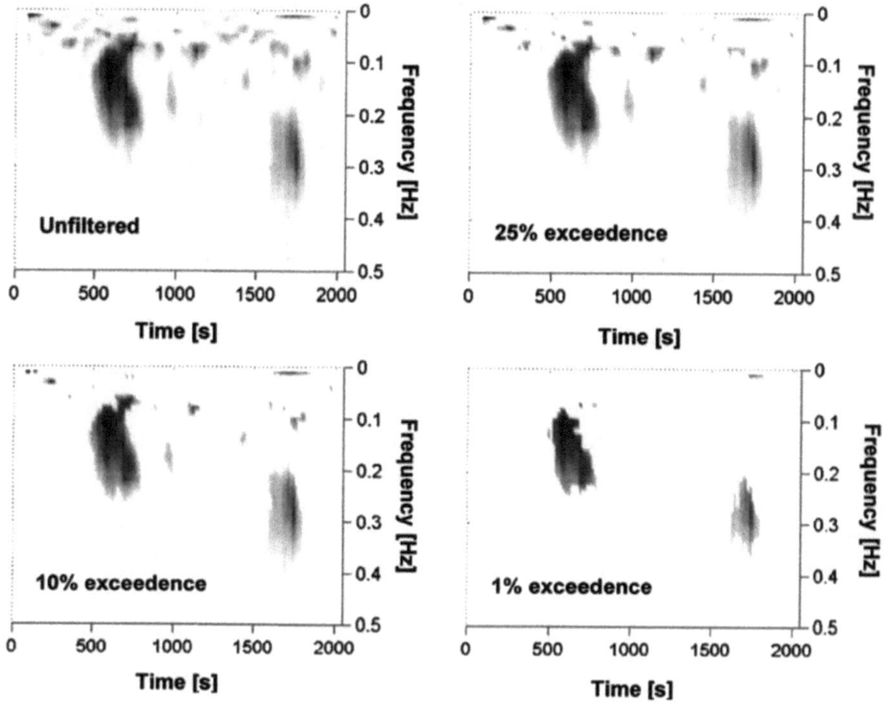

Fig. 10.18 Filtered wavelet coherence map for varying noise exceedence levels (Gurley et al. 2003)

Kijewski 2002). A similar methodology, examples available in Gurley et al. (2003) and Kareem and Kijewski (2002), can be used to identify quadratic interplay between two signals using wavelet bi-coherence. The robustness of the provided thresholding techniques is further established, as it is shown to alleviate the presence of spurious, higher-order coherence, even in cases where variance and leakage are prevalent. Though relatively intensive, this approach facilitates the removal of significant levels of all of the various contributing noise sources.

Wavelet-based system identification and instantaneous spectral analysis offer very valuable tools that permit the tracking of nonlinear characteristics in structural frequency and damping. Such tracking of time-varying frequency content is typically accomplished by monitoring the instantaneous frequency of the signal and extracting ridges from wavelet scalograms to form the Wavelet Instantaneous Frequency Spectra discussed in Kareem and Kijewski (2002) and Kijewski and Kareem (2003). The complex coefficients associated with these ridges can be used directly in a traditional system identification approach based on analytic signal theory to identify the instantaneous frequency, and in the case of free vibration decay or random decrement signatures, damping. Frequency domain perspectives from the wavelet coefficients are also insightful, as the wavelet instantaneous frequency and bandwidth can be respectively tracked to monitor the mean frequency and its deviation as they evolve in time. This basic analysis framework coupled with analysis schemes like SVD and PCA, would help to facilitate delineation of aerodynamic, aeroelastic, or structural parameters.

10.9 Concluding Remarks

This chapter outlines developments in computational stochastic modeling of wind load effects beginning with some historical perspectives and discussing recent developments, improvements, and challenges. The topics discussed include the simulation of univariate and multivariate random processes, the simulation of Gaussian, non-Gaussian, and non-stationary data, conditional simulation, and computational modeling tools. These methodologies introduced are becoming increasingly important in the analysis of wind excited structures that are exposed to a wide range of loads with attendant dynamic effects. Time and frequency domain approaches are discussed, in addition to time-frequency schemes, which are particularly important for non-stationary and non-homogenous processes/fields. Applications of these computational techniques to the analysis of buildings and bridges are highlighted, including the role of stochastic simulation in providing inflow conditions for CFD wind engineering studies.

Acknowledgements The support of the work reported has been made possible in part by the NSF grants CMMI 03-24331, CMMI 06-01143, CMMI 09-28282, and earlier grants including other sponsors over the years. The author is indebted to the contributions of his former students and post-doctoral fellows on the subject of this paper.

References

Aas-Jakobsen K, Strommen E (2001) Time domain buffeting response calculations of slender structures. J Wind Eng Ind Aerodyn 89:341–364

Adrian RJ (1994) Stochastic estimation of conditional structure: a review. Appl Sci Res 53:291–303

Amin M, Ang AHS (1968) Nonstationary stochastic model of earthquake motions. J Eng Mech, ASCE 94:559–583

Armitt J (1968) Eigenvector analysis of pressure fluctuations on the west instrumented cooling tower. Central Electricity Research Laboratories, UK

Billings SA, Tsang KM (1989) Spectral analysis for non-linear systems, part I: parametric non-linear spectral analysis. Mech Sys Sig Process 3:319–339

Borgman LE (1969) Ocean wave simulation for engineering design. J Waterways Harbors Div, ASCE 95:557–583

Borgman LE (1990) Irregular ocean waves: kinematics and forces. In: Le Mehaute B, Hanes DM (eds) Ocean engineering science, part A. Wiley, New York

Cakmak AS, Sherif RI, Ellis G (1985) Modelling earthquake ground motions in California using parametric time series methods. Int J Soil Dyn Earthquake Eng 4:124–131

Carassale L, Solari G (2006) Monte Carlo simulation of wind velocity fields on complex structures. J Wind Eng Ind Aerodyn 94:323–339
Carassale L, Piccardo G, Solari G (2001) Double modal transformation and wind engineering applications. J Eng Mech, ASCE 127:432–439
Chang MK, Kwiatkowski JW, Nau RF, Oliver RM, Pister KS (1979) ARMA models for earthquake ground motions. University of California, Berkeley
Chay MT, Albermani F, Wilson R (2006) Numerical and analytical simulation of downburst wind loads. Eng Struct 28:240–254
Chen X, Kareem A (2001) Aeroelastic analysis of bridges under multicorrelated winds: integrated state-space approach. J Eng Mech, ASCE 127:1124–1134
Chen X, Kareem A (2002) Discussion on 'time domain buffeting response calculations of slender structures' by K. Aas-Jakobsen, E. Strommen. J Wind Eng Ind Aerodyn 90:639–642
Chen X, Kareem A (2005) Proper orthogonal decomposition-based modeling, analysis, and simulation of dynamic wind load effects of structures. J Eng Mech, ASCE 131:325–339
Chen L, Letchford CW (2005) Proper orthogonal decomposition of two vertical profiles of full-scale nonstationary downburst wind speeds. J Wind Eng Ind Aerodyn 93:187–216
Chen L, Letchford CW (2007) Numerical simulation of extreme winds from thunderstorm downbursts. J Wind Eng Ind Aerodyn 95:977–990
Chen S, Billings SA, Luo W (1989) Orthogonal least squares methods and their application to non-linear system identification. Int J Control 50:1873–1896
Chen X, Matsumoto M, Kareem A (2000) Time domain flutter and buffeting response analysis of bridges. J Eng Mech, ASCE 126:7–16
Chen Y, Kopp GA, Surry D (2003) The use of linear stochastic estimation for the reduction of data in the NIST aerodynamic database. Wind Struct 6:107–126
Conner DA, Hammond JL (1979) Modelling of stochastic system inputs having prescribed distribution and covariance functions. Appl Math Model 3:67–69
Conte JP (1992) Effects of earthquake frequency nonstationarity on inelastic structural response. In: Proceedings of the 10th world conference on earthquake engineering, A. A. Balkema
Conte JP, Peng BF (1997) Fully nonstationary analytical earthquake ground-motion model. J Eng Mech, ASCE 123:15–24
Daubechies I (1992) Ten lectures on wavelets. Society for industrial and applied mathematics. Capital City Press, Montpelier, VT
Deodatis G (1996) Simulation of ergodic multivariate stochastic processes. J Eng Mech, ASCE 122:778–787
Deodatis G, Micaletti RC (2001) Simulation of highly skewed non-Gaussian stochastic processes. J Eng Mech, ASCE 127:1284–1295
Deodatis G, Shinozuka M (1988) Auto-regressive model for nonstationary stochastic processes. J Eng Mech ASCE 114:1995–2012
Der Kiureghian A, Crempien J (1989) An evolutionary model for earthquake ground motion. Struct Safety 6:235–246
Deutsch R (1962) Nonlinear transformations of random processes. Prentice-Hall, Englewood Cliffs
Deville J, Lamballais E, Bonnet JP (2000) POD, LODS and LSE: their links to control and simulations of mixing layers. ERCOFTAC Bull 46:29–44
Di Paola M (1998) Digital simulation of wind field velocity. J Wind Eng Ind Aerodyn 74–76:91–109
Di Paola M, Gullo I (2001) Digital generation of multivariate wind field processes. Probabilistic Eng Mech 16:1–10
Ding Q, Zhu L, Xiang H (2006) Simulation of stationary Gaussian stochastic wind velocity field. Wind Struct 9:231–243
Elishakoff I, Ren YJ, Shinozuka M (1994) Conditional simulation of non-Gaussian random fields. Eng Struct 16:558–563
FCMP (2011) Florida coastal monitoring program. http://fcmp.ce.ufl.edu

Filho JCSS, Yacoub MD (2008) Coloring non-Gaussian sequences. IEEE Trans Sig Process 56:5817–5822

Fujita TT (1981) Tornadoes and downbursts in the context of generalized planetary scales. J Atmospheric Sci 38:1511–1534

Fujita TT (1985) The downburst: microburst and macroburst. Report of projects NIMROD and JAWS, Satellite and Mesometeorology Research Project. University of Chicago Department of the Geophysical Sciences, Chicago

Ghanem R, Spanos PD (1991) Stochastic finite elements: a spectral approach. Springer, New York

Goβmann E, Waller H (1983) Analysis of multi-correlated wind-excited vibrations of structures using the covariance method. Eng Struct 5:264–272

Grigoriu M (1984) Crossings of non-Gaussian translation processes. J Eng Mech ASCE 110:610–620

Grigoriu M (1993) On the spectral representation method in simulation. Probabilistic Eng Mech 8:75–90

Grigoriu M (1998) Simulation of stationary non-Gaussian translation processes. J Eng Mech ASCE 124:121–126

Gurley KR, Kareem A (1997) Analysis interpretation modeling and simulation of unsteady wind and pressure data. J Wind Eng Ind Aerodyn 69–71:657–669

Gurley KR, Kareem A (1998a) A conditional simulation of non-normal velocity/pressure fields. J Wind Eng Ind Aerodyn 77&78:39–51

Gurley KR, Kareem A (1998b) Simulation of correlated non-Gaussian pressure fields. Meccanica 33:309–317

Gurley KR, Kareem A (1999) Applications of wavelet transforms in earthquake, wind and ocean engineering. Eng Struct 21:149–167

Gurley KR, Kareem A, Tognarelli MA (1996) Simulation of a class of non-normal random processes. Int J Non-Linear Mech 31:601–617

Gurley KR, Tognarelli MA, Kareem A (1997) Analysis and simulation tools for wind engineering. Probabilistic Eng Mech 12:9–31

Gurley KR, Kijewski T, Kareem A (2003) First- and higher-order correlation detection using wavelet transforms. J Eng Mech ASCE 129:188–201

Haan FL, Kareem A (2009) Anatomy of turbulence effects on the aerodynamics of an oscillating prism. J Eng Mech ASCE 135:987–999

Heredia-Zavoni E, Santa-Cruz S (2000) Conditional simulation of a class of nonstationary space-time random fields. J Eng Mech ASCE 126:398–404

Holmes JD (1992) Optimized peak load distributions. J Wind Eng Ind Aerodyn 41–44:267–276

Hoshiya M (1995) Kriging and conditional simulation of a Gaussian field. J Eng Mech ASCE 121:181–186

Huang G, Chen X (2009) Wavelets-based estimation of multivariate evolutionary spectra and its application to nonstationary downburst winds. Eng Struct 31:976–989

Iyama J, Kuwamura H (1999) Application of wavelets to analysis and simulation of earthquake motions. Earthquake Eng Struct Dyn 28:255–272

Johnson GE (1994) Constructions of particular random process. Proc IEEE 282(2):270–285

Kameda H (1975) Evolutionary spectra of seismogram by multifilter. J Eng Mech Div ASCE 101:787–801

Kameda H, Morikawa H (1992) An interpolating stochastic process for simulation of conditional random fields. Probabilistic Eng Mech 7:243–254

Kareem A (1999) Analysis and modeling of wind effects:numerical techniques. In: Larsen, A, Larose GL, Livesey FM (eds) Wind engineering into the 21st century: Proceedings of the tenth international conference on wind engineering, Copenhagen, 21–24 June 1999. A A Balkema, Rotterdam

Kareem A, Cermak JE (1984) Pressure fluctuations on a square building model in boundary-layer flows. J Wind Eng Ind Aerodyn 16:17–41

Kareem A, Kijewski T (2002) Time-frequency analysis of wind effects on structures. J Wind Eng Ind Aerodyn 90:1435–1452

Kareem A, Li Y (1988) On modelling the nonlinear relationship between random fields by means of high-order spectra. In: Spanos PD (ed) Probabilistic methods in civil engineering. ASCE, New York, pp 384–387

Kareem A, Mei G (2000) Stochastic decomposition for simulation and modal space reduction in wind induced dynamics of structures. In: Stewart MG, Melchers RE (eds) Applications of statistics and probability. A A Balkema, Rotterdam, pp 757–764

Kareem A, Tognarelli MA, Gurley KR (1998) Modeling and analysis of quadratic term in the wind effects on structures. J Wind Eng Ind Aerodyn 74–76:1101–1110

Kijewski T, Kareem A (2003) Wavelet transforms for system identification in civil engineering. Comput-Aided Civil Infrastruct Eng 18:339–355

Kijewski-Correa T, Kareem A (2006) Efficacy of Hilbert and wavelet transforms for time-frequency analysis. J Eng Mech ASCE 132:1037–1049

Kim J, Ho TCE, Hangan H (2005) Downburst induced dynamic responses of a tall building. In: 10th Americas conference on wind engineering, Baton Rouge

Kondo K, Murakami S, Mochida A (1997) Generation of velocity fluctuations for inflow boundary condition of LES. J Wind Eng Ind Aerodyn 67–68:51–64

Krige DG (1966) Two-dimensional weighted moving average trend surfaces for ore-evaluation. J South Afr Inst Min Metall 66:13–38

Kung SY (1993) Digital neural networks. Prentice-Hall, Englewood Cliffs

Kwon H-H, Lall U, Khalil A (2007) Stochastic simulation model for nonstationary time series using an autoregressive wavelet decomposition: applications to rainfall and temperature. Water Resour Res 43(W05407):1–15

Kwon DK, Kijewski-Correa T, Kareem A (2008) e-Analysis of high-rise buildings subjected to wind loads. J Struct Eng ASCE 134:1139–1153

Lee BE (1975) The effect of turbulence on the surface pressure field of a square prism. J Fluid Mech 69:263–282

Levy R, Kozin F, Moorman RBB (1971) Random processes for earthquake simulation. J Eng Mech ASCE 97:495–517

Li Y, Kareem A (1990a) ARMA systems in wind engineering. Probabilistic Eng Mech 5:50–59

Li Y, Kareem A (1990b) On stochastic decomposition and its application in probabilistic dynamics. In: Proceedings of the 5th international conference on structural safety and reliability (ICOSSAR). The Society

Li Y, Kareem A (1991) Simulation of multivariate nonstationary random processes by FFT. J Eng Mech ASCE 117:1037–1058

Li Y, Kareem A (1993a) Parametric modelling of stochastic wave effects on offshore platforms. Appl Ocean Res 15:63–83

Li Y, Kareem A (1993b) Simulation of multivariate random processes: hybrid DFT and digital filtering approach. J Eng Mech ASCE 119:1078–1098

Li Y, Kareem A (1995) Stochastic decomposition and application to probabilistic dynamics. J Eng Mech ASCE 121:162–174

Li Y, Kareem A (1997) Simulation of multivariate nonstationary random processes: hybrid DFT and digital filtering approach. J Eng Mech ASCE 123:1302–1310

Liang J, Chaudhuri SR, Shinozuka M (2007) Simulation of nonstationary stochastic processes by spectral representation. J Eng Mech 133:616–627

Liu B, Munson Jr. DC (1982) Generation of a random sequence having a jointly specified marginal distribution and autocovariance. IEEE Trans Acoust, Speech, Sig Process ASSP-30:973–983

Loeve M (1963) Probability theory I. Van Nostrand, Princeton

Lumley JL (1970) Stochastic tools in turbulence. Academic, New York

Mann J (1998) Wind field simulation. Probabilistic Eng Mech 13:269–282

Maruyama T, Morikawa H (1995) Digital simulation of wind fluctuation in a turbulent boundary layer using experimental data. In: Proceedings of the 9th international conference on wind engineering, pp 659–670

Masters F, Gurley KR (2003) Non-Gaussian simulation: cumulative distribution function map-based spectral correction. J Eng Mech 129:1418–1428

McCarthy J, Wilson JW, Fujita TT (1982) The joint airport weather studies project. Bull Am Meteorological Soc 63:15–22

McCullough M, Kareem A (2013) Testing stationarity with wavelet-based surrogates. J Eng Mech ASCE 139(2):200–209

Mignolet MP, Spanos P (1992) Simulation of homogeneous two-dimensional random fields Part I. AR and ARMA models. J Appl Mech 59:260–269

Morikawa H, Kameda H (2001) Conditional random fields containing non-stationary stochastic processes. Probabilistic Eng Mech 16:341–347

Morlet J, Arens G, Fourgeau E, Giard D (1982) Wave propagation and sampling theory; Part I, Complex signal and scattering in multilayered media. Geophysics 47:203–221

Murakami S (1997) Current status and future trends in computational wind engineering. J Wind Eng Ind Aerodyn 67&68:3–34

Myronenko A, Song X (2010) Point-set registration: coherent point drift. IEEE Trans Pattern Anal Mach Intell 32:2262–2275

Narayana Iyengar R, Sundara Raja Iyengar KT (1969) A nonstationary random process model for earthquake accelerograms. Bull Seismologic Soc Am 59:1163–1188

Orwig KD, Schroeder JL (2007) Near-surface wind characteristics of extreme thunderstorm outflows. J Wind Eng Ind Aerodyn 95:565–584

Orwig-Gast KD, Schroeder JL (2005) Extreme wind events observed in the 2002 thunderstorm outflow experiment. In: Proceedings of the tenth Americas conference on wind engineering, Baton Rouge, LA

Papoulis A, Pillai SU (2002) Probability, random variables, and stochastic processes. McGraw-Hill, New York

Priestley MB (1965) Evolutionary spectra and non-stationary processes. J R Stat Soc Ser B (Methodologic) 27

Rocha MM, Cabral SVS, Riera JD (2000) A comparison of proper orthogonal decomposition and Monte Carlo simulation of wind pressure data. J Wind Eng Ind Aerodyn 84:329–344

Samaras E, Shinozuka M, Tsurui A (1985) ARMA representation of random processes. J Eng Mech ASCE 111:449–461

Saragoni GR, Hart GC (1974) Simulation of artificial earthquakes. Earthquake Eng Struct Dyn 2:249–267

Scherer RJ, Riera JD, Schueller GI (1982) Estimation of the time-dependent frequency content of earthquake accelerations. Nucl Eng Des 71:301–310

Schetzen M (1980) The Volterra and Wiener theories of nonlinear systems. Wiley, New York

Schreiber T, Schmitz A (1996) Improved surrogate data for nonlinearity tests. Phys Rev Lett 77:635–638

Schreiber T, Schmitz A (2000) Surrogate time series. Physica D 142:346–382

Seong SH, Peterka JA (1998) Digital generation of surface-pressure fluctuations with spikey features. J Wind Eng Ind Aerodyn 73:181–192

Shinozuka M (1971) Simulation of multivariate and multidimensional random processes. J Acoust Soc Am 49:357–367

Shinozuka M, Deodatis G (1991) Simulation of stochastic processes by spectral representation. Appl Mech Rev 44:191–203

Shinozuka M, Jan CM (1972) Digital simulation of random processes and its applications. J Sound Vib 25:111–128

Shinozuka M, Sato Y (1967) Simulation of nonstationary random process. J Eng Mech ASCE 93:11–40

Shinozuka M, Zhang R (1996) Equivalence between Kriging and CPDF methods for conditional simulation. J Eng Mech ASCE 122:530–538
Shinozuka M, Yun C-B, Seya H (1990) Stochastic methods in wind engineering. J Wind Eng Ind Aerodyn 36:829–843
Solari G, Carassale L, Tubino F (2005) POD methods and applications in wind engineering. In: Proceedings of the 6th Asia-Pacific conference on wind engineering (APCWE-VI), Seoul, Korea
Spalart PR, Rumsey CL (2007) Effective inflow conditions for turbulence models in aerodynamic calculations. Am Ins Aeronautics Astronautics J 45:2544–2553
Spanos PD, Donley MG (1991) Equivalent statistical quadratization for nonlinear systems. J Eng Mech ASCE 117:1289–1309
Spanos P, Failla G (2004) Evolutionary spectra estimation using wavelets. J Eng Mech ASCE 130:952–960
Spanos P, Mignolet MP (1992) Simulation of homogeneous two-dimensional random fields. Part II. MA and ARMA models. J Appl Mech 59:270–277
Spanos P, Roesset J, Donley M (1987) Evolutionary power spectrum estimation of September 19, 1995 Mexico earthquake accelerograms. In: Proceedings of the stochastic approaches in earthquake engineering: U.S.-Japan joint seminar, Springer, New York
Tamura T, Ono Y (2003) LES analysis on aeroelastic instability of prisms in turbulent flow. J Wind Eng Ind Aerodyn 91:1827–1846
Tamura T, Suganuma S, Kikuchi H, Hibi K (1999) Proper orthogonal decomposition of random wind pressure field. J Fluids Struct 13:1069–1095
Tamura T, Okuda Y, Okada H (2002) LES estimation of wind characteristics in the surface layer over various grounds - urban roughness effects and terrain effects. In: Proceedings of the UJNR panel on wind and seismic effects (Task Committee D)
Theiler J, Eubank S, Longtin A, Galdrikian B, Doyne Farmer J (1992) Testing for nonlinearity in time series: the method of surrogate data. Physica D 58:77–94
Vanmarcke EH, Fenton GA (1991) Conditioned simulation of local fields of earthquake ground motion. Struct Safety 10:247–264
Vasta M, Schueller GI (2000) Phase space reduction in stochastic dynamics. J Eng Mech ASCE 126:626–632
Wang L (2007) Stochastic modeling and simulation of transient events, civil engineering and geological sciences. University of Notre Dame, Notre Dame
Wang L, Kareem A (2004a) Modeling of non-stationary winds in gust-fronts. In: Proceedings of the 9th ASCE joint specialty conference on probabilistic mechanics and structural reliability, PMC, Albuquerque, NM
Wang L, Kareem A (2004b) Simulation of earthquake ground motion based on wavelet and hilbert transforms. In: Proceedings of the 9th ASCE joint specialty conference on probabilistic mechanics and structural reliability, PMC, Albuquerque, NM
Wang L, Kareem A (2005a) Modeling and simulation of transient winds in downbursts/hurricanes. In: Proceedings of the 10th Americas conference on wind engineering, Baton Rouge, LA
Wang L, Kareem A (2005b) Simulation of multi-variate non-stationary random processes based on wavelet and Hilbert transforms. In: Proceedings of the 9th international conference on structural safety and reliability (ICOSSAR), Rome, Italy
Wang L, Kareem A (2006) Description and simulation of gust front wind field. In: Proceedings of the 4th international symposium on computational wind engineering, Yokohama, Japan
Wittig LE, Sinha AK (1975) Simulation of multicorrelated random processes using the FFT algorithm. J Acoust Soc Am 58:630–634
Yamazaki F, Shinozuka M (1988) Digital geneartion of non-Gaussian stochastic fields. J Eng Mech 114:1183–1197
Yamazaki F, Shinozuka M (1990) Simulation of stochastic fields by statistical preconditioning. J Eng Mech ASCE 116:268–287

Chapter 11
CFD Applications in Structural Wind Engineering

Shuyang Cao

Abstract Computational Fluid Dynamics (CFD) is basically a numerical approach to simulating or predicting phenomena and quantities of a flow by solving the equations of motion of the fluid at a discrete set of points. CFD has greatly improved recently and has become an application tool for wind engineering problems. This paper first explains the characteristics of NS equations, and then describes in detail the discretization schemes, turbulence models, inflow turbulence and numerical procedures. Finally, it provides an example of calculation of strong wind in hilly area by utilizing Large Eddy Simulation.

Keywords CFD • CWE • LES • Turbulence model • Wind load • Wind resistance design

11.1 Introduction

Computational Fluid Dynamics (CFD) is basically a numerical approach to simulating or predicting phenomena and quantities of a flow by solving the equations of motion of the fluid at a discrete set of points. It has wide applications in flow-related engineering fields including aeronautical, mechanical and civil/architectural fields, although the difficulties in applying it to particular problems in these fields are different. Computational Wind Engineering (CWE) mainly implies the application of CFD to wind engineering problems. The structural part of CWE usually involves the combination of problems of bluff body aerodynamics, inflow turbulence, wake turbulence, grid generation and high Reynolds number (up to the order of 10^7–10^8). All these problems require special attention in numerical simulation.

S. Cao (✉)
Tongji University, 1239 Siping Road, Shanghai 200092, China
e-mail: shuyang@tongji.edu.cn

Y. Tamura and A. Kareem (eds.), *Advanced Structural Wind Engineering*,
DOI 10.1007/978-4-431-54337-4_11, © Springer Japan 2013

According to the book "Guide to numerical prediction of wind loads on buildings" (AIJ 2005), which was written by an AIJ WG on CWE headed by Prof. T. Tamura of Tokyo Institute of Technology, computational wind engineering has improved into a stage where:

1. It is appropriate when considering the influence of inflow turbulence;
2. Improvement of the turbulence model has enabled us to understand how to treat the complicated separated flow and unsteady flow with strong vorticity;
3. Improvement of the computational technique using the generalized curvilinear coordinate system has enabled it to be used to calculate aerodynamic forces on structures with complicated body shapes. The simulation accuracy is improved.

These improvements enable simulation of wind pressure, wind force and wind response of structures with comparatively complicated shapes immersed in boundary layers with characteristics similar to those of natural wind. When these techniques are appropriately utilized, CFD can be applied to some practical wind engineering problems.

Present wind load codes/standards specify the wind resistance design process as a successive estimations of basic wind speed, design wind speed, wind load on structure frame, wind load on claddings, evaluation of aerodynamic instability and habitability to building vibration, and sometimes evaluation of noise and wind environment. The basic wind speed, which is determined by considering probability of occurrence of strong wind, can be estimated by analyzing the wind climate data and conducting typhoon simulations using the Monte Carlo method. It does not have to be predicted by CFD. The design wind speed is the wind speed profile at a certain location or area that is influenced by local terrain, roughness conditions and other conditions. CFD has been shown to be very powerful in predicting wind profiles in high-density high-rise building areas (Tamura and Nozu 2003) and terrain areas with different slopes (Ishihara and Hibi 2002; Cao and Tamura 2005).The feasibility of CFD in simulating the pressure distribution around a low-rise building (Nozawa and Tamura 2002) and a high-rise building (Nozawa and Tamura 2003) immersed in a rough-wall turbulent boundary layer has also been presented. Three-dimensional flow around a rectangular cylinder in unstable oscillations has been investigated numerically (Tamura and Itoh 1997). These simulations have demonstrated that CFD can be utilized to solve some practical wind resistance design problems. However, when utilizing CFD as a general tool for predicting wind loads on structures, we must realize that, due to differences in discretization methods, boundary conditions, turbulence modeling and numerical procedures, different simulation results are inevitable. Therefore, it is necessary to understand the fundamentals of the numerical simulation in order to make the simulation reliable and robust. The reality is that, although CFD has been widely used and its use has been permitted in solving wind environmental problems such as pedestrian wind, air pollution and others, it is still only a trial method when utilized for wind resistance design because it has not been adequately validated until now.

Although both experimental and numerical approaches are available for investigating flow phenomena, their performances in studying the flow are different. Experimental tools such as wind tunnel and water channel are traditional methods for studying real flow phenomena by utilizing a scaled model. Current knowledge of fluid dynamics has been mainly obtained from experimental and analytical studies. The numerical method has developed along with computer technology. It studies a flow in continuity at a discrete set of points by solving the equations obtained from the experimental or theoretical analysis. Thus, experimental results are closer to real flow phenomena than CFD results. Even if the numerical results are obtained by utilizing an appropriate numerical approach, it cannot be concluded that the characteristics of the governing equations are perfectly simulated. However, an experiment usually uses a scale model and has some restrictions of experimental conditions. For instance, difficulties are very often encountered in arranging pressure taps on the model. However, CFD solves non-dimensional equations without the scale problem, and it is possible to deal with special situations like extremely high pressure or high temperature because it is virtual. CFD can supply temporal and spatial information of the quantities of the flow that enables understanding of the flow field through digital flow visualizations, which is usually very difficult in an experiment. In addition, CFD saves the time and effort of making experimental models and conducting experiments. CFD has been anticipated as an alterative, not only supplementary, to the wind tunnel for practical wind resistant design problems for a long time. However, unfortunately CFD is not as well established as experimentation. In the structural wind engineering field, all national wind load codes/standards are formulated under the premise that wind loads on structures are simulated in wind tunnels. The wind speed profile is even classified into only a few categories in order to make it easy to conduct wind tunnel experiments. With the CFD approach, it is possible to predict the wind speed profile on the construction site and consequently to simulate the wind load on the structure numerically. We may anticipate that improvement of CFD will bring about a revolution to wind resistant design codes.

11.2 CFD Fundamentals

11.2.1 Characteristics of the Governing Equations

The governing equations of the flow are:

Mass conversation equation

$$\frac{\partial u_i}{\partial x_i} = 0 \tag{11.1}$$

Navier Stokes equation

$$\frac{\partial u_i}{\partial t} + u_j \frac{\partial u_i}{\partial x_j} = -\frac{1}{\rho}\frac{\partial p}{\partial x_i} + \frac{1}{\mathrm{Re}}\frac{\partial^2 u_i}{\partial x_j \partial x_j} + f_i \tag{11.2}$$

where u, p, t and Re denote velocity, pressure, time and Reynolds number, respectively.

The second term on the left side of Eq. (11.2) is an advection term, which is nonlinear. The difficulties of solving the NS equation are mainly due to this term. The second term on the right side is the diffusion term, which has the role of smoothing the velocity field and thus stabilizing the simulation. The first term on the right side is the pressure term, which is determined by coupling with the mass conservation equation because the mass conversion equation does not contain the variable t.

The performance of NS equation can be easily investigated if it is simplified into a one-dimensional equation (Burgers equation) as follows:

$$\frac{\partial u}{\partial t} + u\frac{\partial u}{\partial x} = \nu \frac{\partial^2 u}{\partial x^2} \tag{11.3}$$

Burgers equation has an exact solution so it is often used to check the numerical method. Equation (11.3) can be further linearized to an equation of advection and diffusion, as given by Eq. (11.4), by assuming the advection velocity and diffusion coefficient constant.

$$\frac{\partial u}{\partial t} + c\frac{\partial u}{\partial x} = \alpha \frac{\partial^2 u}{\partial x^2} \tag{11.4}$$

The performance of the equation of advection and diffusion can be observed by looking at three simpler equations which contain only two terms of Eq. (11.4):

$$\frac{\partial u}{\partial t} + c\frac{\partial u}{\partial x} = 0 \tag{11.5}$$

$$\frac{\partial u}{\partial t} = \alpha \frac{\partial^2 u}{\partial x^2} \tag{11.6}$$

$$c\frac{\partial u}{\partial x} = \alpha \frac{\partial^2 u}{\partial x^2} \tag{11.7}$$

Equation (11.5) is the advection equation, which is hyperbolic type. It has an exact solution such that

$$u = u(x,t) = \varphi(x - ct) \tag{11.8}$$

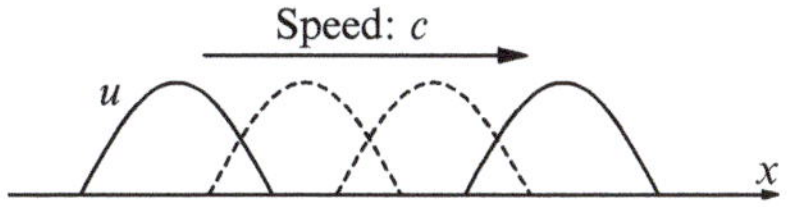

Fig. 11.1 Properties of equation of advection

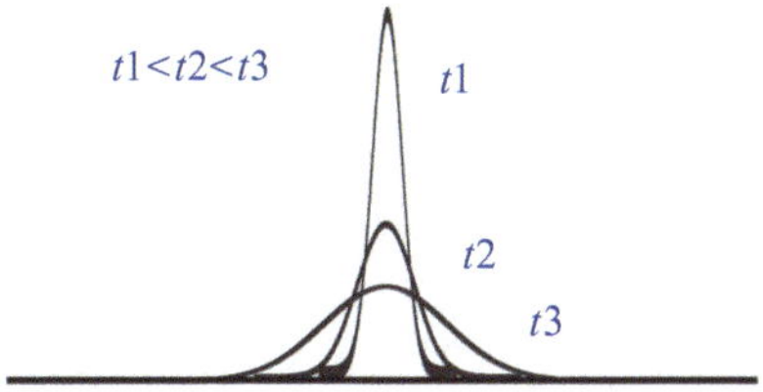

Fig. 11.2 Properties of equation of diffusion

which has the property that

$$u(x + c, t + 1) = u(x, t) \tag{11.9}$$

which means that the flow moves downstream at constant speed c with the initial values maintained as a stationary wave, as illustrated in Fig. 11.1.

Equation (11.6) is an equation of diffusion, which is parabolic type. One of its analytical solutions is:

$$u = u(x, t) = \frac{1}{2\sqrt{\pi t}} e^{-x^2/4t} \tag{11.10}$$

It has the characteristics that $\partial u/\partial t < 0$ when $\partial^2 u/\partial x^2 < 0$ and $\partial u/\partial t > 0$ when $\partial^2 u/\partial x^2 > 0$. This means that the velocity field is always smoothed, in other words, stabilized. The initial value influences the field in the future and the values at $t = 0$ influence the values at points far from the origin. Figure 11.2 shows this most important feature of the diffusion equation.

Equation (11.7) is the steady equation of advection and equation, which is elliptic type influenced by advection term and diffusion term respectively. How it performs depends on which term is dominant.

11.2.2 Difference Scheme

CFD solves the NS equations at a discrete set of points. Generally the error caused by difference depends on the grid size. A finer grid system or high order accuracy scheme leads to smaller error. For a partial differential equation, there are many possible expressions to differentialize it. It is necessary to pay attention to two important issues when considering the error of difference (Kajishima 1999). The first is the performance of the error. The numerical result is smoothed compared to

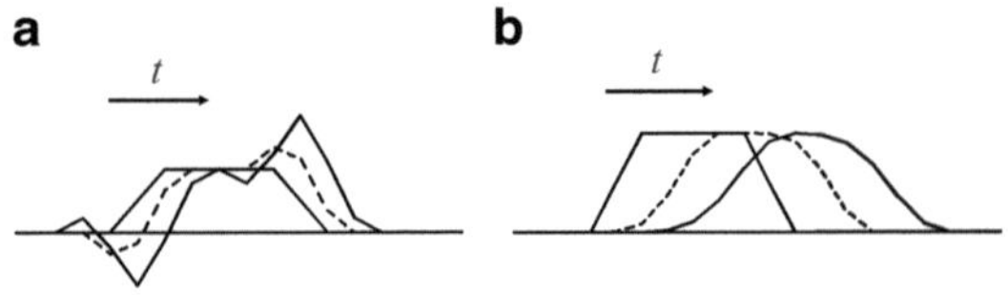

Fig. 11.3 Different behavior of central and upwind difference schemes. (**a**) Central difference, (**b**) Upwind difference

the real phenomenon if the error is diffused. Otherwise, the unphysical fluctuation caused by the error develops and leads to calculation failure if the error is dispersed. Another is that the fundamental characteristics of the partial differential equation must be maintained. The difference expression cannot be accepted if it cannot maintain the original differential relations, even though the difference expression utilized is of high order accuracy.

Central finite difference and upwind finite difference are two widely used difference schemes in CFD. As shown in Eqs. (11.11) and (11.12), the principal term of the truncation error of central difference is the odd (third) order differentiation term, implying that the central finite difference brings dispersed error. Conversely, the principal term of the truncation error of upwind difference is the even (second) order differentiation term, which brings diffused error.

Center difference:

$$u_i' = \frac{-u_{i-1} + u_{i+1}}{2 \times h} - \frac{h^2}{6} \times u_i{}^{(3)} + o(h^4) \tag{11.11}$$

Upwind difference:

$$u_i' = \frac{-u_i + u_{i+1}}{h} - \frac{h}{2} \times u_i{}^{(2)} + o(h^2) \tag{11.12}$$

Figure 11.3 illustrates the performances of the central difference and upwind difference in causing error when solving the equation of advection. As mentioned above, the exact solution is that the original distribution is maintained while the flow moves downstream. It can be seen in Fig. 11.3 that, while moving downstream, irregularity occurs in the distribution for the central difference case, implying an unstable calculation, while the distribution is over-smoothed for the upwind difference case.

An example is shown to present the performance of the upwind difference scheme. If the 1st order upwind scheme is utilized to solve the equation of advection and diffusion, the advection term can be expressed as:

$$\text{if } u > 0 \quad u\frac{\partial f}{\partial x} = u\frac{f_i - f_{i-1}}{\Delta x} \tag{11.13}$$

$$\text{if } u < 0 \quad u\frac{\partial f}{\partial x} = u\frac{f_{i+1} - f_i}{\Delta x} \tag{11.14}$$

These equations can be re-arranged into:

$$u\frac{\partial f}{\partial x} = u\frac{f_{i+1} - f_{i-1}}{2\Delta x} - \frac{|u|}{2}\frac{f_{i-1} - 2f_i + f_{i+1}}{\Delta x} \tag{11.15}$$

It can be noticed that the first term on the right-hand side is the expression of the central difference form, and the second term is just the expression of the diffusion term. Thus, compared with the central difference form, the upwind scheme actually solves the equation of advection and diffusion by solving:

$$\frac{\partial f}{\partial t} + u\frac{\partial f}{\partial x} - (a + \frac{u\Delta x}{2})\frac{\partial^2 f}{\partial x^2} = 0 \tag{11.16}$$

Compared to Eq. (11.4), it is easily seen that there is an additional term ($\frac{u\Delta x}{2}$) that functions in a similar manner to the molecular viscosity, which is called numerical viscosity. The numerical viscosity is proportional to Δx. The grid must be fine enough to make $\frac{u\Delta x}{2} << \alpha$ in order to obtain good simulation at a specified Reynolds number. Otherwise, the results obtained actually correspond to the flow at a Reynolds number that is much smaller than expected. In fundamental research on fluid dynamics, it is necessary to avoid numerical viscosity. However, for simulation of practical engineering problems, sometimes adding numerical viscosity is inevitable. The commercial software utilizes the upwind scheme initiatively in order to make the software applicable to many situations. When using the software to solve practical wind engineering problems, it is necessary to discuss the simulation results carefully (Kajishima 1999).

11.2.3 Turbulence Model

The oncoming flow to the structures is always turbulent because they are immersed in a turbulent atmospheric boundary layer, and the Reynolds number of the structure is usually very high. The flow phenomena of vortex formation, interaction between oncoming turbulence and separated shear layer and dynamic wake formation are under turbulent conditions. In the cascade-down process of turbulence energy, the turbulence energy generated at low frequency transfers to the higher frequency region gradually, i.e., first to the energy preservation region and then to the inertia region. The high frequency fluctuation needs to be considered in the simulation because its role in the energy cascade down process cannot be disregarded.

It is mainly molecular viscosity that works in the process of turbulence energy cascade for real flow phenomena. In the numerical simulation at high Reynolds numbers for wind engineering problems, an extremely fine grid system is necessary in order to determine the turbulence dissipation on molecular viscosity only. In addition, the calculations at high Reynolds numbers usually fail if stabilization measures are not taken. To overcome these difficulties, numerical viscosity or turbulence viscosity is usually added into the simulation. The upwind scheme was once widely used when computer capacity was not large enough to calculate bluff body flows where large scale vortex shedding was the dominate flow phenomenon. With the improvement of computer capacity and CFD technique, the turbulence model is expected to be very useful for solving high Reynolds number problems because they are developed according to the physical characteristics of the flow.

The turbulence model is used to dissipate the turbulent energy at high frequency by means of turbulence viscosity, in addition to molecular viscosity. Generally, there are two kinds of turbulence models: Reynolds Averaged Navier–Stokes (RANS) and Large Eddy Simulation (LES). RANS solves the NS equation by estimating the turbulence effect (turbulence viscosity) through solving supplementary equations of turbulence statistics. The results obtained are averaged over time. In order to reflect the turbulence effect appropriately, very complicated forms sometimes need to be used. RANS is a very efficient method if only the mean or rms values of the flow quantities are required. On the other hand, LES solves the large scale (grid scale) components of the flow directly with the small scale (sub-grid) components modeled. This enables time dependent calculation. LES is very time-consuming because it usually conducts three-dimensional calculations, and long time series data of the flow quantities are required in order to achieve reliable analysis. The formula of LES is simpler than that of RANS, its dependence on the flow phenomenon is less and few parameters need to be decided for the calculation. With LES, the time series data of the flow quantities at all grid points are obtained, allowing evaluation of the mean and peak values as well as the power spectra of the fluctuation at all points. Thus, LES is more advantageous than RANS for wind resistant design. However, the filter operation of LES is actually an averaging process over space. It is necessary to note that the peak value obtained in LES has been subject to a filter operation.

In practical numerical simulation, we need to choose the turbulence model appropriate to the purpose of the calculation.

In wind engineering, multi-scale turbulence exists very commonly because of interaction between structures and atmospheric turbulence. However, how the averaging operation functions on different turbulence scales is not clear in RANS, because its averaging operation is over time. It is available to distinguish which frequency range is numerically operated in LES. The numerical turbulence results obtained by LES can be explained more physically than those obtained by RANS.

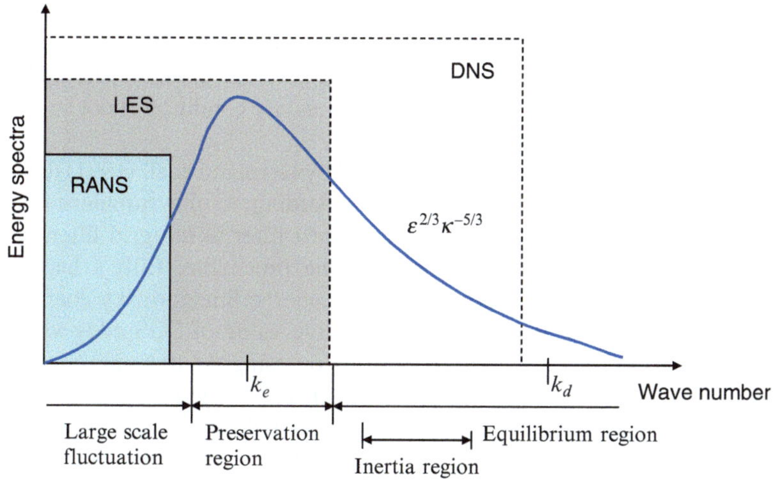

Fig. 11.4 Turbulence spectrum

Figure 11.4 illustrates the distribution of the power spectrum of turbulence. DNS solves the turbulence up to the equilibrium region. It is almost perfect for turbulence simulation. However, it usually requires the grid number to be proportional to $\mathrm{Re}^{9/4}$, which makes simulation of practical wind engineering problems impossible. RANS simulates large scale fluctuation of turbulence only. The frequency range of LES is less than that of DNS, with thc high frequency (sub-grid) component to be modeled. It can be found from Fig. 11.4 that turbulence modeling is more difficult in RANS, because it needs to model a very broad frequency range of turbulence which varies according to flow phenomena.

The standard Smagorinsky model is a widely used LES model based on the assumption of local equilibrium between turbulence production and dissipation. Turbulence viscosity can be estimated from:

$$\nu_e = (C_s\Delta)^2|\overline{D}| \tag{11.17}$$

where $|\bar{D}|$ is a grid scale velocity tensor and C_s is the Smagorinsky constant, which is usually taken as 1.5 for shear flow. The Smagorinsky model has a wide application because of its robustness. It cannot express backward energy cascade. For boundary layer flow, the van Direst function is usually utilized to satisfy the boundary condition:

$$\nu_e = (C_s f_s\Delta)^2|\overline{D}| \tag{11.18}$$

where
$$f_s = 1 - \exp\frac{-y^+}{A^+}$$

$$y^+ = yu_\tau/\nu,\ A^+ = 25.0$$

Fewer parameters are necessary for LES calculation than for RANS calculation, and the simulation results of LES are not greatly influenced by the parameters as they are for RANS because grid scale turbulence is solved directly. The study on the best damping function for the wall or corner boundary condition is not so meaningful (Kajishima 1999).

The dynamic Smagorinsky model proposed by Germano et al. (1991) determines the Smagorinsky coefficient dynamically according to the turbulence field by introducing a test filter. The size ratio of the test filter to the grid filter, which is optimized to 2.0, is the only parameter in the procedure. Lilly's least-squares technique (1992) is often incorporated to estimate the Smagorinsky coefficient. In order to avoid numerical instability, the negative value of SGS eddy viscosity is usually truncated to zero. Large Eddy Simulation using the dynamic Smagorinsky model has been widely utilized as a relatively affordable and easy-to-use approach to simulate separated flow and other complicated flow (Kravchenko and Moin 2000). The procedure for estimating SGS eddy viscosity is shown below.

$$\tau_{ij} - \frac{1}{3}\delta_{ij}\tau_{kk} = -2C \cdot \overline{\Delta}^2 \, |\bar{S}|\overline{S_{ij}} \tag{11.19}$$

$$L_i^k = \bar{u}_i \tilde{\bar{U}}^k - \tilde{\bar{u}}_i \tilde{\bar{U}}^k \tag{11.20}$$

$$M_{ij} = \overline{\Delta}^2 \left(\alpha^2 \left| \tilde{\bar{S}} \right| \tilde{\bar{S}}_i^k | - |\bar{S}| \tilde{\bar{S}}_i^k \right) \tag{11.21}$$

$$C = -\frac{1}{2} \frac{\langle L_{ij} M_{ij} \rangle_z}{\langle M_{ij} M_{ij} \rangle_z} \tag{11.22}$$

The scale Similarity model is based on the assumption that the largest scale component in SGS turbulence is similar to the smallest scale component in GS turbulence. It is a non-viscosity model, so its applicability to turbulence should be broader than the standard and dynamic Smagorinsky model. However, it has few applications to wind engineering problems now.

11.2.4 Inflow Turbulence

As LES is a time-advancing calculation, it is necessary to impose the inflow turbulence on the inlet boundary of the computation domain at each time step. The periodic boundary condition in the streamwise direction, which is often used to simulate turbulent channel flow, cannot be used for practical wind engineering problems. Making a matrix of random signals by considering power spectrum, profiles of turbulence statistics and spatial correlations of wind speed is possible, but it does not satisfy the governing equations of the fluid. When the analytically generated data are imposed at the inlet boundary, the turbulence characteristics

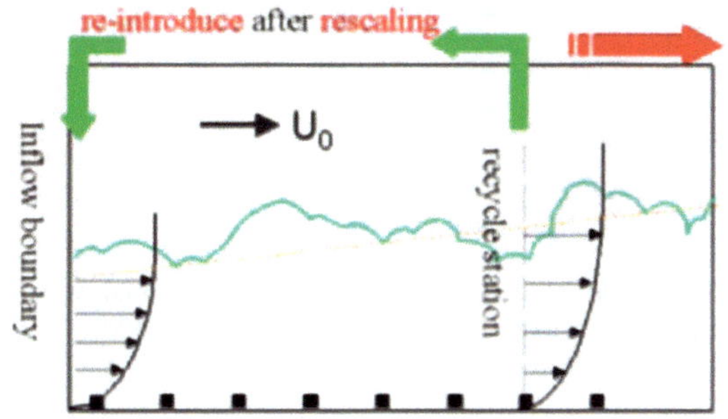

Fig. 11.5 Pseudo periodic boundary condition for simulating spatially-developing boundary layer

change dramatically downstream. It is difficult to control the turbulence statistics at a certain location inside the computational domain.

It is possible to simulate a turbulent layer developing on a smooth or rough flat plate (half channel flow) and to use it as inflow turbulence. However, for wind engineering problems, it is hoped to obtain very long time series pressure/velocity data to evaluate wind load. This requires the properties of the boundary layer imposed at the inlet to be maintained, which is difficult to achieve by the method of half channel flow.

Pseudo-periodic boundary conditions may be the best method for generating inflow turbulence. The ideal atmospheric boundary layer is a spatially-developing one. A very long computational domain is needed to simulate a spatially-developing turbulent boundary layer. The method proposed by Lund et al. (1998) is to rescale the velocity field at a downstream station, and re-introduce it as a boundary condition at the inlet, to allow for the calculation of a spatially developing boundary layer in conjunction with pseudo-periodic boundary conditions applied in the streamwise direction, as illustrated in Fig. 11.5. Lund's method was expanded into rough surface conditions by Nozawa and Tamura (2002), and thermally-stratified conditions by Korean researchers.

11.2.5 Some Numerical Details

11.2.5.1 2D or 3D Calculation

Of course, it is necessary to conduct three-dimensional calculations for three-dimensional structures. However, due to the limitation of computer capacity, 2D simulation for a bridge or cylinder section is often conducted. Although the difference between the statistics of 2D and 3D simulations are sometimes not great, we must realize that all the vortex and dynamic wake structure behind a cylinder are three dimensional. It is thus necessary to conduct three-dimensional calculations if one expects to "simulate" the flow, and not to simply

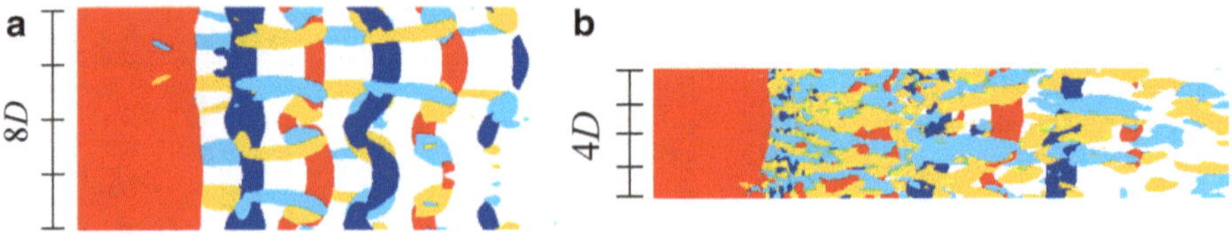

Fig. 11.6 Mode change of three-dimensional wake structure with increase in Re number. (**a**) Mode A (Re = 200) (**b**) Mode B (Re = 500)

"demonstrate" some flow characteristics. Figure 11.6 illustrates the mode change of a three-dimensional wake structure with increase in Re number, which cannot be seen in 2D simulation.

11.2.5.2 Boundary Condition

It is necessary to impose the boundary conditions artificially because the computational domain can not be infinite. An inappropriate boundary layer will produce unphysical numerical results, and even lead to calculation breakdown. The convective boundary condition is the widely accepted outlet boundary condition in the wind engineering field, to let the vortex behind the structure flow smoothly out of the outlet. It is possible to check the boundary condition at the outlet by comparing two calculations (Yoshida et al. 1995). One has a long calculation domain, while the other one has a short calculation domain with one kind of boundary condition. By comparing the time series of pressure at the same point inside the domain, it is possible to identify how much the boundary condition influences the pressure inside, and to judge whether the boundary condition is appropriate or not.

11.2.5.3 Resolution

The maximum scale of turbulence available in the simulation depends on the size of the computational domain, while the minimum scale depends on the grid scale. It is impossible to obtain the flow phenomenon with a scale smaller than the grid size. The maximum wave number available in a grid system with a grid size of Δ is up to $k_c = \pi/\Delta$ (cutoff wave number).

Simulation of interaction between the inflow turbulence and the separated shear flow and vortex behind the structure is important for studying vortex-induced response. Figure 11.7 shows the power spectra of turbulence simulated by a grid system with 64 $\times$ 64 $\times$ 64 grids. The high frequency component is not simulated well enough to be used as inflow turbulence. A denser grid system is necessary to increase the turbulence energy at high frequency. Kaneda (2003) showed that it is possible to simulate the inertia area with 2,000 $\times$ 2,000 $\times$ 2,000 grids.

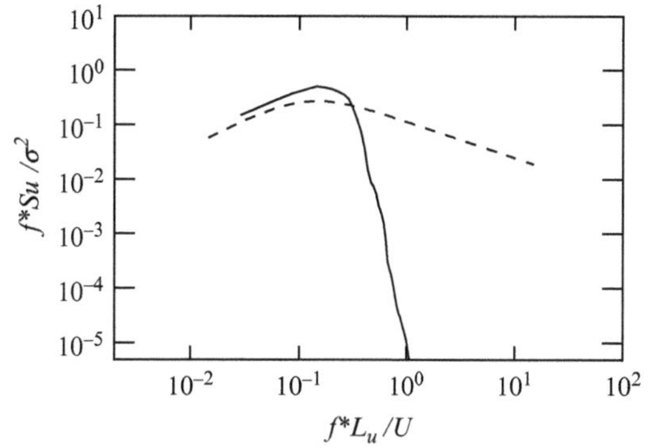

Fig. 11.7 Power spectrum of simulated turbulence

11.2.5.4 Measures for High Reynolds Number Flow

For high Reynolds number flow, it is possible to stabilize the calculation by partially adding the numerical viscosity, as shown by:

$$u\frac{\partial u}{\partial x} = u_i\frac{-u_{i+2} + 8(u_{i+1} + u_{i-1}) + u_{i-2}}{12\Delta x} + \alpha\frac{|u_i|(\Delta x)^3}{12} \times \frac{u_{i+2} - 4u_{i+1} + 6u_i - 4u_{i-1} + u_{i-2}}{(\Delta x)^4} \quad (11.23)$$

where $0 \leq \alpha \leq 1.0$, $\alpha = 0$ corresponds to central difference scheme, $\alpha = 1$ corresponds to 3rd upwind scheme.

11.2.6 Validation

Validation is the process of determining the degree to which a model is an accurate representation of the real world from the perspective of the intended uses of the model.

Validation data sets for CFD fundamentals and wind engineering problems include the database developed by ERCOFTAC (European Res. Com. on Flow, Turb. & Comb.), TTU field data (Texas Tech Univ., USA), AIJ (Architectural Institute of Japan) Working Group for Numerical Prediction and TPU (Tokyo Polytechnic University) aerodynamic database.

Many journals refuse to publish articles that do not meet minimum standards to quantify CFD uncertainty. ASME Journal of Fluids Engineering Editorial Policy is:

Numerical method and truncation error described
At least 2nd order accuracy in space

Artificial viscosity must be assessed and minimized
Grid independence or convergence established
Iterative convergence must be addressed
For transient calculations phase error must be minimized
Accuracy of boundary and initial conditions explained
Existing code must be fully cited in accessible references
Benchmark solutions may be used for validation.

11.3 An Example of Calculation: Flow Around Terrain

With the increased concern over the wind energy and wind loading problems in terrain regions, extensive investigations have been conducted in recent years to study the complicated turbulent boundary layer flow over an isolated or wavy hill by using numerical approaches. Although the RANS based numerical methods are still widely applied (for instance, Ishihara and Hibi 2002, Lun et al. 2003 and Ying and Canuto 1997), Large Eddy Simulation (LES) has become an increasingly used method for predicting the turbulence structures and statistics because it enables the time-dependent calculation of the unsteady flows including large-scale vortices, which are usually established in the wake of a steep hill. There are several reported LES researches that aim to study some aspects of the turbulent flow over hills in neutral flow condition. Gong et al (1996) performed LES of the separated turbulent flow over rough wavy hills to investigate the spanwise instability. Tamura et al. (2002) carried out LES study of the turbulent boundary layer over a three dimensional rough hill by using the simplified generation technique for inflow turbulence. Allen and Brown (2002) compared the numerical results obtained from a LES model proposed by Brown et al. (2001) with the experimental data. However, the systematic numerical research on the effects of hill shape and roughness on the turbulent boundary layer flow over an isolated two dimensional hill is very few.

It is known that many factors, such as the inflow turbulence, hill surface condition and the ratio of boundary layer depth to hill height, have great effects on the turbulent structure and consequently on the turbulent statistics of the flow over the hills. It is because, for instance, the behavior of the separated shear layer from a hill and the vortex motions are sensitively affected by the oncoming flow turbulence and, the roughness blocks on the hill surface influence the onset of the separation and reattachment. It is necessary to achieve a good numerical simulation of the flow field in order to realize the roles of the hill shape and surface roughness. In this study, we carried out Large Eddy Simulation to study the turbulence characteristics of the turbulent boundary layer over two-dimensional steep and low hills with and without surface roughness, i.e. a smooth steep hill in smooth wall inflow, a smooth low hill in smooth wall inflow, a rough steep hill in rough wall inflow and a rough low hill in rough wall inflow. The numerical results

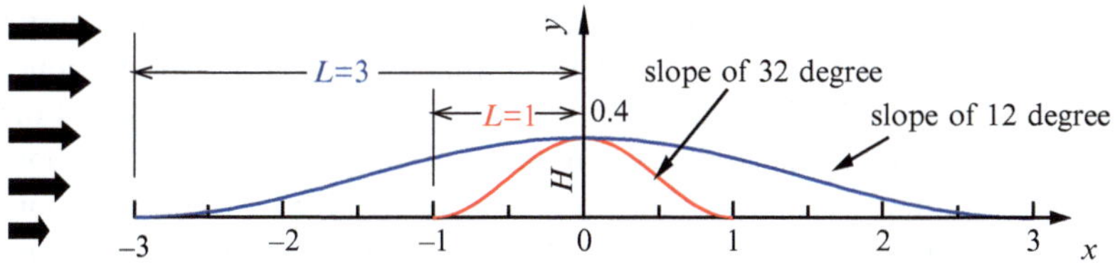

Fig. 11.8 Hill shape

are validated by comparing with the experimental data that were obtained at same Reynolds number. Also, Comparisons of the turbulence statistics obtained at different hill shapes and surface conditions are carried out to make clear the effects of hill shape and surface roughness. During the simulation work, the methods proposed by Lund et al. (1998) and its extended version proposed by Nozawa and Tamura (2002) are applied to simulate the spatially developing boundary layers over smooth and rough walls individually to generate inflow turbulence. Another feature of the simulation is that the rough wall condition is modeled by placing small rectangular cylinders on the computational domain rather than setting roughness length. Although acceptable statistical results were obtained with the latter method of modeling the rough wall, we believe that the former method is closer to the real rough condition because the naturally occurring roughness blocks are trees and buildings, which are not densely arranged as Nikusadse's hypothesis.

11.3.1 Numerical Details

The hill concerned has a sinusoidal geometrical shape defined in Eq. (11.24),

$$y = H\cos^2\left(\frac{\pi x}{L}\right) \tag{11.24}$$

where H is the height of the hill and L is the length as shown in Fig. 11.8. Two hill shapes, $L = 2.5H$ and $L = 7.5H$, which have maximum slope angle of 32° and 12° respectively, are considered. The critical angle to establish the stable separation in the wake is roughly 16° (Finnigan 1988), so these two hill shapes respond to two completely different flow patterns. In the rough hill condition, the cubic cylinders with a side length of $0.125H$ are placed in a staggered pattern whose roughness density equals to 4.1 %, where roughness density is defined as the total roughness front area per unit ground area. In this study we shall denote the streamwise, wall-normal, and spanwise velocity components by u, v, and w, with the corresponding coordinates being x, y, and z.

The simulation consists of two parts actually. The main simulation is carried out for the turbulent flow over a hill, while an auxiliary simulation is for the generation of inflow turbulence. Both simulations are carried out synchronously and the data numerically obtained in the auxiliary simulation are imposed at the inflow boundary of the main simulation at each time step. The generalized curvilinear coordinate system and the Cartesian coordinate system are used to formulate the numerical models for these two simulations respectively. The governing equations, the filtered incompressible Navier–Stokes and the continuity equations, are discretized by finite difference schemes with staggered mesh of Maliska's type in the main simulation and conventional staggered mesh in auxiliary simulation. Equations (11.25) and (11.26) show the filtered continuity and Navier–Stokes equations of the main simulation for the turbulent flow over a hill, where u, p, t and Re denote velocity, pressure, time and the Reynolds number, respectively. J is the Jacobian, U_m is the contravariant velocity component in the generalized coordinate system, and G^{mn} is the mesh skewness tensor. The last term in the right hand of Eq. (11.26) is the external force term, expressed in Eq. (11.27), to bring the flow rest on the body surface.

$$\frac{\partial \overline{u}_i}{\partial x_i} = J\frac{\partial \overline{U}_m}{\partial \xi_m} = 0 \tag{11.25}$$

$$\frac{1}{J}\frac{\partial \overline{u}_i}{\partial t} + \frac{\partial}{\partial \xi_m}\left(\overline{U}_m \overline{u}_i\right) = -\frac{\partial}{\partial \xi_m}\left(\frac{1}{J}\frac{\partial \xi_m}{\partial x_i}\overline{p}\right) + \frac{\partial}{\partial \xi_m}\left(\nu G^{mn}\frac{\partial \overline{u}_i}{\partial \xi_n}\right) - \frac{\partial}{\partial \xi_m} \times \left(\frac{1}{J}\frac{\partial \xi_m}{\partial x_j}\tau_{ij}\right) + \delta(s)\frac{1}{J}f_i \tag{11.26}$$

$$f_i(x_s, t) = \alpha \int_0^t \overline{U}_i(x_s, t')dt' + \beta \overline{U}_i(x_s, t) \tag{11.27}$$

The dynamic procedure based on the Smagorinsky model is used to determine the model coefficient in the main computational domain. The anisotropic part of the SGS stress can be expressed in Eq. (11.28),

$$\tau_{ij} - \frac{\delta_{ij}}{3}\tau_{kk} = -2\nu_\tau \overline{S_{ij}} \tag{11.28}$$

where $\nu_\tau = C\overline{\Delta}^2|\overline{S}| = CJ^{2/3}|\overline{S}|$ and $\overline{S_{ij}} = \frac{1}{2}\left(\frac{\partial \xi_m}{\partial x_j}\frac{\partial \overline{u}_i}{\partial \xi_m} + \frac{\partial \xi_m}{\partial x_i}\frac{\partial \overline{u}_j}{\partial \xi_m}\right)$

In the dynamic procedure to determine the viscosity coefficient, averaging of the viscosity coefficient is performed in the homogeneous spanwise direction,

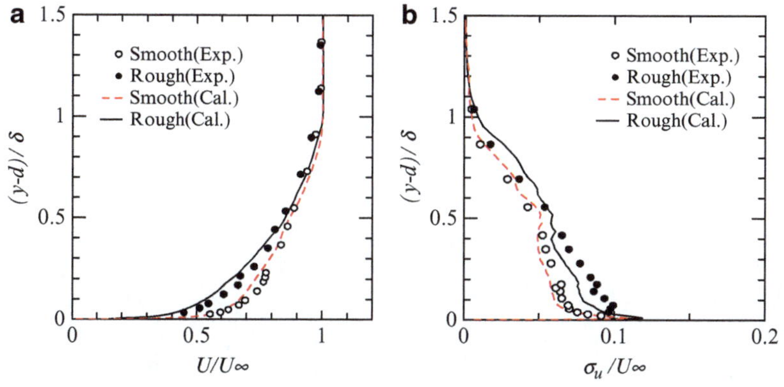

Fig. 11.9 Statistics of the generated inflow turbulence. (**a**) Mean velocity (**b**) Turbulent intensity

in conjunction with Lilly's least squares technique. The ratio of the test scale to grid scale is given a value of 2.0. Averaging and test filter operation are performed only over the spanwise direction where the flow is homogenous. 2nd order central differencing scheme is used for both the convective term and confusion term. As for time marching, explicit Adams–Bashforth differencing scheme is applied for the convention term, and semi-explicit Crank–Nicolson formulation is applied for the diffusion term.

The computational domain is $40H$, $15H$ and $8H$ in the streamwise, wall-normal and spanwise directions respectively. The grid system, which has 240, 129, 81 nodes, is fitted to the hill shape and normalized near the hill surface. Grid size varies from $x^+ = 5.5-66, y^+ = 0.81-70$(boundary layer height)$-170, z^+ = 30$(constant). Initial flow field is from a calculation at lower Reynolds number.

Wind tunnel studies were carried out to supply the experimental data to validate the simulation results. For the rough hill case, small cubes with size of 5 mm each side, or 0.125H, are used to cover the wind tunnel floor and the hill surface to model physically the rough condition. The experiment was carried out at Reynolds number Re = 5,000, close to current numerical study.

11.3.2 Inflow Turbulence

Spatially developing turbulent boundary layers over smooth and rough walls are simulated to generate the inflow turbulence, where the methods of Lund (1991) and Nozawa and Tamura (2002) are applied for the smooth and rough surface conditions respectively. Figure 11.9 compares the vertical distributions of mean

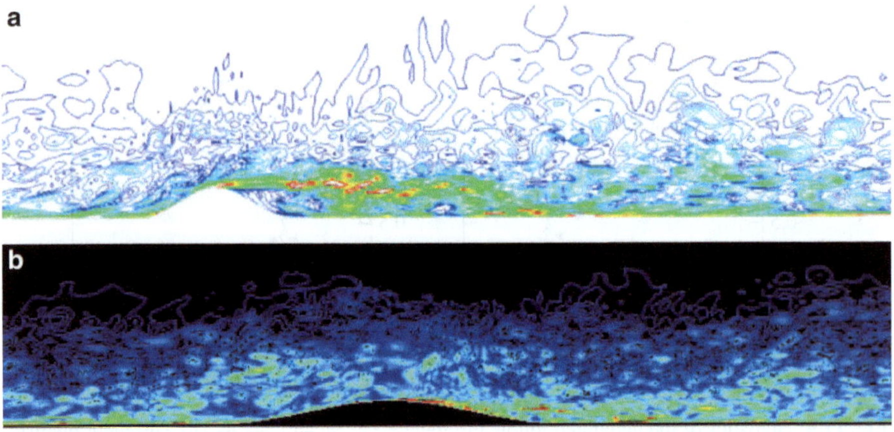

Fig. 11.10 Contours of the instantaneous vorticity magnitude. (**a**) Step smooth hill (**b**) Low smooth hill

velocity and turbulence intensity obtained by the simulation with the experimental data. Overall, the generated inflow turbulent statistics are in not good agreement with the experimental data, perhaps due to the short calculation time. Matching the obtained mean velocity with log law gives $z_0/H = 0.0043$ for the rough case. The roughness Reynolds number, $Re_s(= u_\tau \times z_0/\nu,\ \nu$ is the kinematic viscosity), is 1.18, smaller than the critical value for being fully roughed wall, which is conventionally considered to exceed at least two and perhaps be as much as five before viscous effects are insignificant. However, it has also been reported recently that this critical value can be considered to be $Re^* = 1$ when sharp-edged rectangular cylinder is used as roughness elements (Snyder and Castro 2002). So, however there is undoubtedly Reynolds number effect on the obtained results of this study, it is considered not too significant. In addition, this rough surface condition can be clarified into k-type rough surface.

When this generated turbulent flow is introduced into the main simulation of the flow over a hill, the Reynolds number based on the wind speed on the hilltop and the hill height is 4,865 and 4,175 for the smooth and rough hills respectively.

11.3.3 Results and Discussions

11.3.3.1 Unsteady Flow Characteristics

We at first glance at the turbulent flow field on the unsteady flow feature. Figures 11.10a, b illustrate the contours of the instantaneous vorticity magnitude at

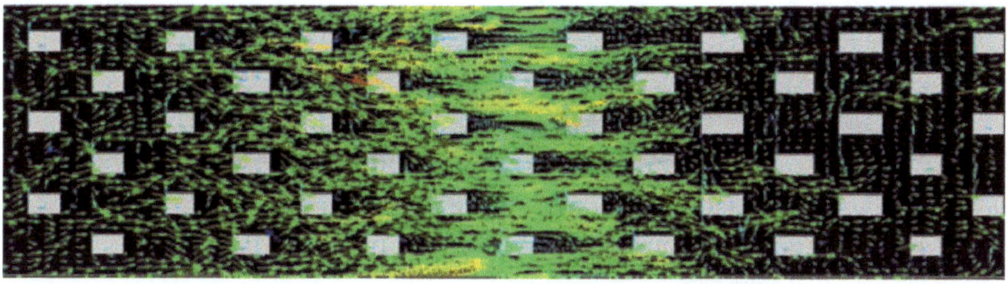

Fig. 11.11 Instantaneous velocity vector around the roughness blocks

the central section of the spanwise direction over the steep smooth hill and low smooth hill. In Fig. 11.10a for the steep hill, there is an apparent roll-up of the shear layer forming a large-scale structure. The separated shear layer extends discontinuously to the reattachment region which is composed of many small, high intensity vortices. However, we cannot find in Fig. 11.10b the similar large scale structure for the low hill. For their counterparts of the rough condition, similar behavior of the separated shear flow can be seen (Figures are not shown), but without recognized difference from those of the smooth case. The obvious difference occurs near the surface where small scale vortices can be recognized locally near the roughness blocks.

Figure 11.11 shows the instantaneous velocity vector near the roughness blocks over the hill. The very small value of the velocity inside the roughness block implies the success of the application of the forcing method. The acceleration of the velocity on the hill top can also be seen in Fig. 11.11.

11.3.3.2 Separation Bubble

The reattachment point, or the length of the flow recirculation zone, downstream of a hill is of interest, for the basic understanding of the separation and reattachment behavior, and for the wind engineering applications. In this study, we investigate the variation of mean reattachment location with the surface condition. The mean reattachment location is determined to be the location where the spanwise-averaged mean velocity $U = 0$ at the nearest grid point to the wall. Figure 11.12 compares the time-averaged streamlines for the steep hills with and without surface roughness. Large separation bubble is formed in the steep hill. In the case of rough surface, the separated shear layer extends longer, and reattaches on the ground at a point further downstream than the smooth surface case. This simulation result agrees qualitatively with the experimental result. Figure 11.13 illustrates the time-averaged streamline over low hills, where the separation bubble cannot be recognized. However, the rough hill has a more obvious tendency to form reversed flow on the leeside.

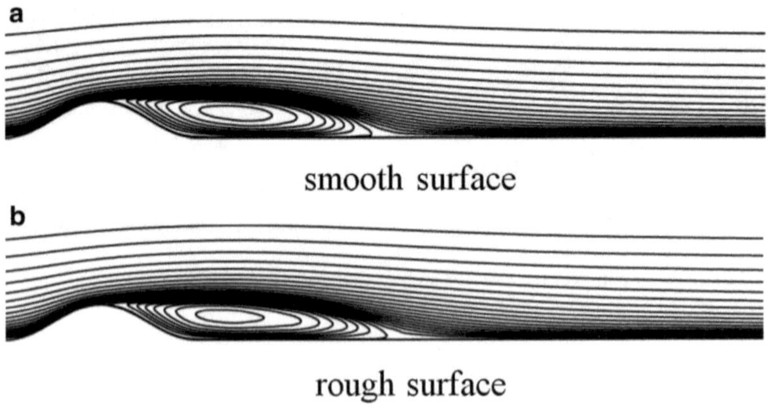

Fig. 11.12 Time-averaged streamlines over steep hill. (**a**) Smooth surface; (**b**) rough surface

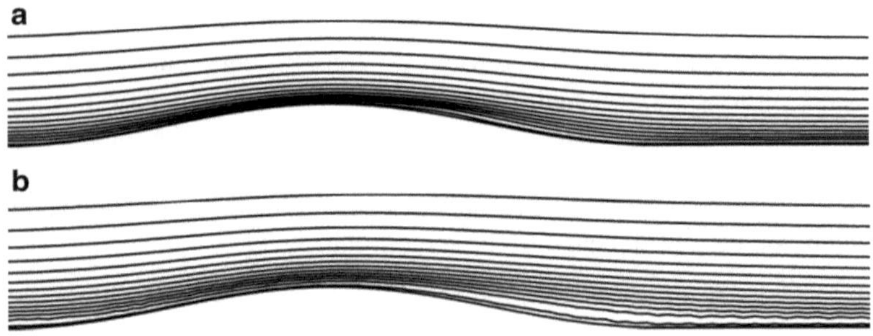

Fig. 11.13 Time-averaged streamlines over low hill. (**a**) Smooth surface; (**b**) rough surface

11.3.3.3 Turbulence Statistics

Figures 11.14, 11.15, 11.16, and 11.17 compare the computed results of turbulence characteristics over the hills with the experimental data. The agreement between them is fair. The discrepancy in turbulence statistics is obvious in the wake region of the steep hill as shown in Fig. 11.14b. The simulation data show acceptable agreements with the experimental data in the low hills, perhaps because of the weaker separated shear layer compared to the steep hill and because it is more appropriate to apply the Smagorinsky type eddy-viscosity models that based on the assumption of equilibrium between dissipation and

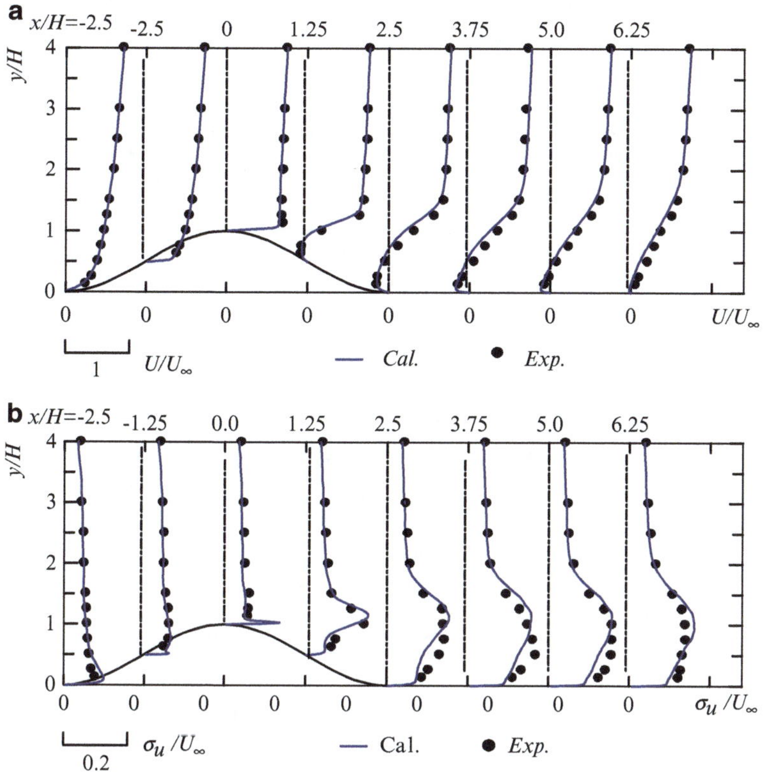

Fig. 11.14 Turbulent statistics over steep smooth hill. (**a**) Mean velocity; (**b**) turbulence intensity

production to the low hill. Turning to the turbulence statistics, for smooth and rough steep hills, mean reversed flow can be seen in the leeward of the hill, and turbulent intensity has a maximum value near the hill height, corresponding to the location of separated shear layer.

This study showed that the Smagorinsky type eddy-viscosity model is more appropriate for the low hill where the assumption of equilibrium between dissipation and production can be anticipated to some degree. Generation of the inflow turbulence plays important roles in the simulation work of the flow over hills. Without the success of the simulation of the inflow turbulence, good results of the turbulent flow field over the hills can not be obtained. Meanwhile, it showed that large scale separation bubble is established on the leeside of the steep hill. The reattachment point extends further downstream when the hill is rough. Stable separation bubble can not be found in the low hill.

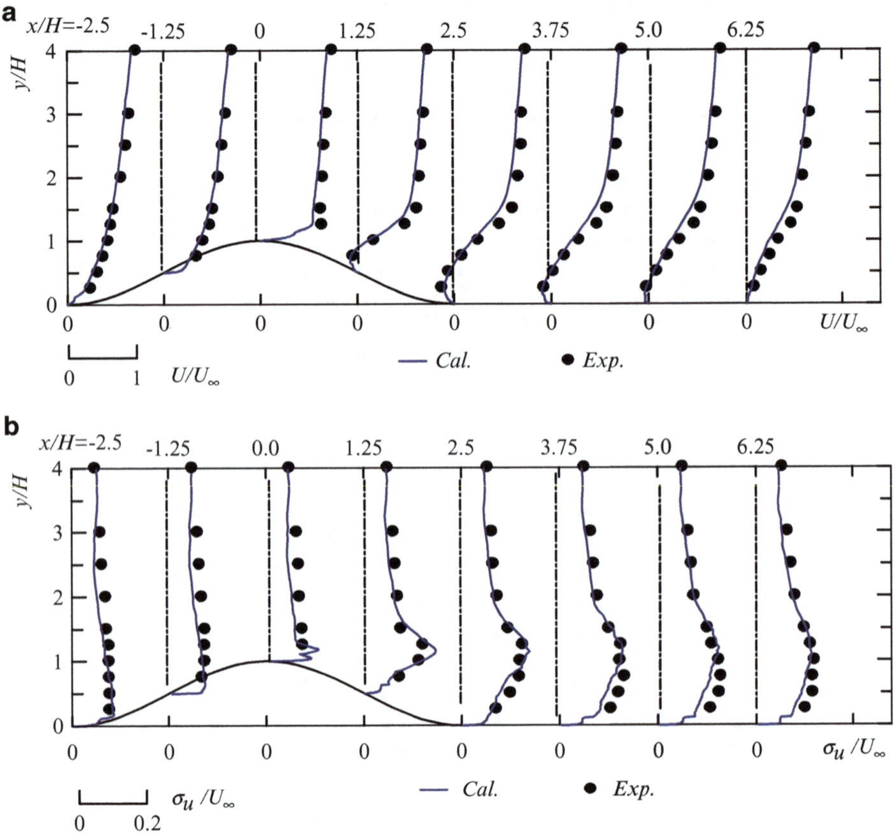

Fig. 11.15 Turbulent statistics over steep rough hill. (**a**) Mean velocity; (**b**) turbulence intensity

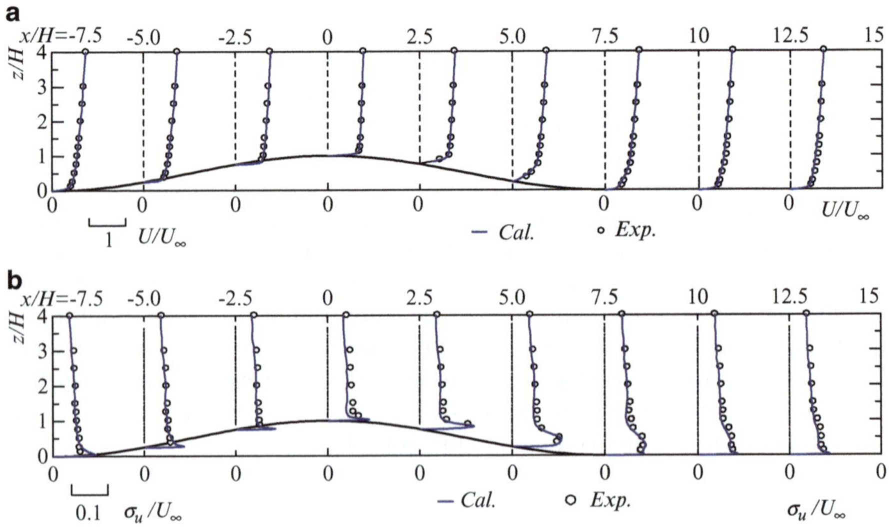

Fig. 11.16 Turbulent statistics over low smooth hill. (**a**) Mean velocity; (**b**) turbulence intensity

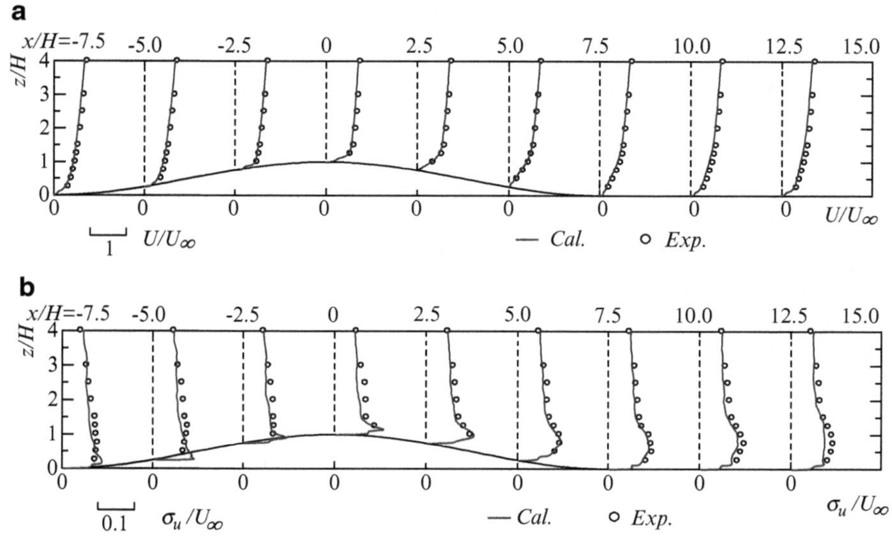

Fig. 11.17 Turbulent statistics over low rough hill. (**a**) Mean velocity; (**b**) turbulence intensity

References

Architectural Institute of Japan (2004) AIJ recommendations for loads on buildings, Maruzen Ltd, Tokyo

Architectural Institute of Japan (2005) Guide for numerical prediction of wind loads on buildings, Maruzen Ltd, Tokyo

Allen T, Brown AR (2002) Large-eddy simulation of turbulent separated flow over rough hills. Bound Layer Meteorol 102:177–198

Brown AR, Hobson JM, Wood N (2001) Large eddy simulation of neutral flow over rough sinusoidal ridges. Bound Layer Meteorol 98:411–441

Cao S, Tamura T (2005) LES analysis of the turbulent boundary layer flow over two dimensional hills. In: Proceedings of 10th Americas wind engineering conferences, CD-ROM, Baton Rouge, USA

Finnigan JJ (1988) Air flow over complex terrain, flow and transport in the natural environment. Springer, Heidelberg, pp 183–229

Germano M, Piomelli U, Moin P, Cabot WH (1991) A dynamic subgrid-scale eddy viscosity model. Phy Fluids A3(7):1760–1965

Gong W, Taylor P, Dornbrack A (1996) Turbulent boundary-layer flow over fixed aerodynamically rough two-dimensional sinusoidal waves. J Fluid Mech 312:1–37

Ishihara T, Hibi K (2002) Numerical study of turbulent wake flow around a three-dimensional steep hill. Wind Struct 5(2–4):317–328

Kajishima T (1999) Numerical simulation of turbulent flows. Yokendo Ltd, Tokyo

Kaneda Y, Ishihara T, Yokokawa M, Itakura K, Uno A (2003) Energy dissipation rate and energy spectrum in high resolution direct numerical simulations of turbulence in a periodic box. Phys Fluids 15:L21–L24

Kravchenko AG, Moin P (2000) Numerical studies of flow over a circular cylinder at Re = 3900. Phys Fluids 12(2):403–417

Lilly DK (1992) A proposed modification of the Germano subgrid-scale closure model. Phys Fluids A 4(4):633–635

Lun YF, Mochida A, Murakami S, Yoshino H, Shirasawa T (2003) Numerical simulation of flow over topographic features by revised k- models. J Wind Eng Ind Aerodyn 91:231–245

Lund TS, Wu X, Squires KD (1998) Generation of turbulent inflow data for spatially- developing boundary layer simulations. J Comput Phys 140:233–258

Nozawa K, Tamura T (2002) Large eddy simulation of the flow around a low rise building immersed in a rough-wall turbulent boundary layer. J Wind Eng Ind Aerodyn 90:1151–1162

Nozawa K, Tamura T (2003) Feasibility study of the LES study on predicting wind loads on high-rise building. In: Proceeding of the 11th international conference on wind engineering, Lubbock, USA

Snyder WH, Castro IP (2002) The critical Reynolds Number for rough-wall boundary layers. J Wind Eng Ind Aerodyn 90:41–54

Tamura T, Itoh Y (1997) Three dimensional vertical flows around a rectangular cylinder in unstable oscillations. J Wind Eng Ind Aerodyn 67/68:141–154

Tamura T, Kido R, Ono Y (2002) LES analysis of the turbulent boundary layer over an insolated three dimensional hill. Summaries of technical papers of annual meeting, AIJ, 95–96

Tamura T, Nozu T (2003) LES analysis of the strong wind turbulence in high-rise building area. J Struct Construction Eng 571:29–38

Ying R, Canuto VM (1997) Numerical simulation of flow over two-dimensional hills using a second-order turbulence closure model. Bound Layer Meteorol 85:447–474

Yoshida M, Watanabe S, Nakamura Y (1995) Numerical study of the outlet boundary of the unsteady incompressible flow. J JSME (B) 61(588):2887–2895

Chapter 12
Human Perception and Tolerance of Wind-Induced Building Motion

Kenny C.S. Kwok

Abstract Wind-induced building motion can interfere with building occupants' daily activities and general well-being. However, human perception of motion and tolerance of wind-induced tall building vibration are essentially a subjective assessment. Hence, there is currently no single internationally accepted occupant comfort serviceability criteria which set a design standard for satisfactory levels of wind-induced vibration in tall buildings. This chapter reviews past studies on human perception of motion and tolerance thresholds of wind-induced tall building motions. Building vibration acceptability, occupant comfort serviceability criteria and assessment methodologies that have been commonly adopted for the assessment of occupant comfort in wind-excited tall buildings are outlined. Occupant comfort evaluation that utilises motion simulator to provide building developers/ owners and design professionals an experience of different levels of simulated wind-induced building vibrations is described, thus facilitating an assessment of the acceptability of the building vibration and the need to adopt vibration mitigation measures. The relevance of two motion sickness theories: Sensory Conflict Theory and Postural Instability Theory on human response to wind-induced building motion is discussed, which may lead to design strategy that aims to minimize the adverse effects of building motion on occupant comfort and well-being and potential degradation of manual task and cognitive performance.

Keywords Building motion perception • Motion sickness • Motion simulator • Occupant comfort • Wind excitation

---

K.C.S. Kwok (✉)
Institute for Infrastructure Engineering, University of Western Sydney,
Locked Bag 1797, Penrith, NSW 2751, Australia
e-mail: k.kwok@uws.edu.au

Y. Tamura and A. Kareem (eds.), *Advanced Structural Wind Engineering*, DOI 10.1007/978-4-431-54337-4_12, © Springer Japan 2013

12.1 Introduction

In tall building design, not only strength and safety requirements need to be satisfied, serviceability criteria are equally important and often govern the design. Wind-induced building motion can cause fear and alarm and interfere with building occupants' daily activities. Prolonged exposure can cause discomfort, affect well-being and task performance, and in extreme cases cause motion sickness type symptons including headache, dizziness and nausea. Hence occupant comfort is a critical design consideration, particularly for wind-sensitive tall and slender buildings constructed in regions of high winds, such as typhoon prone areas in East Asia, coastal region of South China Sea and Gulf of Mexico.

Many studies have been undertaken in the past decades to study huamn response to motion. Many experiments have been conducted using motion simulators to investigate the effects of motion on human. The majority of these experiments has been conducted based on single degree-of-freedom sinusoidal motion, for example to determine perception thresholds of motion. However, human response to motion, particularly wind-induced building motion, is a complex mix of a variety of psychological and physiological factors, including tactile, vestibular, proprioceptive, kinaesthetic, visual and auditory cues, and visual-vestibular interaction. Hence the response of human to sinusoidal motion may be potentially quite different to the narrow-band random motion typical of wind-induced tall building motion. Furthermore, there is little information available to-date on the effects of low-frequency narrow-band random motion on cognitive efficiency and mental performance. With tall and super tall buildings reaching heights of 1,000 m and beyond, proper assessment of occupant perception of motion, comfort and well-being, and task and cognitive performance poses a major challenge due to a lack of information that clearly demonstrate the physiological and psychological responses of occupants at the very low natural frequencies typical of these modern tall buildings.

This chapter reviews past studies on human perception of motion and tolerance thresholds of wind-induced tall building motions. Building vibration acceptability and occupant comfort criteria that have been commonly adopted for the assessment of occupant comfort in wind-excited tall buildings are outlined. Occupant comfort evaluation that utilise a motion simulator to provide building developers/owners and design professionals an experience of different levels of simulated wind-induced building vibrations is described. Two theories on motion sickness and their relevance on human repsonse to wind-induced building motion is discussed.

12.2 Human Perception of Motion and Tolerance Threshold of Wind-Induced Building Motion

There is a vast amount of data available in the published literatures that deal with motion and vibration in a wide range of frequencies, including ship, vehicular and other transportation related vibration, and industrial and manufacturing related vibration, and the effects of these vibrations on passengers and machine operators. These types of vibration are mostly large amplitude vibration and at a frequency range that are outside the range of frequency relevant to building motion. Kwok et al. (2009) reviewed past studies that focus on human perception of motion and tolerance thresholds of wind-induced building motion. These studies generally fall into three categories: field experiments and survey-based studies of building occupants conducted in wind-excited tall buildings; motion simulator and shake table experiments testing human test subjects; and field experiments conducted in artificially excited buildings.

Field experiments and surveys of building occupants conducted in wind-excited tall buildings during wind storms (Hansen et al. 1973; Goto 1983; Lee 1983; Isyumov et al. 1988; Isyumov and Kilpatrick 1996, and others) provide motion perceptibility and acceptability information for various levels of accelerations and at different frequencies. Some surveys were conducted in buildings after or during a storm but with survey results correlated with estimated responses only, usually based on wind tunnel test results and an estimate of the wind speeds during the storm. Better quality and more useful results were derived from experiments and surveys conducted in test buildings that were known to experience unacceptable motions and/or have registered complaints, and with the test buildings properly instrumented to record the incident wind speeds and the resultant accelerations accurately. Repeated tests and surveys conducted in properly instrumented test buildings over multiple wind storm events also produced more reliable data compared with sole event survey data. Unfortunately, despite the availability of modern affordable instrumentation and a general desire by researchers and design professionals to gain an understanding of occupant response in a natural building environment subjected to naturally occurring wind events, field experiments in properly instrumented buildings known to exhibit unacceptable motions remains rare. The lack of progress in this research approach is largely due to unwillingness of building owners and corporate tenants, who are the potential beneficiaries, to actively support such a research endeavour due to a range of financial, commercial and legal reasons. Furthermore, the publication of those good quality data is inevitably hampered by pending or on-going litigations.

Many motion simulator and shake table experiments (e.g. Khan and Parmelee 1971; Chen and Robertson 1972; Irwin 1981, 1983, 1984, 1985, 1988; Kanda et al. 1988, 1990; Goto et al. 1990; Shioya et al. 1992; Shioya and Kanda 1993; Noguchi et al. 1993; Denoon et al. 1999; Denoon 2000; Burton et al. 2004, 2005, 2006; Tamura et al. 2006; Burton 2006; Burton et al. 2011; Denoon and Kwok 2011; Michaels et al. 2013, and others) have been conducted on human test subjects under carefully controlled experimental conditions to overcome the uncertainties posed by field experiments and building occupant surveys. In these motion simulator experiments, motions were generated under controlled conditions by varying frequencies, amplitudes and duration, and with human test subjects tested in different postures. The test signals used in these experiments included uni-directional, bi-directional, and/or yaw (torsion) motions but the vast majority of the experiments were based on sinusoidal signals, with only a limited number based on the narrow-band random signals normally encountered in wind-excited buildings. Thresholds of human perception to highly predictable sinusoidal motion are shown to be quite different than those for unpredictable random motion. Furthermore, most experiments were designed to focus on motion perception without subjecting test subjects to task distraction normally encountered in a work or home environment. Coupled with pre-conditioning, in which the test subjects had a preconceived expectation that they would experience motion, the motion simulator experiments potentially lowered the perception threshold.

Motion perception experiments have also been conducted in artificially excited buildings (Morris et al. 1979; Bouncer et al. 1980, and others). These experiments were conducted in actual buildings excited by counter rotating masses driven by variable speed electric motors of varying frequencies and weights. These tests were parametric studies investigating variables such as fore-aft, lateral and torsional motions, test subject posture, and manual tasks. Although the tests were conducted in an actual building environment unable to be reproduced in a motion simulator and the actual building motions were meant to reproduce feeling of fear in occupants, the excitation method only produced sinusoidal building motions which, as discussed earlier, differ greatly from the narrow-band random building responses normally encountered in wind-excited buildings.

Visual cues experiments were performed by Goto et al. (1996), Goto and Shindo (1999) in which studies on human test subjects were conducted in a motion simulator with an image projection to simulate a moving distant or close visual scene but with no simulated motion and no task distraction for the test subjects. Hence, the experiments delivered purely visual stimuli involving no other physiological factors. Although there have been some other reports of motion perception through external visual cues, with a focus on torsional motion, building occupants experience pure torsional or yaw motion only if located at the centre of rotation of a building, which is likely to be far from a view window and amongst the service part of a building (lift core, fire stair, washrooms). At other positions within a building where most occupants perform their normal activities, particularly far away from the centre of rotation, torsional motion is primarily felt as translational motion. It is noteworthy that during episodes of strong to very strong winds between 1981 and 1987, the highly wind-sensitive Sydney Tower underwent wind-induced motions with a peak translational acceleration as high as 50 milli-g (Kwok and Macdonald 1990), where 1 milli-g = 1,000th of gravity. While standing, the whole body swayed in response to the tower motion and the Tower visually appeared to move, against a background of tall buildings through the view windows. However, once the head was immobilized by being held stationary against a structural member, no deflection or movement was detectable against a far field reference object. This suggests that magnified head movements through the bio-dynamic response of body components in reaction to building motions are potentially linked to creating visual parallax of relative object motion in wind-excited tall buildings. Results to support such theory have been demonstrated in recent studies (Burton et al. 2006).

Human motion perception and tolerance threshold were performed recently utilizing a combination of shake table experiments, field experiments and surveys by Denoon (2000), Denoon and Kwok (2011), Denoon et al. (1999), and Denoon et al. 2000). The field experiments and surveys were conducted in three wind-sensitive towers instrumented with anemometers and accelerometers. Sydney Airport Control Tower has a natural frequency of 0.94 Hz and was installed with motion ranking push button experiments. The air traffic controllers working in the Tower were the test subjects for the motion perception experiments, a long term survey and an exit questionnaire. Brisbane Airport Control Tower (0.54 Hz) was used for motion ranking push button experiments only, while survey questionnaires for limited repeat test subjects were conducted in the Port Operation Control Centre (POCC) in Sydney (0.39 Hz). Occupant perception of motion was found to be frequency-dependent and the associated perception threshold acceleration was in close agreement with ISO6897: 1984 (International Organization for Standardization 1984) when compared on the basis of peak acceleration. While the occupants who had previously experienced motion were more sensitive, those who were distracted by having to perform a primary task recorded a higher motion perception threshold than those for whom the sole purpose was to detect motion. Evidently, habituation and an increase in the tolerance of motion happened over time, based on the results of the long term surveys and exit questionnaires of occupants in the Sydney Airport Control Tower.

Interestingly, while the performances in terms of occupant comfort of both the Sydney and Brisbane Airport Control Towers were "acceptable" according to ISO6897: 1984 (International Organization for Standardization 1984), the Sydney tower registered more complaints. It is believed that this is a result of the frequency of sustained perceptible motion associated with synoptic winds that tend to blow for hours. In contrast, Brisbane tower's responses were characterized by infrequent short bursts of high peak acceleration and high peak factor motion, largely due to a wind climate that is dominated by sporadic thunderstorms with short duration severe winds that typically persist for no more than about 10 min. This suggests that the exposure duration and the motion waveform play an important role in assessing human perception of motion and tolerance threshold, with motion perception and fear of motion triggered by peak acceleration, and standard deviation acceleration, being a more direct measure of sustained motion, more suited to assess occupant comfort. Furthermore, in the assessment of occupant comfort in wind-excited tall buildings, the acceptance criteria should be based on occupant comfort and general well-being, and not occupant perception of motion.

Denoon (2000), Denoon et al. (1999), and Denoon and Kwok (2011) also conducted motion simulator and field experiments on human motion perception and tolerance threshold using the wind-sensitive POCC and a uni-directional shake table. POCC provided the platform for tests using real wind-induced motions at the tower natural frequency of 0.39 Hz, and selected motion data collected in the field experiments at this frequency of 0.39 Hz, one large amplitude with a standard

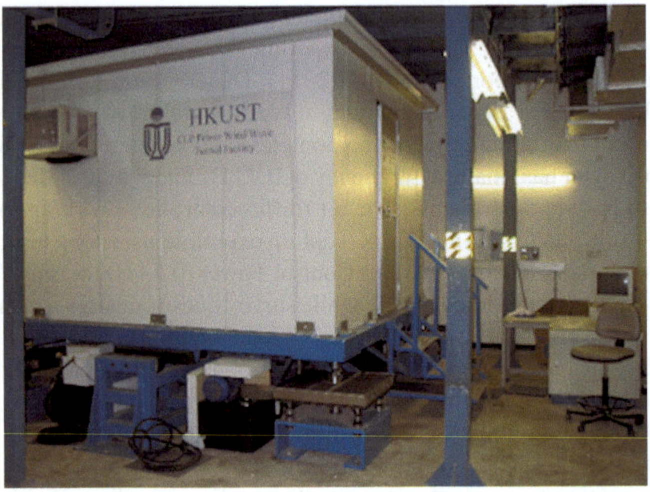

Fig. 12.1 A 3 × 3 m dual axis tall building motion simulator

deviation acceleration of 8.15 milli-g and one small (near zero) amplitude, were used to excite the uni-directional simulated motion of the shake table. In both field experiment and shaker table test setups, cognitive performance tests for both repeated measures tests and prolonged duration tests were conducted. The results collected from these experiments suggested that wind-induced tall building motion has minimal effects on cognitive performance. This would appear to support the rationale that it is not necessary in serviceability acceleration assessment to take into consideration occupant cognitive performance, but the results are far from conclusive. Evidently, measuring cognitive performance is difficult in field experiments and may be more suited to motion simulator experiments.

More recently, Burton et al. (2003, 2006) conducted a series of motion simulator tests on human test subjects using a purpose-built 3 × 3 m dual axis tall building motion simulator, as shown in Fig. 12.1, which is capable of simulating pure and amplitude modulated sine wave and two degree-of-freedom tall building narrow-band random motions with dominant frequencies between 0.1 and 1.2 Hz. In a study of human response to uni-axial horizontal (both lateral and fore-aft) sinusoidal motions at a range of low frequencies (0.15, 0.25, 0.50 and 1.00 Hz) and a constant peak acceleration of 13.5 milli-g, the magnification factor (defined as the head acceleration over the acceleration of the motion simulator) of the head acceleration was found to continuously increase as the frequency of oscillation increases: from about 2 at 0.15 Hz to about 4 at 1.00 Hz. The acceleration of the head is evidently dependent on the frequency of oscillation. Since the head houses the inner ear containing the vestibular organs, the body's primary motion sensors for detecting lateral vibrations, the results suggest that human perception of motion is also dependent on frequency of oscillation.

The results of video image analysis of the test subjects' head movements at a constant simulator peak acceleration of 13.5 milli-g revealed a fairly steady average head displacement of between 20 and 40 mm compared with the peak displacement of the motion simulator: 198, 108, 26 and 6.8 mm for test frequencies of 0.15, 0.25, 0.50 and 1.00 Hz respectively. In wind-excited buildings, the occupants receive cues from their vestibular organs while believing themselves to be stationary relative to the building. However, the observed levels of building motion-induced occupant head movements are sufficient to create an apparent visual cue of building motions when a far-field object such as a tall building or distant scenery is observed against a near-field reference such as a view window or office partition frame. This apparent visual cue often serves as a confirmation and enhances the sensation of building motion.

Burton et al. (2005) conducted motion simulator experiments in which more than 500 test subjects performed manual and cognitive tasks when subjected to sinusoidal, normally distributed or high peak factor waveforms for a range of test frequencies from 0.16 to 0.50 Hz, acceleration amplitudes from 1 to 24 milli-g, and test durations of 12 or 50 min. In addition to demonstrating the frequency dependence of task disruption, the test results showed that test subjects exposed to motions with a normally distributed waveform are more likely to experience significant discomfort, such as nausea, and task disruption than those subjected to motions with a near sinusoidal or high peak factor waveform of similar duration. The shorter 12 min duration test motion, at similar standard deviation accelerations, was found to be less likely to induce discomfort such as nausea or cause task disruption, than the longer 50 min duration test motion. These observations highlight the importance of considering the duration effects of wind events, such as short duration thunderstorms verse long duration synoptic gales and monsoonal winds, when considering the acceptable acceleration level for wind-excited tall buildings. This forms the basis of the wind event/exposure duration based building vibration acceptability criteria proposed by Burton et al. (2007).

In another motion simulator experiment, Burton et al. (2011) investigated the effect of low frequency motion on task performance. Unlike earlier task performance studies that focused on higher frequencies (>1.0 Hz) and higher acceleration levels (>50 milli-g) than those relevant to tall building vibration (e.g. McLeod and Griffin 1989) and the comprehensive studies conducted by Irwin and Goto (1984) on manual dexterity tests in the frequency range of 0.02 to 10.0 Hz, Burton et al. (2011) conducted manual task performance tests in the frequency range of 0.125 to 0.50 Hz and at acceleration levels ranging from 2 to 30 milli-g typically experienced by tall buildings in a short duration strong wind event such as thunderstorm. Fourteen test subjects undertook a manual tracking task in the form of an aim and shoot game in a standing position for 13 randomly presented motion conditions plus one no motion control condition. Overall, the performance, which was measured by completion time and accuracy, was found to exhibit no statistically significant degradation or improvement as a function of frequency and acceleration. It was concluded that a more complex manual task and testing at a higher acceleration level may be necessary to elicit significant degradation in manual task performance. However, even for relatively simple manual tasks such as line tracing and target tracking, a more accurate measurement of performance, and hence a measure of performance degradation or improvement, can be achieved by quantifying error magnitude in terms of degree of excursion from the target and error recovery time in terms of time spent in error zones of different error margins.

Michaels et al. (2013) conducted motion simulator experiments to study occupant comfort and well-being. One experiment was based on independent measures design involving 578 test subjects randomly assigned a test motion condition of bi-directional narrow-band random motion within a matrix of three frequencies, three acceleration levels and three exposure durations. A second experiments was based on repeated measures design involving eight test subjects exposed to 40 min of six different uni-directional motion conditions, including a control no motion condition, randomly presented from a matrix of five frequencies and five levels of acceleration. In addition to rating comfort level using a five-level scale, test subjects also self-assessed their motion sickness susceptibility (MSS). By utilizing a binary logistic regression analysis, predicted probability of discomfort was established as a function of frequency, acceleration, MSS and culture. Not surprisingly, test subjects generally experienced greater discomfort with increasing frequency and acceleration level, and test subjects who reported higher levels of motion sickness susceptibility also felt greater discomfort. Interestingly, test subjects who were susceptible to motion sickness and were Collectivists, who are representative of Asian, African, Central and South American cultures, were more likely to report discomfort than those susceptible Individualists, who are representative of Western cultures. This suggests that societal influences may play a role in defining motion acceptability for and possibly complain behavior of occupants in wind-excited tall buildings.

In summary, significant advances in our understanding of human response to low frequency, low amplitude narrow-band random building motion commonly experienced by occupants in wind-excited tall buildings have been achieved through the considerable amount of research undertaken in the past four decades. One noteworthy finding is human perception threshold at frequencies commonly encountered in tall building vibration. Despite significant differences in test equipment and test methodology used by different researchers, vibration perception thresholds were found to be generally similar and clearly dependent on the frequency of vibrations for both sinusoidal and narrow-band random vibrations, as shown in Fig. 12.2. However, more research efforts utilizing field measurements and survey (e.g. Kijewski-Correa and Pirnia 2009; Lamb et al. 2010, 2012a, 2012b; Selvakumar et al. 2012; Walton et al. 2012) and motion simulator experiments on human test subjects (e.g. Kwok and Wong 2012) are needed to unravel the complex nature of the effects of wind-induced building motion on cognitive performance, task performance, work productivity, occupant well-being and other human responses and behaviors.

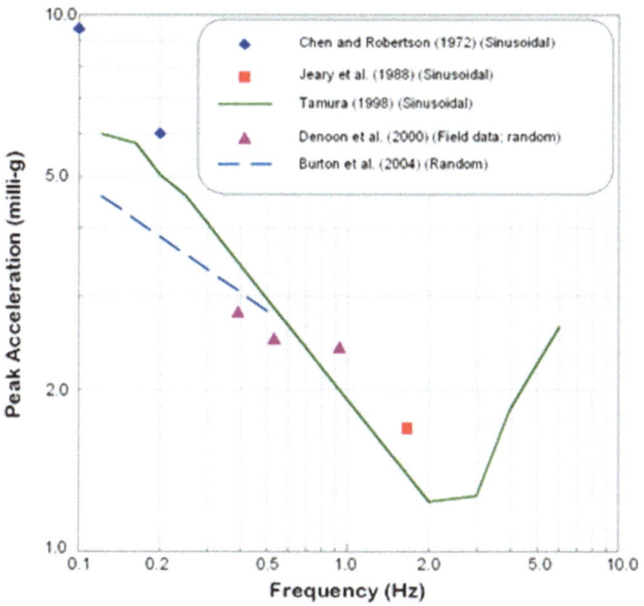

Fig. 12.2 Average vibration perception thresholds for sinusoidal and narrow-band random vibrations (from Kwok et al. 2009)

12.3 Occupant Comfort Serviceability Assessment for Wind-Excited Tall Buildings

In addition to the effects of well-established psychological and physiological factors, which include tactile, vestibular, proprioceptive, kinaesthetic, visual and auditory cues, and visual-vestibular interaction, on human response to building vibration, prior experience, motion expectation, habituation, personality, culture and even job satisfaction also play an important role. This complex mix of diverse factors makes predicting an individual's reaction to building vibration and the assessment of motion acceptability for occupants in wind-excited tall buildings a difficult task. Furthermore, since human perception and tolerance of wind-induced tall building vibration are essentially a subjective assessment, it is not surprising that there are significant differences and uncertainties amongst the few occupant comfort serviceability criteria commonly in use by wind tunnel laboratories and design professionals, which generally reflect country/regional building code requirements, building design professionals' interpretation and preference, and market forces. Kwok et al. (2009) summarised the characteristics and compared the suggested acceptable acceleration levels of these commonly adopted criteria, as shown in Fig. 12.3, including ISO6897: 1984 (International Organization for Standardization 1984), Melbourne and Cheung (1988), Isyumov (1993), AIJ-GEH-2004 (Architectural Institute of Japan Recommendations, 2004), ISO/FDSI 10137:2007 (International Organization for Standardization 2007) and Burton et al. (2007).

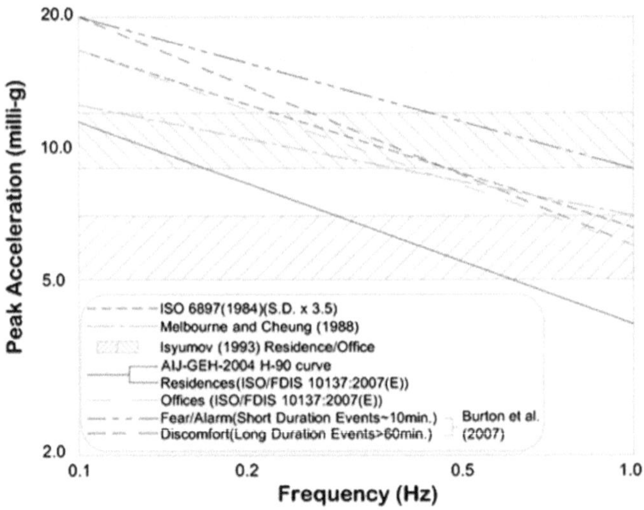

Fig. 12.3 Comparison of occupant comfort serviceability criteria for a 1 year return period wind storm (from Kwok et al. 2009)

By synthesizing laboratory motion simulator investigations (Chen and Robertson 1972; Irwin 1978) and full-scale data (Hansen et al. 1973), Irwin (1978) proposed the design recommendations which led to the development of the ISO6897: 1984 (International Organization for Standardization 1984) guideline for evaluating the acceptability of low-frequency, in the frequency range of 0.063–1.0 Hz, horizontal motion of buildings subjected to wind forces. Important features of the guidelines that establish the benchmark for building motion acceptability are: frequency dependence, acceptable acceleration levels expressed in standard deviation, frequency of occurrence of unacceptable acceleration level expressed in recurrent interval (return period), storm duration, and building/structure types. Although ISO6897: 1984 (International Organization for Standardization 1984) was first published nearly 30 years ago, it remains one of the most comprehensive and popular guidelines currently in use. The guidelines are based on the worst consecutive 10 min of a wind storm with a recurrence interval of 5 years. A conversion factor of 0.72 has been applied, as suggested by the guidelines, to represent a recurrence interval of 1 year for comparison purposes in Fig. 12.3.

Melbourne and Cheung (1988) proposed an occupant comfort criteria based on peak acceleration values. The limiting acceleration values were derived from ISO6897: 1984 (International Organization for Standardization 1984) by applying a frequency and duration dependent peak factor to the standard deviation acceleration values suggested in ISO6897: 1984 (International Organization for Standardization 1984). Melbourne and Palmer (1992) proposed further refinements by including recurrent interval as a variable. The suggested frequency-dependent acceptable peak acceleration values for a recurrence interval of 1 year are shown in Fig. 12.3.

Isyumov (1993) suggested acceptance criteria based on peak acceleration for recurrent intervals of 1 and 10 years. For 1 year return period, two ranges of peak acceleration were specified: 5–7 milli-g for residential buildings and 9–12 milli-g for office buildings. The most obvious difference compared with other criteria is the proposed non-frequency-dependent acceptable acceleration values for two building types. Although applicable within the frequency range of the majority of tall buildings normally encountered, say between 0.15 and 0.30 Hz, these non-frequency-dependent acceptable acceleration values may not be appropriate for buildings with unusually high or low natural frequencies of vibration.

The Architectural Institute of Japan Guidelines for Evaluation of Habitability AIJ-GEH-2004 (Architectural Institute of Japan Recommendations, 2004) offers frequency-dependent guidelines based on probabilistic perception thresholds for evaluating wind-induced building vibration. The basic evaluation curves specify peak accelerations of 1 year recurrence for five levels of perception probability from 10 % to 90 %. The Architectural Institute of Japan Guidelines differ from other occupant comfort serviceability criteria by not specifying an acceptable acceleration level for habitability. Instead, building developers/owners and design professionals need to determine the level of motion perception and occupant comfort that the building will be designed for, probably based on factors including proposed building usage/function, target clientele/tenant and prestige of the building. The 90 % perception probability curve, H-90, is presented in Fig. 12.3 for comparison.

The recently published ISO/FDSI 10137:2007 (International Organization for Standardization 2007) addresses the serviceability design of buildings subjected to wind-induced vibrations based on a recurrence interval of 1 year with two frequency-dependent acceptance curves: one for residences and one for offices. The acceptance curve for residences is similar to the H-90 90 % perception probability curve given in AIJ-GEH-2004 (Architectural Institute of Japan Recommendations 2004). The acceptance curve for offices is factored by 0.7. Since both the AIJ-GEH-2004 (Architectural Institute of Japan Recommendations 2004) and ISO/FDSI 10137:2007 (International Organization for Standardization 2007) are based on probabilistic perception threshold and does not directly address occupant comfort, the acceleration values are, as expected, significantly lower than the acceptable acceleration levels suggested in other criteria.

Burton et al. (2007) proposed wind event/exposure duration based building vibration acceptability criteria, as shown in Fig. 12.3, which recognizes the importance in building design to address occupant fear and alarm as well as occupant comfort during wind events that result in perceptible wind-induced building vibration. Evidently, fear and alarm are primarily caused by wind-induced vibration associated with infrequent extreme wind events such as typhoons and thunderstorms, while discomfort results from sustained and/or frequently occurring vibration associated with long duration regularly occurring wind events such as synoptic gales and monsoonal winds. It was further demonstrated that occupants subjected to a longer duration building vibration are more likely to experience discomfort and hence complain than those subjected to the same vibration magnitude but for a shorter duration. The proposed criteria for long duration wind event/duration is similar in magnitude to peak acceleration values suggested by the ISO6897: 1984 (International Organization for Standardization 1984) guidelines. The fear and alarm criteria for short duration event/exposure were set at higher acceleration levels.

Recognizing the complexity of occupant comfort assessment in wind-excited tall buildings, differences in assessment methodologies current in use, and the variability of acceptable acceleration levels, the recent published wind loading handbook for Australia and New Zealand AWES-HB-001-2012 (Holmes et al. 2012), which serves as a commentary to AS/NZS 1170.2: 2011 Wind Actions (Standards Australia/Standards New Zealand 2011) Wind Actions (Holmes et al. 2012), offers a simple guideline. The handbook suggests that, in general, buildings that frequently exhibit wind-induced vibrations with peak accelerations greater than 10 milli-g are unlikely to be acceptable to most occupants, since a majority of people will perceive motion with a peak acceleration of around 10 milli-g, as shown in Fig. 12.2.

In addition to the deterministic approach adopted by most occupant comfort serviceability criteria, including those described above, a more comprehensive probabilistic approach has been proposed by Kanda et al. (1993), Tamura et al. (2006), Basher and Kareem (2007) and others to consider the uncertainties of a range of governing parameters including design wind speed, building dynamic and aerodynamic properties, and response prediction process, in addition to human response to wind-induced building motion including perception threshold and occupant comfort statistics. These probabilistic evaluation methodologies provide a more advanced and sophisticated platform to assess occupant comfort and evaluate building habitability, and are a practical alternative to the more commonly adopted deterministic approach in the tall building design process.

In summary, a review of current and proposed occupant comfort serviceability criteria and assessment methodologies highlights the major differences between various criteria/guidelines currently in use, in terms of frequency dependency, building types (residences versus offices), assessment methodology (deterministic versus probabilistic), and in particular the acceptable acceleration level for building occupants (perception versus comfort versus fear/alarm). Although deterministic approaches dominate the current assessment of occupant comfort and evaluation of building habitability, most probably due to the ease of application and interpretation and the general acceptance by building developers/owners and design professionals in the assessment processes, probabilistic evaluation methodologies provide a more advanced and sophisticated platform to assess occupant comfort and evaluate building habitability. With appropriate selection, simplification and generalization of the governing parameters and assessment protocol, a probabilistic evaluation approach is a practical alternative in the design of wind-excited tall buildings for occupant comfort. While there has been relatively few major complaints about building performance under wind actions in terms of occupant comfort, these major differences nonetheless cast doubts and uncertainties about the accuracy and the true value of the outcome of the assessment processes and emphasise the need to develop practical occupant comfort serviceability criteria that are internationally applicable and that can be confidently used by building developers/owners and design professionals.

12.4 Occupant Comfort Test Using a Motion Simulator

Previous studies have shown that the majority, up to 80 % or more, of the general population, including building developers/owners and design professionals who are responsible for the design and construction of all tall buildings and structures, has never experienced wind-induced building vibration, either at work or at home. In addition to conducting research on human perception of motion and occupant comfort assessment in wind-excited tall buildings reported in this chapter and elsewhere, tall building motion simulators such as the HKUST's purpose-built dual axis tall building motion simulator shown in Fig. 12.1 have been utilized to conduct specially designed occupant comfort tests for building developers/owners and design professionals to enable them to experience a range of simulated bi-directional narrow-band random wind-induced tall building vibrations typically experienced in wind-sensitive buildings. Modified flight/ship simulators such as the test facility at Centre for Marine Simulations at Memorial University of Newfoundland, Canada, have also been used as a demonstration facility.

Although the objectives of conducting an occupant comfort test may vary, their aims typically involve one or more of the following:

- To experience different levels of simulated wind-induced tall building vibration at different frequencies and amplitudes based on assumed design values for building natural frequencies and structural damping values;
- To assess the acceptability of these vibrations in terms of comfort level and/or frequency of occurrence;
- To assess the need to adopt vibration mitigation measures; and
- To experience and assess the potential benefits of expected reductions in building vibration if selected vibration mitigation measure(s) is adopted.

Furthermore, there are major differences in experimental design and expected outcomes between a research study and an occupant comfort test for demonstration/education purposes. Unlike test subjects who volunteer to undertake tests in the tall building motion simulator for research purposes, this interest group participates in an occupant comfort test purposely to assess the acceptability of predicted accelerations for a specific building. Hence, the participants are preconditioned and have a heightened expectation of motion. Such a test also serves an additional purpose in cases where the building performance exceeds certain occupant comfort serviceability criteria thus necessitates consideration to introduce vibration mitigation measure(s) to bring the building performance to within occupant comfort limit. Some participants, particularly the developers/building owners, are highly market-driven and profit-driven with a primary cost-benefit focus. Hence developers/building owners are very likely to assess the need to adopt costly vibration mitigation measures differently to office workers and/or tall building residents who normally expect a generally comfortable work and living environment which the developers/building owners purported to deliver during the sale and/or lease campaign. This interest group also invariably has a tight timeframe to undertake the tests. Hence, a special test setup and test procedure has to be designed to accommodate this special interest group and to meet the test objectives.

During the occupant comfort test session, a number of test motion conditions, typically six motion conditions including a no-motion control condition are presented in a random order to the participants. Each test motion condition is typically 3–4 min in duration, resulting in a total test session duration of less than 30 min to experience six test motion conditions. An appropriate distraction, such as a presentation on occupant comfort assessment, is delivered for the entire test session such that the participants are distracted from solely sensing motion. Each participant receives a "comfort test questionnaire" to rate their comfort level for each of the test motion conditions experienced during the test session, with ratings ranging from "no motion perceived" to "very uncomfortable". The primary objective of the questionnaire is to gauge the participants' self-assessment of the comfort level of each test motion condition. At the conclusion of the occupant comfort test session, all participants attend a de-briefing session which included a review and discussion of the comfort test questionnaire results.

Based on the results of a number of occupant comfort tests conducted at HKUST by Kwok and Hitchcock (2008), it was found that the percentage of participants who rated a motion test condition as uncomfortable generally decreased as the acceleration magnitude of the test motion condition was decreased, as expected. Evidently, experiencing the accelerations first hand encouraged informed decisions to be made on limiting levels of accelerations expected to be experienced in the subject building. For example, the results from one study suggested that approximately 90 % of the participants found 25 milli-g peak acceleration at a specified frequency of oscillation to be uncomfortable, whereas only about half of the participants found 18 milli-g peak acceleration to be uncomfortable. For lower levels of acceleration (<15 milli-g), more than half of the participants either did not perceive the motion or found the motion to be not uncomfortable.

According to Burton (2006), occupants may tolerate building vibration at sufficiently large amplitudes that cause discomfort provide that it does not occur frequently. Conversely, frequently occurring building vibrations, even at lower perceptible amplitude, may cause greater discomfort. The simulation of expected wind-induced building vibration corresponding to wind speeds of different recurrence intervals, based on wind tunnel test results, may also offer the participants the opportunity to appreciate the relationship between comfort and exposure duration and frequency of occurrence of perceptible building vibration.

By replicating the expected wind-induced building vibration for a specific total damping value (structural damping plus added damping provided by potential vibration mitigation measure), based on wind tunnel test results, participants are able to experience the potential benefits of a reduction in the vibration amplitude and the resultant enhanced occupant comfort. The range of increase in damping value, and the corresponding reduction in wind-induced building vibration, offered by adopting different types or sizes of vibration control devices/strategies can also be assessed. Although conducting occupant comfort tests with a small subject sample is difficult to derive statistically explicit results, motion simulator studies such as the occupant comfort tests described herein encourages building developers/owners and design professionals to make an assessment of the need to adopt vibration mitigation measures and conduct appropriate cost-benefit analysis of available vibration mitigation strategies.

12.5 Etiology of Motion Sickness: Two Theories

Motion sickness has plagued travelers for the last several thousand years (Reason and Brand 1975), and continues to be problematic in a variety of environments such as fun rides, motion simulators, modern video games, and reportedly also in wind-induced high-rise buildings. The typical symptoms of motion sickness consist of headaches, nausea, pallor, and eventually vomiting. Less well-known low-level symptoms of motion sickness are tiredness, fatigue and decrements in hand-eye coordination, often referred to as Sopite Syndrome (Graybiel and Knepton 1976). Despite the long-held concern for the effects of motion sickness, there is no unified theory that explains the etiology of motion sickness. The dominant and widely accepted theory is called Sensory Conflict Theory (Reason and Brand 1975; Reason 1978; Oman 1982), and an alternative theory is called Postural Instability Theory (Stoffregen and Riccio 1991).

Kwok et al. (2010) and Walton et al. (2011) recently reviewed the two theories and examined the role these theories play in occupant response to wind-induced building motion. There are a number of variations of Sensory Conflict Theory, but common to all is the notion that conflicting sensory information results in motion sickness. The main perceptual systems implicated in Sensory Conflict Theory are the visual system (the eyes and the occipital cortex), the vestibular system (otolith organs in the inner ear, responsible for sensing gravity and in the detection of linear accelerations), and non-vestibular proprioceptors (receptors located in the muscles that give the individual a sense of where their limbs are) (Reason and Brand 1975; Reason 1978; Oman 1982). Sensory Conflict Theory suggests that if two or more of these systems are at variance, as determined by a cognitive system that compares all sensory inputs, motion sickness is experienced by the individual. The degree of motion sickness experienced scales to the degree of sensory conflict.

Stoffregen and Riccio (1991) question the approach adopted by Sensory Conflict Theory and offer a comprehensive critique. The three main criticisms are: (1) Sensory Conflict Theory does not adequately predict which environments are nauseogenic; (2) The addition of high-pass and low-pass filters limit the presentation of symptoms to certain ranges of sensory conflict; and (3) The postulation of a cognitive comparative system that must logically be able to determine what information about the world is veridical, or "true", or else the stated conflict cannot logically exist.

Riccio and Stoffregen (1991) offer an alternative theory, Postural Instability Theory. They adopt an ecological approach to psychology where the focus is on the individual's observable interaction with their environment, as opposed to the cognitive approach advocated by Sensory Conflict Theory which concerns "hidden" processes in the mind. Riccio and Stoffregen (1991) define postural stability as the coordinated control of all body segments, and show that the prolonged loss of this coordinated stabilization or control always precedes symptoms of motion sickness. For example, it is well known that drivers are less prone to car-sickness than passengers, which Riccio and Stoffregen (1991) attribute to the driver's control over the car and the associated ability to anticipate vehicle motion, maintaining a higher level of postural control than their passengers.

Low-frequency vibration in the frequency range of natural body sway is thought to cause wave interference and de-stabilize posture. Individuals subject to such environments have been shown to require effort to maintain control over posture resulting in performance degradation (Graybiel and Knepton 1976). Despite these significant criticisms, Sensory Conflict Theory is held by many as a generally accepted theory, and Postural Instability Theory is often overlooked as an alternative theory.

Studies have found that often the only observable symptoms of motion sickness are tiredness and impaired hand-eye coordination (e.g. Graybiel and Knepton 1976). Hence less attention is typically paid to these more subtle symptoms of motion sickness. Occupants of wind-excited tall buildings have been found to display low-level symptoms of motion sickness such as nausea and headaches (e.g. Hansen et al. 1973; Goto 1983). However, there have been no published reports of extreme symptoms such as vomiting, probably because the overwhelming majority of buildings do not exhibit wind-induced motion that is strong enough to elicit motion sickness, very large or extreme building motion is a rare event, and/or motion sickness prone occupants purposely vacating affected buildings in advance of the onset of large or extreme motion or prior to the onset of motion sickness. Nevertheless, wind-excited tall buildings represent a low-dose environment, and induce a corresponding low-response, i.e. typically tiredness and fatigue. These symptoms may be misattributed to reasonably common experiences at work, and therefore are masked by the environment for both occupants and researchers. Postural Instability Theory provides an alternative predictions and understanding about how individuals might behave in low-dose environments. The effects of building sway on occupants needs to be understood by designers in order to mitigate building vibration, thus minimizing the underlying symptoms of motion sickness and the corresponding effects on comfort and well-being and potential degradation of task and cognitive performance.

12.6 Conclusion

Human response to wind-induced building vibration have been investigated by using a variety of test methods including field experiments and surveys, motion simulators and shake tables, and artificially excited buildings. The majority of these investigations utilized motion simulators to study human perception of motion, comfort and well-being, manual task performance and cognitive performance while subjected to low frequency and low acceleration amplitude vibration, mostly as sinusoidal vibration with a limited number of studies utilizing bi-directional narrow-band random vibration typically experienced in wind-excited buildings. Vibration perception thresholds were found to be clearly dependent on the frequency of vibration. Motion simulators have also been utilized to conduct occupant comfort tests for building developers/owners and design professionals to gain an experience of different levels of simulated wind-induced building vibrations and to facilitate an assessment of the acceptability of the building vibration and the need to adopt vibration mitigation measures.

Due to the subjective nature of human perception of motion and tolerance of wind-induced tall building vibration, there are currently no internationally accepted occupant comfort serviceability criteria. A comparison of current and proposed building vibration acceptability and occupant comfort serviceability criteria/guidelines highlighted the significant differences between various criteria/ guidelines and the assessment methodologies currently in use.

The application and relevance of two motion sickness theories: Sensory Conflict Theory and Postural Instability Theory on human response to wind-induced building motion have been discussed. Postural Instability Theory provides new predictions and understanding of building occupant behaviors in low-dose stimulus environments typically found in wind-excited buildings, which may lead to design strategy that aims to minimize the adverse effects of building motion on occupant comfort and well-being and potential degradation of task and cognitive performance.

More research efforts are needed to unravel the complex nature of the effects of wind-induced building motion on cognitive performance, task performance, work productivity, occupant well-being and other human responses and behaviors to facilitate the development of practical occupant comfort serviceability criteria that are internationally applicable and that can be confidently used by building developers/owners and design professionals.

Acknowledgements The author acknowledges the invaluable contributions to research on human perception and tolerance of wind-induced building motion over the past 20 years by his research team, in particular Dr Roy Denoon, Dr Melissa Burton, Dr Peter Hitchcock, Dr Richard Roberts, Ms Marianne Michaels, Professor Vaughan Macefield, Dr Darren Walton, Mr Steve Lamb, Mr Elie Hammam and Mr Kwok Shing Wong. The author also gratefully acknowledges HKUST's Design and Manufacturing Services Facility who designed and built two tall building motion simulators. This research is supported by the Australian Research Council Discovery Project (Project: DP 1096179), Research Grants Council of Hong Kong (Projects: HKUST CA04/05. EG01) and HKUST Research Equipment Funding (REF 2006).

References

Architectural Institute of Japan Recommendations (2004) Guidelines for the evaluation of habitability to building vibration, AIJES-V001-2004, Tokyo

Basher R, Kareem A (2007) Probabilistic performance evaluation of buildings: an occupant comfort perspective. In: Proceedings of 12th international conference on wind engineering, vol 2, Cairns, July 2007, pp 1335–1342

Bouncer TH, Morris RC Tomlinson RW (1980) Perception threshold responses to induced low-frequency motion of tall buildings, Final Report S.R.C. Contract No. GR/B 20938. Plymouth Polytechnic

Burton MD (2006) Effects of low frequency wind induced building motion on occupant comfort. PhD thesis, The Hong Kong University of Science and Technology, Hong Kong

Burton MD, Denoon RO, Roberts RD, Kwok KCS, Hitchcock PA (2003) A motion simulator to investigate wind-induced building motion. In: Proceedings of 11th international conference of wind engineering, Texas, pp 1341–1348

Burton MD, Denoon RO, Kwok KCS, Hitchcock PA (2004) Effects of uni-axial and bi-directional random motion on cognitive performance. In: Proceedings of 11th Australasian wind engineering society workshop, Darwin, July 2004

Burton MD, Kwok KCS, Hitchcock PA, Roberts RD (2005) Acceptability curves derived from motion simulator investigations and previous experience with building motion. In: Proceedings of 10th Americas conference on wind engineering, Baton Rouge, Louisiana, 31 May–4 June 2005, pp 286–287 (Full paper on CD-ROM)

Burton MD, Kwok KCS, Hitchcock PA, Denoon RO (2006) Frequency dependence of human response to wind-induced building motion. J Struct Eng ASCE 132(2):296–303

Burton MD, Kwok KCS, Hitchcock PA (2007) Occupant comfort criteria for wind-excited buildings: based on motion duration. In: Proceedings of 12th international conference on wind engineering, vol 1, Cairns, July 2007, pp 1207–1214

Burton MD, Kwok KCS, Hitchcock PA (November 2011) Effect of low frequency motion on the performance of a dynamic manual tracking task. Wind Struct 14(6):517–536

Chen PW, Robertson LE (1972) Human perception threshold of horizontal motion. J Struct Div ASCE 98(8):1681–1695

Denoon RO (2000) Designing for wind-induced serviceability accelerations in buildings. PhD thesis, The University of Queensland, Australia

Denoon RO, Letchford CW, Kwok KCS, Morrison DL (1999) Field measurements of human reaction to wind-induced building motion. In: Proceedings of 10th international conference on wind engineering, Copenhagen, pp 637–644

Denoon RO, Roberts RD, Letchford CW, Kwok KCS (2000) Field experiments to investigate occupant perception and tolerance of wind-induced building motion, Research Report No. R803. Department of Civil Engineering, University of Sydney, Australia

Denoon RO, Kwok KCS (November 2011) Implications of full-scale motion experience for acceleration design guidelines. Wind Struct 14(6):537–557

Goto T (1983) Studies on wind-induced motion of tall buildings based on occupants' reactions. J Wind Eng Ind Aerodyn 13:241–252

Goto T, Iwasa Y, Tsurumaki H (1990) An experimental study on the relationship between motion and habitability in a tall residential building. In: Proceedings of tall buildings: 2000 and beyond, fourth world congress, Hong Kong, pp 817–829

Goto T, Tsurumaki H, Inoue J (July 1996) Effects of visual views on the perception threshold of building motion. Jap Informal Group Hum Response Vib Shiraoi 13–14:1–15

Goto T, Shindo S (1999) Effect of window view on the human perception of tall building motion. In: Proceedings of the ICTB, Kuala Lumpur

Graybiel A, Knepton J (1976) Sopite syndrome: a sometimes sole manifestation of motion sickness. Aviat Space Environ Med 47:873–882

Hansen RJ, Reed JW, Vanmarcke EH (1973) Human response to wind-induced motion of buildings. J Struct Div ASCE 99(ST7):1589–1605
Holmes JD, Kwok KCS, Ginger JD (2012) wind loading handbook for Australia and New Zealand background to AS/NZS1170.2 wind actions. AWES-HB-001-2012. Australasian Wind Engineering Society, 109 pp
International Organization for Standardization (1984) Guidelines for the evaluation of the response of occupants of fixed structures, especially buildings and offshore structures, to low-frequency horizontal motion (0.063 to 1.0 Hz), ISO 6897. International Organization for Standardization, Geneva
International Organization for Standardization (2007) Bases for design of structures—serviceability of buildings and walkways against vibration. Final Draft International Standard ISO/FDIS 10137:2007(E). International Organization for Standardization, Geneva
Irwin AW (1978) Human response to dynamic motion of structures. Struct Eng 56A(9):237–243
Irwin AW (1981) Perception, comfort and performance criteria for human beings exposed to whole body pure yaw vibration and vibration containing yaw and translational components. J Sound Vib 76(4):481–497
Irwin AW (1983) Relative influence of noise and whole-body vibration on the response of humans, Internoise 83. In: Proceedings of international conference on noise control engineering, vol II, Edinburgh, 13–15 July 1983, pp 921–924
Irwin AW (1984) Human perception, task performance and simulator sickness in single and multi-axis low-frequency horizontal linear and rotational vibration, United Kingdom informal group meeting on human response to vibration, Edinburgh, 21–22 September 1984, pp 289–313
Irwin AW (1985) Assessment methods and case studies on human response to combined influences of noise and mechanical vibration, United Kingdom informal group meeting on human response to vibration, Derby, 17–19 September 1985
Irwin AW (1988) Human response to vibration of structures—criteria and assessment, seminar on vibration control, Eger, 29 September–1 October 1988
Irwin AW, Goto T (1984) Human perception, task performance and simulator sickness in single and multi-axis low-frequency horizontal linear and rotational vibration, United Kingdom informal group meeting on human response to vibration, Edinburgh, 21–22 September 1984, pp 289–313
Isyumov N (1993) Criteria for acceptable wind-induced motions of tall buildings. In: Proceedings of international conference on tall buildings, CTBUH, Rio De Janeiro, 17–19 May 1993
Isyumov N, Mascantonio A, Davenport AG (1988) Measured building motions of tall buildings in wind and their evaluation. In: Proceedings of symposium/workshop on serviceability of buildings (movements, deformations, vibrations), Ottawa, May 1988, pp 181–199
Isyumov N, Kilpatrick J (1996) Full-scale experience with wind-induced motions of tall buildings. In: Proceedings of 67th regional conference council on tall buildings and urban habitat, Chicago, 15–18 April 1996, pp 401–411
Jeary AP, Morris RG, Tomlinson RW (1988) Perception of vibration—tests in a tall building. J Wind Eng Ind Aerodyn 29:361–370
Kanda J, Tamura Y, Fujii K (1988) Probabilistic criteria for human perception of low-frequency horizontal motions. In: Proceedings of symposium/workshop on serviceability of buildings, Ottawa, pp 260–269
Kanda J, Tamura Y, Nakamura O, Uesu K (1993) Probabilistic criteria for serviceability limit of wind response. In: Proceedings of international colloquium, structural serviceability of buildings, IABSE Reports, vol 69, Goteborg, pp 59–66
Kanda J, Tamura Y, Fujii K (1990) Probabilistic perception limits of low-frequency horizontal motions, conference with international participation, serviceability of steel and composite structures proceedings, Pardubice, pp 61–72
Khan FR, Parmelee RA (1971) Service criteria for tall buildings for wind loading. In: Proceedings of 3rd international conference on wind effects on buildings and structures, Tokyo, pp 401–407

Kijewski-Correa T, Pirnia D (2009) "Pseudo-full-scale" evaluation of occupant comfort in tall buildings. In: Proceedings of 11th Americas conference on wind engineering, San Juan, Puerto Rico, June 2009
Kwok KCS, Macdonald PA (1990) Wind-induced response of Sydney Tower. Eng Struct 12(3):153–162
Kwok KCS, Hitchcock PA (2008) Occupant comfort test using a tall building motion simulator. In: Proceedings of 4th international conference on advances in wind and structures, Jeju, 28–30 May 2008 (Full paper on CD-ROM)
Kwok KCS, Hitchcock PA, Burton MD (2009) Perception of vibration and occupant comfort in wind-excited tall buildings. J Wind Eng Ind Aerodyn 97:368–380
Kwok KCS, Michaels MN, Lamb S (2010) Human perception and tolerance of wind-induced building motion. In: Proceedings of 6th international symposium on environmental effects on buildings and people: actions, influences, interactions, discomfort, Krakow, 11–13 October 2010, pp 27–40
Kwok KCS, Wong KS (2012) A dual-axis tall building motion simulator to investigate effects of wind-induced building motions on human. In: Proceedings of 15th Australasian wind engineering society workshop, Sydney, February 2012, pp 77–80
Lamb S, Kwok KCS, Walton D, Carpenter P (2010) A longitudinal approach to measuring occupant comfort and productivity in wind-excited tall buildings. In: Proceedings of 14th Australasian wind engineering society workshop, Canberra, 5–6 August 2010, pp 9–12
Lamb S, Kwok KCS, Walton D (2012) Experience with wind-induced building motion in Wellington, New Zealand: motion sickness, compensatory behaviours and work location preferences". In: Proceedings of 15th Australasian wind engineering society workshop, Sydney, February 2012, pp 73–76
Lamb S, Walton D, Kwok KCS (2012) The effects of exposure to wind-induced tall building motion on occupant comfort in Wellington, New Zealand. In: Proceedings of 1st international conference on performance-based and life-cycle structural engineering, Hong Kong, 5–7 December 2012
Lee BE (1983) The perception of the Wind-induced vibration of a tall building—a personal viewpoint. J Wind Eng Ind Aerodyn 12:379–384
McLeod RW, Griffin MJ (1989) A review of the effects of translational whole-body vibration on continuous manual control performance. J Sound Vib 133(1):55–115
Melbourne WH, Cheung JCK (1988) Designing for serviceable accelerations in tall buildings. In: Proceedings of 4th international conference on tall buildings, Hong Kong and Shanghai, pp 148–155
Melbourne WH, Palmer TR (1992) Accelerations and comfort criteria for buildings undergoing complex motions. J Wind Eng Ind Aerodyn 41–44:105–116
Michaels MN, Kwok KCS, Tung YK (2013) Exploratory analyses and modelling of parameters influencing occupant behaviour due to low frequency random building motion. J Wind Eng Ind Aerodyn 115:82–92
Morris RC, Dennis I, Tomlinson RW, Clarke A (1979) Vibration in tall buildings: performance tests, B.R.E. Contract Report. Plymouth Polytechnic, UK
Noguchi K, Hiwatashi A, Kobayashi A, Tsujita O, Goto T (1993) Human response to horizontal motion of tall buildings. In: Proceedings of international colloquium, structural serviceability of buildings, IABSE Reports, Goteborg, vol 69, Sweden, pp 53–58
Oman CM (1982) A heuristic mathematical model for the dynamics of sensory conflict and motion sickness. ACTA Oto-laryngologica Suppl 392:1–44
Reason JT (1978) Motion sickness adaptation: a neural mismatch model. J R Soc Med 71:819–828
Reason JT, Brand JJ (1975) Motion sickness. Academic, London
Riccio GE, Stoffregen TA (1991) An ecological theory of motion sickness and postural instability. Ecol Psychol 3:195–240

Selvakumar K, Kwok KCS, Lamb S (2012) Assessment of occupant response in wind-excited tall buildings. In: Proceedings of 15th Australasian wind engineering society workshop, Sydney, February 2012, pp 69–72

Shioya K, Kanda J, Tamura Y, Fujii K (1992) Human perception thresholds of two-dimensional horizontal motion. In: Proceedings of the ASCE structures congress, San Antonio, 13–15 April 1992, pp 480–483

Shioya K, Kanda J (1993) Human perception thresholds of horizontal motion. In: Proceedings of international colloquium goteborg, structural serviceability of buildings, IABSE Reports, vol 69, 1993, pp 45–52

Standards Australia/Standards New Zealand (2011) Structural design actions Part 2: wind actions AS/NZS 1170.2:2011, Sydney

Stoffregen TA, Riccio GE (1991) An ecological critique of the Sensory Conflict theory of motion sickness. Ecol Psychol 3:159–194

Tamura Y (1998) Application of damping devices to suppress wind-induced responses of buildings. J Wind Eng Ind Aerodyn 74–76:49–72

Tamura Y, Kawana S, Nakamura O, Kanda J, Nakada S (2006) Evaluation perception of wind-induced vibration in buildings. In: Proceedings of institution of civil engineers, structures & buildings, vol 159(SB5) October 2006, pp 283–293

Walton D, Lamb S, Kwok KCS (2011) Low dose motion sickness and wind-excited tall buildings. Wind Struct 14(6):499–515

Walton D, Burton MD, Kwok KCS, Lamb S, Merwood Y (2012) Motion sickness sensitivity and its relationship to occupant comfort due to tall building motion sway. In Proceedings of 1st international conference on performance-based and life-cycle structural engineering, 5–7 December 2012, Hong Kong

Chapter 13
Damping in Buildings and Estimation Techniques

Yukio Tamura

Abstract This chapter first discusses physical causes of damping in buildings and dynamic properties of buildings, especially amplitude dependency of damping. It is emphasized that there is no evidence of increasing damping ratio in the very high amplitude range within the elastic range of main frames, unless there is damage to secondary members or cladding. Then, damping predictors are introduced based on the Japanese Damping Database, and design values for damping ratio are recommended. Next, damping estimation techniques are introduced and some points to note in damping estimation are discussed. The feasibility and efficiency of two simple and user-friendly but accurate damping estimation techniques are discussed. One is the Frequency Domain Decomposition technique, which uses Singular Value Decomposition of the Power Spectral Density matrix of multiple outputs, and the other is the Multi-mode Random Decrement technique. Both can be applied to ambient excitations such as micro-tremors, enabling easy handling of closely-spaced and even repeated modes. Some full-scale examples demonstrating the damping estimation efficiency of both techniques are also shown.

Keywords Ambient vibration • Amplitude dependency • Damping ratio • Frequency domain decomposition (FDD) • Random decrement (RD) technique

Y. Tamura (✉)
Wind Engineering Research Center, Tokyo Polytechnic University, 1583 Iiyama, Atsugi, Kanagawa, Japan
e-mail: yukio@arch.t-kougei.ac.jp

Y. Tamura and A. Kareem (eds.), *Advanced Structural Wind Engineering*, DOI 10.1007/978-4-431-54337-4_13,

13.1 Introduction

In order to evaluate the responses of buildings, their dynamic characteristics such as natural frequencies, mode shapes and damping ratios should be accurately known. It is well known that a structure can be damped by mechanisms with different internal and external characteristics: friction between atomic/molecular or different parts, impacts, air/fluid resistance, and so on. Combination of different phenomena results in various types of damping. Vibration energy is finally dissipated as thermal energy, or is radiated outside the boundary of the system. Damping in buildings can not only reduce wind- or earthquake-induced vibrations, but also increase the onset velocity for aeroelastic instability. Generally, their mathematical descriptions are quite complicated, and not suitable for vibration analysis of complicated structures. In structural dynamics, damping is described by viscous, hysteretic, coulomb or velocity-squared models. Viscous damping occurs when the damping force is proportional to the velocity, and is mathematically convenient because it results in a linear second order differential equation for engineering structures. Transient decay of a viscously under-damped system is exponential. Fortunately, damping in buildings can be approximated by the viscous damping formula, even if there is no viscous material in it.

As there is no theoretical method for estimating damping in buildings, it should be estimated from full-scale data, which shows significant dispersion (Haviland 1976). Damping is the most important but most uncertain parameter affecting the dynamic responses of buildings. This uncertainty significantly reduces the reliability of structural design for dynamic effects. There are many potential factors influencing dispersion of full-scale damping data including:

- Soil types
- Foundation types
- Structural materials
- Joint types
- Structural frame systems
- Exterior walls
- Interior walls
- Cladding
- Other non-structural members
- Vibration amplitude

These can be taken as parameters for defining the damping value. Although it is very difficult, as shown in Tamura et al. (2000), if full-scale damping data were appropriately categorized based on these parameters, dispersion might be minimized. Thus, these are not necessarily essential sources of dispersion.

One of the major reasons for dispersion of full-scale data is inappropriate use of damping evaluation techniques. If a technique is not appropriate, the evaluated damping ratio would be quite different from the true value (Tamura 2012). However, measurement and evaluation techniques cannot be parameters for the damping value, and if the measurement and evaluation techniques are appropriately chosen and buildings and vibration conditions are efficiently classified, as discussed in Sect. 13.5, the dispersion of full-scale damping data will be minimized.

13.2 Physical Causes of Damping in Buildings

There are various physical causes of damping in buildings as follows (AIJ 2000):

- Internal friction damping
  Energy dissipation due to internal friction of solid materials
- Plasticity damping
  Energy dissipation due to plasticity of solid materials
- External friction damping
  Energy dissipation due to friction between solids (joints, contact surfaces, friction dampers)
- Radiation damping
  Energy transfer and radiation due to soil–structure interaction
- External viscosity damping
  Energy dissipation due to viscosity of liquids such as oil dampers
- Aerodynamic damping
  Aerodynamic interaction between building and air
- Others
  Damping devices, Vibration control

The majority of total damping of general buildings without any damping devices is radiation damping due to soil-structure interaction (SSI) and external friction damping (EFD) due to friction between contact surfaces of elements such as frames, secondary members and cladding.

13.3 Amplitude Dependency of Dynamic Properties of Buildings

13.3.1 General Tendency

Figures 13.1 and 13.2 show typical amplitude dependency of dynamic properties obtained by the Random Decrement (RD) technique. Figure 13.1 is a case of a steel-framed observatory tower 99-m high. It has a 16.5-m square cross section in its lower third, a 12.1-m square cross section in its middle third and an octagonal section in its upper third. The reinforced concrete foundation is supported by caisson-type piles. The exterior walls are made of glass-fiber-reinforced concrete panels (Tamura and Suganuma 1996). It is clearly demonstrated that the fundamental natural frequency decreases and the damping ratio increases with tip acceleration amplitude (Tamura et al. 2000). The increasing tendency ends at around $a_H = 46$ mm/s^2. Considering the natural frequency around $f_1 = 0.66$ Hz, the tip displacement is roughly estimated by $x_H \approx a_H/(2\pi f_1)^2$ as 3 mm. Thus, the tip drift ratio x_H/H is only 3×10^{-5}, which is very small. Another example shown in

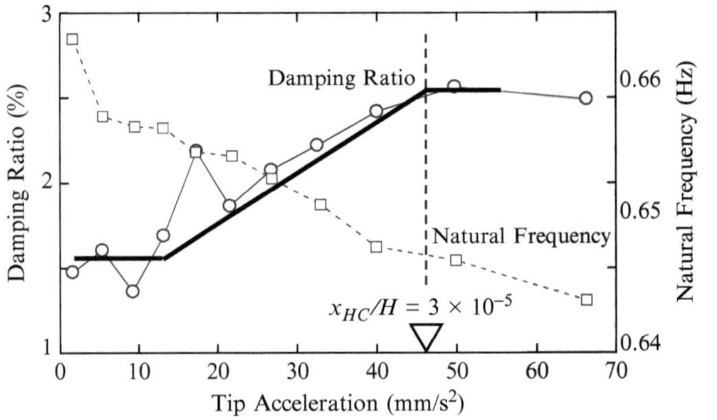

Fig. 13.1 Dynamic properties of 99-m high steel observatory building (Tamura et al. 2000; Tamura 2012)

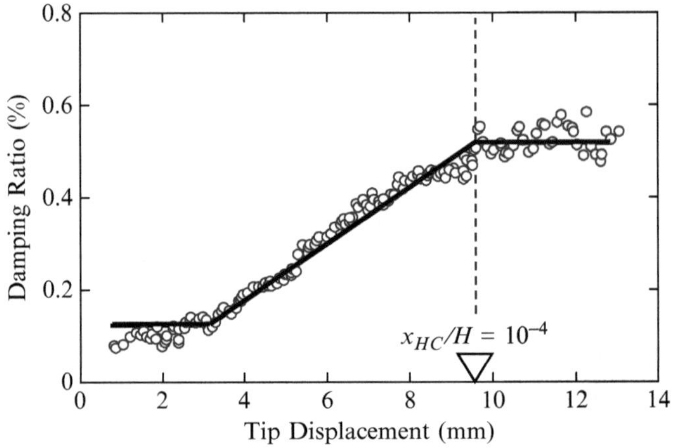

Fig. 13.2 Dynamic properties of 100-m high steel building obtained by RD technique (Jeary 1998)

Fig. 13.2 is a case of a 100-m high steel building (Jeary 1998). Careful observation of Fig. 13.2 also shows the same tendency as Fig. 13.1. Jeary (1998) pointed out the following three parts: a low-amplitude plateau, a mid-amplitude slope, and a high-amplitude plateau. It should be noted that the mid-amplitude slope where the damping ratio increases with amplitude ends at a relatively small amplitude level of around 11 mm in terms of the tip displacement, x_H. As the building height H is 100 m, the tip drift ratio x_H/H is only around 10^{-4}. Figure 13.2 implies that damping ratio does not necessarily increase at higher amplitude levels beyond the amplitude range shown in the figure.

The tip drift ratio where the damping increasing tendency ends is defined as "critical tip drift ratio, x_{HC}/H" by Tamura (2012). It should also be noted that the damping ratio seems to begin to decrease in Figs. 13.1 and 13.2 at the

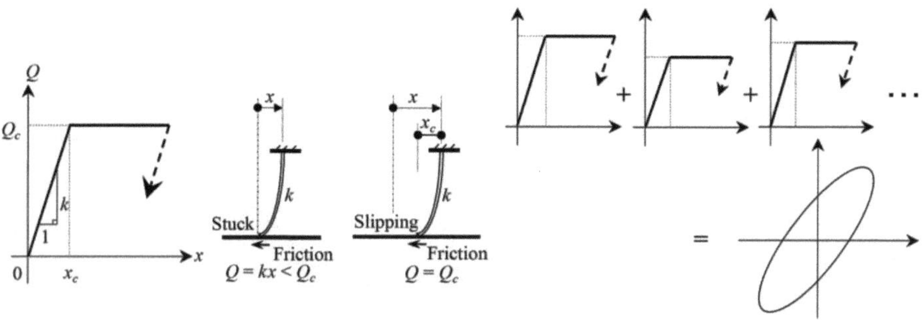

Fig. 13.3 Stick-slip model for damping in buildings and viscous damping model (Tamura 2012)

high-amplitude plateau. The damping ratio does not increase any more with amplitude in the amplitude range above the critical tip drift ratio.

Before reaching the critical tip drift ratio, the number of slipping contact surfaces increases with amplitude, so the damping ratio due to friction damping increases as the natural frequency decreases, as shown in Fig. 13.1. This is expressed by the term "Stick-Slip" in Davenport and Hill-Carroll (1986). This was originally evolved by Wyatt (1977) expressed by the term "STICTION" as the most realistic model of the physical EFD mechanism for built-up structures. STICTION represents a STuck frICTION element, in which a contact surface between members does not move in the very low amplitude range, but begins to slip at a particular amplitude, thus losing its stiffness as illustrated in Fig. 13.3.

Stick-slip contact surfaces are randomly distributed with respect to amplitude, and the integrated effect of a lot of different friction damping elements causes the building to respond as if it had a linear viscous damping system. Therefore, from a practical point of view, equivalent viscous damping, which models the overall damped behaviour of a structural system as being viscous, is often adopted in structural dynamics and is thought to be appropriate.

13.3.2 Damping Ratio at High Amplitude Level

There is not enough data on damping ratio at a high amplitude level, and it is difficult to show general characteristics of dynamic properties in a very high amplitude regime.

However, an example of dynamic characteristics in a very high amplitude level is shown in Fig. 13.4 (Fukuwa et al. 1996), in which tendencies similar to Figs. 13.1 and 13.2 are recognized in the natural frequency and damping ratio. The building is a three-storey steel-framed experimental house with autoclaved lightweight concrete (ALC) panel slabs, ALC panel exterior walls, and plaster board partitions. The ALC panel exterior walls, with a thickness of 100 mm, were designed not to provide shear stiffness under horizontal loads. They are connected to the frames at their top

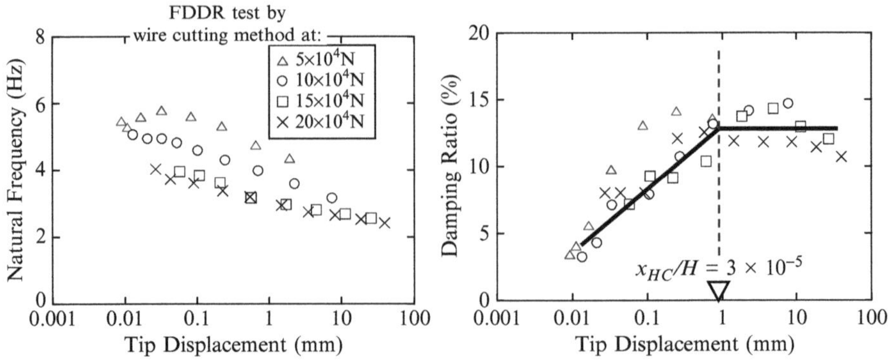

Fig. 13.4 Variations of dynamic properties of 3-storey building with amplitude (Fukuwa et al. 1996; Tamura 2012). (**a**) Natural frequency; (**b**) Damping ratio

only to permit a rocking type movement, although silicone sealing compounds fill the gaps between panels. This exterior wall system contributes significantly to structural damping, so it cannot be said that the results demonstrate the typical dynamic characteristics of general buildings. As the building height is only about 10 m, the effect of SSI is very significant. Additionally, excitation was made up to 4×10^{-3} in terms of the tip drift ratio, and the maximum acceleration was almost 1,000 milli-g. The maximum amplitude was large enough to damage the plaster board partitions, and cracks appeared not only in the ALC slabs but also in the ALC exterior walls. Under the above conditions, the estimated damping ratio was large. However, it reached a maximum value at around 0.3 mm tip displacement. Thus, the "critical tip drift ratio" is estimated at 3×10^{-5} (Fig. 13.4b), which is consistent with the previous results in Figs. 13.1 and 13.2.

Figure 13.5a–c show another very interesting result obtained for a 200-m high steel framed office building with a Hybrid Mass Damper (HMD) installed at its top. Before occupation of this building, the HMD was used as an exciter, and relatively large amplitude vibration tests were conducted (Okada et al. 1993). Figure 13.5a shows an example of a free decay damped response (FDDR) after a sudden stop of excitation by the HMD. From this damped-free oscillation record, step-by-step evaluations of damping ratio and natural frequency were made (Okada et al. 1993). Obtained fundamental natural frequency and damping ratio are shown in Fig. 13.5b, c (after Okada et al. 1993). The fundamental natural frequency, $f_1 \approx 0.26$ Hz, decreases with velocity amplitude and saturates at around 30 mm/s. The damping ratio increases in the low-amplitude regime, and reaches its maximum at around the response velocity $v_H = 6$ mm/s. Thus, the tip displacement $x_H \approx v_H/(2\pi f_1)$ is roughly estimated at 4 mm. The critical tip drift ratio is estimated at $x_{HC}/H = 2 \times 10^{-5}$, which is also very small. In the higher amplitude regime greater than this critical tip drift ratio, the damping ratio clearly decreases with amplitude.

Even after the damping ratio reached its maximum value with increasing amplitude, the natural frequency still decreased, as shown in Figs. 13.1, 13.4a and

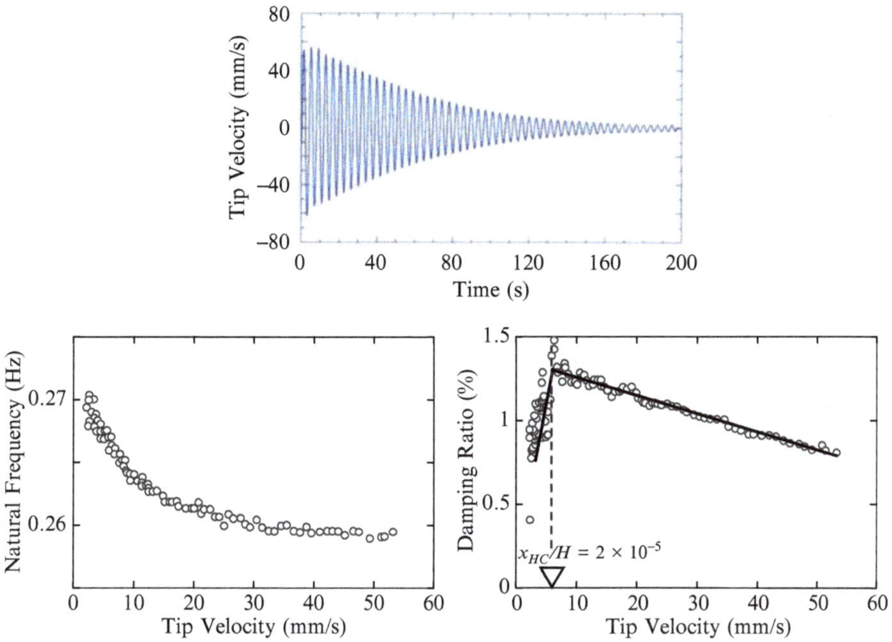

Fig. 13.5 Variation of fundamental natural frequency and damping ratio of 200-m high steel-framed office building (Step by step evaluation from FDDR, based on Okada et al. 1993). (**a**) FDDR; (**b**) Fundamental natural frequency; (**c**) Damping ratio

13.5b. This is also reasonable, because the damping ratio can decrease even when the number of slipping contact surfaces increases, thus reducing the stiffness, if the increasing rate of amplitude is more significant than that of the number of slipping contact surfaces.

13.3.3 Critical Tip Drift Ratio

The critical value of tip drift ratio representing the maximum damping ratio in the non-damaged situation has been defined as "critical tip drift ratio" by Tamura (2012). Table 13.1 summarizes the critical tip drift ratios, x_{HC}/H, obtained for the four buildings shown in Figs. 13.1, 13.2, 13.4 and 13.5. It is very interesting that those ratios for buildings with heights ranging from 10 to 200 m show similar values ranging from 2×10^{-5} to 1×10^{-4}. Although it is dangerous to make a general conclusion, we should note that the damping ratio of a building does not necessarily increase with amplitude continuously, but reaches its maximum at a very low amplitude level.

Table 13.1 Critical tip drift ratio

Buildings	Frame	Height: H (m)	Critical tip drift ratio: x_{HC}/H
Observatory building	Steel	99	3×10^{-5}
Office building	Steel	100	1×10^{-4}
Office building	Steel	200	2×10^{-5}
Residential house	Steel	10	3×10^{-5}

13.3.4 Decrease of Damping at Very High Amplitude Level

One example shown in Fig. 13.5c clearly indicates, and all other examples shown in Figs. 13.1, 13.2, and 13.4b suggest, a "decreasing tendency" of damping at the higher amplitude level. The reason for this is not difficult to understand based on the stick-slip model (Tamura 2012). As already mentioned, increasing amplitude increases the number of slipping contact surfaces, and the damping ratio increases with amplitude. However, after reaching an amplitude where all contact surfaces slip, the total friction force at the slipping contact surfaces causing friction damping does not increase any more with amplitude, i.e., the total friction force remains constant with increasing amplitude. For a simple forced vibration system with a constant damping force F, i.e. Coulomb damping, a steady state amplitude A and a circular natural frequency ω, the equivalent viscous damping coefficient C can be simply expressed as Eq. (13.1):

$$C = \frac{4F}{\pi A \omega} \tag{13.1}$$

The equivalent damping ratio ζ is given as Eq. (13.2):

$$\zeta = \frac{C}{2M\omega_0} = \frac{2F}{\pi A M \omega \omega_0} \tag{13.2}$$

Thus, if the friction damping force F is constant, the "damping ratio" decreases with increasing amplitude A. There has been a misunderstanding that the damping ratio increases with amplitude. However, that is not correct. It decreases with amplitude from the "critical tip drift ratio", which is very small, being in the order of 10^{-5} to 10^{-4}.

There is a possibility of damage to secondary members such as non-load-bearing walls, interior partitions, slabs, and cladding in the large amplitude range. However, if those kinds of damage are "not allowed" in allowable-stress-level design, as in many current codes, a high damping ratio cannot be expected even near the elastic limit of frames.

13.4 Damping Database and Design Value Recommendations

13.4.1 Dynamic Properties of Buildings Based on Japanese Damping Database

Data of dynamic properties of buildings, including damping ratios, have been collected by the Committee on Damping Data organized by the Architectural Institute of Japan in 1993 (Tamura et al. 2000; AIJ 2000). Reliable high-quality full-scale damping data on buildings for various conditions were accumulated from about 40 organizations. Additional damping data have been collected from many journals and proceedings published mainly after 1970. Then, the Committee conducted questionnaire studies by mail to literature authors, designers and building owners to confirm the measured values, and to obtain additional important information, especially on measurement and evaluation techniques and vibration amplitudes as well as detailed building information. Data were excluded if their amplitudes were not clear or could not be deduced from the measuring methods. Consequently, many original non-published data were collected and additional information was added to the published data.

Table 13.2 summarizes information contained in the Japanese Damping Database (Tamura et al. 2000; AIJ 2000). The database includes building information and dynamic properties for 137 steel buildings (Steel), 43 steel-encased reinforced concrete buildings (SRC), 25 reinforced concrete buildings (RC), and 80 tower-like non-building structures, making a total of 285 buildings and structures.

The tip drift ratios x_H/H are approximately in the following ranges:

RC building:	5×10^{-7}–2×10^{-5}
SRC buildings:	10^{-7}–10^{-2}
Steel buildings:	10^{-7}–10^{-4}

Although the upper bound of the amplitude range for SRC buildings is relatively large, the tip drift ratios of the majority of the data are less than 2×10^{-5}.

Figure 13.6 shows relations between the fundamental natural period $T_1(\mathrm{s}) = 1/f_1$ and the building height H (m) (Sasaki et al. 1997; Satake et al. 2003). Figure 13.6a, b suggest a fundamental natural period T_1 proportional to building height H as follows:

$$T_1 = 0.015H \quad \text{for RC/SRC buildings} \tag{13.3a}$$

$$T_1 = 0.020H \quad \text{for Steel buildings} \tag{13.3b}$$

As is well known, Ellis (1980) proposed a simple approximation for the fundamental natural frequency $f_1 = 46/H$; thus, the natural period $T_1 = 0.022H$ for all structural types of buildings, which is slightly longer than those given by Eqs. (13.3a) and (13.3b).

Table 13.2 Contents of the Japanese Damping Database as of 1999

Number of buildings and structures			
285			
Steel buildings (Steel) 137	Steel encased reinforced concrete buildings (SRC) 43	Reinforced concrete buildings (RC) 25	Tower-like non-building structures 80
15.5 m $\leq H \leq$282.3 m (Ave. 101 m)	11.6 m $\leq H \leq$167.4 m (Ave. 60 m)	10.8 m $\leq H \leq$129.8 m	9.1 m $\leq H \leq$226.0 m (Ave. 124 m)
Office: 99 Hotel: 25 Others: 13	Apartment: 35 Others: 20 School: 4 Others: 9		Chimney: 26 Lattice: 24 Tower: 23 Others: 7
Contained information			
Building information	Location Time of completion Building usage Shape Height Dimensions Number of stories	Structural type Cladding type Foundation type Embedment depth Length of foundation piles Soil conditions Reference	
Dynamic properties	Damping ratio (up to the 6th mode) Natural frequency (up to the 6th mode) Time of measurement	Excitation type Experimental & measurement method Evaluation technique Amplitude etc.	

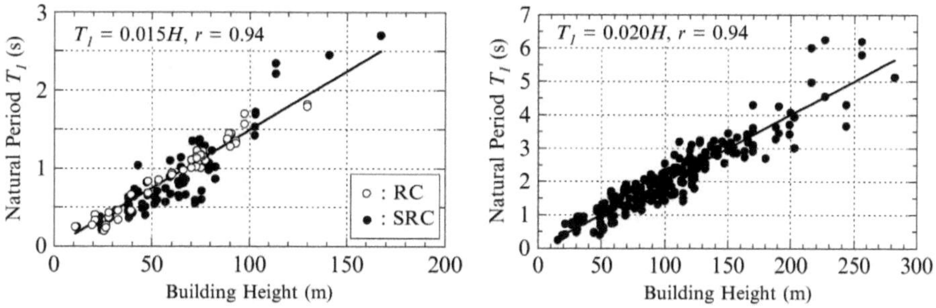

Fig. 13.6 Fundamental natural periods and building heights for data in Japanese Damping Database (Sasaki et al. 1997; Satake et al. 2003). (**a**) RC/SRC Buildings; (**b**) Steel Buildings

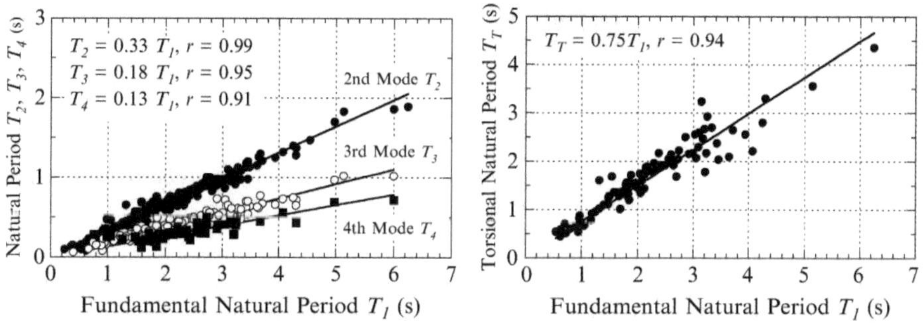

Fig. 13.7 Relation of natural periods of higher translational modes and torsional mode with fundamental natural period for Steel buildings in Japanese Damping Database (Sasaki et al. 1997; Satake et al. 2003) (**a**) Higher Translational Modes; (**b**) Torsional Mode

Higher mode natural periods are proportional to the 1st mode natural period (Sasaki et al. 1997; Satake et al. 2003) as shown in Fig. 13.7a, b for Steel buildings. Thus, the natural frequencies of the higher modes and the torsional mode are well expressed by the fundamental natural frequency f_1 as shown in Table 13.3 for Steel buildings.

Satake et al. (1997) reported that the measured fundamental natural periods are generally 20 % smaller than their design values for Steel buildings, as shown in Fig. 13.8 for the low-amplitude regime. Shioya et al. (1993) reported almost the same result for thirteen RC apartment buildings whose heights are in the range of 60.05–113.6 m. Measurements were basically made under low-amplitude conditions, in which almost all contact surfaces were stuck and the secondary members contributed to stiffness. On the other hand, the design values were estimated based on Finite Element Method (FEM) models considering only main frames, and it is difficult to include the contributions of secondary members. The difference between the design values and the full-scale values of the natural periods in the low-amplitude regime may be attributed to the contributions of secondary members such as furring strips, studs, partitions, and cladding.

Table 13.3 Higher mode natural frequencies

Mode	2nd mode	3rd mode	4th mode	Torsion
Steel buildings	$f_2 = 3.05\,f_1$	$f_3 = 5.46\,f_1$	$f_4 = 7.69\,f_1$	$f_T = 1.34\,f_1$
RC buildings	–	–	–	$f_T = 1.36\,f_1$

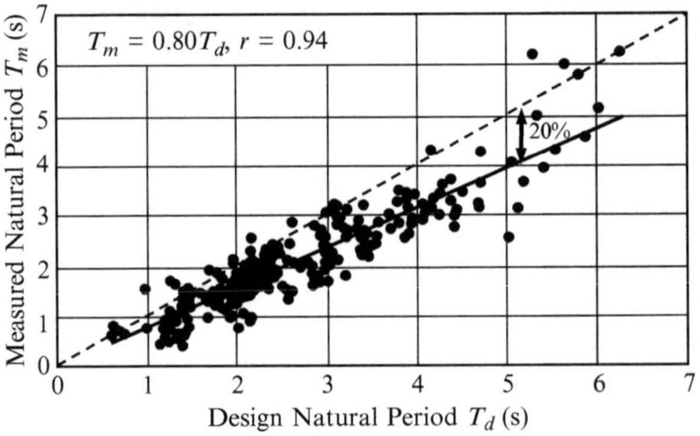

Fig. 13.8 Difference between measured and design values of natural periods based on Japanese Damping Database (Tamura et al. 2000; Satake et al. 2003)

Consequently, the design natural periods are 20 % higher than the measured values, as shown in Fig. 13.8. Therefore, the natural period approximated by Eqs. (13.3a) and (13.3b) shall be considered as the value for the low-amplitude regime.

13.4.2 Proposed Natural Frequencies Based on Japanese Damping Database

Based on the above results shown in Figs. 13.6 and 13.8, Tamura et al. (2000) proposed fundamental natural frequencies for the low amplitude habitability design level and the large amplitude safety design level as follows:

$$f_1 = \begin{cases} \dfrac{1}{0.015H} = \dfrac{67}{H} \quad \text{(for habitability level)} \quad \text{(a)} \\ \dfrac{1}{0.018H} = \dfrac{56}{H} \quad \text{(for safety level)} \quad \text{(b)} \end{cases} \quad : \text{RC Buildings} \qquad (13.4)$$

$$f_1 = \begin{cases} \dfrac{1}{0.020H} = \dfrac{50}{H} \quad \text{(for habitability level)} \quad \text{(a)} \\ \dfrac{1}{0.024H} = \dfrac{42}{H} \quad \text{(for safety level)} \quad \text{(b)} \end{cases} \quad : \text{Steel Buildings} \qquad (13.5)$$

For the large amplitude safety level design, almost all contact surfaces are assumed to be slipping, and the natural frequencies are based on the FEM model, which is almost 20 % lower than the measured low amplitude values without consideration of stiffness contribution by secondary non-load bearing members.

Based on Table 13.3 and Eqs. (13.5a) and (13.5b), the higher mode natural frequencies are given by the following equations for Steel buildings:

$$
\left.
\begin{aligned}
f_2 &= \begin{cases} \dfrac{150}{H} & \text{(for habitability level)} \quad \text{(a)} \\[2ex] \dfrac{130}{H} & \text{(for safety level)} \quad \text{(b)} \end{cases} \\
f_3 &= \begin{cases} \dfrac{270}{H} & \text{(for habitability level)} \\[2ex] \dfrac{230}{H} & \text{(for safety level)} \end{cases} \quad \text{(c)} \\
f_4 &= \begin{cases} \dfrac{380}{H} & \text{(for habitability level)} \\[2ex] \dfrac{320}{H} & \text{(for safety level)} \end{cases} \quad \text{(d)} \\
f_T &= \begin{cases} \dfrac{67}{H} & \text{(for habitability level)} \quad \text{(e)} \\[2ex] \dfrac{56}{H} & \text{(for safety level)} \quad \text{(f)} \end{cases}
\end{aligned}
\right\} : \text{Steel Buildings} \qquad (13.6)
$$

13.4.3 Damping Predictor

Tamura et al. (2000) proposed the following damping predictors as empirical regression equations within the range of the tip drift ratio $x_H/H \leq 2 \times 10^{-5}$, for building heights in the range $10\,\text{m} < H < 100\,\text{m}$ for RC buildings and $30\,\text{m} < H < 200\,\text{m}$ for Steel buildings based on the Japanese Damping Database.

$$\zeta_1 = 0.014 f_1 + 470 \frac{x_H}{H} - 0.0018 \; : \; \text{RC Buildings} \qquad (13.7\text{a})$$

$$\zeta_1 = 0.013 f_1 + 400 \frac{x_H}{H} + 0.0029 \; : \; \text{Steel Buildings} \qquad (13.7\text{b})$$

It should be noted that the frequency dependent terms in Eqs. (13.7a) and (13.7b) are not actually the effect of the natural frequency. There is no clear reason for the frequency effects. They may be attributed to the effect of SSI. A low-rise building with a high natural frequency shows a significant contribution of the SSI effect. Thus, it tends to have higher total damping. Therefore, it is better to replace the frequency-dependent term with a height-dependent term as follows based on Eqs. (13.4a), (13.5a), (13.7a) and (13.7b) (Tamura 2012):

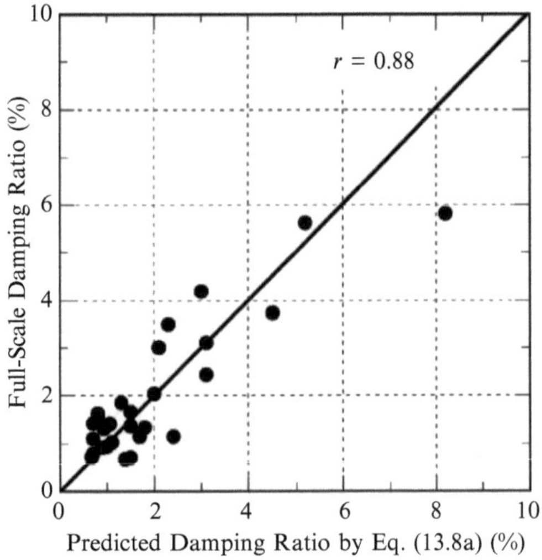

Fig. 13.9 Comparison of known amplitude full-scale damping data with those predicted by Eq. (13.8a) (Tamura et al. 2000)

$$\zeta_1 = \frac{0.93}{H} + 470\frac{x_H}{H} - 0.0018 \; : \; \text{RC Buildings} \tag{13.8a}$$

$$\zeta_1 = \frac{0.65}{H} + 400\frac{x_H}{H} + 0.0029 \; : \; \text{Steel Buildings} \tag{13.8b}$$

The correlation coefficient of the damping values predicted by Eq. (13.8a) and the full-scale values whose amplitudes are known, for example, was 88 %, as shown in Fig. 13.9. There have been several proposals for damping predictors such as Jeary (1986) and Lagomarsino (1993), but the predictors given by Eqs. (13.7a) and (13.7b), or Eqs. (13.8a) and (13.8b) are believed to be reliable.

13.4.4 Recommended Design Values of Natural Frequency and Damping Ratio

Damping design values commonly used for general buildings in various countries range from almost 0.3 % to 5 % for Steel buildings and from 0.5 % to 5 % for RC buildings, showing significant dispersion of design damping values (Tamura et al. 2000). EUROCODE (1994) and AIJ (2000) propose frequency and amplitude dependent damping ratios, and ÖNORM B4014 (1993) adopts somewhat complicated damping formulae depending upon materials, construction types, and foundation types. However, the majority adopt simple definitive values regardless of building height, amplitude or frequency (Tamura et al. 2000).

Based on the proposed predictors given in Eqs. (13.4a), (13.4b), (13.5a), (13.5b), (13.8a) and (13.8b) and discussions on amplitude levels, Tamura et al. (2000) proposed to use the recommended values given in Tables 13.4 and 13.5 for designing RC buildings and Steel buildings, respectively. The damping ratios for

Table 13.4 Design damping ratios for RC buildings (Tamura et al. 2000)

	Habitability			Safety (elastic range)		
	Natural frequency (Hz)	Damping ratio ζ_1 (%)		Natural frequency (Hz)	Damping ratio ζ_1 (%)	
Building height H (m)	$f_1 = 67/H$	Rec.	Standard	$f_1 = 56/H$	Rec.	Standard
30	2.2	2.5	3	1.9	3	3.5
40	1.7	1.5	2	1.4	2	2.5
50	1.3	1.2	1.5	1.1	2	2.5
60	1.1	1.2	1.5	0.93	1.5	2
70	0.95	0.8	1	0.79	1.5	2
80	0.83	0.8	1	0.69	1.2	1.5
90	0.74	0.8	1	0.62	1.2	1.5
100	0.67	0.8	1	0.56	1.2	1.5

– "Rec." means "Recommended" values
– The "Standard" values or larger values (max. × 1.2) can be used for buildings such as hotels and apartment houses, which have many interior walls
– Less than the "Recommended" values should be used for buildings that have few interior walls
– The value for $H = 30$ m should be used for buildings lower than 30 m
– The tabulated values represent the damping ratios including SSI effects. Thus, they should not be used for structural damping of only the upper structure

Table 13.5 Design damping ratios for Steel buildings (Tamura et al. 2000)

	Habitability			Safety (elastic range)		
	Natural frequency (Hz)	Damping ratio ζ_1 (%)		Natural frequency (Hz)	Damping ratio ζ_1 (%)	
Building height H (m)	$f_1 = 50/H$	Rec.	Standard	$f_1 = 42/H$	Rec.	Standard
30	1.7	1.8	2.5	1.4	2	3
40	1.3	1.5	2	1.0	1.8	2.5
50	1.0	1	1.5	0.83	1.5	2
60	0.83	1	1.5	0.69	1.5	2
70	0.71	0.7	1	0.60	1.5	2
80	0.63	0.7	1	0.52	1	1.5
90	0.56	0.7	1	0.46	1	1.5
100	0.50	0.7	1	0.42	1	1.5
150	0.33	0.7	1	0.28	1	1.5
200	0.25	0.7	1	0.21	1	1.5

Same notes as in Table 13.4

evaluation of building habitability performance in the low amplitude regime, such as human perception threshold, and the damping ratios for evaluation of structural safety in the large amplitude regime were recommended. However, the safety evaluation level does not mean at the ultimate limit state where the frames would be partially damaged, but in the elastic range. It has not been clarified how buildings would behave in the non-elastic region under extremely strong wind conditions, which have a static component and act for a rather long duration comparing to seismic actions. Plastic behavior tends to reduce the natural frequency, which suggests more wind excitation. Because of these reasons, wind resistant design of buildings is generally made in the range where the main frames behave mostly elastically. Even though the main frames are in the elastic range, the dynamic behaviour of buildings depends on the amplitude as shown in Sect. 13.3. However, it is too complicated to give an amplitude dependent damping model in the wind-induced response analysis. The damping ratio is generally given as a constant viscous damping for response analysis of a building.

Damping ratios recommended in Tables 13.4 and 13.5 are for wind-induced response analyses for amplitude levels lower than the elastic limit of main frames under the assumption of no damage to secondary members and cladding. However, even in a very high amplitude range exceeding the elastic limit of the main frames, as the contribution of plastic-hysteretic damping becomes significant, it is recommended to use these values as the structural damping ratio.

For higher-mode damping, the ratios of mode damping values such as ζ_2/ζ_1 and ζ_3/ζ_2 were examined based on the Japanese Damping Database (Tamura et al. 2000), and average values were obtained as follows:

$$\frac{\zeta_{n+1}}{\zeta_n} = 1.4 \ : \ \text{RC Buildings} \tag{13.9a}$$

$$\frac{\zeta_{n+1}}{\zeta_n} = 1.3 \ : \ \text{Steel Buildings} \tag{13.9b}$$

for $n = 1$ and 2. On average, 30 – 40 % larger damping ratios were obtained for the next higher mode.

However, it is noteworthy that the above relations do not imply stiffness-proportional damping characteristics, which is often assumed in structural design. It is also important that some data suggest higher mode reducing damping ratios for medium-rise and low-rise buildings, and the scope of application of Eqs. (13.9a) and (13.9b) should be limited to tall buildings.

13.5 Damping Estimation Techniques

Modal identification techniques and damping estimation techniques are summarized in Table 13.6. Here, quality of estimated damping, points to note of damping estimation, and the efficiency of two user-friendly damping evaluation techniques are discussed.

13.5.1 Quality of Estimated Results

The dynamic characteristics of an actual full-scale structure can be obtained by traditional experimental modal analysis techniques. However, these techniques have the following limitations. Artificial excitation or input information is also necessary

Table 13.6 Modal identification methods (Maia and Silva 1997; He and Fu 2001; Tamura et al. 2005)

Time domain methods		
Least squares	LS	– FDDR – MDOF curve fitting method
Complex exponential	CE	– MDOF method in SISO – IRF and complex poles and residues – Sensitive to noise
Least squares complex exponential	LSCE	– MDOF method in SIMO – IRF
Poly-reference complex exponential	PRCE	– MDOF method in MIMO – IRF
Ibrahim time domain	ITD	– MDOF method in SIMO – Eigenvalue problem from IRF data
Eigensystem realization algorithm	ERA	– MDOF method in MIMO – SVD of Markov parameters matrix
Autoregressive moving average	ARMA	– MDOF method in SISO – Selection of the optimal order
Direct system parameter identification	DSPI	– MDOF method in MIMO – Generalization of CE, PRCE and ITD
Random decrement	RD	– ACF – SDOF method extracting FDDR
Multi-mode random decrement	MRD	– ACF – MDOF method extracting FDDR – LS
Frequency domain methods		
$\hat{H}_1(\omega)$ model	IFRF	– FRF – No input noise assumption – Cross-PSD of input and output & Input Auto-PSD
$\hat{H}_2(\omega)$ model	OFRF	– FRF – No output noise assumption – Output Auto-PSD & Cross-PSD of input and output
$\hat{H}_v(\omega)$ model and $\hat{H}_s(\omega)$ model	VFRF SFRF	– FRF – Both output noise and input noise – LS
Frequency domain decomposition	FDD	– PSD matrix – SVD
Hilbert transform	HT	– FRF – Detecting system non-linearity – No domain change

ACF, auto-correlation function; FDDR, free decay damped response; FRF, frequency response function; IRF, impulse response function; MDOF, multi-degree of freedom; MIMO, multi-input multi-output; PSD, power spectral density; SDOF, single-degree-of-freedom; SIMO, single-input multi-output; SISO, single-input single-output; SVD, singular value decomposition to measure FRFs or IRFs for subsequent modal parameter extraction. These response functions are very difficult to obtain under field conditions or for large structures. Some components, instead of the complete system, can be tested in the laboratory environment, but it is not easy to precisely simulate the boundary conditions.

Test procedures and estimation techniques significantly affect the quality of the obtained modal parameters, i.e. natural frequencies, damping ratios, and mode shapes. In order to obtain high quality estimates, it is necessary to conduct careful and high quality measurements. Especially in damping estimations, great care is required for high quality data acquisitions. In the building engineering field, chances to obtain full-scale measurements are quite rare, and obtained data are very valuable. Therefore, modal parameters obtained by full-scale measurements have been treated as a kind of precious heritage. This is true for natural frequencies and mode shapes obtained under field conditions, which have only 10 – 20 % errors. However, low quality damping data can result in significant errors of 100 – 200 % or even more than 1,000 %. Short term measurements at only a few points for Output-Only Modal Estimation (OOME) occasionally have almost no meaning in damping estimations and can provide nonsense results in some cases. Furthermore, it cannot be denied that many full-scale data have over-estimated damping ratios because of inappropriate measurements and damping estimation methods. Thus, it is very important to carefully examine the existing damping data and to have the courage to reject such contaminated damping data, even though almost no data would then be available.

13.5.2 Damping Estimation Techniques

Damping estimation techniques obtained from FDDR, IRF, FRF and ambient vibration data are shown in Table 13.7.

Many civil engineering structures can be adequately excited by ambient excitations. Ambient modal analysis based on response measurements has two major advantages compared to traditional analysis. One is that no expensive and heavy excitation devices are required. The other is that all (or part) of the measurements can be used as references, and MIMO techniques can be used for modal analysis, thus enabling easy handling of closely-spaced and even repeated modes. This is most valuable for long-span roof structures or complex and huge systems such as power transmission line systems.

Under these conditions, two simple and user-friendly but accurate damping estimation techniques are discussed in Sects. 13.5.4 and 13.5.5. One is the MRD technique (Tamura et al. 2002) and the other is the FDD technique (Brincker et al. 2000) using SVD of the PSD matrix. Both can be applied to ambient excitations such as wind, turbulence, traffic, and/or micro-tremors.

Table 13.7 Damping estimation techniques

Time domain techniques		
Logarithmic decrement	LD	– FDDR – SDOF curve fitting method
Random decrement	RD	– SDOF method extracting FDDR – LD
Multi-mode random decrement	MRD	– MDOF method extracting FDDR – LD
Hilbert transform	HT	– LD – FRF
Frequency domain techniques		
Half power	HP	– Auto-PSD – SDOF
Transfer function	TF	– FRF – SDOF curve fitting method
1/√2 Method	1/√2	– FRF
Phase gradient	PG	– FRF
Frequency domain decomposition	FDD	– PSD matrix – SVD – Peak picking
Wavelet transform	WT	– Time-scale decomposition – TRF

TRF, time reponse function

13.5.3 Points to Note in Damping Estimation

Measurements of a FDDR of a SDOF system can directly provide the system's TRF. Here, TRF includes IRF, FDDR and the RD signature. The LD method is commonly used to give a quick estimate of the damping ratio from the TRF, represented by an exponentially decaying harmonic motion. Under field conditions, the FDDR signal is generally polluted by noise. More amplitude points are chosen and plotted on semi-logarithmic plane, and a straight line is obtained by linear regression. This technique is available for linear systems or structures. However, the dynamic properties of buildings generally have non-linearity, as shown in Figs. 13.1, 13.2, 13.4 and 13.5. For actual buildings, the straight line approximation can only provide the average tendency; for accurate values, step-by-step estimation of the damping ratio for each amplitude level is necessary, as demonstrated in Fig. 13.5c.

When dealing with a MDOF system, one can filter out the effect of other modes in the frequency domain and transform the signal back to the time domain. The accuracy of the method breaks down when there are two close modes.

In dynamic tests, where both input and output signals can be measured, FRF can be estimated by Auto-PSD and Cross-PSD through Fast Fourier Transform (FFT). The simplest technique for estimating damping ratios is the HP method directly using the Auto-PSD of the responses. Actually, the bandwidth and the damping ratio can be more accurately evaluated from the imaginary or real part of FRF, the phase gradient, and the Nyquist plot in the complex plane.

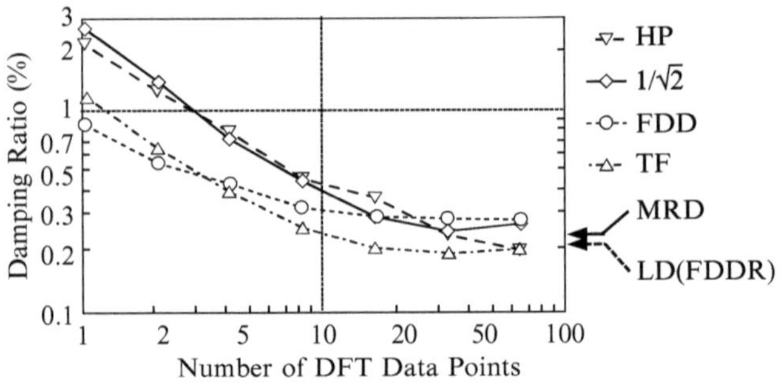

Fig. 13.10 Variations of damping ratios evaluated by PSD-based frequency domain techniques with DFT data points demonstrating leakage effects (Tamura et al. 2004)

It should be noted that damping estimation techniques in the frequency domain are more or less based on output PSDs, which require the following conditions for accurate estimation:

- Stationarity of building responses
- Sufficient frequency resolution near resonant frequency in FFT analysis

The first condition is closely related to the inherent nonlinearity of dynamic properties of buildings, shown in Figs. 13.1, 13.2, 13.4 and 13.5. Mixing of different levels of building responses results in a deformed or broader resonant peak providing inaccurate estimation of the damping ratio (Jeary 1992).

The second condition comes from the leakage error in PSD estimation due to data truncation of Discrete Fourier Transform (DFT), which always takes place in practice. This generally results in a larger damping estimation, and occasionally it becomes very large. It has been emphasized by Jeary (1986), Tamura et al. (2002) and others, but there have been many papers reporting unrealistically large damping values without any consideration of this point. This is because natural frequencies and mode shapes are not significantly affected by insufficient frequency resolution.

Figure 13.10 demonstrates the importance of fine frequency resolution in damping estimation, say a sufficient length of sample or a sufficient number of DFT data points (Tamura et al. 2004). This is the result obtained from laboratory tests on ambient vibrations of a simple four-storey lumped-mass model. Damping ratios were obtained by four different frequency domain estimation methods, where the Auto-PSDs or FRF was calculated via DFT. The abscissa indicates the number of data points used for DFT calculation. The evaluated damping ratios decrease with increasing number of DFT data points and converge to a constant value around 0.25 %, which was thought to be the accurate damping value. If the number of DFTs were only set at 1024, as the fundamental estimation of PSD by FFT often does, the evaluated damping ratio can be 3 %, which is more than ten times the accurate damping value, thus producing a 1,000 % error. It is not appropriate to say "error" for a discrepancy of 100 % or more: it should be called a "mistake".

Leakage is a kind of bias error that cannot be eliminated by windowing, e.g. by applying a Hanning window. Furthermore, it is harmful to the damping estimation accuracy, which relies on the PSD measurements. The bias error caused by leakage is proportional to the square of the frequency resolution (Bendat and Piersol 1986). Therefore, increasing frequency resolution is a very effective way to reduce leakage error.

It is noted that the number of DFT data points should be more than 16,384 in this case, which is almost 600 times the natural period of the model, for accurate identification of the damping ratio. This means that the sample should have enough length. An ensemble averaging operation with a sufficient number of samples is also necessary to obtain a correct Auto-PSD. A very long total record length is required to accurately evaluate the damping ratio by using a sufficient number of samples for ensemble averaging with enough length for each sample for fine frequency resolution. It is also noted that stationarity has to be satisfied in those records, which also requires careful selection of vibration records with similar conditions, especially in amplitude level.

In summary, difficulties and errors in damping evaluation of buildings and structures based on OOME due to ambient excitations mainly result from the following:

1. Non-stationarity of excitation phenomena and amplitude dependency of damping in buildings
2. Non-well-separated modes in terms of natural frequencies
3. Failure in local white noise assumption of excitation
4. Leakage error in PSD estimation
5. Insufficient number of ensemble averaging for accurate PSD estimation
6. Inappropriate width of a band-pass filter for RD technique
7. Insufficient number of superimpositions to obtain RD signature

Clearly, accurate damping can only be estimated by careful and time consuming efforts. As there is no theoretical means for estimating damping ratios in buildings, consulting high quality field data is the only way to make a good prediction of design damping value. It is thus necessary to establish a reliable database consisting of a statistically sufficient number of data sets of various structure types, building dimensions, cladding, soil and foundation conditions, amplitudes and so on. It is necessary to accumulate more high-quality damping data.

Table 13.8 summarizes the important points and necessary conditions to be satisfied in damping evaluation. The damping data should be evaluated and reported considering these points.

13.5.4 Random Decrement (RD) Technique and Multi-Mode Random Decrement (MRD) Technique

RD techniques for superimposing sub-samples of responses with an initial amplitude can be used to extract an ACF eliminating randomly excited components (Cole 1973; Jeary 1986; Tamura and Suganuma 1996). The RD technique assuming a SDOF system can efficiently evaluate the damping ratios and the natural

Table 13.8 Points to note for evaluation of damping in buildings using ambient vibrations (OOME)

Techniques	Required conditions	Measures
Time domain – RD – MRD	– Sufficient number of superimpositions	– Long enough data – Error ratio (<3%) based on theoretical SDOF FDDR
	– Appropriate bandwidth of band-pass filter (not too narrow and wide enough)	– Sufficient number of superimpositions – Application of MRD
	– Local white-noise assumption	– Careful check of excitation condition[a]
	– Well-separated modes	– Application of MRD
Frequency domain	– Stationarity	– Run tests
– HP – FDD	– Fine frequency resolution (Minimal leakage error)	– Convergence check of damping ratio by increasing sample length – Long enough data
	– Sufficient ensemble averaging of PSD	– High quality PSD – Long enough data
	– Local white-noise assumption	– Careful check of excitation condition[a]
	– Well-separated modes	– Application of FDD
Other necessary information	– Amplitude information	– Tip drift ratio x_H/H – Initial amplitude (RD/MRD)
	– Building information	– Height and stories – Plan shape and dimensions – Material of structural frames – Structural type – Building usage – Type of foundation – Soil condition – Cladding type
	– Measurement/excitation information	– Excitation type – Length of samples

[a]A narrowband excitation near the target natural frequency such as vortex resonance is inappropriate

frequencies only for well-separated vibration modes. It is very efficient for evaluating the amplitude dependency of the dynamic characteristics of buildings. However, if there are closely located predominant frequency components, a beating phenomenon is observed in the RD signature. In these cases, traditional RD cannot be used to estimate the damping ratio. In order to evaluate multiple closely located vibration modes, the MRD technique has been proposed (Tamura et al. 2002), where the FDDRs of multiple SDOF systems with different dynamic characteristics are superimposed, and the LS curve fitting approximates the RD signature with the beating phenomenon.

The advantage of MRD is in application to closely located modes, which almost no traditional methods can do, and actual buildings and structures often have vibration modes that are not well-separated.

13.5.5 Frequency Domain Decomposition (FDD)

The FDD technique is a recently developed powerful tool that can be used in the frequency domain. Instead of using Auto-PSD of the building responses directly, as in the classical frequency domain technique, the PSD matrix is decomposed at each frequency line via SVD. SVD has a powerful property of separating noisy data from disturbance caused by unmodeled dynamics and measurement noise. For the analysis, Singular Value (SV) plots, as functions of frequencies, calculated from SVD can be used to determine modal frequencies and mode shapes. It has been proved that the peaks of an SV plot indicate the existence of structural modes (Brincker et al. 2000). The singular vector corresponding to the local maximum SV is unscaled mode shape. This is exactly true if the excitation process in the vicinity of the modal frequency is white noise. One of the major advantages of the FDD technique is that closely-spaced modes, even repeated modes, can be easily dealt with. The only approximation is that mode shape orthogonality is assumed.

The basic idea of the FDD technique is as follows (Brincker et al. 2001). The SV in the vicinity of the natural frequency is equivalent to the PSD function of the corresponding mode (as a SDOF system). This PSD function is identified around the peak by comparing the mode shape estimate with the singular vectors for the frequency lines around the peak. As long as a singular vector is found that has a high Modal Amplitude Coherence (MAC) value with the mode shape, the corresponding SV belongs to the SDOF function. If at a certain line none of the SVs has a singular vector with a MAC value larger than a certain limit value, the search for matching parts of the PSD function is terminated. From the fully or partially identified SDOF PSD function, the natural frequency and the damping ratio can be estimated by taking the PSD function back to the time domain by inverse FFT as an ACF of the SDOF system. From the ACF, the natural frequency and the damping ratio are found by LD or other methods.

13.6 Practical Applications of MRD and FDD Techniques

13.6.1 Application of MRD and FDD to Tall Chimney

Figure 13.11a shows an elevation and plan of the tested chimney, which consists of steel trusses and a concrete funnel, and has an octagonal cross section. Servo-type accelerometers were installed on three different levels. Two horizontal components (x, y) and one vertical component (z) were measured at each level.

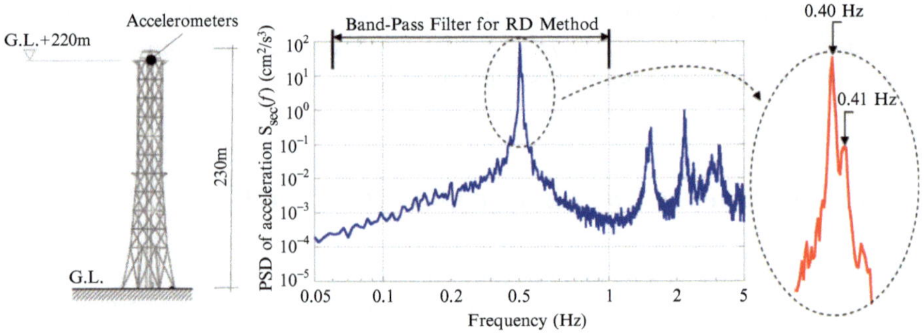

Fig. 13.11 Auto-PSD of tip acceleration of 230-m high chimney (Tamura et al. 2002). (**a**) Elevation; (**b**) Auto-PSD of tip acceleration (*y*-dir)

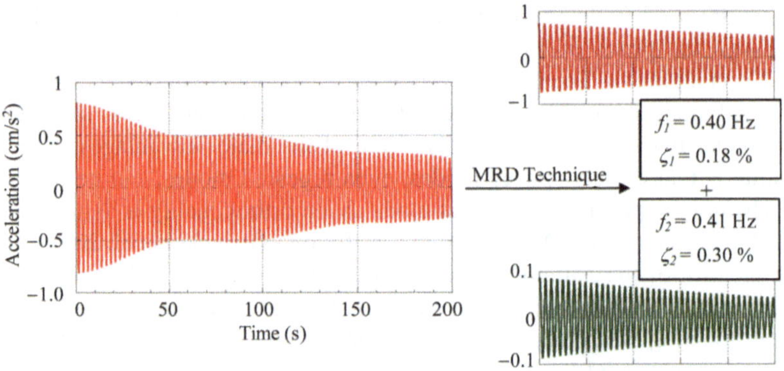

Fig. 13.12 RD signature of tip acceleration (*y*-dir.) of 230-m high chimney and dynamic properties estimated by MRD technique (Tamura et al. 2002)

The sampling rate of the acceleration records was set at 100 Hz, and ambient responses were measured for 90 min in total. Figure 13.11b shows the Auto-PSD of acceleration (*y*-dir.) at the top of the chimney. Peaks corresponding to several natural frequencies are clearly shown. From such a beautifully isolated peak, it is generally easy to estimate the damping ratio, as well as the natural frequency, by the traditional RD technique.

The traditional RD technique assuming a SDOF system was applied for system identification using the ambient accelerations in the y direction at the top: GL + 220 m. By processing with a numerical band-pass filter with a frequency range of 0.06 – 1.0 Hz, only the frequency components around the lowest peak near 0.4 Hz depicted in Fig. 13.11b were extracted. The initial amplitude of the acceleration to get the RD signature was set at the standard deviation. From the clear peak near 0.4 Hz in the PSD shown in Fig. 13.11b, it was believed that the dynamic property of the first mode would be easily obtained with appropriate accuracy, but this was not correct. Figure 13.12 shows the obtained RD signature, where a beating phenomenon is

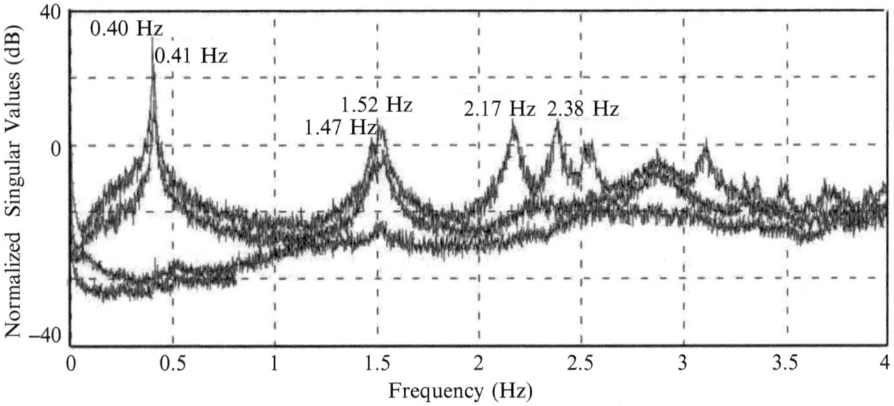

Fig. 13.13 SV distributions of PSD matrix of ambient vibrations of 230-m high chimney (Tamura et al. 2002)

observed, suggesting two closely located dominant frequency components. By careful examination of a close-up view of the peak near 0.4 Hz, it is seen that there are actually two peaks: at 0.40 and 0.41 Hz. It is very difficult to identify whether the peak at 0.41 Hz seen in Fig. 13.11b is meaningful or not from PSD due to FFT, and the traditional RD method based on SDOF fails to accurately estimate the damping ratio in such cases. However, the MRD technique can overcome this shortcoming of the traditional RD method.

The MRD technique (Tamura et al. 2002) was thus applied to approximate the beating RD signature, and the damping ratio and the natural frequency of the chimney were estimated at 0.18 % and 0.40 Hz for the 1st mode, and 0.30 % and 0.41 Hz for the 2nd mode, as shown in Fig. 13.12. The dynamic characteristics of the 3rd and 4th modes were also estimated by the MRD technique.

The FDD technique was also applied to the six horizontal components of the acceleration responses at the three different heights to evaluate the chimney's dynamic characteristics in detail. Figure 13.13 shows the frequency distribution of the SV plots obtained by the FDD technique from the PSD matrix of the acceleration responses. The 3rd and 4th natural frequencies are also closely located, and it is also very difficult to estimate damping ratios for these two modes by the HP method or the traditional RD technique. However, as shown, the FDD technique can estimate damping ratios even in such cases.

Figure 13.14 shows the damping ratios of the lowest two modes obtained by the FDD technique. The figure shows the effects of the number of data points used for PSD calculation on the estimated damping ratios. The damping ratios converge to certain values with increase in the number of data points, suggesting that it is necessary to take 8,192 data points in this case. Table 13.9 shows the dynamic characteristics of a 230-m high chimney obtained by MRD and FDD. They agree well except for the third mode damping ratio, and it is obvious that both techniques are useful even for closely located modes.

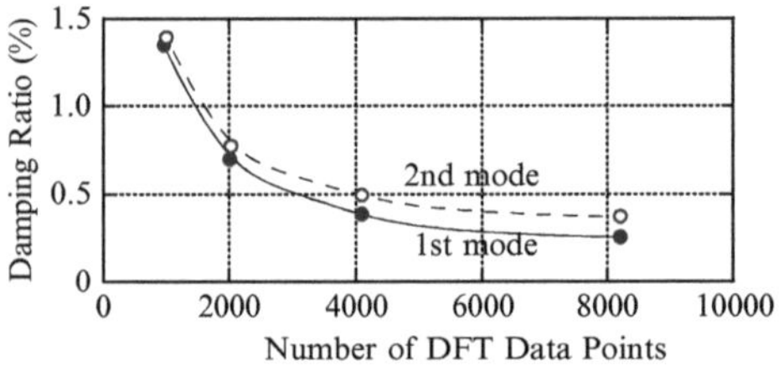

Fig. 13.14 Variations of damping ratios of 230-m high chimney with DFT data points (Tamura et al. 2002)

Table 13.9 Dynamic characteristics of 230-m high chimney (Tamura et al. 2002)

	Natural frequency (Hz)		Damping ratio (%)	
Mode	MRD	FDD	MRD	FDD
1	0.40	0.40	0.18	0.24
2	0.41	0.41	0.30	0.39
3	1.47	1.47	0.83	0.3
4	1.53	1.52	0.85	0.91
5	2.17	2.17	0.55	0.65
6	2.38	2.38	0.42	0.39
7	–	2.87	–	–
8	–	3.10	–	0.77

13.6.2 Application of MRD and FDD to 15-Storey Office Building

A series of field measurements were made of ambient vibrations of a medium-rise 15-storey office building, and its dynamic characteristics were evaluated (Tamura et al. 2002, and Miwa et al. 2002). The building extends from 6 m underground to 59 m above basement level, as shown in Fig. 13.15a. The columns are concrete-filled tube and the beams are wide-flange steel. The exterior walls of the first floor are of pre-cast concrete. The walls from the second floor to the top are of ALC. Fourteen servo-type accelerometers were used for one setup with two at the 15th floor as references, and 53 components were measured in total. By assuming that the floor was subject to lateral rigid body motion, the measured vibration was translated into equivalent motions at the desired corners.

The FDD technique was applied to evaluate the dynamic characteristics of the building based on the measured ambient vibrations. Cross-PSDs were estimated using the full data of 53 components.

Figure 13.15b shows the SV plots of the office building. There were many peaks of less than 5 Hz corresponding to the natural frequencies, and it was possible to obtain up to the 9th mode below 5 Hz. Figure 13.16 depicts the corresponding 9 mode shapes. Figure 13.17 shows a typical "bell" of the SDOF system: the 1st and 2nd modes of the 15-storey building. From the identified SDOF PSD function, the

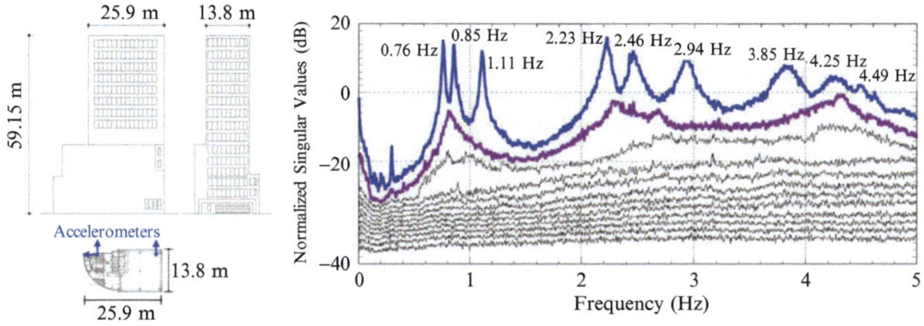

Fig. 13.15 15-storey office building and SV distributions of PSD matrix of ambient vibrations (Tamura et al. 2002) (**a**) Elevation and general plan; (**b**) SV plots

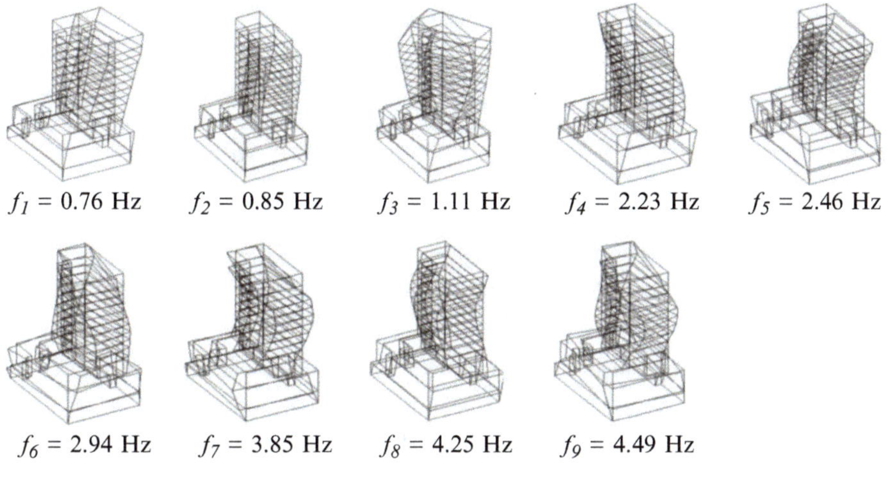

Fig. 13.16 Mode shapes of 15-storey office building obtained by FDD (Tamura et al. 2002)

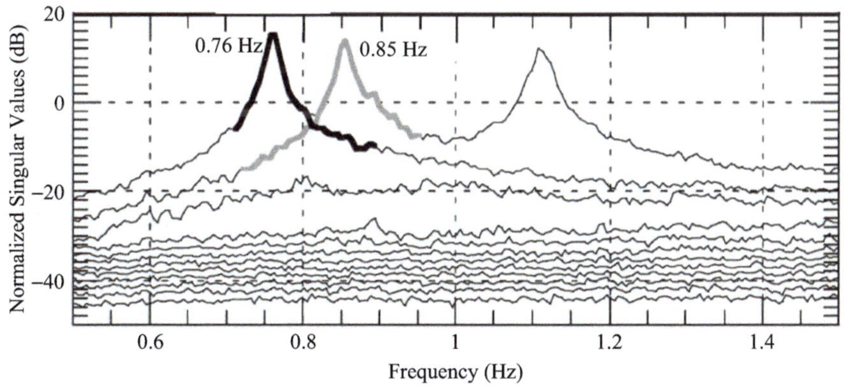

Fig. 13.17 Close-up view of SV plots near lowest mode of 15-storey building (Tamura et al. 2002)

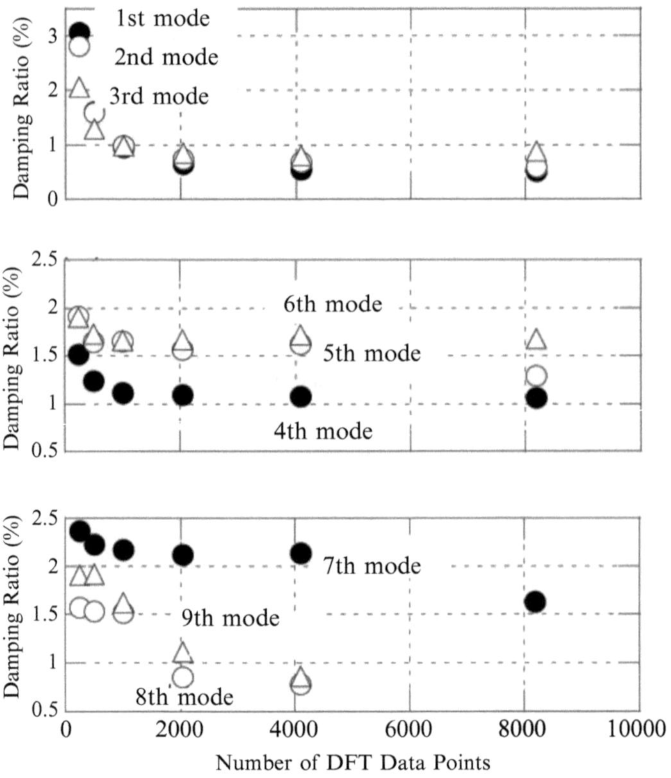

Fig. 13.18 Variations of damping ratios of 15-storey office building with DFT data points

modal frequency and the damping ratio can be estimated by taking the PSD function back to the time domain by inverse FFT as the ACF of the SDOF system. The modal frequency and the damping ratio are found from the ACF.

Figure 13.18 presents the variations of the damping ratios with the number of data points used for DFT calculation. The damping ratios of all modes decrease, while the number of data points, or frequency resolution, increases. It appears that damping estimates converge when the number of data points is large enough (up to 4,096 or 8,192).

Table 13.10 shows the dynamic characteristics of the 15-storey building obtained by the FDD technique, where 4,096 data points were used for DFT calculation. Corresponding natural frequencies obtained by FEM analysis are also shown in Table 13.10. In the original FEM analysis, the building's stiffness was estimated only for the members of the main structure, and the natural frequencies by FEM analysis were evaluated as slightly smaller than the full-scale values. The stiffness of the building's secondary members was therefore added so that the first-mode natural frequency was identical to the full-scale value. As a result, satisfactory agreement with less than 5 % error was automatically obtained up to the 9th vibration mode, as shown in Table 13.10.

Table 13.10 Dynamic characteristics of 15-storey office building

Mode	Natural frequency (Hz)			Damping ratio (%)
	FEM model	FDD (full-scale)	Error (%)	FDD (full-scale)
1	0.76	0.76	0	0.65
2	0.87	0.85	2.35	0.74
3	1.15	1.11	3.60	0.84
4	2.14	2.23	4.04	1.10
5	2.53	2.46	2.84	1.56
6	3.02	2.94	2.72	1.67
7	3.85	3.85	0	2.12
8	4.26	4.25	0.24	0.85
9	4.67	4.49	4.01	1.11

References

AIJ (2000) Damping in buildings, Committee on damping evaluation (chaired by Y. Tamura), Architectural Insitute of Japan, Maruzen

Bendat J, Piersol A (1986) Random data: analysis and measurement procedures. Wiley, New York

Brincker R, Zhang LM, Anderson P (2000) Modal identification from ambient response using frequency domain decomposition. In: Proceedings of the 18th international modal analysis conference (IMAC), San Antonio, Texas, February 7–10, 2000

Brincker R, Ventura CE, Andersen P (2001) Damping estimation by frequency domain decomposition. In: Proceedings of the 19th international modal analysis conference (IMAC)

Cole HA (1973) On-line failure detection and damping measurement of aerospace structures by the random decrement signatures. NASA CR-2205

Davenport AG, Hill-Carroll P (1986) Damping in tall buildings: its variability and treatment in design. In: Building motion in wind, ASCE Spring Convention, Seattle, pp 42–57

Ellis BR (1980) An assessment of the accuracy of predicting the fundamental natural frequencies of buildings and the implications concerning the dynamic analysis of structures. In: Proceedings of the institution of civil engineers, Part 2, 69, pp 763–776

EUROCODE (1994) Basis of design and actions on structures, Part 2–4: Wind actions, CEN

Fukuwa N, Nishizawa R, Yagi S, Tanaka K, Tamura Y (1996) Field measurement of damping and natural frequency of an actual steel-framed building over a wide range of amplitude. J Wind Eng Ind Aerodyn 59:325–347

Haviland R (1976) A study of the uncertainties in the fundamental translational periods and damping values for real buildings. Massachusetts Institute of Technology, PB-253, 188

He JM, Fu ZF (2001) Modal analysis. Butterworth-Heinemann, USA

Jeary AP (1986) Damping in buildings—a mechanism and a predictor. J Earthquake Eng Struct Dyn 14:733–750

Jeary AP (1992) Establishing non-linear damping characteristics of structures from non-stationary response time-histories. Struct Eng 4(70):61–66

Jeary AP (1998) The damping parameter as a descriptor of energy release in structures. In: Proceedings on CD-ROM, T193-4, Structural engineers world congress, San Francisco, p 8

Lagomarsino S (1993) Forecast models for damping and vibration periods of buildings. J Wind Eng Ind Aerodyn 48:221–239

Maia NMM, Silva JMM (1997) Theoretical and experimental modal analysis. Research Studies Press, Ltd., England

Miwa M, Nakata S, Tamura Y, Fukushima Y, Otsuki T (2002) Modal identification by FEM analysis of a building with CFT columns. In: Proceedings of the 20th international modal analysis conference (IMAC)

Okada K, Nakamura Y, Shiba K, Hayakawa T, Tsuji E, Ukita T, Yamaura N (1993) Forced vibration tests of ORC200 Symbol Tower, Part 1: test methods and results. Summaries of technical papers of annual meeting of architectural institute of Japan, Structures 1, pp 875–876

ÖNORM B4014, Teil 1 (1993) Code for Austria

Sasaki A, Satake N, Arakawa T, Suda K, Ono J, Morita K (1997) Damping properties of buildings using full-scale data. In: Proceedings of the symposium on damping technology of the 21st century, Dynamics and design conference, The Japan Society of Mechanical Engineers Centennial Grand Congress, pp 25–28 (in Japanese)

Satake N, Arakawa T, Morita K (1997) Characteristics of damping in steel buildings based on full-scale data: differences due to foundation types, measuring methods and damping evaluation techniques. In: Proceedings of the 67th research meetings, 1996, Kanto Chapter, AIJ, Structures, materials & construction, 77–80 (in Japanese)

Satake N, Suda K, Arakawa T, Sasaki A, Tamura Y (2003) Damping evaluation using full-scale data of buildings in Japan. J Struct Eng ASCE 129(4):470–477

Shioya K, Uesu K, Tamura Y, Kanda J (1993) Dynamic characteristics of reinforced concrete high-rise apartment buildings. Summaries of technical papers of annual meeting, AIJ, I, pp 481–482 (in Japanese)

Tamura Y (2012) Amplitude dependency of damping in buildings and critical tip drift ratio. Int J High-Rise Build 1:1–13

Tamura Y, Suganuma S (1996) Evaluation of amplitude-dependent damping and natural frequency of buildings during strong winds. J Wind Eng Ind Aerodyn 59:115–130

Tamura Y, Suda K, Sasaki A (2000) Damping in buildings for wind resistant design. In: Proceedings of the international symposium on wind and structures for the 21st century, 26–28 January 2000, Cheju, pp 115–130

Tamura Y, Zhang LM, Yoshida A, Nakata S, Itoh T (2002) Ambient vibration tests and modal identification of structures by FDD and 2DOF-RD technique. In: Proceedings of the structural engineers world congress (SEWC), Yokohama, 9–12 October, T1-1-a-1, 8

Tamura Y, Yoshida A, Zhang LM (2004) Evaluation techniques for damping in buildings. In: International workshop on wind engineering & science, October 2004, Wind Engineering & Science, New Delhi

Tamura Y, Yoshida A, Zhang L (2005) Damping in buildings and estimation techniques. In: Proceedings of the sixth Asia-Pacific conference on wind engineering, 12–14 September 2005, Seoul, CD-ROM

Wyatt TA (1977) Mechanisms of damping. In: Proceedings of the symposium on dynamic behaviour of bridges, 19 May 1977, Transport and Road Research Laboratory, Crowthorne

Chapter 14
Control of the Wind Induced Response of Structures

Ahsan Kareem, Enrica Bernardini, and Seymour M.J. Spence

Abstract In recent years, a drive towards the construction of buildings of increasing heights and bridges of increasing span lengths has generated the most fascinating challenges in the civil engineering field. Indeed, the particular sensitivity to wind loads of such extreme structures has to be faced when designing structural systems that have to possess adequate stiffness and damping characteristics to ensure acceptable performance for survivability, serviceability and habitability. In order to meet these needs, several routes can be taken that involve different design aspects. In addition to the modification of the structural system, which is the first and foremost option for the structural designers, these can be divided in the adoption of shape tailoring measures, aimed at improving the aerodynamics/aeroelasticity of the structure, and the introduction of auxiliary motion control devices. In this chapter these two aspects are treated, with the intent of providing the reader with a general overview of the engineering solutions that may be considered for the control of the wind induced response of structures. In particular, discussion on the aerodynamic/aeroelastic shape tailoring is provided, alongside with meaningful examples of application of this strategy to real structures. The possibility of adopting motion control devices is also investigated. The most important typologies of devices for the control of wind induced vibrations are classified according to the principle on which they are based and their main characteristics are illustrated.

Keywords Motion control devices • Shape tailoring • Wind induced response

---

A. Kareem (✉) • E. Bernardini • S.M.J. Spence
NatHaz Modeling Laboratory, University of Notre Dame, Notre Dame, IN 46556, USA
e-mail: kareem@nd.edu; ebernard@nd.edu; sspence@nd.edu

Y. Tamura and A. Kareem (eds.), *Advanced Structural Wind Engineering*,
DOI 10.1007/978-4-431-54337-4_14, © Springer Japan 2013

14.1 Introduction

Globally, high-rise developments in urban settings have become venue of choice for housing both businesses and residences. This has brought about trends toward structures with increasing heights, lighter and slenderer construction whose performance is sensitive to complex wind environments. A drive towards more extreme structures is seen not only for buildings, for which the record of 828 m of height established by the Burj Khalifa in 2010 will most likely be broken soon, but also for bridges. The current trend is indeed towards spanning ever longer bodies of water, as for example in the case of the proposed Strait of Messina Bridge, Italy (3,300 m), using a cable supported solution. Even in the design of high-rise buildings, where the resistance to lateral wind loads can be achieved with simple and classical structural systems (that are carefully chosen and designed), serviceability and habitability issues may pose major challenges. In particular, it is the documented human sensitivity to acceleration and jerk, occasionally causing various symptoms of discomfort in the building's occupants, that has a critical influence and imposes stringent requirements on the structural response. On the other hand, with the increase in span length, bridges are becoming more sensitive to the action of wind and phenomena such as flutter instability and vortex induced vibration of the deck, as well as cable vibration, are becoming sources of major concern. New challenges are therefore being faced in designing structural systems with adequate stiffness and damping characteristics that are able to ensure acceptable performance for survivability, serviceability and habitability.

In order to meet these needs, several routes can be taken that involve different design aspects. In addition to the modification of the structural system, which is the first and foremost option for the structural designers, these possible mitigative routes can be divided in the adoption of shape tailoring measures, aimed at improving the aerodynamics/aeroelasticity of the structure, and the introduction of auxiliary motion control devices (Fig. 14.1). These topics are discussed in this chapter; additional details can be found among others in Tamura (1998), Kareem et al. (1999).

14.2 Aerodynamic/Aeroelastic Shape Tailoring

The tailoring of the aerodynamic shape is a powerful means to achieve optimal performance of buildings and bridges under winds which is used routinely in other fields, such as in aerospace and automobile engineering. The external shape of a building, as well as the geometric profile of a bridge deck, plays a remarkably important role in determining the entity of the wind loads acting on the structure. Modifications of the geometry can be introduced at the cross-section level with the aim of defining a more aerodynamic profile with the consequent reduction of the drag forces or to change the detachment point of the vortices, or at the

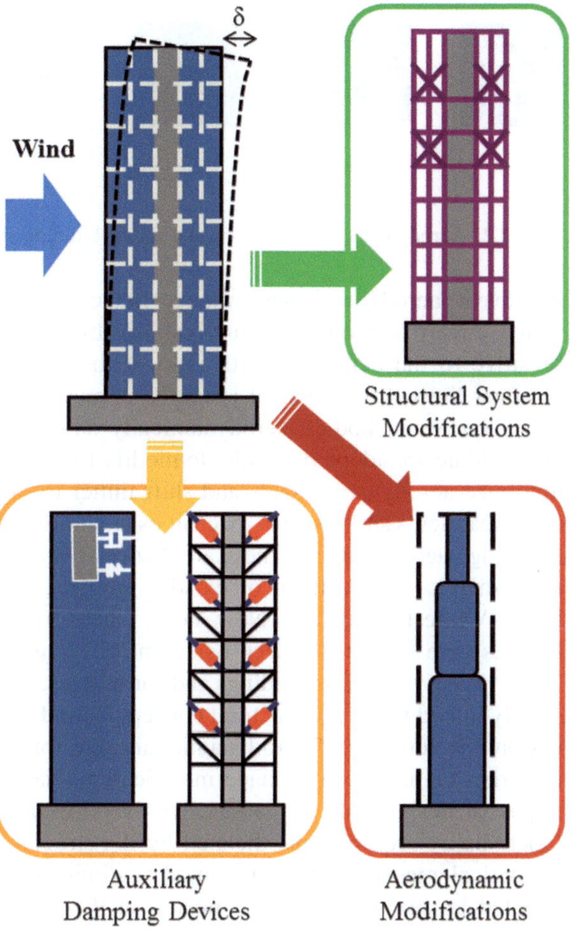

Fig. 14.1 Possible strategies for motion mitigation control

longitudinal profile level (i.e. over the height of the building or over the length of the bridge) mainly in order to disrupt the synchronization between the vortices and therefore reducing the vortex-shedding loads. In the case of buildings, modifications of the corner geometry, as for example the inclusion of chamfered corners or horizontal slotted corners (Fig. 14.2) (Kareem et al. 1999), have been found to considerably reduce the response of buildings, in comparison to the performance of a basic square cross-section, with these improvements becoming more marked as the corners are progressively rounded. As for the longitudinal modifications, the inclusion of set-backs and tapering, the cutting of corners or the progressive elimination of corners with height not only are characteristics of some of the most graceful and notable structures, but also have been demonstrated to have a practical aerodynamic purpose (e.g. Tanaka et al. 2012). Similarly, in bridge aerodynamics, it is the shape of a bridge deck section that plays an important role in its behavior. The reduction of the height of the deck edge is usually the easiest way to

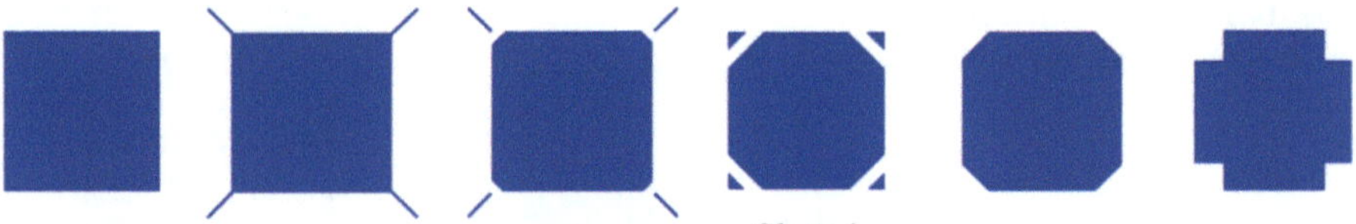

Fig. 14.2 Aerodynamic modifications to a square building shape [after Kareem et al. (1999)]

improve its aerodynamic performance. In fact, flutter and vortex-induced vibrations, which govern not only the survivability but also the serviceability design, are sensitive to the height-to-width ratio of the deck. This measure of reduction is however limited by practical issues. Commonly, in the case of bridges spanning over large water bodies, aerodynamically tailored box sections with openings in their middle are adopted in order to modify the flow around the deck in an effort to minimize aerodynamic loads and shift flutter to higher wind speeds. In addition, edge treatments such as deck extensions and fairings can be used to further enhance the bridge aerodynamics. These are devices commonly adopted also in the case of aerodynamic enhancement of existing bridges. A recent example of this is the Bronx-Whitestone Bridge, New York, where triangular fiber reinforced polymer fairings were added to the deck's sides in 2007 after the removal of stiffening trusses.

Obviously, both for buildings and for bridges, any choice regarding the shape is strictly linked to the architectural concept behind the structure's aesthetics. For this reason, the adoption of aerodynamic tailoring for the reduction of the wind action requiring highly incisive shape modifications has to be embedded in the earliest stages of design, where the shape is developed considering both aesthetic and functionality requirements dictated by the structure's ultimate use. For the same aesthetical reasons, retrofitting modifications of bridge deck's edges are often rejected by bridge designers. Indeed, in the case of buildings, recent trends in joint architect-engineer involvement in the overall evolution of a building project has led to building shapes that benefit from aerodynamic profiling to help deter formation of flow fields generating higher loads. It is also true that some minor building shape modifications (e.g. rounding of corners, change of inclination of a slot cutting through the building) can be introduced in a more advanced design stage and prove to be extremely beneficial to the wind-induced behavior even if not drastic from an aesthetic point of view.

Some recent examples have shown that geometric choices that create aerodynamically efficient forms can be integrated into the design of tall buildings without sacrificing their appearance. For example, in the case of the Taipei 101 in Taipei, Taiwan, completed in 2004 (Fig. 14.3a), every group of eight floors constitutes a tapered segment, creating a pattern that recalls important cultural symbolism (an Asian pagoda, a bamboo stem, symbol of learning and growth, and a stack of traditional Chinese money boxes, symbol of abundance) enhancing at the same time the aerodynamic performance. Similarly, in the Petronas Towers, Kuala Lumpur, Malaysia, completed in 1998 (Fig. 14.3b), the repeated use of multi-level setbacks

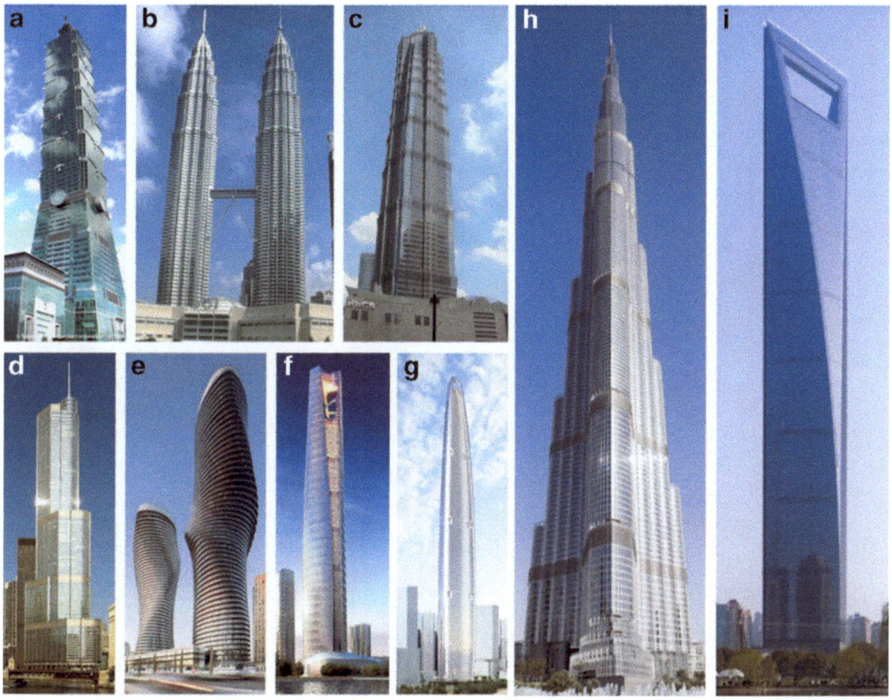

Fig. 14.3 (**a**) Taipei 101 (Taipei, Taiwan), (**b**) Petronas Towers (Kuala Lumpur, Malaysia), (**c**) Jin Mao Building (Shaghai, China), (**d**) Trump International Hotel & Tower (Chicago, USA), (**e**) Absolute World Buildings (Mississauga, Canada), (**f**) Dalian Greenland Center (Dalian, China), (**g**) Wuhan Greenland Center Tower (Wuhan, China), (**h**) Burj Khalifa (Dubai, United Arab Emirates), (**i**) Shanghai World Financial Center (Shanghai, China) (all images from the website of the Council on Tall Buildings and Urban Habitat, http://www.ctbuh.org/)

to taper the towers to reduce the plan with height provides an elegant appearance while still allowing the integration of Islamic geometric patterns in the cross-sectional shape thus merging cultural imagery with modern architecture. Further evidence of the aesthetic benefits of aerodynamic modifications is provided by the Jin Mao building in Shanghai, China (1999, Fig. 14.3c), which uses setbacks to gently taper its façade, gradually reducing and redefining the shape of the structure at the upper levels, where the effects of wind are most critical, transforming the cross-section from a square to a complex cruciform shape. The breathtaking structure not only pays tribute to the Chinese culture, with its ornate tiers reminiscent of the ancient pagodas, but also withstands typhoon winds, in part due to judicious choice of lateral system and the use of aerodynamic tapering. More recent structures with setbacks along the height include the Trump International Hotel & Tower in Chicago, USA (2009, Fig. 14.3d) and Burj Khalifa, Dubai, UAE (2010, Fig. 14.3h), which with its 828 m of height holds the record of the tallest building in the world. In particular, in this last case, the influence of vortex-induced excitations was minimized by deterring the formation of a coherent wake structure, through

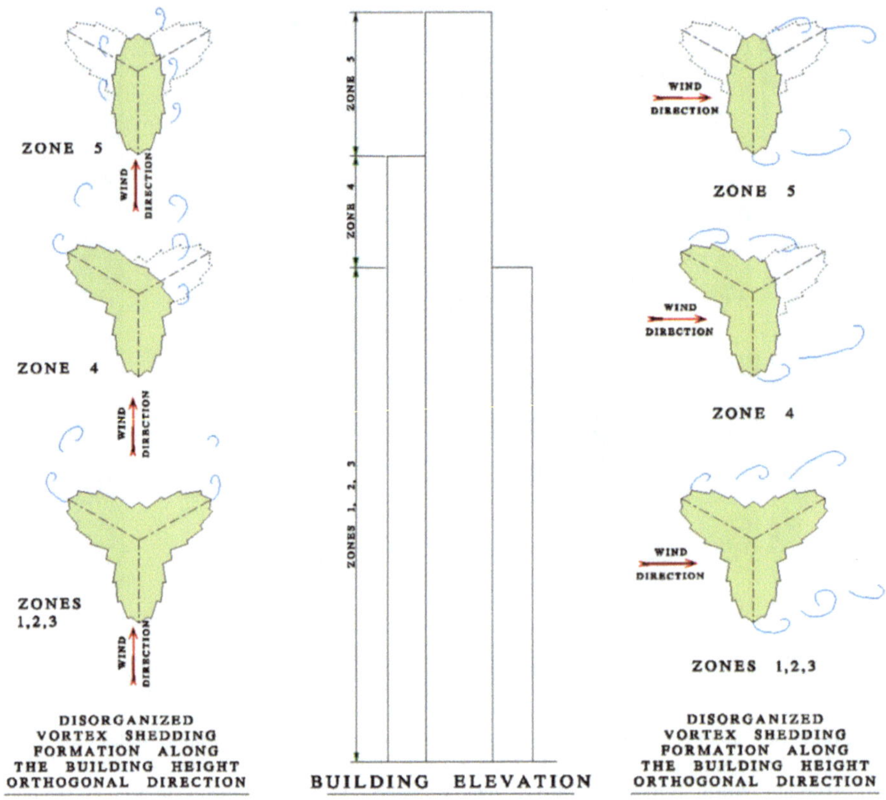

Fig. 14.4 Example of aerodynamic shape tailoring to disrupt vortex shedding along the height: Tower Palace III, Seoul, Korea [after Abdelrazaq et al. (2005)]

frequent and drastic changes in its cross-section which, at the base, has a triple-lobed shape, an abstraction of the Hymenocallis flower, and over the height is progressively modified with setbacks that follow a spiral along the height, further vitiating the impact of any potentially coherent formation of vortices. Earlier, this favorable aerodynamic behavior was also achieved for the Tower Palace III (Seoul, Korea, 2004) with a similar strategy, as demonstrated conceptually in Fig. 14.4. Indeed, the variation of the building cross-section over the height generates a building with longitudinal profile which is aerodynamically less conducive to vortex shedding and dynamic interaction of vortices from the sides of the cross-section (Abdelrazaq et al. 2005). In some cases, the change in geometry over the building height is not achieved through setbacks, i.e. sudden changes of shape, but through smoother modifications. This is the case of many of the last generation of buildings, with architectures characterized by twisting movements (as for example in the case of the Infinity Tower in Dubai, currently under construction, which is twisted by 90°) or free-form shapes (e.g. the Absolute World Building in Mississauga, Canada, completed in 2012, shown in Fig. 14.3e). The presence of

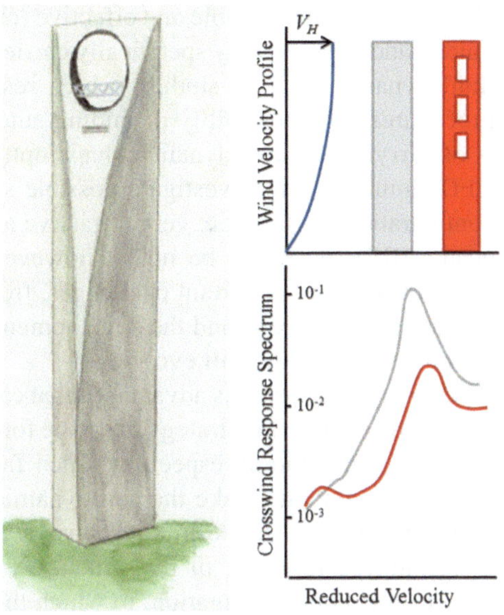

Fig. 14.5 Buildings with openings and their effectiveness (sketch courtesy of T. Kijewski)

openings extending across the building, in particular near the top, can also significantly improve the aerodynamic response, considerably reducing the wake induced forces (Fig. 14.5). Such a design strategy has been utilized in several buildings, as for example in the Shanghai World Financial Center (Fig. 14.3i), completed in 2008, which features a diagonal face that is shaved back with an aperture at the top of the building which measures 51 m in width. Other noteworthy examples are the Dalian Greenland Center, China (Fig. 14.3f), currently under construction, which is planned to be characterized by an opening of notable dimensions near the building top and the Wuhan Greenland Center Tower, China (Fig. 14.3g), also under construction, in which the distinctive floor slots at different elevations help reduce the vortex-shedding effect (Fu et al. 2012).

In addition to these passive forms of aerodynamic shape tailoring, possibilities exist to introduce active measures such as bleeding or suction of air in the separated flow regions adjacent to the building surfaces. Such addition or extraction of air causes a virtual modification of the building surface profile, manipulating the aerodynamics therefore allowing the minimization of the wind loads. Initial studies related to such measures can be found in Gad-el-Hak (2000) and Zheng and Zhang (2012).

As a final remark, it should be kept in mind that the relation between the aerodynamic characteristics of a structure and the resulting intensity of wind excitation is not straightforward. This means that the beneficial influence that geometric modifications can have on the flow field, therefore on the wind loads, is always difficult to predict, even in simple cases (e.g. introduction of chamfered corners in an originally square shape) and for minor alterations. For this reason, the choice of the most suitable and effective (optimal) changes in shape must always be guided and supported by specifically carried out wind tunnel tests or computational fluid dynamics (CFD) studies, which result in more costly and lengthy design procedures. The possibility of finding automatized solutions to this problem, that is of carrying out aerodynamic shape optimization procedures through the use of CFD simulations to investigate possible shape changes and discover the optimal configuration, is therefore very attractive and has recently gained attention (Bobby et al. 2012). It should be noted, however, that the solution to such a complex problem presents important challenges, from a modeling, computational and engineering point of view, and the development of CFD-based optimization techniques for civil structures is still evolving.

Despite the enormous advantages that can be obtained from a smart choice of the shape, and even if this strategy has to be for the designers the first direction in which to look for a solution, especially when facing a particularly challenging project, inherent limitations make the aerodynamic/aeroelastic shape tailoring not always sufficient to reach the desired response level. In a number of cases, the looked-for wind induced response of the structure is obtained through the incorporation of devices for motion mitigation, to which the following sections are dedicated.

14.3 Devices for Motion Mitigation: Generalities and Classification Principles

14.3.1 Introduction

Over the years, a relevant number of devices for motion mitigation have been studied and applied to buildings, towers, chimneys and bridges in an effort to exploit the numerous advantages these devices can provide. In (Kareem et al. 1999), an overview of the measures that reduce the structural response of buildings including a summary of auxiliary damping devices and their applications to buildings around the world is presented. Other interesting examples of applications of devices for motion mitigation to real structures can be found in Holmes (1995), Tamura (1998), Nishitani and Inoue (2001), Spencer and Nagarajaiah (2003), Tamura (2003). It is in general possible to group the various design solutions into several main typologies, typically based on the physical principle at the basis of their action. Nevertheless, it must be kept in mind that in many cases the purpose the device is required to fulfill and the peculiarity of the application make the device a one-of-its-kind. The aim of this section therefore is not, and could not be, to describe the entire spectrum of possible applications that have been studied and/or applied over the years, but to give a comprehensive overview of the main typologies, highlighting their principal characteristics and providing some significant examples.

14.3.2 Manipulation of Structural Properties for Motion Mitigation

The dynamic response of a structure is governed by its mass, stiffness and damping properties: in principle, a reduction of the response level under a certain external excitation can be achieved through suitable manipulation of any of these three properties during the design stage. Nevertheless, the mass distribution characterizing a building or a bridge, while being fairly easy to estimate, is very difficult to modify in a significant way in order to obtain benefits in terms of reduction of the structural response.

Stiffness can also easily be assessed, and relatively easily changed by means of suitable modifications in the structural system, but if the stiffening of a structure can be taken into account as a possibility to reduce the motion amplitude it will not in general benefit the acceleration responses (Kareem 1979; Tamura 1998) which, as anticipated, are fundamental in the design of slender structures for occupant comfort (and fatigue, in the case of bridges). Also, the significant stiffening of a structure often comes at a high cost in terms of material and constructability issues. A cleverer way to exploit structural stiffness modification for the mitigation of motion does however exist and is represented by the adoption of variable stiffness devices, that will be treated in the following. A specially configured structural system may offer higher stiffness without any penalty in terms of added material cost, as discussed in Kareem et al. (1999) and Taranath (2010).

On the other hand, damping plays a fundamental role in the structural response and enormous benefits can be attained by increasing the damping of the structure. Differently from mass and stiffness characteristics, however, the estimation of damping characterizing a civil structure is in general very difficult (Kareem and Gurley 1996). This can be ascribed to the complex nature of the mechanisms that are at the root of this phenomenon. Indeed, it is well known that a structure will be dynamically damped by a number of energy dissipation mechanisms with different origin and characteristics such as material damping, originating in complex molecular interaction, frictional damping between structural and/or non-structural members and connections, damping due to energy dissipation mechanisms occurring in the foundation, and aerodynamic damping. To face the enormous difficulties in the estimation of this parameter, the importance of databases collecting damping values estimated over the years for real structures is universally recognized. These collections of data are however characterized by considerable scatter, partly due to the amplitude dependence of damping, and partly due to possible errors in its identification (Haviland 1976; Davenport and Hill-Carroll 1986; Jeary 1986; Lagomarsino 1993; Satake et al. 2003). Also, much of the full-scale data results from ambient excitations and therefore the system identification has to be carried out without knowledge of the input. This restriction is often problematic and leads to a host of possible errors. Commonly, the random decrement technique and spectral analysis are the system identification techniques adopted for this problem (e.g. Jeary 1986; Kareem and Gurley 1996; Kareem et al. 1999) and, in spite of a well-established theory regarding the bias and variance errors of power spectra and random decrement signatures, the theoretical error formulae give only approximate indications of errors inherent to the estimated dynamic properties. Kijewski and Kareem (2002) utilized a bootstrapping approach to assess the quality of system identification by providing surrogate estimates of damping and natural frequencies to generate useful statistics and confidence intervals.

Being extremely difficult to estimate, damping originating from the aforementioned sources, often termed inherent damping, is consequently difficult to engineer. If however controlling the structural response level through the governing of the inherent damping is nearly an impossible task, this is not true for damping provided by auxiliary devices. The adoption of these devices, suitably designed to assist in controlling the amplitude of vibration of the structure, has had wide diffusion over the years to tame the behavior of particularly lively structures whenever the inherent damping is not sufficient to obtain the desired response level, and undoubtedly constitutes a relatively predictable and reliable method of imparting additional damping to the system.

14.3.3 Classification and Basic Principles

In the present and in the following section the main typologies of motion control devices adopted for the protection of structures, in particular against wind, are briefly described. Table 14.1 presents a classification of the various devices that have been adopted in buildings and bridges. The various categories of devices are organized considering the principle on which they are based. It should be noticed that the base isolation system and the metallic dissipators, included in the table for the sake of completeness, are only used for the protection against seismic loads, therefore will not be treated in detail in this chapter.

A large number of the devices for motion mitigation fall into the broad category of the so-called auxiliary damping devices. This encompasses a vast series of devices which dissipate energy through various physical mechanisms and have different properties which make each typology unique and particularly suitable for a specific problem. In addition to these devices which have in common the introduction of additional damping in the system, other motion control devices can be considered. These last are able to help in controlling the structural response through real-time modification of the stiffness properties of the structure (as in the case of variable stiffness devices), direct application of forces to the structure (distributed actuator systems) or variation of the geometry with consequent load reduction (variable aerodynamics devices).

Table 14.1 Motion control device classification

Category	Principle	Typology			Notes
		Passive	Active or semi-active	Hybrid[a]	
Inertial dampers	Vibration absorption through inertial effects	• Tuned mass dampers (TMD) • Tuned liquid dampers (TLD) among which tuned sloshing dampers (TSD) and tuned liquid column dampers (TLCD) • Impact dampers among which hanging chain dampers (HCD)	• Active mass dampers (AMD) • Semi-active TSDs and TLCDs • Semi-active impact dampers • Active gyrostabilizers	• Hybrid mass dampers (HMD)	Nearly exclusively for wind induced response
Viscous fluid dampers (VFD)	Development of a viscous resistant force in the fluid in opposition to the movement of the structure	• Oil dampers (OD) • Damping walls (DW), also indicated as viscous wall dampers (VWD)	• Magneto-rheological (MR) and electro-rheological dampers (ER) • Variable-orifice ODs	–	MR and ER are often used in conjunction with other devices within hybrid systems
Viscoelastic dampers (VED)	Dissipation of energy due to the hysteretic behavior of the viscoelastic material	• Single layer VED • Constrained layer VED • "Sandwich" VED	–	–	Particularly efficient against strong winds and moderate earthquakes

(continued)

Table 14.1 (continued)

Category	Principle	Typology			Notes
		Passive	Active or semi-active	Hybrid[a]	
Friction dampers (FD)	Dissipation of energy due to hysteresis of the frictional forces	• Braced frame friction dampers • Rigid frame friction dampers	• Semi-active variable friction dampers	–	Simple configuration, low cost Rigid frame only for seismic control
Metallic dissipators[b]	Dissipation of energy due to hysteretic behavior of the metallic components	• Steel dampers (SD) • Lead dampers (LD) • Unbonded braces • Added damping and stiffness devices (ADAS)	–	–	Only for seismic response (to be replaced after the seismic event)
Variable stiffness devices	Modification of the stiffness in order to prevent resonance occurring		• Active variable stiffness (AVS) system • Intelligent stiffener for bridge (ISB)	–	These systems are by nature only active
Variable aerodynamics devices	Modification of the geometry in order to reduce		• Active aerodynamic appendages • Active flaps	–	

	the aerodynamic loads				
Distributed actuator systems	Direct application of forces to the structure to counteract the action of the external excitation		• Active tendons • Active bracings	–	
Base isolation systems[b]	Shifting of the structure's frequency of vibration to avoid resonance with the input or limitation of the maximum shear forces transmitted to the structure	• Elastomeric bearings • Lead/rubber bearings • Sliding friction pendulum	• Active isolation system (passive system + control actuators) • Semi-active (passive system + semi-active damping devices)	–	Only for seismic response

[a]Hybrid devices can be in principle obtained by combining different types of passive/active/semi-active devices. These possible combinations are numerous and various and, in principle, any kind of motion control system can be combined in a hybrid typology. In the table only the hybrid mass dampers are indicated, which are the most "standard" typology

[b]Metallic dissipators and base isolation systems, included in the table for the sake of completeness, are only used against seismic loads, therefore will not be treated in detail in this chapter

Table 14.1 divides the devices into the three typologies of passive, active (or - semi-active) and hybrid. Systems for motion control are defined to be passive when their action can be initiated and maintained without the necessity of providing any power source therefore resulting, in general, inexpensive and reliable. Passive systems are also characterized by the fact that their parameters are fixed, which constitutes a limitation because their behavior cannot be adjusted to changing excitation or system characteristics. A system is instead defined as active when a controller is introduced that, based on the information provided by sensors, triggers actuators, i.e. devices able to produce forces, which in turn act on the system (in the case of the active mass dampers, the actuators initiate the movement of the additional mass introduced in the system). Although an active system can be constructed that renders better performance than the best possible passive system, the application of active systems can also be limited by higher initial and maintenance costs, higher complexity and consequently a comparatively lower reliability (Karnopp et al. 1974). The fact that an active system needs a power supply is itself a limitation, as shortage of power is likely to occur in the case of a severe event like an earthquake or wind storm. The actuators present in active motion control systems can be of various types, depending on the particular application; commonly they are hydraulic or electro-mechanical. A semi-active device can be generally defined as a device in which actuators do not operate directly on the principal system but modify the parameters of a passive motion control device introduced in the original system. As a consequence, the power required is generally less than in the case of active systems representing the fundamental advantage of this typology of devices. It should be observed, however, that the definition of semi-active devices in the field of structural control is somewhat vague and it is not uncommon to find the term semi-active control used to indicate a control system in which the actuator needs power only on specific occasions, not throughout the entire control operation (e.g. Nishitani and Inoue 2001) or when the power required is much lower than the power required by an analogous active control system (e.g. Housner et al. 1997).

For both active and semi-active devices, sensors that provide information to the controller can be installed to monitor either the structural response or the structural input. In the first case the control algorithm is closed-loop, or feedback, otherwise it is open-loop or feedforward; if the controller receives information from both the excitation and the response the algorithm it is defined as feedback-feedforward.

The term hybrid is instead adopted for all the devices that behave partially actively and partially passively. These two behaviors can occur in parallel, as for example when the device features passive control in one direction of motion and active in another (e.g. Saito et al. 2001), or sequentially, when the device is initially active, then switches to passive behavior when the excitation or the motion reaches a predefined level. Because multiple control devices are operating, hybrid control systems can allow overcoming some of the limitations that exist when each system is acting alone, therefore potentially allowing higher performance levels.

14.4 Devices for Motion Mitigation: Review of the Main Typologies

14.4.1 *Inertial Dampers*

Inertial dampers are vibration absorbers which utilize inertial effects. The additional damping provided to the principal system does not result from a dissipation of energy, as in all other typologies of auxiliary damping sources, but it originates from the modification of the building motion, or more specifically from the modification of the building frequency response (Kareem 1983). This results from the dynamic interaction between the building and the damper masses, which is tailored to be out of phase to manifest response reduction.

14.4.1.1 Tuned Mass Dampers

The basic typology of mass damper is the (passive) tuned mass damper (TMD). The principle at the base of its efficacy can be simply explained by schematizing the main structure as a single degree of freedom (DOF) linear system. The TMD consists of an additional mass that is attached to the main system through a spring and a dashpot, therefore the total system can be easily schematized as a 2-DOF system. The frequency associated with the secondary mass is tuned to a particular structural frequency: when that frequency is excited, the damper resonates out of phase with the structural motion. In the frequency domain, the presence of the TMD modifies the structural transfer function, whose original resonance peak is shifted (the new transfer function presents two peaks, corresponding to the two modal frequencies of the 2-DOF system, one higher and one lower than the original system's frequency). Different kinds of TMDs have been designed for various structures. They can be divided into translational TMDs, where the mass rests on bearings that allow it to translate laterally, and pendulum TMDs, where the mass is instead supported by cables. In this last case, it is more frequent the adoption of compound pendulum schemes, which allow a shorter effective length of the pendulum: this can be extremely advantageous when there are limitations on the space available. Bidirectional translational TMDs are configured with springs/dampers in two orthogonal directions and therefore are capable of controlling the structural motion in two orthogonal planes. For the application to buildings, the mass weighs around 0.25–0.75 % of the building weight, corresponding approximately to 1–2 % of the first modal mass (Kareem 1983). Notable examples, also for historical reasons, of the application of translational TMDs are the John Hancock Building (241 m, Boston, USA, 1977), where two dampers of 270 ton each (consisting of lead-filled steel boxes) are placed at the opposite ends of the 58th story, in order to mitigate both sway and twisting of the building, and the Citycorp Center (278 m, New York, USA, 1978), where a TMD of around 370 ton, made out of concrete, is located in the crown of the structure. A more recent example, relevant for the weight of the TMD (600 ton), is the Trump Tower in New York City (262 m, New York, USA, 2001). In the case of the Bronx-Whitestone Bridge, New York, which opened to traffic in 1939, a 94 ton TMD was installed in 1987 to improve aerodynamic stability. Among the TMDs with a pendulum scheme, it is worth mentioning the Crystal Tower (157 m, Osaka, Japan, 1990), where a tuned pendulum mass damper had to be included in the early phase of the design to decrease the wind-induced response level by about 50 %. To this end, six air cooling and heating ice thermal storage tanks, each weighing 90 ton, have been hung and used as pendulum masses (four working in one direction and two in the opposite direction). By using masses that were already present in the building, the cost of the motion control system was significantly reduced. Another well-known example of a pendulum TMD is the case of the Taipei 101 (508 m, Taipei, Taiwan, 2004), particularly remarkable for its 740 ton of weight, that make it to date the heaviest TMD in the world. The TMD mass consists of a large steel ball hung through a mezzanine level in the buildings making it attractive for the visitors and the bar customers.

Multiple TMDs (Kareem and Kline 1995) have also been studied and implemented to overcome the inherent limitation of the classic single TMD systems which can only act on a single natural frequency even though multiple frequencies will in general be excited by the external input. The aim of multiple TMDs is therefore to distribute the natural frequencies of a number of TMDs around the fundamental natural frequency of the structure in order to have a more effective control of the structural response.

14.4.1.2 Tuned Liquid Dampers

Since 1950s liquid dampers have been used to stabilize marine vessels or to control motion of satellites, but only in the 1980s they started to be considered as possible motion control devices in civil structures (Kareem and Sun 1987). Tuned liquid dampers (TLD) encompass tuned sloshing dampers (TSDs) and tuned liquid column dampers (TLCDs). These devices are basically similar to TMDs where the secondary system (mass, spring and dashpot) is replaced by a partially liquid-filled (usually with water) rigid tank. By only partially filling the tank the liquid can slosh from one side to another inside the tank therefore acting analogously to the mass in the TMD systems. Also, additional damping is provided through dissipation of energy generated by the friction between the boundary layers of the fluid and wave breaking. This dissipation mechanism can be enhanced through the inclusion of screens or partially closed sluices, or the addition of floating particles to the fluid. TSDs can be classified into deep and shallow types, depending on the ratio of the water depth to the wave length in the direction of motion (this classification is also important from a modeling perspective) (Kareem 1990). TLCDs are a special type of TLDs relying on the motion of a column of liquid in a U-tube-like container, where damping is introduced in the oscillating liquid column through an orifice in the liquid passage. An enhancement of the robustness of these devices can be achieved through the possibility of adjusting the natural frequency of the mass

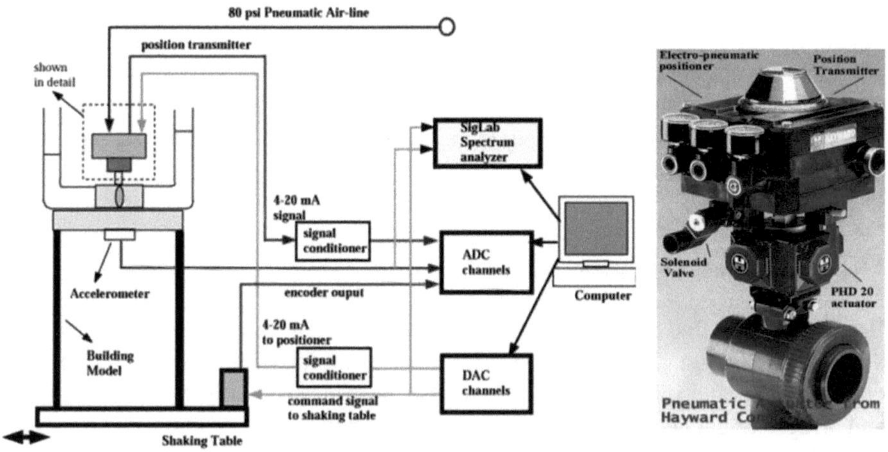

Fig. 14.6 Semi-active control of TLCD (*left*) and electro-pneumatic valve (*right*) (Yalla and Kareem 2003)

damper systems in order to tune it to the structure's frequency (or re-tune it, due to possible changes in the structural characteristics occurring during the structure lifetime) utilizing period-adjustable TLDs (Shimizu and Teramura 1994). In contrast to the TMDs, the damping of TLDs is amplitude dependent. This makes these devices governed by non-linear dynamics and in general more difficult to model. The liquid sloshing has been modeled with different levels of sophistication, i.e. it ranges from simplistic models based on the tuned mass damper analogy to the solution of shallow water equation representing sloshing. Despite these advances, most models need some tweaking to account for large damping at wave breaking. Yalla and Kareem (1999) developed a mechanical analog of the sloshing action named sloshing-slamming damper, or S2 damper. The model consists of a linear sloshing model at low amplitudes of base excitation, whereas at higher amplitudes of motion a part of its mass transforms from sloshing action to rolling between the tank walls, thus acting as an impact damper (see below). The impact action adds damping, thus explaining the high level of damping at large amplitudes. The non-optimal performance of TLCDs at a wide range of wind speeds is enhanced by introducing a semi-actively controlled electro-pneumatic valve (Yalla and Kareem 2003) (Fig. 14.6). Analogously to the case of TMDs, multiple tuned liquid dampers can also be considered (Fujino and Sun 1993). Despite the complexity in modeling, TLDs present notable advantages among which low initial and maintenance cost and, most importantly, the fact that the water can be utilized as water supply for the building (e.g. for fire extinguishing purposes), unlike a TMD where the dead weight of the mass has in general no other functional use. Among the first civil structures with applications of TLDs for reduction of wind-induced vibrations, where the efficiency of these devices was proved through full-scale monitoring, are the Yokohama Marine Tower (105 m, Yokohama, Japan), where a TLD formed by 39 cylindrical vessels was installed in 1987, the Shin-Yokohama Prince Hotel

(149 m, Yokohama, Japan, 1992, 30 cylindrical vessels), and the Tokyo International Airport Tower (78 m), where a number as high as 1,400 vessels containing floating polyethylene pieces in addition to water was installed in 1992. It is interesting to observe that the majority of TLDs have been installed for protection against wind while only a small percentage have also the aim of protecting against seismic forces (Tamura 1998).

14.4.1.3 Impact Dampers

Impact dampers find their main field of application in mechanical engineering, with the aim of reducing vibrations of systems such as turbine blades, light poles, printed circuit boards and robot arms, but have also been occasionally used for buildings and bridges (e.g. Ogawa et al. 1997) when their oscillations occur mainly in a single plane. The basic typology of impact damper consists of a rigid mass which is able to slide, or oscillate, according to the direction of oscillation of the structure and whose movement, which is triggered by the structure's vibration, is confined within a container. When the mass reaches the container's walls, alternatively on one side or on the other, it impacts inelastically on the container's walls, therefore dissipating energy. Due to the fact that the system has to be tweaked to the frequency of the system, the gap distance is the fundamental parameter that allows the tuning of the damper. Hanging chain dampers (HCD) are a particular kind of impact damper, which is mainly used to control the dynamic response of tower-like structures, as spires or pinnacles atop buildings, features used in many iconic constructions.

14.4.1.4 Active, Semi-Active and Hybrid Mass Dampers

The idea of using active control has been adopted in a full-scale application for the first time by the Kajima Corporation in 1989 (Kobori et al. 1991), when an active mass damper (AMD) was applied to the 11-story Kyobashi Seiwa building in Tokyo, Japan, for the seismic protection of the building. As previously anticipated, an AMD is simply a passive TMD whose movement is initiated by actuators, according to a control law. In this way, the coupling between the tuned mass damper and the building is actively controlled and consequently higher damping (or the same damping level with however a smaller mass) is provided together with a fast tuning to the building motion (therefore higher efficiency). While a conventional TMD may provide an additional damping of 3–4 % of critical damping, an active system can add more than 10 %. On the other hand, the additional equipment required for an active system increases its cost (Kwok and Samali 1995). AMDs are usually designed for wind protection; in some cases, seismic action of small entity can also be considered. As for the TMDs, there exist interesting cases of AMDs exploiting masses already present in the building to save damper space and avoid adding mass which otherwise places a heavy premium on the structural system.

Among these, it is worth mentioning the 400 ton heat storage tanks used as a mass of AMD in LTC Bank (Tokyo, Japan), and the 480 ton heliport deck used as a mass of AMD in Hankyu Chayamachi Building (Osaka, Japan).

Semi-active mass dampers can be obtained introducing a control of the characteristics of a classic TMD or TLD in order to tune in real-time the frequency of the device and to increase the range of effective application. In the case of semi-active TMDs, the control regulates the stiffness of the additional single degree of freedom system or its damping. In the case of TLDs different semi-active control strategies have been studied, such as the regulation of the natural period of the water tank through control of the orientation of a set of rotatable baffles in the tank or, in the case of TLCD, the insertion of an electropneumatic valve to provide optimal damping at a wide range of structural motion amplitudes (Yalla and Kareem 2003).

Over the years, active, semi-active and hybrid mass dampers (combination of active/semi-active and passive devices) experienced wider acceptance all over the world. Many of the devices that have been installed feature peculiar characteristics, being specifically designed for the particular structure in which they are applied, and it is virtually impossible to report all the typologies which have been engineered for different applications (Tamura 1998). Among the most interesting cases it is worth mentioning the semi-active technology of TMD coupled with magneto-rheological dampers, variable stiffness TMDs and multiple TMDs (Nagarajaiah and Sonmez 2007), adaptive pendulum TMDs (Nagarajaiah 2009), semi-active tuned liquid column dampers (Yalla and Kareem 2003) and the multi-story semi-active tuned mass damper building system (Chey et al. 2010).

14.4.2 Viscous Fluid Dampers

Viscous dampers exploit the properties of highly viscous materials, such as silicone oils, to develop resisting forces proportional to the body's velocity that oppose the movement of the body.

14.4.2.1 Oil Dampers (OD)

The most broadly used viscous damper consists of a cylinder-piston system containing the viscous fluid, which is forced through orifices, as the piston head moves, from one side to the other side converting the upstream pressure into kinetic energy. These devices, usually termed oil dampers (ODs), were extensively tested at the beginning of the 1990s, when they started to be used for structures subjected to earthquakes. Successively, their use was extended for wind-induced vibration control (McNamara and Taylor 2003), even if this application remains much rarer. In the case of buildings, ODs are generally inserted in braced systems distributed over the structure. The configuration of the braces (diagonal, chevron...) can be important, because the damper will be more

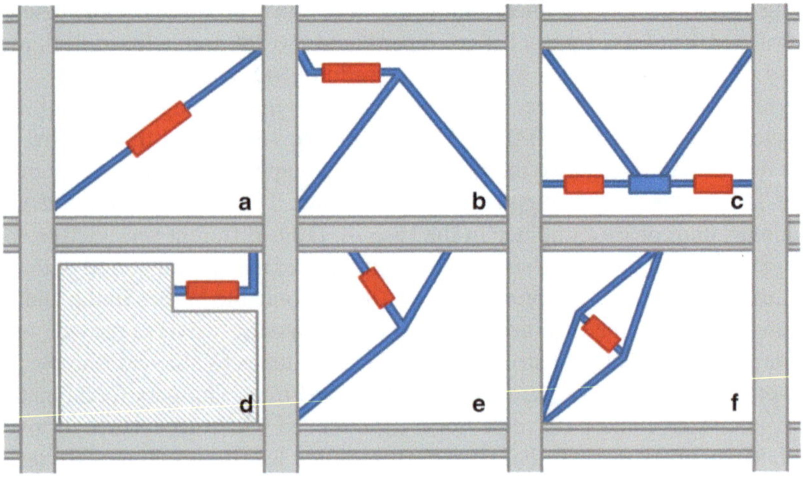

Fig. 14.7 Examples of installation schemes for dampers, (**a**) diagonal brace, (**b**) chevron brace, (**c**) V brace, (**d**) rigid stud wall system, (**e**) toggle system, (**f**) scissor jack system

effective depending on the ratio of the displacement absorbed by the damping device to the inter-story drift displacement. For this reason the choice of non-classical bracing configurations, such as the toggle or the scissor-jack configuration (Fig. 14.7), has been the subject of several studies (e.g. Constantinou et al. 2001). It should be observed that the bracing configuration is even more important in the case of wind-induced vibrations, where the devices have to be activated at a low displacement level (McNamara and Taylor 2003). One of the advantageous characteristics of this kind of device lies in the fact that they provide forces which are out of phase with respect to the motion of the structure and therefore reach their maximum values when the forces in the members due to the deformation of the structure are at their minimum. Viscous fluid dampers of this typology have been used in many structures, also in retrofitting damaged or old structures. A notable example of this is the Solomon R. Guggenheim Museum in New York, opened in 1959, where 54 viscous dampers with 30 mm stroke and 20 kN of limit for the actuating force have been installed in order to control wind and traffic vibrations in 2008. Another interesting example is the recently-built Manhattan office tower 250 West 55th Street, a 39-story building completed in 2012. For its initial design, the level of the wind-induced movement predicted based on wind tunnel test results was unacceptable. The problem was fixed by means of the incorporation of viscous dampers (7 with 100-mm stroke and 1,690 kN of limit actuating force) embedded in the outrigger system. In the case of bridges, for example, the cable-stayed Pomeroy-Mason Bridge, West Virginia, was provided in 2007 with fluid viscous dampers that were connected to the stay cables in order to reduce wind- and rain-induced vibrations.

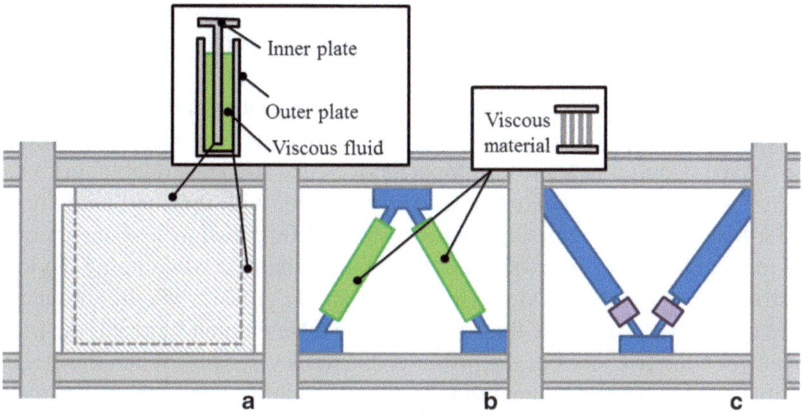

Fig. 14.8 Example of installation scheme for various typologies of dampers: (**a**) viscous damping wall, (**b**) viscoelastic dampers, (**c**) friction damper

14.4.2.2 Damping Walls

A damping wall (DW, sometimes also termed viscous wall damper or VWD) (Lu et al. 2008) is a kind of viscous fluid damper firstly developed by the Japanese engineer Mitsuo Miyazaki and his co-workers (Arima et al. 1988; Miyazaki and Mitsusaka 1992). A damping wall occupies an entire panel of a frame, as illustrated in Fig. 14.8a, and consists of a steel wall plate (that does not carry any loads) contained in an external steel case with a viscous fluid injected in the gap between the two. The middle plate is connected to the upper floor, while the external case is connected to the lower floor, so that a relative movement of the floors is opposed by the resistance operated by the viscous fluid filling the gap. Viscous damping walls have been installed in the Sato Building in Tokyo in 1992 (Miyazaki and Mitsusaka 1992) and in many other buildings in Japan, both for wind and earthquake protection, while the use of these devices is not as common in other countries. Indeed, the first application in the United States of this technology is the Cathedral Hill Hospital of San Francisco, California, currently under construction, where the presence of the DWs is envisaged to benefit the wind-induced response, even if the devices have been primarily designed as a means of seismic protection (Brown 2011).

14.4.2.3 Electro-Rheological and Magneto-Rheological Dampers

Starting from the 1980s, and more intensively at the beginning of the 1990s, two typologies of semi-active oil dampers which use electro-rheological (ER) and magneto-rheological (MR) fluids have been investigated and gained a wide diffusion in the field of civil structures. ER and MR fluids can change from liquid to semi-solid state under an electrical or magnetic field. This change can take place within milliseconds and is completely reversible. Although the configurations of semi-active ER and MR dampers can be various, the main idea is that the variation of the viscosity of the fluid, which is opposing the relative motion of two parts of the device, therefore contrasting the motion of the structure, changes the stiffness of the device (Coulter et al. 1993). Most devices that use controllable fluids can be classified as having either fixed poles (valve mode, with the fluid contained in a cylinder-piston system and subjected to tension/compression and shear) or relatively moveable poles (direct-shear mode), but in the case of damping devices for civil structures the valve mode is adopted (Carlson et al. 1995). In absence of an electric or magnetic field, the fluid behaves like a Newtonian fluid and the pressure drop between the two sides of the piston is due to its viscosity. When the electric or magnetic field is applied, the fluid develops a yield stress which produces an additional pressure drop. The total damping force developed by the device is given by the summation of these two force components and can therefore be controlled by varying the degree of polarization of the fluid, regulating the voltage source that controls the field. Several phenomenological models have been attempted for the description of the ER and MR dampers' complex behavior (e.g. Spencer et al. 1997; Sims et al. 1999). The accuracy of these models is fundamental for the device sizing and in the design of suitable control systems.

14.4.2.4 Semi-Active Variable-Orifice Oil Dampers

The semi-active variable-orifice oil dampers are equipped with a device controlling the damping coefficient by adjusting the opening of a valve to govern the flow of confined oil. Analogously to other kinds of semi-active devices, these dampers allow the regulation of a large reaction force with a small amount of external energy. It has been demonstrated that semi-active variable-orifice oil dampers can dissipate twice as much energy as an ordinary passive damper (Kurino et al. 2003).

14.4.3 Viscoelastic Dampers

Viscoelastic dampers (VEDs) are among the earliest devices included in structural systems in order to reduce the amplitude of the dynamic structural response and the acceleration level in the case of wind forces (Samali and Kwok 1995). They exploit the viscous characteristics of some polymeric materials, such as neoprene, which have the property of dissipating energy by means of heat loss when subjected to shear. VEDs can be classified in three distinct types: single layer (where the viscoelastic layer is deformed axially), constraint layer (again with a single layer of viscoelastic material, but subjected to both shear and axial deformation) and "sandwich" type (consisting in one or more layers that, constrained between rigid metal plates, are deformed nearly exclusively for shear). The sandwich type can be considered in general more efficient and therefore is more widely used. Analogously to the case of oil dampers, a damping system involving VEDs consists of a number of devices installed at different points of the structure where a relative displacement during a wind or seismic event is expected. Generally VEDs are placed in bracing systems (Fig. 14.8b) but in some cases different systems have been designed, such as the World Trade Center Towers, New York, built in the 1970s, where VEDs were distributed between the lower chords of the floor joists and the columns belonging to the peripheral wall, at approximately 10,000 locations for each of the two buildings. In this way, the dampers' action was activated by the relative movement of the floor and the external walls, when the structure was subjected to horizontal wind loads. Another notable example of application of VEDs for wind motion reduction is the case of the Columbia Seafirst Center of Seattle, built in 1984, where 260 dampers are attached to the bracing systems. This represents the first case in which multilayer VEDs were used in a tall building. After these pioneering applications for reduction of wind-induced vibrations, VEDs have been widely studied and applied, and, starting from the 1990s, the possibility of using these devices also for seismic protection has been investigated. Current research is devoted at exploring efficient ways to determine, by means of numerical analysis and test results, the optimal parameters, optimal location and innovative installation schemes of viscoelastic devices, among which it is worth mentioning the application of VEDs between adjacent buildings (Hejazi et al. 2010) or as coupling beams between shear walls of concrete buildings.

14.4.4 Friction Dampers

The direct dissipation of energy through frictional mechanisms providing a plastic behavior in selected regions of a structure (Fig. 14.8c) is another widely used means of structural response reduction. Friction dampers (FDs) are designed to have a fixed slipping load, dependent on the clamping force applied, below which they are able to provide high stiffness. When the activation load is reached, two surfaces dissipate energy by slipping on each other. The simple mechanisms, low required maintenance and powerful energy dissipation capability together with low installation/maintenance costs represent the main advantages of the friction dampers over other kinds of passive dissipative devices. It should be observed that in many cases these devices are intended for seismic protection; after the event, i.e. after plastic hinges are formed, their original configuration is not recovered (the devices have to be substituted). Friction dampers can however also be applied for wind-induced vibration control (Patil and Jangid 2009).

In order to improve the performance of passive dampers, various semi-active FDs have been proposed in the literature. Basically, a semi-active FD is able to adjust its slip force by controlling its clamping force in real-time, according to the motion of the structure. An example of semi-active FD is the piezoelectric FD (Xu et al. 2001). This device exploits the properties of a piezoelectric material, in which an electric charge generates a strain state, to control the clamping force, therefore the slip force of the damper, based on the feedback on the structural motion.

14.4.5 Variable Stiffness Devices

The active variable stiffness (AVS) system was first developed by Kobori and coworkers at the beginning of the 1990s (Kobori et al. 1993) as a seismic response control device and was first implemented in a case study 3-story building built in a research facility, the Kajima Research Laboratory. This system has the aim of producing a non-stationary, non-resonant condition in the building during an earthquake by actively controlling its stiffness. This is done by means of variable stiffness devices (VSDs), installed between steel braces and beams, which can be engaged or released, becoming ineffective, in accordance with the feedforward control algorithm's decision. The VSDs consist in hydraulic cylinders whose valves connecting the two chambers can be in open or close position therefore unlocking or locking the relative movement between brace and beam. In this way, the stiffness of the system can be instantaneously changed by modifying the condition of the VSDs according to different combinations (e.g. only the VSDs of the first floor, only the first and second, and so on) in order to obtain a host of various possible stiffness values for the building. The control algorithm chooses the stiffness corresponding to the modal frequency of the structure which gives instantaneously the least building response level. An important advantage of this system compared to other active control systems is the low energy power required for the valve control, which makes a small generator sufficient to maintain the system operative even during blackouts. Other studies on the possible implementation and the benefits of variable stiffness devices were conducted by Nemir et al. (1994), where controllable spring elements that can be instantaneously clamped or released are considered. Many years after the first applications, research on this kind of devices is still open. To the knowledge of the authors, apart from the laboratory tests, applications of this system to real structures have yet to be carried out.

A system based on a similar principle that has been applied to a real bridge is the intelligent stiffener for bridges (ISB). This is an adjustable stiffness system that was attached to an in-service interstate bridge (the Walnut Creek Bridge, in the vicinity of Purcell, Oklahoma). The stiffener is equipped with an adjustable hydraulic link which is used to regulate its stiffness in order to reduce the damage generated by vehicle-induced vibrations. The system was monitored between 1997 and 1999 and its effectiveness was proved (Patten et al. 1999).

As a final remark, it should be observed that the concept of variable stiffness is often used in the literature in conjunction with other kinds of devices, typically mass dampers, to obtain a semi-active control by changing the stiffness of the

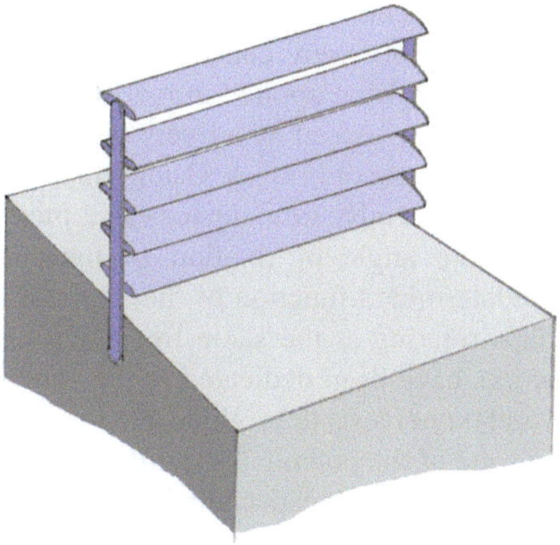

Fig. 14.9 Example of aerodynamic appendage design

passive device itself. Such cases have to be considered, for the purpose of this classification, as semi-active mass dampers and not as variable stiffness devices, which instead act directly on the stiffness of the structures.

14.4.6 Variable Aerodynamics Devices

The idea of actively controlling the wind-induced response of tall buildings through variable aerodynamic devices was first introduced by Klein et al. (1972). The proposed aerodynamic appendage consisted essentially in a metallic plate placed on the building roof, which is completely folded when the velocity at the building top is along the main wind direction and fully deployed when the velocity is in the opposite direction to the main wind direction (see for example Fig. 14.9). The main objective of this device was to use the power generated by the wind to produce control forces while using only a limited amount of external energy for the deployment/disposal of the appendages. In the following years, studies have been dedicated to the development of efficient control algorithms for this device, and experiments in wind tunnel tests have been carried out (Abdel-Rohman and Al-Zanaidi 1987; Soong 1990). The appendage has been found efficient only for wind blowing perpendicularly to its plane and for very flexible buildings. In addition, the feasibility of such a system is questionable, because of the large actuator force needed for small intervals of time (Abdel-Rohman 1987). A different proposal for suppressing aerodynamic vibrations in the case of tall buildings with a square cross-section is the moving surface boundary-layer control (MSBC) (Kubo et al. 1993, 1996). This consists in the insertion of cylinders at the corners of a building. These cylinders extend for a certain length over the building height and, actively rotating about their vertical axes, help in reducing the wind-induced loads. Wind tunnel tests have been conducted to study the efficacy of these devices where several different configurations were considered. To the knowledge of the authors, similar devices have never been applied in real structures as they pose challenges that even dwarf the concerns posed by a lively building.

In the case of bridges, Ostenfeld and Larsen (1992) first proposed the concept of variable aerodynamic devices as a means of flutter control, suggesting to actively vary the angle of rotation of flaps attached to the sides of the girder as a predetermined function of the angle of rotation of the girder, in order to have the flaps rotating at the same frequency but out of phase. After this proposal, other works have been dedicated to the study of this kind of devices, also supported by wind tunnel tests (e.g. Wilde and Fujino 1998; Nissen et al. 2004; Kirch et al. 2009). In spite of the potentially significant advantages that could be obtained by this kind of active control, especially in the case of suspension bridges of exceptional length, the problem is characterized by high complexity, and the possibility of application to real bridges would still seem distant.

14.4.7 Distributed Actuator Systems

A structure with distributed actuators is designed in order to have a certain number of actuators generating forces on selected structural members in order to compensate the effects of the external forcing functions such as impact, wind and earthquake forces (Adeli and Saleh 1997; Saleh and Adeli 1998). Obviously the design of such a system is challenging and this field is still under development, especially for what concerns the definition of feasible and effective control strategies for complex systems (many elements, many actuators) subject to different load typologies, and for the optimization of the actuator arrangement over the structure. A state-of-the-art on these "smart" structures can be found in Fisco and Adeli (2011).

Within the family of the structures with distributed actuator systems it is possible to identify two main typologies, which consider the placement of actuators in tendons and in bracing systems, respectively. The idea of using an active tendon control system was first theoretically considered during the 1970s (Roorda 1975; Yang and Giannopoulos 1978). Tendons are typical structural elements that have always been used to provide stiffness to a structure and therefore to reduce the amplitude of oscillations. However, the addition of stiffness tends to increase the frequency of the structure and consequently can generate problems of material fatigue and discomfort. The idea of using actuators to actively operate the tendons comes therefore from the need of a trade-off between advantages and disadvantages in the use of these structural elements (Leipholz and Abdel-Rohman 1986) and has been used in a number of applications. The active tendon system exploits a system of pulleys and tendons to transmit the forces generated by hydraulic actuators to the structure (Fig. 14.10); this allows to control various vibrational modes with the applied forces modifying both structural stiffness and damping. In cable-stayed bridges, active tendons are placed at the end of cables. The development of this kind

Fig. 14.10 Schematic diagram of an active tendon system

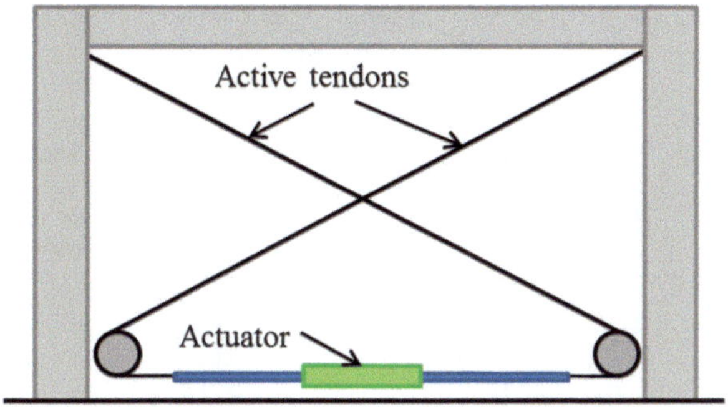

of system owes a lot to the numerous tests that have been carried out especially throughout the 1980s; by 1990 they had already been installed in full-scale structures (Soong 1990). Active bracing systems (ABSs) are in principle similar to the active tendon systems. They consist of solid diagonal tube braces attached at the building frame which can be extended or shortened through hydraulic actuators, according to the input provided by the control system. Finally, the aforementioned systems can use pulse generators, therefore pneumatic mechanisms, instead of hydraulic actuators to generate the control forces.

14.5 Recent Trends

Due to the nature of steady state wind excitation found in most wind conditions, control devices are generally efficient in controlling wind-induced motions, but their efficiency under seismic actions may sometimes be compromised, especially for extremely strong earthquakes. As a result, the urgency of the early 1990s to install AMDs or HMDs seems to have faded. For protection against strong earthquakes, metallic dissipators may currently be the most commonly chosen devices in Japan, considering their cost and maintenance as well as their efficiency and their promising performance. Some typologies of dampers, such as the VFDs and the VEDs, can be used for applications to both earthquakes and winds, but the recent trend is instead to use multiple devices to target different design requirements. This is the case, for example, of the 193 m tall Nittele Tower, Tokyo, Japan, built in 2003, where four different types of damping devices have been installed (Fig. 14.11): two TMDs for habitability under wind-induced vibrations, 32 ODs for wind-induced vibrations and weak/medium-class earthquakes, 64 unbonded bracing dampers and 312 link beam dampers (both types belonging to the category of the metallic dissipators) for extremely strong earthquakes (Tamura 2003). During the same year, AMDs for habitability under wind-induced vibrations and semi-active ODs

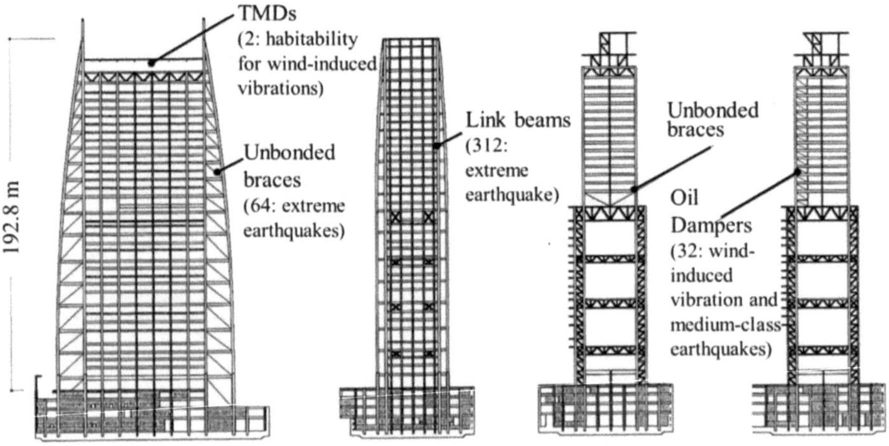

Fig. 14.11 Nittele Tower, Tokyo, Japan: placement of the four different types of devices installed for protection against wind and earthquake (courtesy of Y. Tamura)

for wind-induced and extremely strong earthquakes were installed in the Royal Park Shiodome Tower, Tokyo, Japan (173 m). AMDs and metallic dissipators (honeycomb dampers) were installed in the Kyodo News Service building (172 m, 2003) for habitability under wind-induced vibrations and extremely strong earthquakes, respectively (Fig. 14.12). In Roppongi Hills, Tokyo, Japan (238 m, 2003), 192 semi-active ODs were installed for winds and extremely strong earthquakes and 356 unbonded braces for extremely strong earthquakes. In the case of the 634 m tall Tokyo Sky Tree Tower (Tokyo, Japan, 2012), a number of ODs are used to connect the reinforced concrete central pillar to the external steel structure between the heights of 125 and 375 m, with the main scope of controlling the response under earthquake excitation. These devices are activated by the relative movement between the internal core and the exterior structural system (Kumar and Kareem 2011). In addition, two inverted pendulum TMDs have been installed at the heights of 620 and 625 m to control the wind induced vibrations. These combinations of different types of devices allow a more efficient and reliable damper design by focusing on the target performance of each damping system, thus enhancing their effectiveness and robustness (Tamura 2003).

Another interesting example concerns the joint application of damping devices and damage monitoring devices. Harumi Island Triton Square (2001, Tokyo, Japan) consists of three tall buildings, Offices X (195 m), Y (175 m) and Z (155 m), each with a square plan. These buildings have unbonded bracing dampers made of low yield strength steel for extremely strong earthquakes (Asano et al. 2002). In Office Y, 2,840 unbonded braces are arranged in the outside frames from the third story up. These devices were implemented for seismic responses only, and not for wind-induced responses. The main structural frames were designed not to reach their full-plastic moment capacity even under extremely strong earthquakes with a maximum velocity of 50 cm/s. A host of computations and experimental results

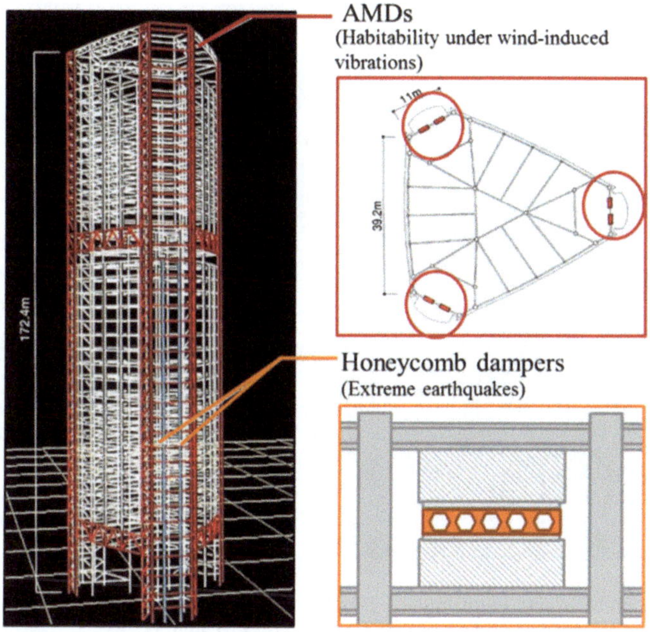

Fig. 14.12 Kyodo News Service Building, Tokyo, Japan: placement of the different types of devices installed for protection against wind and earthquake (courtesy of Y. Tamura)

were conducted to ensure the performance of the damping devices. However, in considering unexpected situations within the building's lifetime or reliability of the design values, it was necessary to capture the exact condition of the damping devices after completion of the buildings. Then, damage monitoring sensors or devices were attached to all the unbonded bracing dampers. The buildings are also characterized by the presence of two active damping bridges for habitability under wind-induced vibrations (Koike et al. 2001). Each active damping bridge connects two adjacent buildings and acts as an active damper with acceleration feedback control when the two buildings experience relative displacements, therefore reducing wind-induced vibrations based on the acceleration (Fig. 14.13).

Another approach could be to isolate the building cladding from the structural system therefore avoiding that the dynamic loads be transferred to this last. This approach can also be followed by spatially dividing the cladding into regular and isolated sections over the height of the building. In this manner isolated panels could be responsible for breaking up the wind load correlation, hence resulting in reduced overall loads (Kareem 1994).

A new technology recently proposed by Kareem and Zuo (2012) consists in the harvesting of the vibration energy created by wind acting on tall buildings while simultaneously reducing the wind induced structural response. This is accomplished by using TMDs with embedded high-energy-density electromagnetic

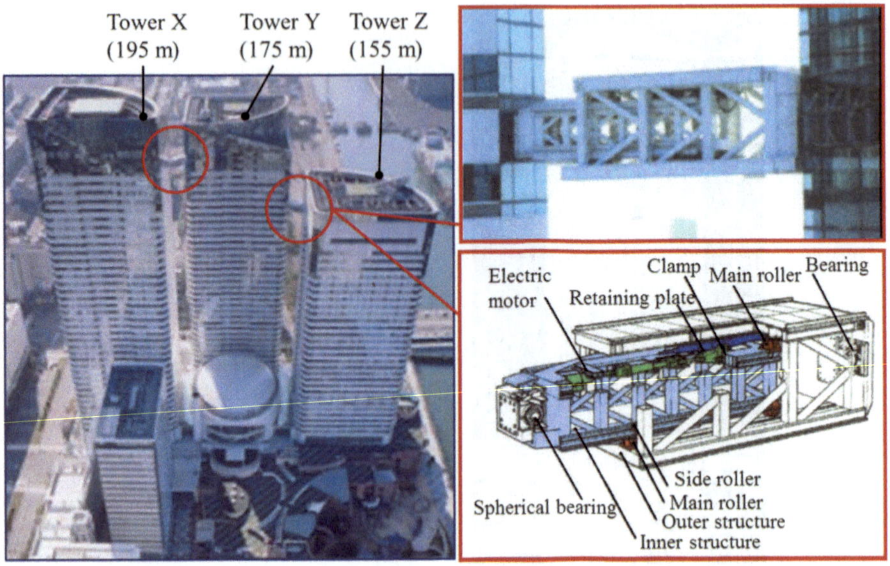

Fig. 14.13 Harumi Island Triton Square Buildings, Tokyo, Japan: active damping bridges for habitability under wind-induced vibrations (courtesy of Y. Tamura)

transducers that allow the conversion of the energy of the low frequency building motion to electricity. This green source of energy may give a fundamental contribution to improve the building's sustainability. This concept can also be considered for the isolated cladding panels discussed above.

Finally, among the most recent trends in the structural motion control field, it is worth mentioning the interest in the use of variable stiffness devices exploiting the use of "smart" materials, in particular shape memory alloys (SMAs). These materials are characterized by polymorphism, i.e. they have more than one crystal structure, and the reversible change from one to the other depends on both temperature and external stress. Although they were discovered in the 1930s, only recently have researchers started to study ways of exploiting the different electric, thermal and above all mechanical properties that are associated to the various phases for the application in motion control devices.

While the incorporation of such technologies may permit contemporary structures to reach even greater heights, it is interesting to reflect and notice that these concepts are not new. Nearly 1,200 years ago, the Japanese utilized many of these concepts summarized here in the design of their famous pagodas, making these structures resistant to both the action of typhoons and earthquakes. The secret of their enduring strength and stability lies in their tapered configuration, the variation of their cross-section with height, and the fact that the energy dissipation occurs at each level, since the levels are not attached to one another and may independently slide back and forth. This is achieved using the "shinbashira", the central pillar attached to the ground, which serves as a pivot, constraining each level from swinging too far in any direction. As the independent levels impact this fixture, energy is introduced, which is then dispersed through soil radiation damping. Thus the concept of secondary inertial system, friction and impact dampers, and aerodynamic tailoring are not so revolutionary. In fact, the same strategies exploited in modern times for urban skyscrapers, today's counterpart of pagodas, have been ingeniously tapped by the ancient Japanese builders for centuries (Kareem et al. 1999). Accordingly, a short journey into the past may be invaluable in crafting a brighter outlook for the future!

Acknowledgements The support for this study was in part provided by NSF Grants CMMI 0324331, CMMI 06-01143, CMMI 09-28282 and the Global Center of Excellence at Tokyo Polytechnic University, funded by MEXT. The author would also like to thank Prof. Yukio Tamura for his collaboration on this study: without his many contributions this study would be incomplete.

References

Abdel-Rohman M (1987) Feasibility of active control of tall buildings against wind. J Struct Eng 113(2):349–362

Abdel-Rohman M, Al-Zanaidi M (1987) Design of appendages for tall building control. J Struct Eng 113(2):397–408

Abdelrazaq A, Kijewski-Correa T, Song Y, Case P, Isyumov N, Kareem A (2005) Design and full-scale monitoring of the tallest building in Korea: Tower Palace III. In: Proceedings of the sixth Asia-Pacific conference on wind engineering, Seoul

Adeli H, Saleh A (1997) Optimal control of adaptive/smart bridge structures. J Struct Eng 123 (2):218–226

Arima F et al (1988) A study on building with large damping using viscous damping walls. In: Proceedings of the 9th conference on earthquake engineering, Tokyo-Kyoto

Asano M, Yamano Y, Yoshie K (2002) Damage controlled design for the office towers in Harumi Island Triton Square. In: Proceedings of the structural engineers world congress. Yokohama, pp T1–6–d–5

Bobby S, Spence SMJ, Wei D, Bernardini E, Kareem A (2012) A cohesive performance-based optimization framework for the design of tall buildings subject to wind loads. In: 3rd American association for wind engineering workshop, Hyannis

Brown JL (2011) Viscous wall dampers enhance hospital's seismic performance. Civil Eng ASCE 81(6):32–33

Carlson JD, Catanzarite DM, Clair KAS (1995) Commercial magneto-rheological fluid devices. In: Proceedings of the 5th international conference on electro-rheological, magneto-rheological suspensions and associated technology. Sheffield

Chey MH, Mander JG, Carr JB (2010) Semi-active tuned mass damper building systems: application. Earthquake Eng Struct Dyn 39(1):69–89

Constantinou MC, Tsopelas P, Hammel W, Sigaher AN (2001) Toggle-brace-damper seismic energy dissipation systems. J Struct Eng 127(2):105–112

Coulter JP, Weiss KD, Carlson JD (1993) Engineering applications of electrorheological materials. J Intell Mater Syst Struct 4(2):248–259

Davenport AG, Hill-Carroll P (1986) Damping in tall buildings: its variability and treatment in design. In: Isyumov N, Tschanz T (eds) Building motion in wind. ASCE convention, Seattle, Washington, pp 42–57

Fisco NR, Adeli H (2011) Smart structures: part I—active and semi-active control. Scientia Iranica Trans A: Civil Eng 18(3):275–284

Fu G, Betancur J, Poon D, Dannettel M (2012) Wuhan Greenland Center Main Tower: seamlessly integrating structure and architecture. In: Wood A, Johnson T, Li G-Q (eds) Proceedings of the CTBUH 9th world congress. Shanghai, China: council on tall buildings and urban habitat, pp 399–406

Fujino Y, Sun LM (1993) Vibration control by multiple tuned liquid dampers (MTLDs). J Struct Eng 119(12):3482–3502

Gad-el-Hak M (2000) Flow control: passive, active and reactive flow management. Cambridge University Press, London

Haviland R (1976) A study of the uncertainties in the fundamental translational period and damping values for real buildings. Massachusetts Institute of Technology Report, Cambridge

Hejazi F, Dalili M, Noorzaei J, Jaafar MS, Abdullah AAA (2010) Coupling parallel buildings by supplemental viscous damper device. In: Tiziani W (ed) Proceedings of the international conference on computing in civil and building engineering, Nottingham

Holmes JD (1995) Listing of installations. Eng Struct 17(9):676–678

Housner GW, Bergman LA, Caughey T, Chassiakos AG, Claus RO, Masri SF, Skelton RE, Soong TT, Spencer BF, Yao JTP (1997) Structural control: past, present, and future. J Eng Mech ASCE 123(9)

Jeary AP (1986) Damping in tall buildings—a mechanism and a predictor. Earthquake Eng Struct Dyn 14:733–750

Kareem A (1979) Reduction of wind-excited motion of high-rise buildings. In: Proceedings of the 3rd ASCE/EMD specialty conference. Austin

Kareem A (1983) Mitigation of wind induced motion of tall buildings. J Wind Eng Ind Aerodyn 11(1–3):273–284

Kareem A (1990) Reduction of wind induced motion utilizing a tuned sloshing damper. J Wind Eng Ind Aerodyn 36:725–737

Kareem A (1994) Methods to control wind-induced building motions. In Baker NC, Goodno B (eds) Proceedings of the XII structures congress, Atlanta, pp 654–659

Kareem A, Gurley K (1996) Damping in structures: Its evaluation and treatment of uncertainty. J Wind Eng Ind Aerodyn 59(2):131–157

Kareem A, Kijewski T, Tamura Y (1999) Mitigation of motions of tall buildings with specific examples of recent applications. Wind Struct 3(2):201–251

Kareem A, Kline S (1995) Performance of multiple mass dampers under random loading. J Struct Eng 121(2):348–361

Kareem A, Sun W-J (1987) Stochastic response of structures with fluid-containing appendages. J Sound Vibration 119(3):389–408

Kareem A, Zuo L (2012) Energy harvesting and motion control of tall buildings using tuned mass dampers. In: Wood A, Johnson T, Li G-Q (eds) Proceedings of the CTBUH 9th world congress. Shanghai, China: council on tall buildings and urban habitat, pp 351–357

Karnopp D, Crosby MJ, Harwood RA (1974) Vibration control using semi-active force generators. J Eng Ind, pp 619–626

Kijewski T, Kareem A (2002) On the reliability of a class of system identification techniques: insights from bootstrap theory. Struct Safety 24:261–280

Kirch A, Peil U, Borri C (2009) Limits for the control of wind-loaded slender bridges with movable flaps. Part I: aerodynamic modelling, state-space model and open-loop characteristics of the aeroelastic system. In: Proceedings of the 5th European African conference on wind engineering, Florence

Klein RE, Cusano C, Slukel JV (1972) Investigation of a method to stabilize wind induced oscillations in large structures. In: 1972 ASME winter annual meeting, New York

Kobori T, Koshika N, Yamada K, Ikeda Y (1991) Seismic response controlled structure with active mass driver system. Part 1: design. Earthquake Eng Struct Dyn 20:133–149

Kobori T, Takahashi M, Nasu T, Niwa N, Ogasawara K (1993) Seismic response controlled structure with active variable stiffness system. Earthquake Eng Struct Dyn 22(11):925–941
Koike Y, Inoue Y, Kimura T, Yamamoto Y, Enokuchi M, Yoshie K (2001) Active-damping bridge for buildings of Harumi Island Triton Square, Part 3 Outline of active-damping bridge. Architectural Institute of Japan Report
Kubo Y, Yasuda H, Kotsubo C, Hirata K (1993) Active control of super-tall structure vibrations under wind action by a boundary-layer control method. J Wind Eng Ind Aerodyn 50:361–372
Kubo Y, Modi VJ, Kotsubo C, Hayashida K, Kato K (1996) Suppression of wind-induced vibrations of tall structures through moving surface boundary-layer control. J Wind Eng Ind Aerodyn 61:181–194
Kumar D, Kareem A (2011) Dynamic response of tall building with embedded dampers in the structural system and multiple TMDs. In: Proceedings of the 11th international conference on wind engineering. Amsterdam, The Netherlands
Kurino H, Tagami J, Shimizu K, Kobori T (2003) Switching oil damper with built-in controller for structural control. J Struct Eng 129(7):895–904
Kwok KCS, Samali B (1995) Performance of tuned mass dampers under wind loads. Eng Struct 17 (9):655–667
Lagomarsino S (1993) Forecast models for damping and vibration periods of buildings. J Wind Eng Ind Aerodyn 48:221–239
Leipholz HHE, Abdel-Rohman M (1986) Control of structures. Nijhoff, M, Dordrecht
Lu X, Zhou Y, Yan F (2008) Shaking table test and numerical analysis of RC frames with viscous wall dampers. J Struct Eng 134(1):64–76
McNamara RJ, Taylor DP (2003) Fluid viscous dampers for high-rise buildings. Struct Des Tall Spec Build 12:145–154
Miyazaki M, Mitsusaka Y (1992) Design of a building with 20% or greater damping. In: Proceedings of the 10th world conference on earthquake engineering, Madrid
Nagarajaiah S (2009) Adaptive passive, semiactive, smart tuned mass dampers: identification and control using empirical mode decomposition, Hilbert transform, and short term Fourier transform. Struct Control Health Monit 16(7–8):800–841
Nagarajaiah S, Sonmez E (2007) Structures with semiactive variable stiffness single/multiple tuned mass dampers. J Struct Eng 133(1):67–77
Nemir DC, Lin Y, Osegueda RA (1994) Semiactive motion control using variable stiffness. J Struct Eng 120(4):1291–1306
Nishitani A, Inoue Y (2001) Overview of the application of active-semiactive control to building structures in Japan. Earthquake Eng Struct Dyn 30:1565–1574
Nissen HD, Sørensen PH, Jannerup O (2004) Active aerodynamic stabilisation of long suspension bridges. J Wind Eng Ind Aerodyn 92:829–847
Ogawa K, Ide T, Saitou T (1997) Application of impact mass damper to a cable-stayed bridge pylon. J Wind Eng Ind Aerodyn 72:301–312
Ostenfeld KH, Larsen A (1992) Bridge engineering and aerodynamics. In: Proceedings of the frst international symposium on aerodynamics of large bridges, Copenhagen
Patil VB, Jangid RS (2009) Double friction dampers for wind excited benchmark building. Int J Appl Sci Eng 7(2):95–114
Patten WN, Sun J, Li G, Kuehn J, Song G (1999) Field test of an intelligent stiffener for bridges at the I-35 Walnut Creek Bridge. Earthquake Eng Struct Dyn 28:109–126
Roorda J (1975) Tendon control in tall structures. J Struct Div ASCE 101:505–521
Saito T, Shiba K, Tamura K (2001) Vibration control characteristics of a hybrid mass damper system installed in tall buildings. Earthquake Eng Struct Dyn 30(11):1677–1696
Saleh A, Adeli H (1998) Optimal control of adaptive/smart multistory building structures. Comput-Aided Civil Infrastruct Eng 13(6):389–403
Samali B, Kwok KCS (1995) Use of viscoelastic dampers in reducing wind- and earthquake induced motion of building structures. Eng Struct 17(9):639–654

Satake N, Suda K, Arakawa T, Sadaki A, Tamura Y (2003) Damping evaluation using full-scale data of buildings in Japan. J Struct Eng 129(4):470–477

Shimizu K, Teramura A (1994) Development of vibration control system using U-shaped tank. In: Proceedings of the 1st international workshop and seminar on behavior of steel structures in seismic areas, Timisoara, Romania, pp 7.25–7.34

Sims ND, Stanway R, Johnson AR (1999) Vibration control using smart fluids: a state-of-the-art review. Shock Vib Digest 31(3):195–203

Soong TT (1990) Active structural control: theory and practice. Longman, London

Spencer BFJ et al (1997) Phenomenological model of a magnetorheological damper. J Eng Mech ASCE 123(3):230–238

Spencer BFJ, Nagarajaiah S (2003) State of the art of structural control. J Struct Eng 129 (7):845–856

Tamura Y (1998) Application of damping devices to suppress wind-induced responses of buildings. J Wind Eng Ind Aerodyn 74–76:49–72

Tamura Y (2003) Design issues for tall buildings from accelerations to damping—tribute to Hatsuo Ishizaki and Vinod Modi. In: Proceedings of the 11th international conference on wind engineering, Lubbock

Tanaka H, Tamura Y, Ohtake K, Nakai M, Kim YC (2012) Experimental investigation of aerodynamic forces and wind pressures acting on tall buildings with various unconventional configurations. J Wind Eng Ind Aerodyn 107–108:179–191

Taranath BS (2010) Reinforced concrete design of tall buildings. CRC, Boca Raton

Wilde K, Fujino Y (1998) Aerodynamic control of bridge deck flutter by active surfaces. J Eng Mech ASCE 124(7):718–727

Xu YL, Qu WL, Chen ZH (2001) Control of wind-excited truss tower using semiactive friction damper. J Struct Eng 127(8):861–868

Yalla SK, Kareem A (2003) Semiactive tuned liquid column dampers: experimental study. J Struct Eng 129(7):960–971

Yalla SK, Kareem A (1999) Sloshing-slamming dampers. In: Proceedings of the 10th international conference on wind engineering, Copenhagen

Yang J-N, Giannopoulos F (1978) Active tendon control of structure. J Eng Mech ASCE 104 (3):551–568

Zheng C-R, Zhang Y-C (2012) Computational fluid dynamics study on the performance and mechanism of suction control over a high-rise building. Struct Des Tall Spec Build 21 (7):475–491

MIX
Papier aus verantwortungsvollen Quellen
Paper from responsible sources
FSC® C105338

If you have any concerns about our products, you can contact us on
ProductSafety@springernature.com

In case Publisher is established outside the EU, the EU authorized representative is:
Springer Nature Customer Service Center GmbH
Europaplatz 3, 69115 Heidelberg, Germany

Printed by Libri Plureos GmbH
in Hamburg, Germany